Earth History and Palaeogeography

Palaeogeography is the challenging yet fascinating study of changing geography and geomorphology through deep time, in response to tectonic plate movements. This full-colour volume presents our latest knowledge of the Earth's dynamic evolution over the last 540 million years, making it an invaluable reference for researchers, graduate students, professional geoscientists, and anyone interested in the geological history of the Earth.

Using full-colour palaeogeographical maps from the Cambrian to the present, this interdisciplinary volume explains how plate motions and surface volcanism are linked to processes in the Earth's mantle, and to climate change and the evolution of the Earth's biota. These new and very detailed maps provide a complete and integrated Phanerozoic story of palaeogeography. They illustrate the development of all the major mountain-building orogenies, both those that have ended (such as the Caledonide and Variscan) and those continuing (such as the Andean and Himalayan). Old lands, seas, ice caps, volcanic regions, reefs, and coal beds are highlighted on the maps, as well as faunal and floral provinces. Many other original diagrams show sections from the Earth's core, through the mantle, and up to the lithosphere, and how large igneous provinces (LIPs) are generated, helping to understand how plates have appeared, moved, and vanished through time.

Supplementary resources are available online, including software, data files, operating instructions, and extended descriptions of continental plates and terranes, enabling readers to make their own reconstructions at any given time over the past 540 million years.

Trond H. Torsvik is the Founding Director of the Centre for Earth Evolution and Dynamics (CEED), University of Oslo, Norway, and Honorary Professor at Wits University, Johannesburg. He is a Member of the Norwegian Academy, and was awarded the prestigious Arthur Holmes Medal (European Union of Geosciences) in 2016 and the Leopold von Buch Medal (German Geological Society) in 2015 for outstanding achievements in the geosciences, among various other awards and prizes. He has written over 200 publications in refereed journals and books.

L. Robin M. Cocks OBE TD is a Scientific Associate in the Department of Earth Sciences at the Natural History Museum, London, UK, where he was formerly Keeper of Palaeontology. He has been President of the Geological Society of London, the Palaeontological Association, the Geologists' Association, and the Palaeontographical Society. In 1995 he was awarded the Geological Society's Coke Medal, and in 2010 the Lapworth Medal, by the Palaeontological Association, its highest honour.

Earth History and Palaeogeography

Trond H. Torsvik
University of Oslo

and

L. Robin M. Cocks
The Natural History Museum, London

Shaftesbury Road, Cambridge CB2 8EA, United Kingdom

One Liberty Plaza, 20th Floor, New York, NY 10006, USA

477 Williamstown Road, Port Melbourne, VIC 3207, Australia

314–321, 3rd Floor, Plot 3, Splendor Forum, Jasola District Centre, New Delhi – 110025, India

103 Penang Road, #05–06/07, Visioncrest Commercial, Singapore 238467

Cambridge University Press is part of Cambridge University Press & Assessment, a department of the University of Cambridge.

We share the University's mission to contribute to society through the pursuit of education, learning and research at the highest international levels of excellence.

www.cambridge.org
Information on this title: www.cambridge.org/9781107105324

First published 2017

A catalogue record for this publication is available from the British Library

ISBN 978-1-107-10532-4 Hardback

Additional resources for this publication at www.cambridge.org/torsvik

For Stephanie and Elaine

Contents

Preface

Although Trond and Robin had met previously, our first significant interaction was at the Europrobe meeting at St Petersburg in 1999, with the result that Robin was invited by Trond to the Norwegian Geological Winter Meeting at Trondheim in 2000, after which our collaboration into unravelling global palaeogeography began in earnest. Some years and many joint papers later, we thought that a summary and extension of our work would be timely, and this book is the result. Our specialities are complementary: Trond is a geophysicist specialising in palaeomagnetism and mantle dynamics, whilst Robin specialises in Palaeozoic stratigraphy and faunas. However, both of us had previously published on global and regional palaeogeography, alone and with other colleagues, and we both have first degrees in geology, which has formed the essential common language for appropriate discussion, which has been both challenging and fun. In addition and most importantly, Trond has developed the software to generate flat maps from a spherical Earth, and which can move the lithospheric units through time with kinematic objectivity, as explained in Chapter 2. That means that he has constructed virtually all the diagrams in this book, whilst Robin has written slightly more of the words. Nevertheless, we are under no illusions that this book is a final summary of the Earth's changing geography; merely a progress report.

Acknowledgements

We are indebted to a host of friends, colleagues, and acquaintances from around the world, with whom we have separately and together interacted over many years. It is impossible to name them all here, but we particularly thank (in alphabetical order) Lew Ashwal, Kevin Burke, Mat Domeier, Pavel Doubrovine, Richard Fortey, Carmen Gaina, Morgan Jones, Paul Kenrick, the late Stuart McKerrow, Adrian Rushton, Grace Shephard, Bernhard Steinberger, the late Brian Sturt, Henrik Svensen, Rob Van der Voo, and Douwe van Hinsbergen.

1 Introduction

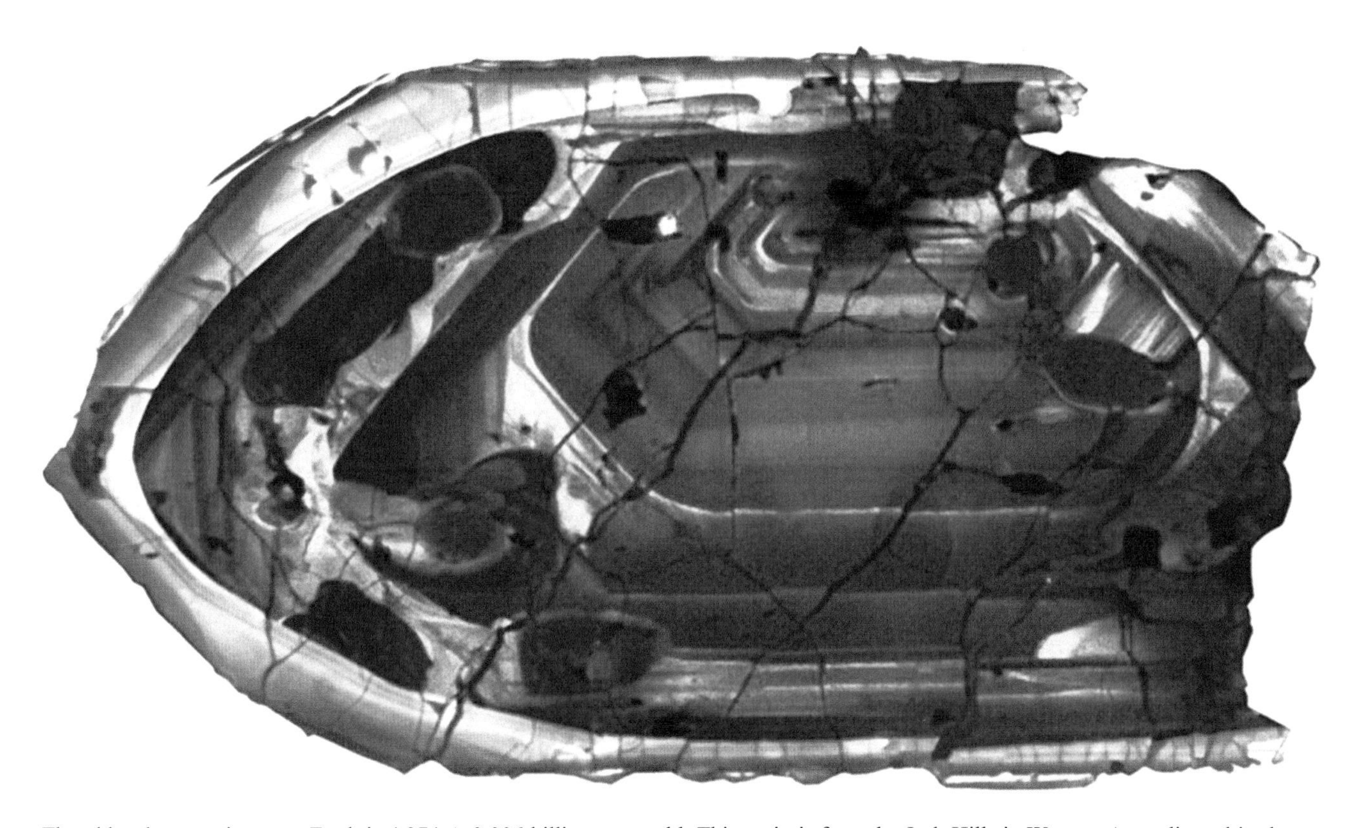

The oldest known zircon on Earth is 4.374 ± 0.006 billion years old. This grain is from the Jack Hills in Western Australia and is about 200 million years younger than the age of the Earth, and is probably a remnant of the oldest continental crust. Credit: John Valley, University of Wisconsin/Nature Geoscience.

When the Earth is viewed from space on a cloudless day, all that can be seen are the edges of lands, seas, and ice caps, all of which can be objectively mapped. From further geophysical exploration, the identities and margins of the oceans and the continental lithosphere which lie beside and below them today can also be discovered. However, how and when all those margins have changed through geological time becomes progressively less easy to discover and also less objective, since an uncertain number of the plates and their included oceanic and continental lithosphere have disappeared by subduction into the Earth's interior. In addition, much of the lithosphere has been distorted through tectonic processes, in many places very heavily.

Our chief aim in writing this book is to interpret, decipher, and describe the complex history of our planet over the most recent half-billion years and the processes through which it has changed, and to compile maps of the distribution of the many tectonic plates through that time, and also show where the lands and seas were situated over that long period. As is usual with narration, we start at the beginning and carry on progressively through time as it elapsed, but the result of that natural sequence is to commence by discussing the periods over which we have the fewest quantitative constraints on Earth's old geography, and thus our geographical reconstructions gradually become more accurate as time continues up to the present day.

The periods into which geological time is divided are shown on the endpapers within this book's covers. The history of the Earth falls naturally into two very unequal divisions: the Precambrian, in which there are no fossils of use in determining the positions of the former continents, and which, including the origin of the planet, is only summarised here in Chapter 4. The Precambrian was followed by the Phanerozoic at 541 million years ago (Ma), and the latter started with the Palaeozoic, from which there is no old *in situ* ocean crust preserved, but this is when the biota was distributed in faunal and floral provinces which are very relevant in assessing oceanic separations in the absence of much useful geophysical data (apart from palaeomagnetism). The boundary between the Palaeozoic and the overlying Mesozoic to the present day was at 252 Ma, after which the ocean-floor magnetic stripes and other useful geophysical data become progressively more abundant and objective, but when the biota, although interesting in its evolutionary development, is again of no primary help in deciphering the palaeogeography. Thus there are separate chapters here for each of the main geological systemic periods from the Cambrian to the Quaternary (Chapters 5 to 15).

But what are the stepping stones, which cover many geological and geophysical disciplines, by which we can achieve our aims? After this brief introduction, in Chapter 2 we describe the varied and often independent methods that we have used to reconstruct old lands and seas. In Chapter 3 we list the 268 unit areas among the many making up our planet which are the ones we have used in the construction of our kinematic computer-generated palaeogeographical maps through time, with a very brief sketch of their geological constitutions. Each of those units, which vary in size from large continents to small terranes, can be downloaded digitally from www.earthdynamics.org/earthhistory, together with a digital rotation file and various other files, which can be used by anyone to make their own reconstructions with GPlates (www.gplates.org) for any area and at any given time for the past 540 million years, with no fees involved but acknowledgements requested.

Over the past billion years, our planet's climate has fluctuated wildly between hot and cold temperatures, some so extreme that any life has been scarcely possible. Thus, as well as mentioning those climates during the individual periods in Chapters 4 to 15, the final Chapter 16 brings together the many factors which affect and support the Earth's climate, and also describes how and why that climate has changed so much during the half-billion years and how it has come to be what it is today. Unfortunately, that deeper time perspective appears to be lacking in many modern-day climate scientists and politicians.

But underlying all that, our book and all of geological thinking depends on knowing how long ago each past event occurred. So that we can comprehend and evaluate the number of years over which the many changes of the Earth's surface and interior have taken place since its origin, and from that the rates of those changes, it is essential to know objectively the amount of time available for their progression. Thus a reliable primary geochronology is critical for underpinning our work.

Since the pioneer work by Arthur Holmes in the early twentieth century, rocks have been dated by radiometric methods, using a great variety of the longer-lived radiogenic isotopes, some of whose half-lives extend over billions of years (Torsvik & Cocks, 2012). The most useful elements have been found to be carbon for the most recent 30,000 years, and a variety of others, including argon–argon ($^{40}Ar/^{39}Ar$) and uranium–lead (U/Pb), for older rocks. All radiometric ages have errors calculated individually, which are given in the original papers, and most are published with dates including proportions within a million years, but, so that this text can flow relatively unimpeded, we have rounded all ages earlier than the Cenozoic (66 Ma) to the nearest one million years, and do not quote the published error ranges here. But, although lacking the objective numbers which have come from geochronology, in much of the Phanerozoic finer time divisions exist through the use of quickly evolving animals and plants from which biozones have been defined, and which have been used in the correlation of rocks. An example of the latter is graptolites in the Silurian, some of whose biozones are less than 100,000 years long, in contrast to the radiometric ages for that period, which are not accurate to within about one-third of a million years.

The overall time scale on which our work depends is inside the endpapers of this book, where the dates for the bases of the major time units are shown, most of which have now been standardised by the International Union of Geological Sciences (IUGS) Commission on Stratigraphy (Cohen et al., 2013).

It has been difficult to know how many references should best be cited. Many textbooks are frustrating in their relative lack, or sometimes even complete absence, of references, whilst research papers usually include at least one reference to support every fresh statement, and often far too many more, particularly to papers written by friends of the authors. This has led us to compromise, and we apologise both to current workers for the omission of precise citation of much invaluable work, and also to those earlier scientists on whose shoulders we all stand, particularly since we have tended to refer to summary articles in many places here, rather than to the many papers which underpin those works.

2 Methods for Locating Old Continents and Terranes

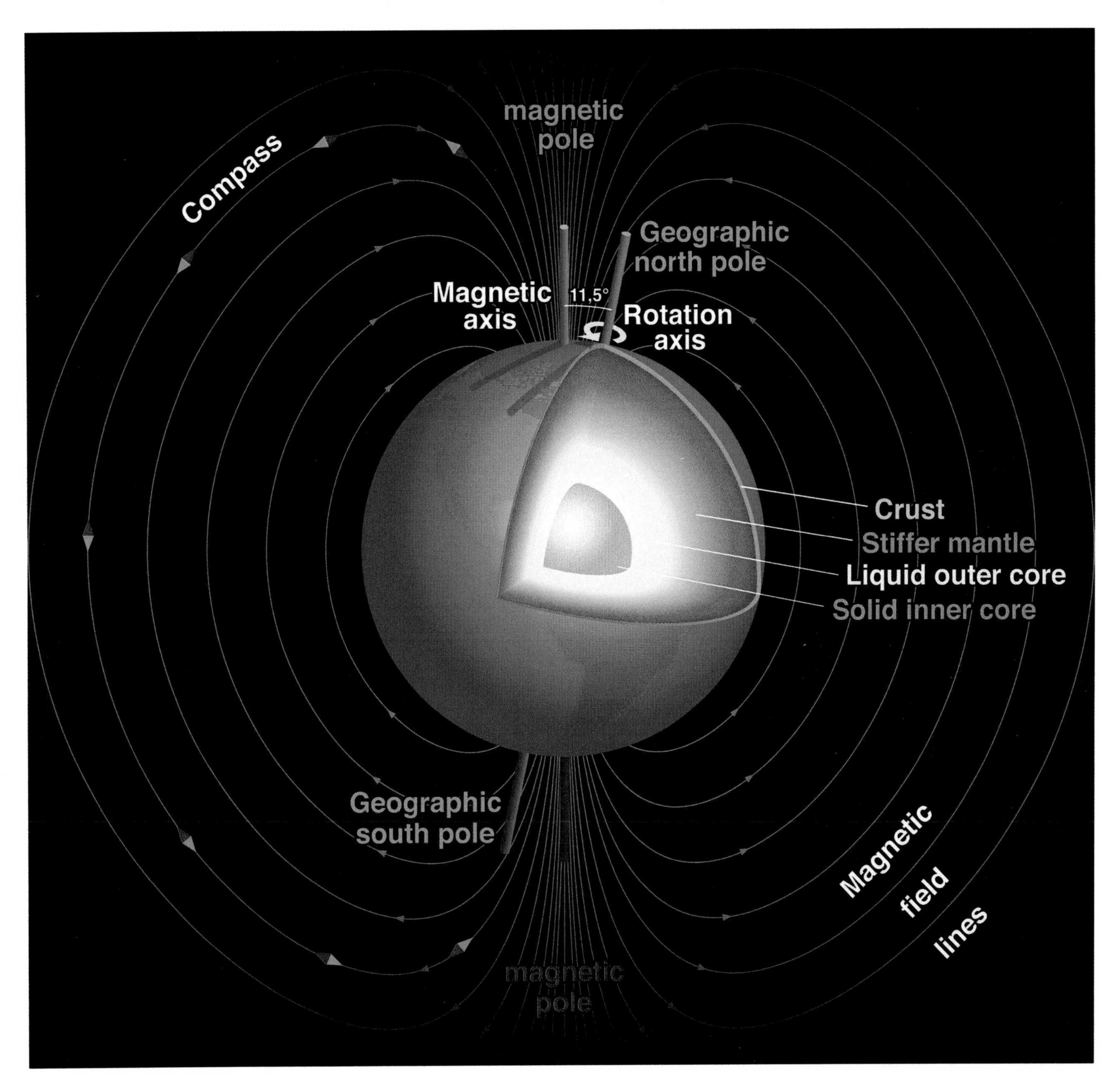

From the vantage point of geological time, the Earth is best seen as a giant heat engine. The decay of radioactive nuclides in the deep interior provides energy for its most fundamental dynamical processes: convection in both the liquid outer core and the stiffer (solid) but slowly deforming mantle. Under the influence of the Earth's rotation, the electrically conductive outer core generates the Earth's magnetic field. This field shelters us from cosmogenic radiation, modulates atmospheric escape, and provides guidance for many migratory species. The Earth's ancient magnetic field provided one of the fundamental markers used to document the motion of the continents and evolution of the Earth. Changes in ancient magnetic polarity at irregular intervals are recorded in the surface rock record, and over some fifty years such palaeomagnetic data have been used to create the geomagnetic time scale, to firmly document sea-floor spreading, to validate plate tectonics, and to reconstruct vanished supercontinents. The magnetic poles differ from the geographical poles because the magnetic axis is inclined relative to the geographical (rotation) axis (today ~11.5°). The magnetic axis, however, is rotating slowly around the geographical axis and, over a period of a few thousand years, the averaged magnetic poles have corresponded reasonably well with the geographical poles. Credit: Furian/Shutterstock.

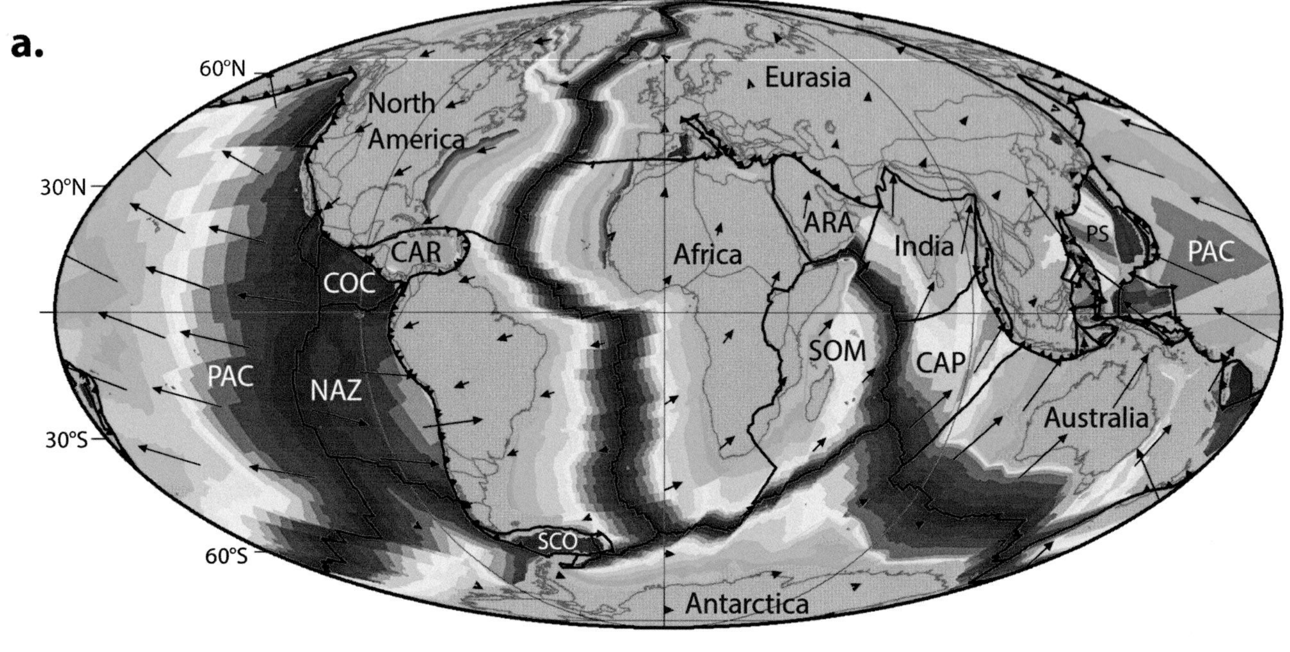

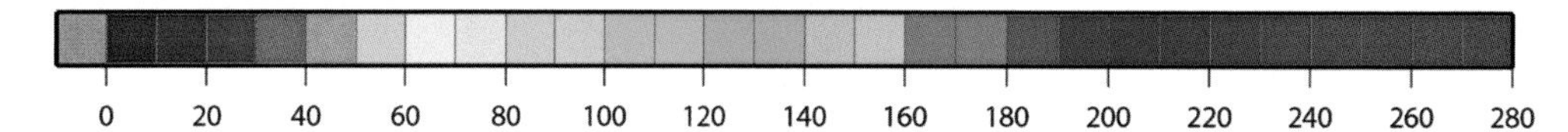

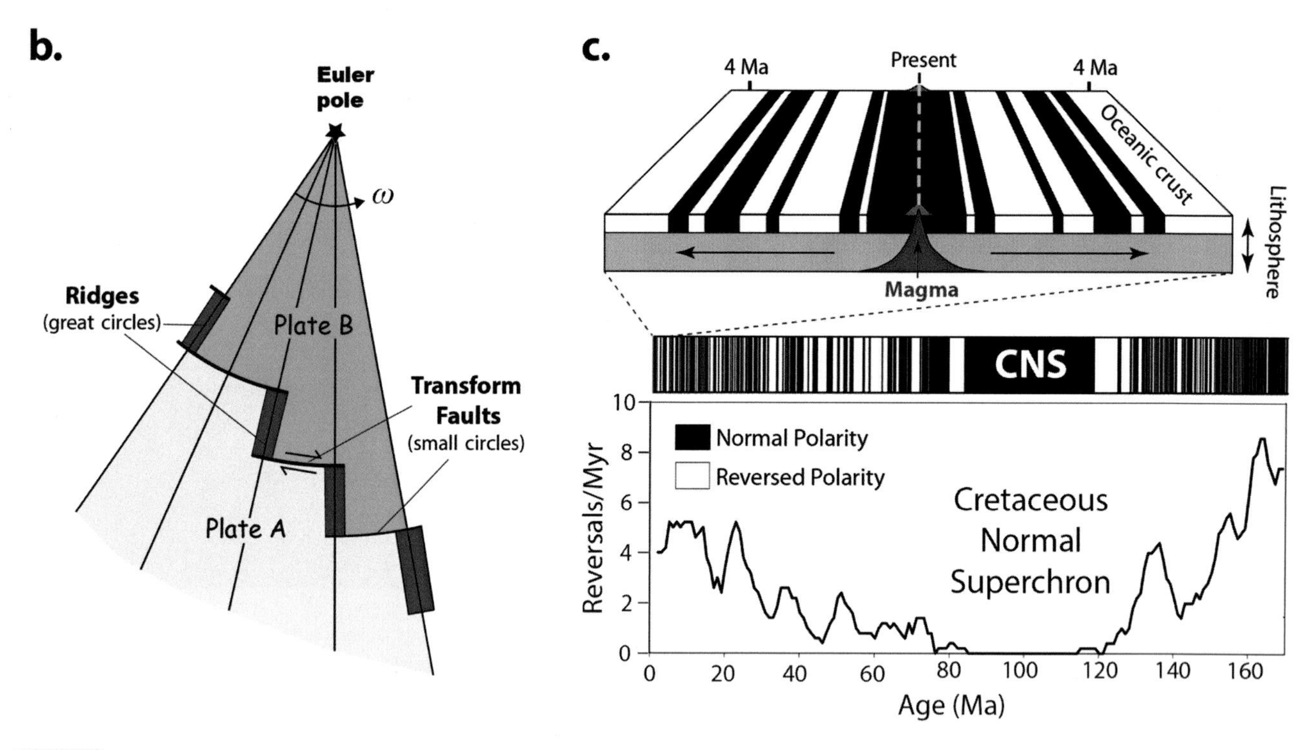

Fig. 2.1 (a) Present-day distribution of tectonic plates. Base map shows the modelled age of oceanic crust; warm (red) colours indicate young sea floor (close to the spreading axis), while cold (blue/violet) colours indicate old oceanic crust. The oldest preserved oceanic crust (possibly 195 million years old: Early Jurassic) is found, among other places, in the Central Atlantic between North America

Over the last century, our portrayal of the movement and deformation of the Earth's outer layer evolved from the hypothesis of Continental Drift (1912) into Sea-Floor Spreading (1962) and then to the paradigm of Plate Tectonics (1967). The word 'tectonics' comes from Greek and means 'to build', and plate tectonics is as fundamentally unifying to the Earth Sciences as Darwin's (1859) Theory of Evolution is to the Life Sciences. Most of this book describes how the geography of the Earth has changed over geological time. However, this chapter sets out the various methods by which the positions of old continents can be deduced, most of them completely independent from each other. It is important to bear in mind that the geographical areas (Chapter 3) in older times shown on our reconstructions (Chapters 4–15) rarely reflect the real shape of those continents at those individual times in the past, since their margins have been much changed by subsequent tectonic events.

Plate reconstructions at successive intervals in geological time are the result of a reiterative process using a wealth of methods. The relative positions of continents are commonly determined from ocean-floor magnetic anomalies (since the Jurassic) and fracture-zone geometries, analysis of continent–ocean boundaries, palaeomagnetic poles, and other geological and geophysical data. Continents and terranes are then reconstructed to their ancient positions on the globe using hotspot trails (since the Cretaceous) or palaeomagnetic data, and by identifying and discriminating between the distributions of various fauna and flora in their various provinces at different times. These biological distributions can indicate if terranes with similar biota were close to one another or were separate. The distribution of key sediments, such as glacial deposits, coal, and evaporites, can also be useful, but they are largely latitudinally determined rather than terrane-specific.

Plates and Plumes

Plate tectonics describes how the Earth's lithosphere (the crust and upper mantle) is constructed from a dozen large and many smaller rigid blocks that move in relation to each other (Fig. 2.1a). It is a captivating story of oceanic and thicker continental plates that move across the Earth's surface, and how they glide apart forming new oceanic crust (divergent plate boundaries), collide to form mountain belts (convergent boundaries), or move sideways in relation to each other (transform boundaries). The motions of rigid plates can be described by Euler rotations (Fig. 2.1b). Plate velocities range from about 1 to 15 cm/yr, and earthquakes and volcanic eruptions close to the plate boundaries are key elements of the plate tectonic paradigm. Under the influence of the Earth's rotation, the electrically conductive outer core generates the Earth's magnetic field. The history of that ancient magnetic field provides one of the fundamental clues used to document the motion of the continents and evolution of the Earth. Changes in ancient magnetic polarity at irregular intervals are recorded in the surface rock record, and those palaeomagnetic data have been used to create the geomagnetic time scale (Fig. 2.1c).

Hotspots can be referred to as volcanism unrelated to plate boundaries and rifts, and many intra-plate hotspots such as the one beneath Hawaii (Wilson, 1963) have been suggested to overlie plumes that originated at the core–mantle boundary. A few hotspots also lie at the ends of plume trails (chains of volcanic islands) which are connected to large igneous provinces (LIPs), e.g. the Tristan (Paraná–Etendeka) and Reunion (Deccan) hotspots. The Hawaii Hotspot was probably also linked to a starting LIP, which was long subducted (Chapters 13 and 14), whilst the New England Hotspot lies at the end of a plume trail that was connected with Jurassic kimberlite volcanism in continental north-eastern America. Many hotspots are clearly intra-plate (e.g. in the Pacific and Africa, Fig. 2.2a) and therefore unrelated to plate tectonics, but some are located on or near a plate tectonic boundary.

The lowermost mantle is characterised by two large heterogeneities where shear-wave velocities are up to three per cent slower than in the surrounding mantle. Those thermochemical piles are near antipodal and equatorially centred

Fig. 2.1 (*cont.*) and North-West Africa, and linked to the very earliest breakup of the supercontinent of Pangea. Red lines denote subduction zones, black lines denote mid-ocean ridges and transform faults. Plate motions (moving hotspot frame) are shown as black arrows. ARA, Arabia; CAP, Capricorn; CAR, Caribbean; COC, Cocus; NAZ, Nazca; PAC, Pacific; PS, Philippine Sea; SCO, Scotia; SOM, Somalia (data from EARTHBYTE). (b) On a sphere the relationship between two plates (A and B in this example) is described by a rotation (Euler) pole (latitude, longitude, and angle) and the relative speed (ω) is commonly expressed in °/Myr. Spreading ridges define great circles, while ocean fracture zones (transform faults) define small circles. (c) Cartoon illustrating the formation of magnetic anomalies at a mid-ocean ridge and the marine magnetic anomaly polarity record and reversal frequency (10 Myr running mean based on data in Biggin et al., 2012). Reversal frequency peaked in the Jurassic (~150–170 Ma), but the most extreme geomagnetic behaviour was in the Cretaceous Normal Superchron (CNS) from 121 to 84 Ma (Late Jurassic to Mid–Late Cretaceous), when the field was of single polarity for almost 40 Myr. Two older Phanerozoic superchrons have also been suggested: the Permian–Carboniferous Kiaman Reverse Superchron (~265–310 Ma) and an Ordovician Reversed Superchron between about 460 and 490 Ma.

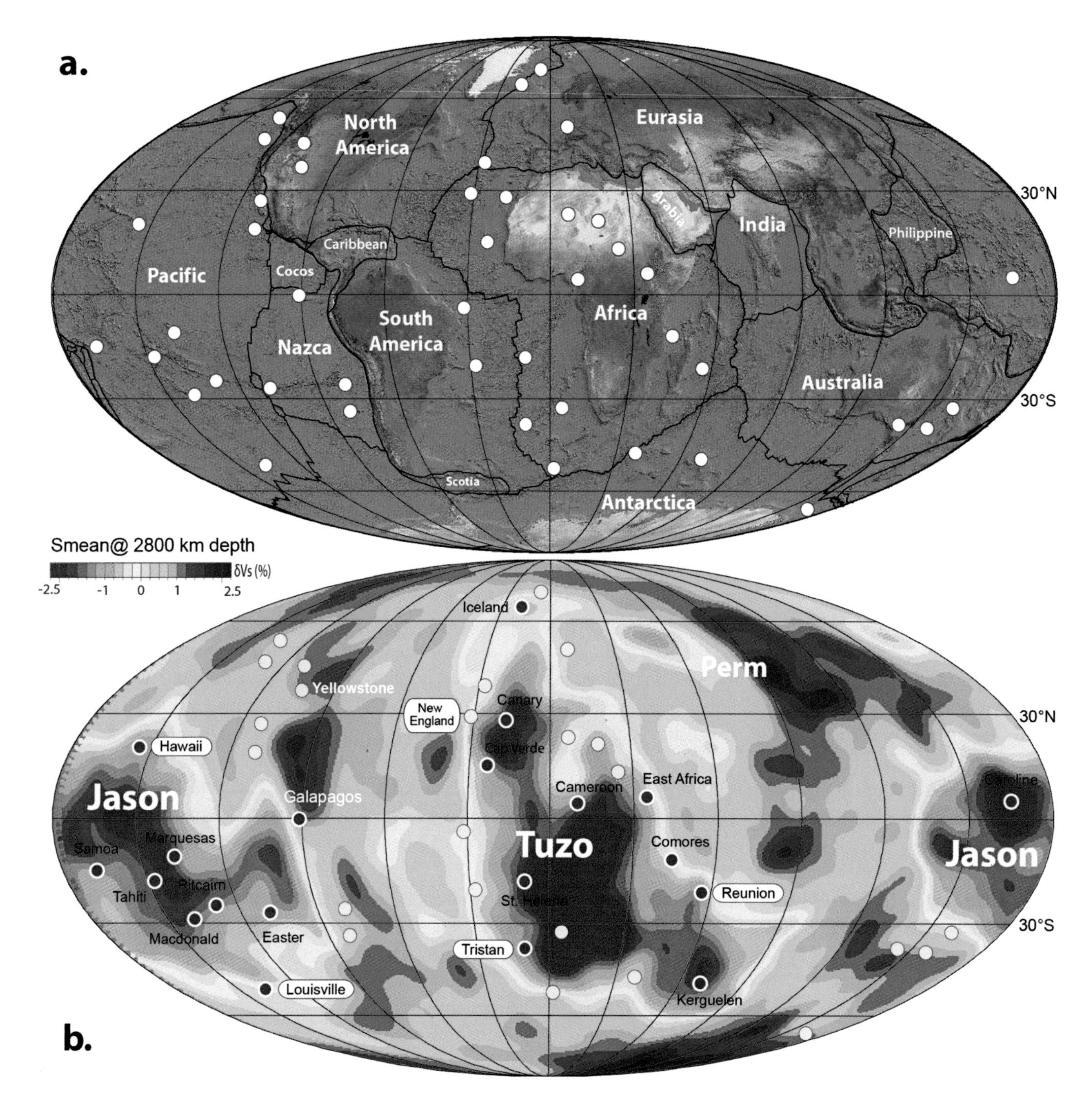

Fig. 2.2 (a) Main plate boundaries (black thick lines) and the distribution of hotspots (red circles with white filling; Steinberger, 2000) draped on a global map of topography and bathymetry. In plate tectonic theory, volcanism, deformation, and earthquakes should be confined mainly to regions near the plate boundaries. However, many hotspots are found within the plates (e.g. Pacific and African plates) which are not readily explained by plate tectonic processes. (b) Plot of hotspots as in (a) but draped on the SMEAN shear-wave velocity anomaly model at 2,800 km depth (Becker & Boschi, 2002). Velocity anomalies (δV_s) in percentage and red denote regions with low velocity. The lower mantle is characterised by two main zones of low shear-wave velocities, mainly beneath Africa (Tuzo) and the Pacific (Jason). In addition, a smaller one named Perm (which may alternatively be part of Tuzo) is located beneath Siberia. Many hotspots appear to overlie regions of slower than average shear-wave velocities (notably those associated with Tuzo) but there are clear exceptions (e.g. Yellowstone in North America). Those hotspots, thought to be sourced by deep plumes from the core–mantle boundary (primary and clearly resolved plumes in French & Romanowicz, 2015), are shown as white circles with red filling, whilst others of unknown origin are shown as yellow dots.

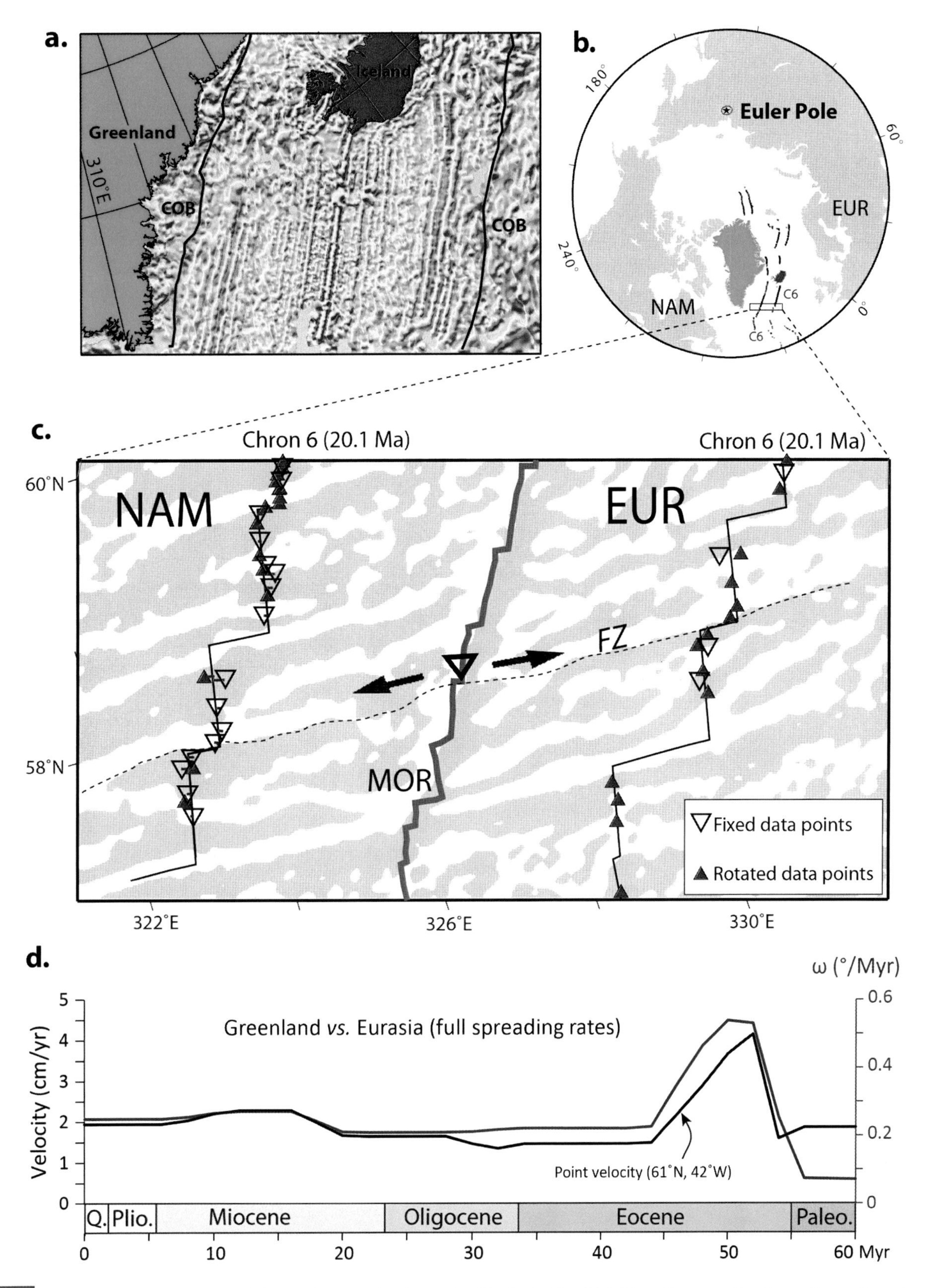

Fig. 2.3 (a) Magnetic anomaly grid of the north-east Atlantic south of Iceland. The thick black lines are Continent–Ocean Boundary transitions (COBs). (b) Example of marine magnetic anomaly interpretations (700 data points of Chron 6, C6) and resulting Euler pole (EP) (black star) and uncertainty ellipse (red contour around EP; enlarged three times so as to be visible on the map) for relative motion between

(Fig. 2.2b), and are termed large low shear-wave velocity provinces (LLSVPs) (Garnero et al., 2007) or more simply 'Tuzo' (beneath Africa) and 'Jason' (beneath the Pacific) by Burke (2011). Courtillot et al. (2003) and Ritsema and Allen (2003) concluded that only eight hotspots (Afar, Easter Island, Hawaii, Iceland, Louisville, Réunion, Samoa, and Tristan) potentially had a deep plume origin. Montelli et al. (2006) added three extra hotspots (Azores, Canary, and Tahiti) to those of deep origin, and Torsvik et al. (2006) pointed out that all hotspots of potential deep origin plot on or very near the margins of Tuzo or Jason. In a recent study using full-wave tomography, French and Romanowicz (2015) identified 20 primary or clearly resolved plumes (Fig. 2.2b) that all overlie Tuzo and Jason.

Relative Positions of Continents and Ocean Floor Magnetic Stripes

As long ago as the 1950s, scientists noticed the regular pattern of positive and negative magnetic anomalies across the ocean floors (Fig. 2.3a), and it was realised subsequently that the lines of those magnetic anomalies are arranged in sub-parallel stripes on both sides of each of the mid-ocean ridges (Vine & Matthews, 1963). That was explained by the realisation that, as magma is erupted at the mid-ocean ridges, it flows laterally and displaces the rock erupted previously. That causes the ocean floor to spread out relatively symmetrically from both sides of the spreading ridges (Figs. 2.1a and 2.3a). When the magma cools to form rock, magnetic minerals in the oceanic crust acquire a magnetisation aligned with the Earth's magnetic field (Fig. 2.9). That magnetisation in the rock reflects the polarity of the magnetic field (Fig. 2.1c), which has reversed from north to south and back again at irregular intervals many times in the past. The Late Jurassic witnessed high reversal frequencies but in the Cretaceous the field was of a single normal polarity lasting for almost 40 Myr (Cretaceous Normal Superchron, ~84–121 Ma), and thus no magnetic stripes exist for that time interval. On average, there has been a reversal every 200,000 years (five reversals per million years) during the past 10 Myr.

The matching of magnetic anomalies and fracture zones of the same age, corresponding to patterns of palaeo-ridge (great-circle) and palaeo-transform (small-circle) segments at any particular reconstruction time (Fig. 2.1b), is used to determine the relative motions between tectonic plates (Fig. 2.3b and c). The Hellinger (1981) method is most commonly used to derive best-fit rotations (Euler poles) from conjugate magnetic anomalies and fracture zone data (Fig. 2.3c), which are in turn used to derive spreading velocities between two plates. Figure 2.3 shows gridded magnetic anomalies in the north-east Atlantic from Iceland and southwards, with an example of fitting magnetic anomalies and fracture zones at around Chron 6 time (20.1 Ma) along the Reykjanes Ridge. Applying this method to all the identified Chrons in the north-east Atlantic (Gaina et al., 2002) one can calculate the relative speed back to the initial opening of the north-east Atlantic at around 54 Ma. Velocities peaked during the Early Eocene opening stage (about 0.5°/Myr or 4 cm/yr for a location along the southern Greenland margin), but otherwise have been quite stable at around 0.2°/Myr or 2 cm/yr. Spreading velocities are often reported as half-spreading rates in the literature; the values calculated in Fig. 2.3d are full spreading rates and the Reykjanes Ridge, and the other ridges in the north-east Atlantic and the high Arctic, are considered as slow to ultra-slow spreading ridges.

Absolute Plate Reconstructions Using Hotspot Tracks

After Wilson (1963) suggested that linear chains of seamounts and volcanoes are caused by hotspots, Morgan (1971) proposed that hotspots may be caused by mantle plumes up-welling from the lower mantle, and constructed the first hotspot reference frame. In that and later models

Fig. 2.3 (*cont.*) North America (NAM) and Eurasia (EUR). (c) Detailed image shows a subset of the Chron 6 (20.1 Ma) interpretation in the North Atlantic and illustrates Hellinger's (1981) criterion of fit. Fixed data points are represented by inverted triangles; rotated data points are red triangles. The background shows the vertical gradients of free air gravity that allow identification of fracture zones (FZ) and offsets between spreading segments. Great circles were fitted for data points in each individual spreading segment. For a given rotation the measure of fit represents the sum of squares of the weighted distances (short blue line segments perpendicular to the great circle segment shown as an example on the NAM isochron). The thick grey line shows the present-day mid-ocean ridge (MOR); the arrows indicate the direction of spreading on NAM and EUR plates (simplified from Torsvik et al., 2008b). (d) Relative velocity between the Greenland and Eurasian Plates expressed in °/Myr (ω, blue curve) and cm/yr (black curve) for a point in southern Greenland (61° N, 42° W). Those velocities are calculated from Euler poles determined by the Hellinger (1981) method detailed in Gaina et al. (2002). Sea-floor spreading between Greenland and Eurasia started at around 54 Ma, but before then there was a period of Late Cretaceous–Paleocene pre-drift extension, and therefore velocities are not zero before breakup. Velocities are calculated over a 5 Myr window and shown at 2 Myr intervals. Paleo., Paleocene; Plio., Pliocene; Q., Quaternary.

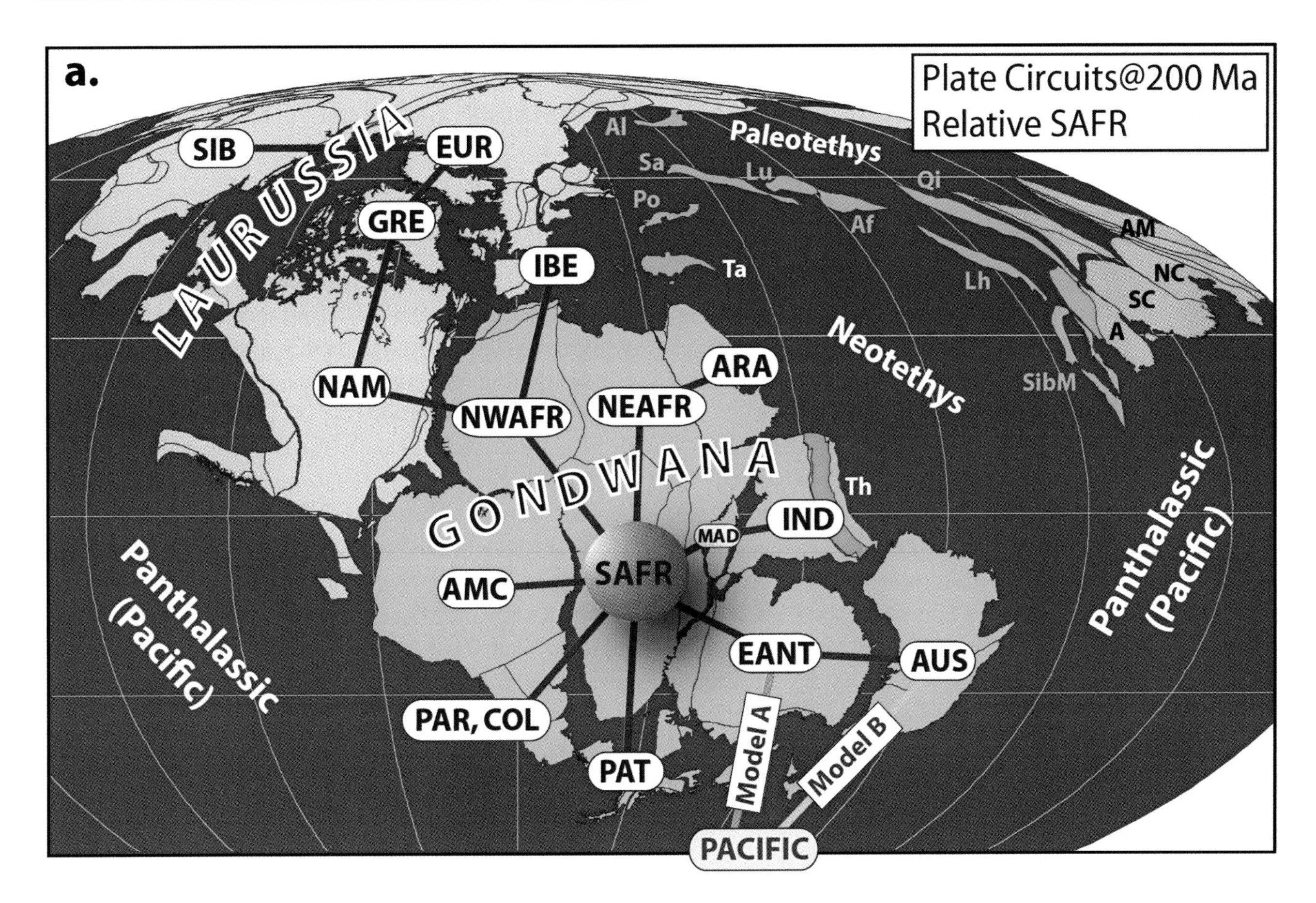

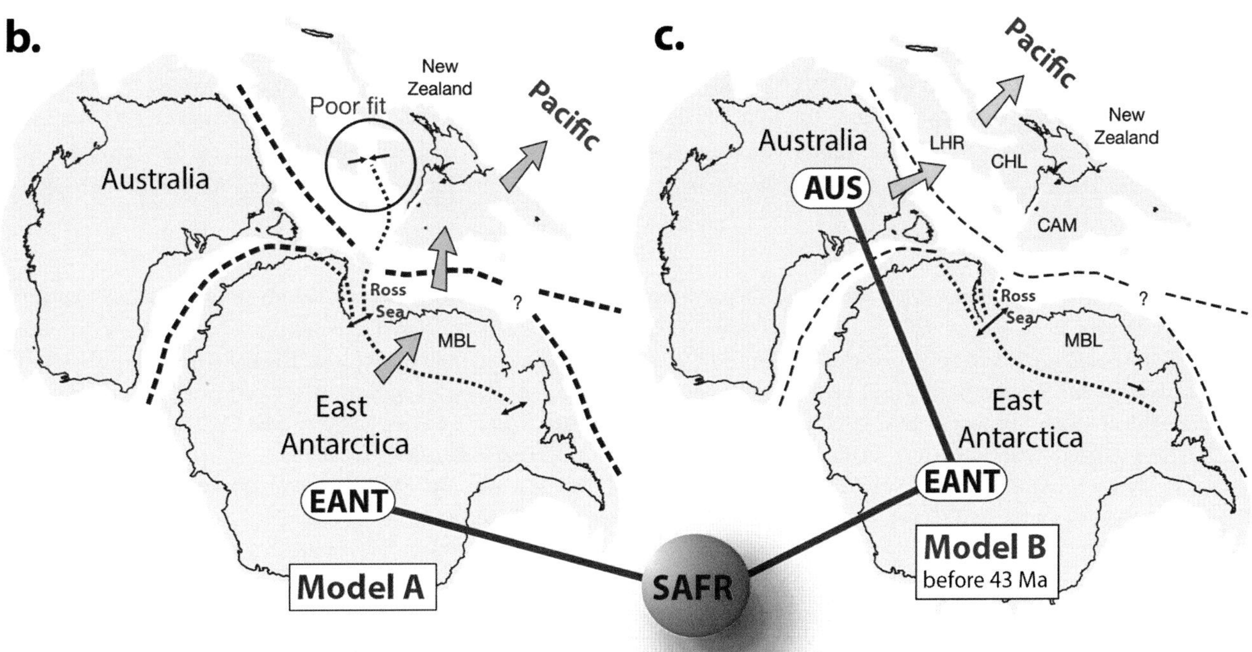

Fig. 2.4 (a) When building global plate motion models before the Cretaceous it is important to choose a plate as an initial reference that has undergone a small amount of longitudinal movement. Africa is the ideal candidate because it has remained extremely stable for the past 200 Myr, having mostly been surrounded by spreading centres since the Jurassic. Africa has the added advantage of being nearly in the

(e.g. Müller et al., 1993), plumes remained fixed relative to each other over long periods of time (fixed hotspot hypothesis). However, there is abundant evidence that hotspots are moving relative to each other; for example, the fixed Pacific and African hotspot reference frames do not agree with each other (Fig. 2.5a), and there is sound palaeomagnetic evidence (Tarduno et al., 2009) that the Hawaii Plume underwent considerable southward drift in Late Cretaceous to Paleogene times (Fig. 2.7). Fixed hotspot reference frames must therefore (at least before 40 Ma) be replaced by a mantle (moving hotspot) frame in which motions of hotspots in a convecting mantle are assumed (Steinberger et al., 2004).

Hotspot reference frame reconstructions are absolute in the sense that they constrain the ancient positions of continents (described by latitudes, longitudes, and orientations). There are three key elements that define a moving hotspot reference frame: (i) relative plate reconstructions (Fig. 2.4), (ii) the ages and geometries of hotspot tracks (Figs. 2.5 and 2.6), and (iii) the motions of mantle plumes that rely on backward advection of present mantle density structures. Incorporating data from hotspot tracks formed on different plates into a common reference frame requires estimates of relative plate motion. That is achieved by reconstructing coeval locations from all the tracks that are present at a certain age relative to a selected 'anchor' plate, which is most commonly South Africa (Unit 701, Chapter 3). A global absolute plate motion model is therefore made in such a way that all the tracks were on the southern African Plate (using plate circuits, Fig. 2.4) and then averaged to define a global moving hotspot reference frame (GMHRF).

The GMHRF of Steinberger et al. (2004) was based on four hotspot tracks, two in the Pacific (Hawaiian and Louisville), one in the South Atlantic (Tristan), and the fourth in the Indian Ocean (Réunion). All those hotspots probably initiated as LIPs (catastrophic melting of the upper mantle) at around 125–120 Ma (Ontong Java: Louisville), 134 Ma (Paraná–Etendeka: Tristan), and 65 Ma (Deccan: Réunion). Indo-Atlantic Plate circuits are quite robust but relating the Pacific Plate to the Indo-Atlantic system is not straightforward. The Steinberger et al. (2004) model of hotspot motion predicted a southward motion (up to a few cm/year) of the Hawaii Hotspot, and in combination with a plate motion chain that connects Africa and the Pacific via East Antarctica and Marie Byrd Land (West Antarctica) allowed a fit of hotspot tracks globally for times after the age of the Hawaiian–Emperor bend. In this model, no motion occurred between East Antarctica and Marie Byrd Land prior to 43.8 Ma (model A in Fig. 2.4b). Before 43.8 Ma, an east–west misfit between predicted and observed Hawaiian hotspot tracks remains (Fig. 2.5a). Steinberger et al. (2004) thus explored alternative plate circuits linking Africa and the Pacific via East Antarctica–Australia–Lord Howe Rise in older times (Model B, Fig. 2.4c), and with this model they were able to achieve a reasonable fit to hotspot tracks globally back to around 65 Ma. Prior to that, a notable misfit between predicted and observed Hawaiian tracks remains (Fig. 2.5a).

The GMHRF of Torsvik et al. (2008b) also used the model B plate motion chain of Steinberger et al. (2004), but relative plate motions are slightly different and the South Africa *versus* East Antarctica and Australia *versus* Lord Howe Rise rotations were smoothed. The transition between models A and B was also smoothed. By doing that, at 75 Ma (Campanian), the difference between observed and predicted

Fig. 2.4 (*cont.*) centre of the plate circuit tree, thus limiting error propagation. Abbreviations in our 200 Myr reconstruction keeping Africa fixed: SIB, Siberia; EUR, Europe; GRE, Greenland; NAM, North America; IBE, Iberia; NWAFR, North-West Africa; NEAFR, North-East Africa; SAFR, South Africa; AMC, Amazonia Craton; PAR, COL, Paraná, Colorado sub-plates; PAT, Patagonia sub-plate; IND, India; ARA, Arabia; MAD, Madagascar; EANT, East Antarctica; AUS, Australia; Th, Tethyan Himalaya. Once peri-Gondwanan terranes (dark greenish) and not part of Pangea include Ta (Taurides, Turkey), Po (Pontides, Turkey), Sa (Sanand, Iran), Lu (Lut, Iran), Al (Alborz, Iran), Af (Afghanistan), Qi (Qiantang, North Tibet), Lh (Lhasa, South Tibet) and SibM (Sibumasu). China block not part of Pangea (separated from Asia by the Mongol–Okhotsk Ocean) includes A (Annamia), SC (South China), NC (North China), and AM (Amuria, Central Mongolia). We also show two different relative plate circuit models between Indo-Atlantic (Africa) and Pacific hotspots before the Middle Eocene (Chron 20, 43 Ma). After Chron 20, models A and B follow the same plate motion chain through East Antarctica and Marie Byrd Land. The models were originally named model 1 and 2 (Steinberger et al., 2004). (b) and (c) Late Cretaceous (Maastrichtian) South Pacific reconstructions, the standard model A and the alternative plate chain model B, which link Africa and the Pacific via East Antarctica, Australia, and the Lord Howe Rise (LHR). For the South-West Pacific Plate motion chain, model B predicts intra-Antarctic motion prior to 43.8 Ma with extension in the Ross Sea area, whereas model A does not involve movements between East and West Antarctica before 43.8 Ma. The large arrows show paths of plate motion chains, thick stippled lines are divergent plate boundaries with sea-floor spreading, and dotted lines are conditional intra-continental plate boundaries necessary to accommodate the two different plate circuit models. The standard model A demonstrates a New Zealand misfit whilst model B requires about 300 km extension in the Ross Sea. CHL, Challenger Plateau; CAM, Campbell Plateau; MBL, Marie Byrd Land.

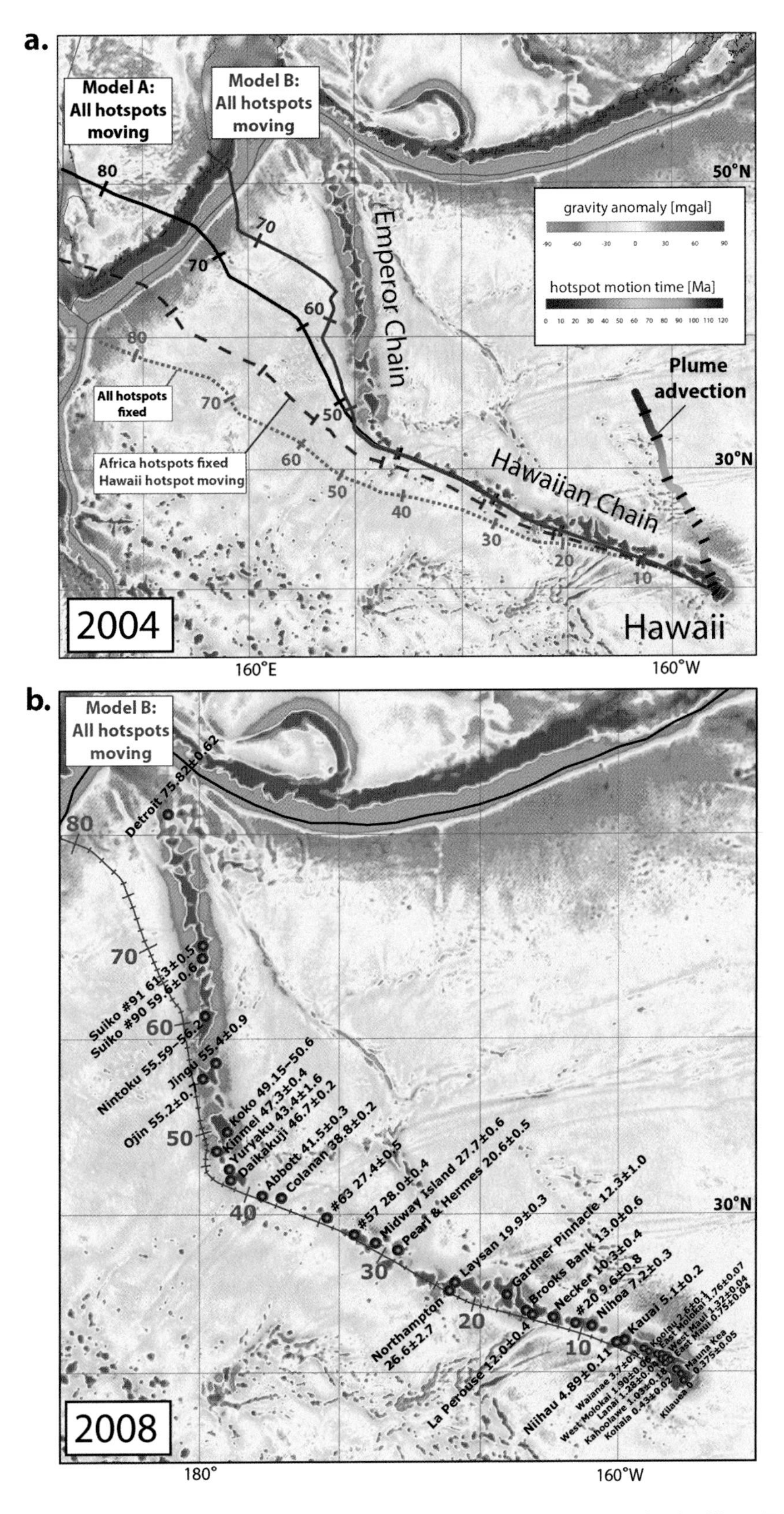

Fig. 2.5 (a) Dotted line: Computed fixed hotspot track for Hawaii in a reference frame based on fixed African (Tristan and Réunion) hotspots and plate chain model A (Fig. 2.4b). Dashed line: Same reference frame, but Hawaiian hotspot motion (due to plume advection, as shown in the figure) considered. Neither mimics the trend of the Emperor–Hawaii Chain. The black line (track computed in a reference frame that best fits four hotspot tracks – Hawaii, Louisville, Tristan and Réunion – and considers computed hotspot motion using plate chain model A) partly mimics the bend but plots west of the Emperor Chain. A model B plate chain (blue line) clearly improves the fit to the Emperor Chain and captures the Emperor–Hawaii Chain for the past 65 Myr quite well. (b) Predicted tracks for the Hawaii Hotspot in the global moving hotspot reference frame of Torsvik et al. (2008b). Hawaiian hotspot motion is the same as in (a) with model B plate circuits, but the transition between model A and B (46.3–43.8 Ma) and the plate circuits between East Antarctica–South Africa and Australia–Lord Howe Rise are smoothed compared with (a).

Hawaii–Emperor hotspot tracks was reduced to 300 km or less (Fig. 2.5b). Torsvik et al. (2008b) also extended the GMHRF from about 83 Ma to 130 Ma (Africa) and 150 Ma (Pacific), using rotation rates relative to fixed hotspots, but the extension was done separately for the Pacific and Africa.

A more recent GMHRF (Doubrovine et al., 2012) also included the New England hotspot track and was therefore based on a total of five hotspot tracks. A starting LIP for the New England Hotspot (located near Great Meteor Seamount, south of the Azores) has not been identified, but the New England track started with Triassic kimberlite intrusions in north-eastern America (e.g. Zurevinski et al,, 2011). Individual tracks were resampled with spherical splines (Fig. 2.6a), errors were assigned to track locations, a spherical regression and goodness of fit test was included, and an iterative model convergence scheme was implemented to derive the final GMHRF (Fig. 2.6b). Doubrovine et al. (2012) also tested fixed hotspot frames, model A *versus* B plate chains as in Fig. 2.5a and also alternative plate circuits between Australia and Lord Howe Rise. They concluded that model B (with Australia–Lord Howe Rise fits as in Steinberger et al., 2004; Torsvik et al., 2008b) produced the statistically best GMHRF. The difference between observed and predicted Hawaii–Emperor hotspot tracks is much reduced compared with earlier GMHRFs (compare Figs. 2.5 and 2.6).

The GMHRF of Doubrovine et al. (2012) showed that Africa has been dominated by northward movement since the Cretaceous and the net longitude component for a specific location in Africa (15° N, 20° E) is less than 10° for the past 120 Myr (Fig. 2.8a). For Africa, the Doubrovine et al. (2012) GMHRF shows gross similarities (statistically overlapping) with that of Torsvik et al. (2008c), except for 60 and 70 Ma. For the past 100 million years we also show the movement of Africa based on the Indo-Atlantic moving hotspot (O'Neill et al., 2005; the 'rotation motor' of Seton et al., 2012) and 'slab-fitting' (van der Meer et al., 2010) frames. These frames are compatible with the GMHRFs for only the past 40 million years; the slab-fitted (subduction) frame is a direct derivation of the Indo-Atlantic hotspot frame in which longitudes of all the Earth's continents were calibrated so that the location of subducted material (identified in the mantle by seismic tomography) was consistent with the plate tectonic model. The subduction frame consistently predicts more westerly longitudes (Fig. 2.8a) than any other reference frame, and is to a large extent based on positioning the Farallon–Mexican–Caribbean–South American slabs near (below) the Andean-type continental margin of the Americas.

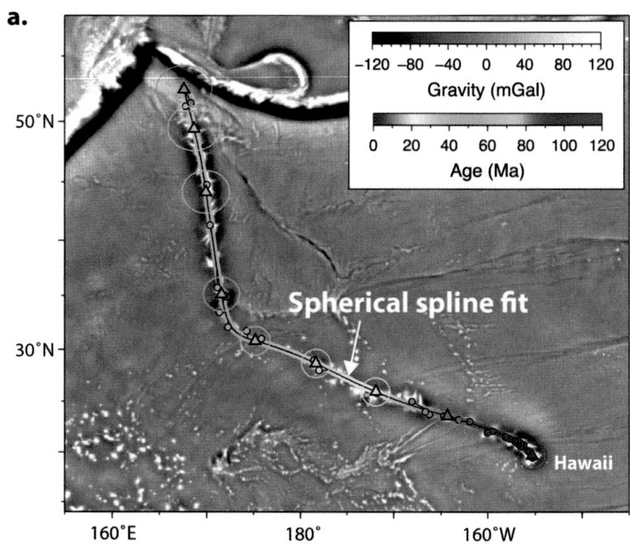

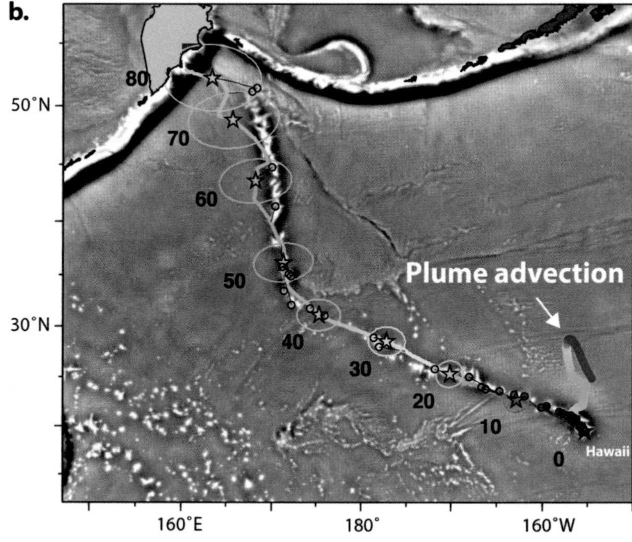

Fig. 2.6 (a) Smoothed spherical spline with 95% uncertainty circles (10 Myr increments) fitted to the age data for the Hawaiian hotspot track. Small dots with tie-lines to the spline are locations of radiometrically dated samples. (b) Computed motion and track for a moving hotspot and plate circuit model B (with Australian plate circuit model 1 in Doubrovine et al., 2012). Model track calculated by combining absolute plate motions and hotspot motions is shown as rainbow-coloured paths (colour-coded according to age) with crosses at 10 Myr increments (ellipses show the 95% uncertainty regions). Thicker rainbow-coloured swaths represent the surface motion of the Hawaii Hotspot estimated from numerical models of plume conduit advection. The global moving hotspot model of Doubrovine et al. (2012) was calculated from five hotspot tracks, four as in Fig. 2.5, and also the New England seamount chain. In modelling the Réunion hotspot track, Doubrovine et al. (2012) excluded Site 707 (erupted close to a ridge and about 500 km from the Réunion plume conduit) and Chagos (Site 713) was modelled to belong to the African (and not Indian) Plate at ~50 Ma (Torsvik et al., 2013).

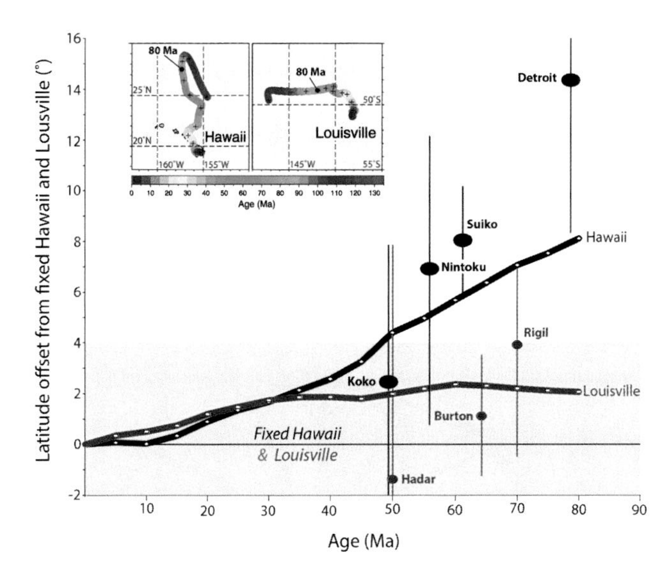

Fig. 2.7 Palaeomagnetically derived latitudes for seamounts along the Emperor (Detroit, Suiko, Nintoku, and Koko; Tarduno et al., 2003; Doubrovine & Tarduno, 2004) and Louisville Chains (Rigil, Burton, and Hadar; Koppers et al., 2012). Latitudes are shown as latitude offsets from zero (observed latitude minus latitude of Hawaii and Louisville) and compared with latitudinal estimates of plume advection (Doubrovine et al., 2012). For a system with fixed plumes (and no true polar wander) all the latitudes should be on the zero line, i.e. same latitude as Hawaii and Louisville today. Inset diagrams show the modelled surface motion (colour coded) of the Hawaii and Louisville hotspots back to 120 Ma. Hawaii is characterised by southward motion for the past 100 Myr whilst Louisville is dominated by eastward motion. This is also clearly reflected by palaeomagnetic data where the Emperor seamounts show large latitude deviations whereas latitudes of the Louisville seamounts are within error of the present-day location of Louisville (but show a small systematic southward component from Rigil to Hadar). The Emperor seamounts indicate more advection of the Hawaii plume conduit than that estimated from numerical modelling (Chapter 14), but the latitudes estimated for Koko, Nintoku, and Suiko are clearly within errors.

The GMHRF of Doubrovine et al. (2012) shows that absolute velocities for 'Africa' peaked at 5.5 cm/yr in the Upper Cretaceous (Campanian), declined during most of the Cenozoic, and for the past 20 Myr (Neogene) have averaged 1.4 cm/yr (Fig. 2.8b). North-easterly directed (as seen today) motion of the African Plate is recognised over the past 40 Myr. Absolute rotations for the remaining plates are calculated by adding the relative plate motions to the absolute rotations of southern Africa. Two examples of this are shown in Fig. 2.8c: Europe (Oslo) has not moved much in the last 120 Myr and has been extremely stable for the past 30 Myr, with absolute velocities below 1.5 cm/yr (Fig. 2.8b). Conversely, North America has moved about 50° westwards in the past 120 Myr; the dominant longitude change (Fig. 2.8c) is the cause of the so-called Cretaceous apparent polar wander 'still-stand' in North America (Torsvik et al., 2008b).

Palaeomagnetism

The Earth's magnetic field is described by its inclination (angle with respect to the local horizontal plane), declination (angle with respect to the local north–south meridian), and field strength. The inclination of the Earth's magnetic field varies systematically with latitude (Fig. 2.9), which is of prime significance for palaeomagnetic reconstructions. For example, at the north magnetic pole, the inclination of the field is +90° (vertically down), and at the Equator the field inclination is zero (horizontal). The magnetic north and south poles normally differ from the geographical north and south poles because the magnetic axis is inclined relative to the geographical (rotation) axis (today ~11.5°). The magnetic axis, however, is slowly precessing around the geographical axis and, over a period of a few thousand years, it is probable that the averaged magnetic poles have corresponded reasonably well with the geographical poles. That is known as the geocentric axial dipole (GAD) hypothesis. We can therefore imagine that a magnetic dipole is placed at the centre of the Earth and aligned with the Earth's rotation axis (Fig. 2.9b).

There are various ways in which a rock can acquire a remanent (permanent) magnetisation parallel to the Earth's magnetic field at a given location (Fig. 2.9a), and the cooling of a basaltic lava flow from high temperature is an example. The most important magnetic mineral in basaltic rocks is titanomagnetite with maximum Curie temperatures (i.e. temperature above which a magnetic material loses its magnetism because of thermal agitation) near 580 °C (pure magnetite). During a basaltic volcanic eruption, the temperature of a lava is about 1,200 °C, but, when cooled below the Curie temperature, the magnetic minerals acquire a *thermo-remanent magnetisation* (TRM) aligned with the Earth's magnetic field (Fig. 2.9b). The declination, inclination, and magnetisation intensity (proportional to the strength of the field), can be measured in the laboratory. In a comparable way, during the deposition of sediments, magnetic mineral grains settle on average in the direction of the Earth's magnetic field and a *detrital remanent magnetisation* (DRM) is acquired. DRMs, however, are commonly associated with inclination shallowing errors, which are latitude dependent (Tauxe & Kent, 2004; Kent & Tauxe, 2005; Kodama, 2009;

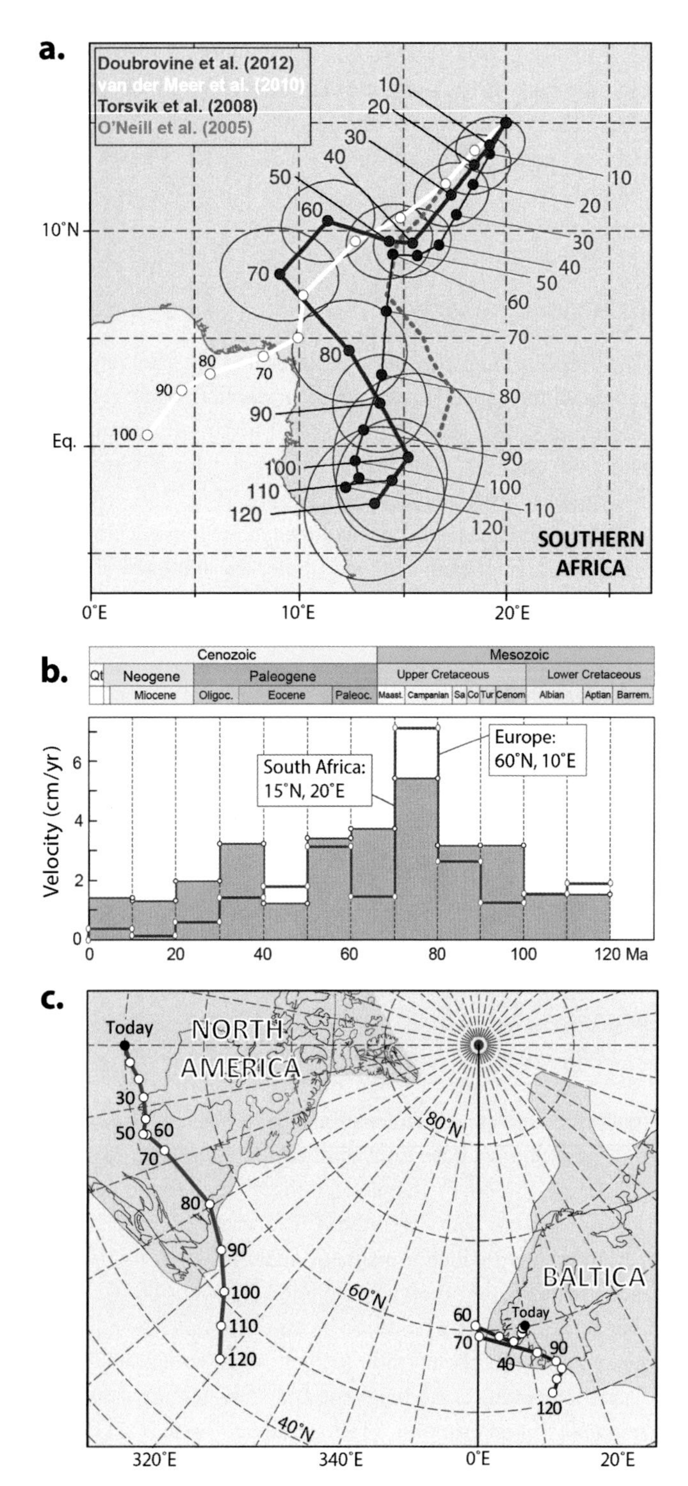

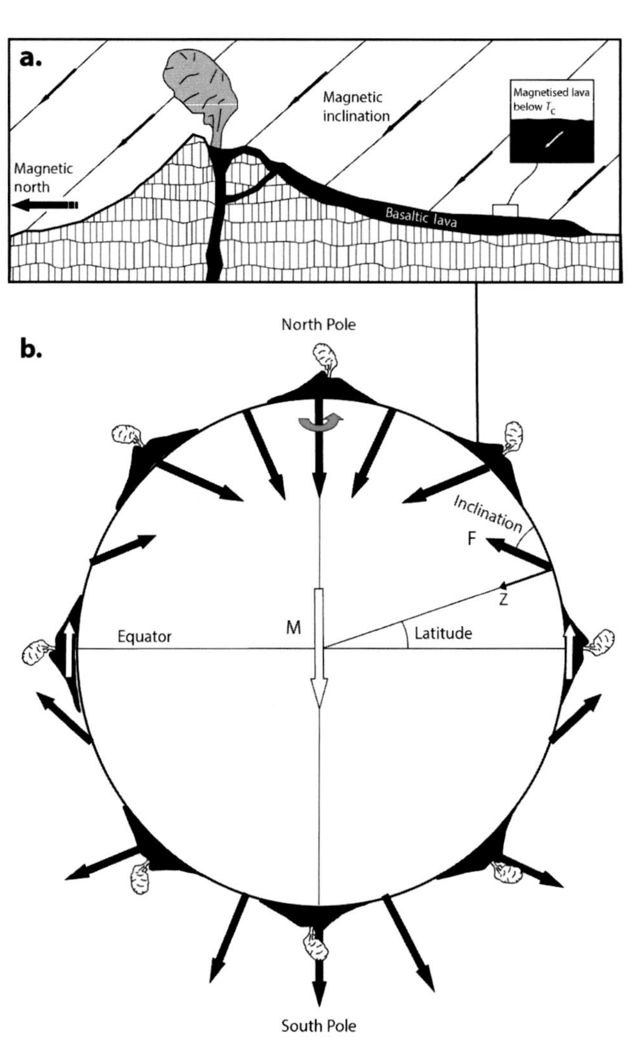

Fig. 2.8 (a) Africa motion (15° N, 20° E) in two different global hotspot reference frames (GMHRFs) at 10 Myr increments. Red dots and lines correspond to the GMHRF of Doubrovine et al. (2012). Blue is based on an earlier GMHRF by Torsvik et al. (2008b). For the Cretaceous (120–80 Ma) and the past 50 Myr, the two reference frames show gross similarities, but there are notable differences at 70 and 60 Ma. We also compare these GMHRFs with an Indo-Atlantic reference frame for the past 100 Myr (stippled brown line; O'Neill et al., 2005) and a slab-fitting frame (white thick lines; van der Meer et al., 2010) derived from the former (only longitude shifts). (b) Absolute velocities for a location in Africa and Europe (Oslo). (c) North America and Europe (Oslo) plate motion based on Doubrovine et al. (2012).

Fig. 2.9 (a) Example of acquisition of a thermoremanent magnetisation at intermediate northerly latitudes (acquired in a normal polarity field similar to today's). A lava will acquire a thermoremanent magnetisation upon cooling below the Curie temperature (T_C), and the inclination will parallel the inclination of the external field and have declination due north. (b) Field lines at the Earth's surface for a geocentric axial dipole. At the Equator the inclination is flat (zero), and at the north and south poles the inclination is vertical (+90° and −90° respectively). The inclination recorded in volcanoes formed on the Earth's surface is dependent on the latitude. Declinations in a normal polarity field, such as today's, should point to the north. Note that longitude cannot be determined because volcanoes erupted at different longitudes will always have the same inclination and declination.

Torsvik et al., 2012). Inclination shallowing is commonly predicted from:

$$\tan(\text{inclination}_{\text{observed}}) = f \times \tan(\text{inclination}_{\text{field}})$$
(f is flattening factor; $f = 1$, no flattening)

Based on the measurement of the remanent inclination, we can calculate the ancient latitude for an outcrop on a continent when the rock formed:

$$\tan(\text{inclination}) = 2 \times \tan(\text{latitude})$$

In addition, the remanent declination, which deviates from 0° or 180° (depending on the polarity of the Earth's magnetic field), provides information about the subsequent rotation of a continent.

The inclination and declination change with the position of the sampled rock on the globe as its host terrane unit moves with time (Fig. 2.9b), but the direction of the magnetic pole of a GAD is independent of the locality at which the rock acquired its magnetisation. Thus it is practical to calculate pole positions so as to compare results from various sites and to provide objective data to underpin plate tectonic reconstructions. Ideally, as a time average, a palaeomagnetic pole (calculated from the declination, the inclination, and the geographical site location) for a newly formed rock will correspond to the geographical pole. If a continent moves later, the palaeomagnetic pole moves with the continent. To create a reconstruction with palaeomagnetic poles, we therefore have to calculate the rotation (Euler) pole and angle that will bring the palaeomagnetic pole back to the geographical north or south pole, and then rotate the continent by the same amount. In our example (Fig. 2.10a), a Late Permian palaeomagnetic pole from Oslo, Norway, will position Baltica at latitudes between 15° and 50° N, causing Oslo to have been located at 24° N. Therefore, because the current latitude of Oslo is 60° N, Baltica must have drifted northwards since the Permian.

Because its palaeolongitude is unknown from palaeomagnetic measurements alone, we can position Baltica at any longitude we wish, subject to other geological constraints. In addition to that, we cannot tell in old rocks whether a palaeomagnetic pole is a south or north pole. In Fig. 2.10a, we assumed that the pole was a north pole, but, if we used a south pole, Baltica would plot in the southern hemisphere but in a geographically inverted orientation (Fig. 2.10b). Hence, there is freedom to select north or south poles when producing reconstructions, placing the continent in an opposite hemisphere and rotated by 180°. The hemispheric choice is simple in this case, since Baltica had formed part of Pangea from 320 Ma, but in the deeper past it can be a problem.

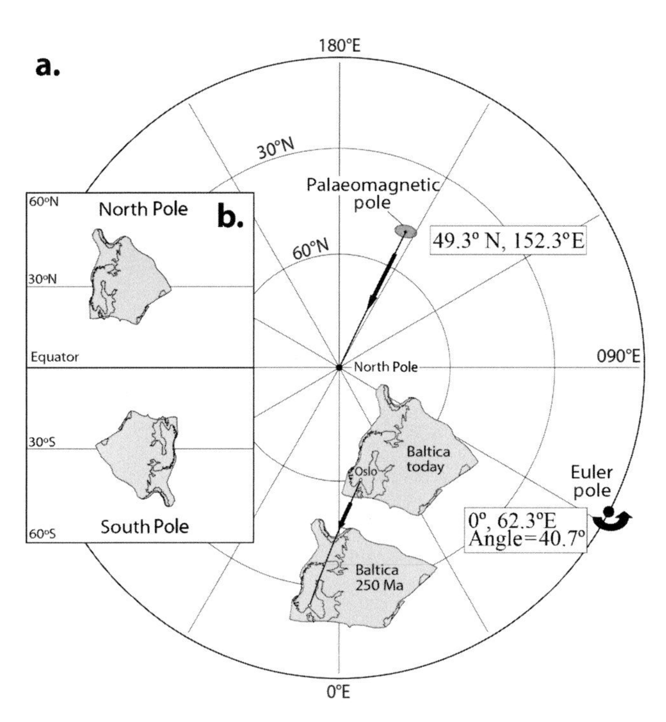

Fig. 2.10 (a) The reconstruction of a continent, in this diagram Baltica, is achieved as follows. Determine the Euler pole needed to rotate a palaeomagnetic pole (in this case a 250 Ma pole: 49.3° N, 152.3° E from Oslo, Norway) to the geographical North Pole (we calculate 0°, 62.3° E and a rotation of 40.7°). This Euler pole is then used to rotate the continent by the same amount. Thus Baltica today is rotated back about this pole to the position it occupied in Late Permian times at 250 Ma. (b) In (a), we assumed that the palaeomagnetic pole was a north pole. If we assume that it was a south pole, then the continent will have been situated in the opposite hemisphere, and geographically inverted (after Torsvik & Cocks, 2005).

Apparent Polar Wander (APW) Paths

Palaeomagnetic results can be expressed in terms of palaeopoles that are calculated using the GAD model. Those palaeopoles can in turn be used to construct APW paths. In this way, the motion of the polar axis relative to the continent (held fixed) is visualised. Many APW paths have been published for Baltica and its younger incarnations (Stable Europe) and the currently accepted path is shown in Fig. 2.11a (Torsvik et al., 2012). By keeping Baltica/Europe fixed, mean south poles for the Cambrian are located in Arctic Siberia (not shown in diagram), followed by a drift of the pole over Arabia and Central Africa in the Ordovician to the Atlantic off the north-east corner of Brazil in the Silurian–Devonian cusp. Subsequent southward movement brought the south pole near

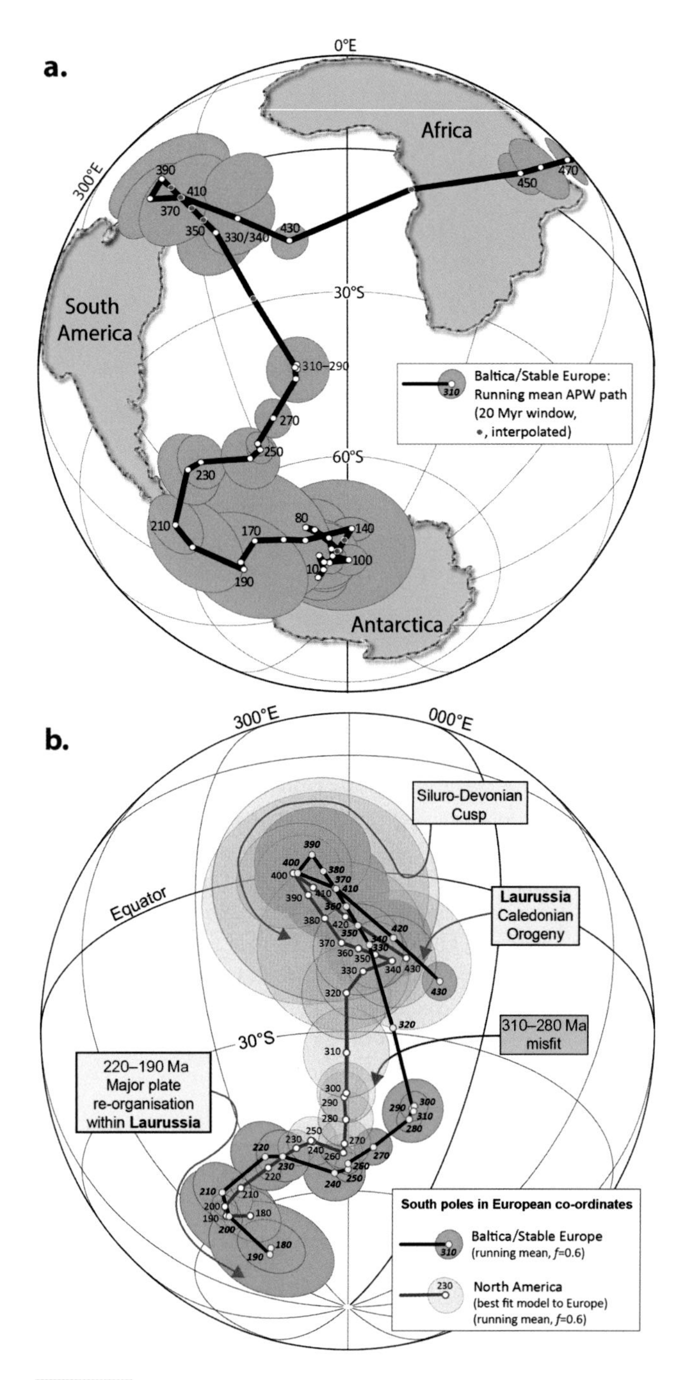

Fig. 2.11 (a) Running mean APW path (20 Myr window) for Baltica/Stable Europe shown with 95% confidence ovals at 10 Myr intervals. The APW path shows the movement of the south pole relative a fixed Baltica/Europe (grey shaded circles are interpolated). (b) Best fit of North American and Baltic/Stable Europe APW paths (running means) from 430 Ma (formation of Laurussia) to 190 Ma. Late Carboniferous–Early Permian poles (310–280 Ma) have been excluded in the fitting procedure, and between 220 and 190 Ma the Euler fit changes from 78.7° N, 161.9° E (angle = −31.0°) to 69° N, 154.8° E (angle = −23.6°). Palaeomagnetic input poles listed in Torsvik et al. (2012).

Patagonia in the Triassic, after which the pole remained close to Antarctica in the Jurassic.

From palaeomagnetic data, we calculate APW rates, the latitude for a specific location through time, north–south latitudes (longitude unknown), and the speed and change in angular rotation of a continent (Fig. 2.12). In our Baltica example, Oslo has drifted northwards from 60° S to 60° N during the Phanerozoic. North–south velocities were notably high (about 12 cm/yr) during the Late Ordovician–Early Silurian (just prior to and during collision with Laurentia and the subsequent formation of Laurussia), and the Late Carboniferous (Pangea assembly). Angular rotations peaked in the Early Ordovician, when Baltica was rotating counter-clockwise at rates of up to 5° /Myr.

In Fig. 2.12a we show an example of reconstructing Baltica's changing position over the past 500 million years based on the palaeomagnetic data (Fig. 2.11a). In Late Cambrian and Early Ordovician times (500 and 470 Ma reconstructions) Baltica was strongly rotated (compared with today) and was located in the southern hemisphere. Since Late Devonian times, Baltica has largely undergone northward drift. The westerly longitudinal movement of Baltica (with decreasing time in Fig. 2.12a) is entirely subjective, but there is an analytical trick that can constrain longitude semi-quantitatively for the times after Pangea formed at about 300 Ma. But that trick requires that we identify the continent that has moved the least in longitude, and, after reconstructing that particular continent, all the other continents must be reconstructed relative to it in order to minimise longitudinal uncertainty. We therefore need to build a global APW path through the continent's history. Taking relative plate motions into account, palaeomagnetic data from all the major continents can be combined in a global APW path. However, the choice of an appropriate reference plate for that global APW is critical. Judicious selection of that reference plate results in reduced longitudinal uncertainty in palaeogeographical reconstructions. The 'zero-longitude' Africa approach (Burke & Torsvik, 2004; Torsvik et al., 2008c) makes use of the observation that Africa has remained the continent that has moved least in longitude since the breakup of Pangea. To build a global APW path (GAPWaP) we must first compile all reliable palaeomagnetic data worldwide (from stable cratons) and rotate all the raw poles to South Africa co-ordinates to calculate a running mean GAPWaP (alternatively use spherical splines). Next we calculate Euler poles from the GAPWaP to reconstruct South Africa, and then the remaining plates are reconstructed with plate chains relative to South

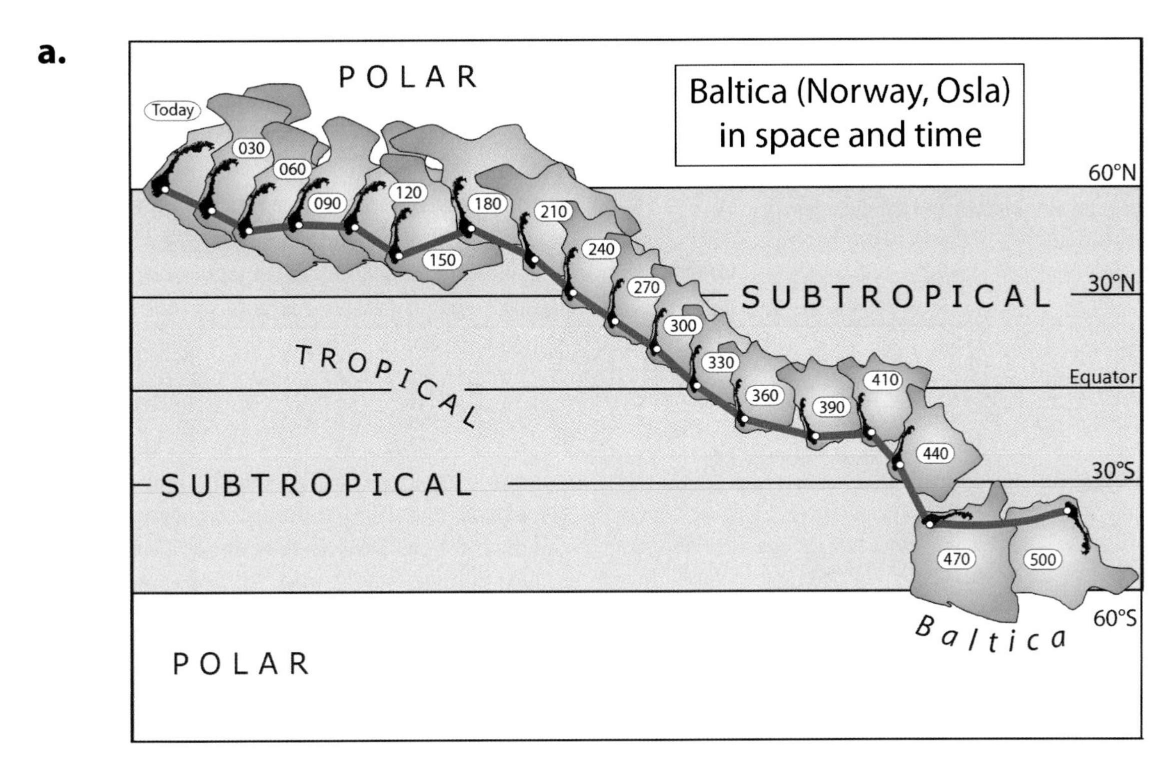

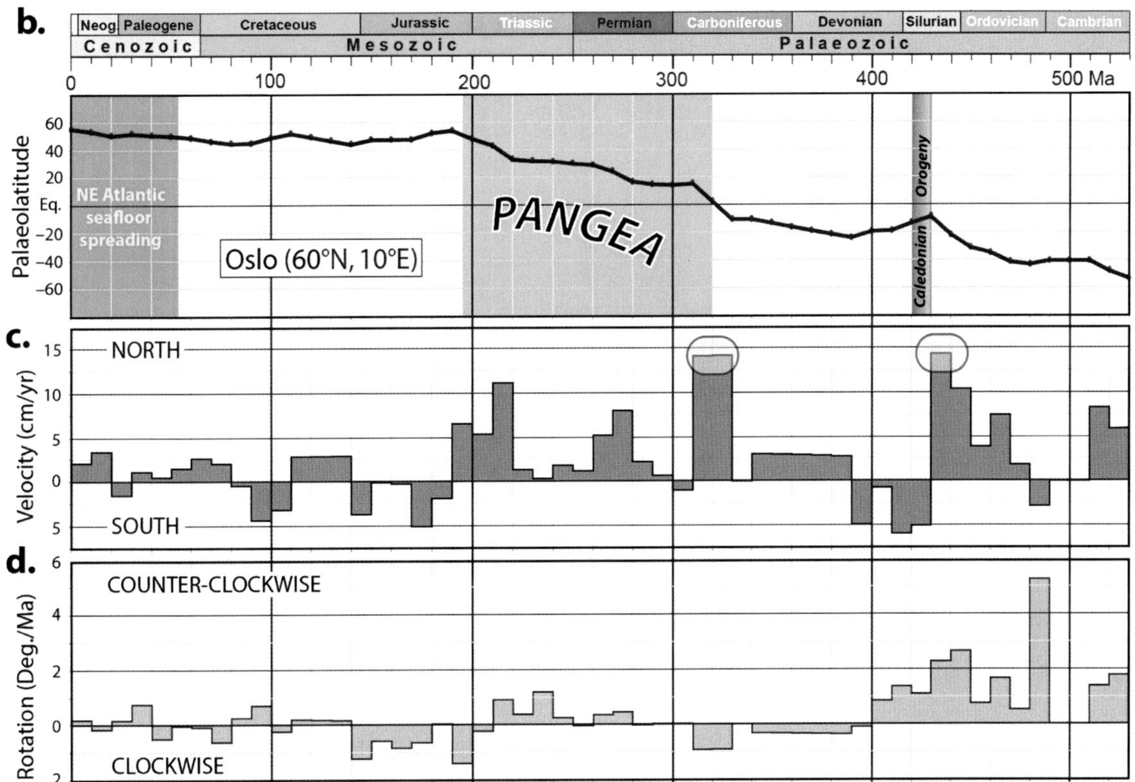

Fig. 2.12 (a) The drift history for Baltica (Oslo) at 30 Myr intervals based on the APW path in Fig. 2.11a. We also indicate the collision with Laurentia (Caledonian Orogeny), the time interval Pangea existed, and sea-floor spreading in the north-east Atlantic when Baltica separated from Laurentia (Greenland), which is still ongoing. (b) Palaeolatitudes for Oslo in 10 Myr intervals. (c) Palaeolatitudes for Oslo separated into northward and southward motion (minimum velocities since palaeolongitude change are 'unknown'). (d) Rotation rate of Oslo.

(1) Compile **Palaeomagnetic** poles worldwide

(2) Rotate all poles to Southern Africa coordinates using plate circuits (Fig. 2.4)

(3) Construct a **G**lobal **A**pparent **P**olar **W**ander **P**ath (GAPWaP) using the running mean method (e.g. 20 Myr window and 10 Myr interval)

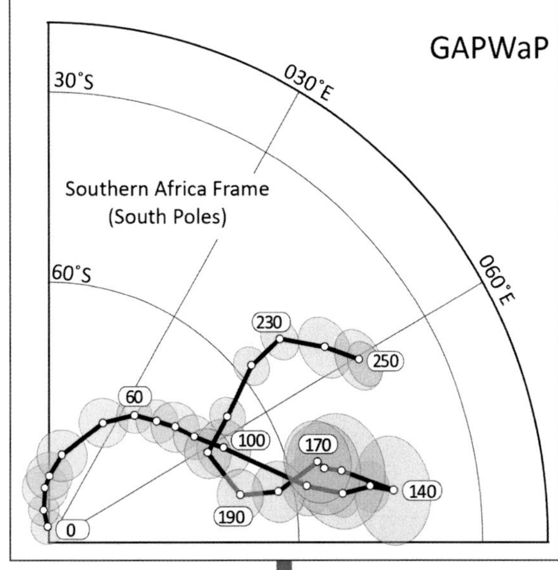

(4) Calculate 'absolute' Euler Rotation Poles for (i) Southern Africa and (ii) rotations for the remaining plates by adding relative plate motions as in Step (2) to the absolute rotations of Southern Africa.

(5) **Global Palaeomagnetic Reference Frame**
'Zero longitude' Africa approach

Fig. 2.13 The workflow to build a global palaeomagnetic reference frame since Pangea formed.

Africa. This is termed a palaeomagnetic reference frame, the 'zero longitude' Africa approach (Fig. 2.13).

APW Paths and Plate Circuits

Relative plate circuits between plates are determined by magnetic anomalies and fracture zones (Figs. 2.3 and 2.4), but only after about 180 Ma, which is the oldest preserved *in situ* sea floor. Before sea-floor spreading there can be hundreds of kilometres of stretching of the continental crust (here referred to as pre-drift extension) and therefore plate fits before breakup should be much tighter. As Euler poles and their angles change during sea-floor spreading, pre-drift Euler poles can also change through geological history. Theoretically, Euler poles can be derived from a comparison of APW paths from conjugate continents (Fig. 2.11b), but their quality is often not good enough for this purpose, and thus additional information is needed, such as estimates of stretching from basin analysis or crustal thickness estimates from seismic refraction data or gravity inversion. We assume an average continental thickness of say 35 km: anything less tells us something about the amount of stretching but not the timing. But on passive margins we normally consider that pre-drift extension stopped when sea-floor spreading started. As an example, we show a very tight fit between Greenland and Norway (Europe) in Fig. 2.14b. At 55 Ma and just prior to sea-floor spreading in the NE Atlantic (Fig. 2.14a), the current continent–ocean boundary transitions (COBs) for Greenland and Europe almost coincide (as they should), but in the Late Triassic they show a huge overlap, up to almost 400 km (Fig. 2.14b). Bullard et al. (1965) developed the first computer-generated fit by matching 500-fathom contours of conjugate margins in the Atlantic realm. Their fit for the North Atlantic matches North American and European palaeomagnetic poles reasonably well from Middle Palaeozoic to Early Mesozoic times (Van der Voo, 1993; Torsvik et al., 1996; 2012). For that reason, many North Atlantic reconstructions use the Bullard et al. (1965) fit despite the somewhat problematic geological implications this reconstruction creates. Beck and Housen (2003) attempted to create an alternative to the Bullard fit by fitting 300–200 Ma palaeomagnetic poles from North America and Europe, but the amount of continental overlap of Europe and Greenland (over 1000 km) is beyond any geological acceptance.

On the basis of this unrealistic continental overlap, Beck and Housen (2003) therefore concluded that mathematically fitting APW paths from North America and Europe does not provide satisfactory results. But, by excluding Late Carboniferous to Early Permian poles (Fig. 2.11b), or substituting the North America data with Gondwanan data (Torsvik et al., 2006), we can reconcile the 430–330 Ma poles (defining the so-called Silurian–Devonian cusp) and 290–220 Ma poles with the same Euler pole. However, in order to depict the change in APW wander in the Late Triassic to Early Jurassic (more open cusp) we must gradually change the Euler pole between 220 and 190 Ma. The amount of extension predicted from our plate model compares much better with extension predicted from gravity inversion (or seismic data) than the classic Bullard fit (Fig. 2.14c). The implications of the plate model changeover between 220 and 190 Ma are discussed in Chapter 11.

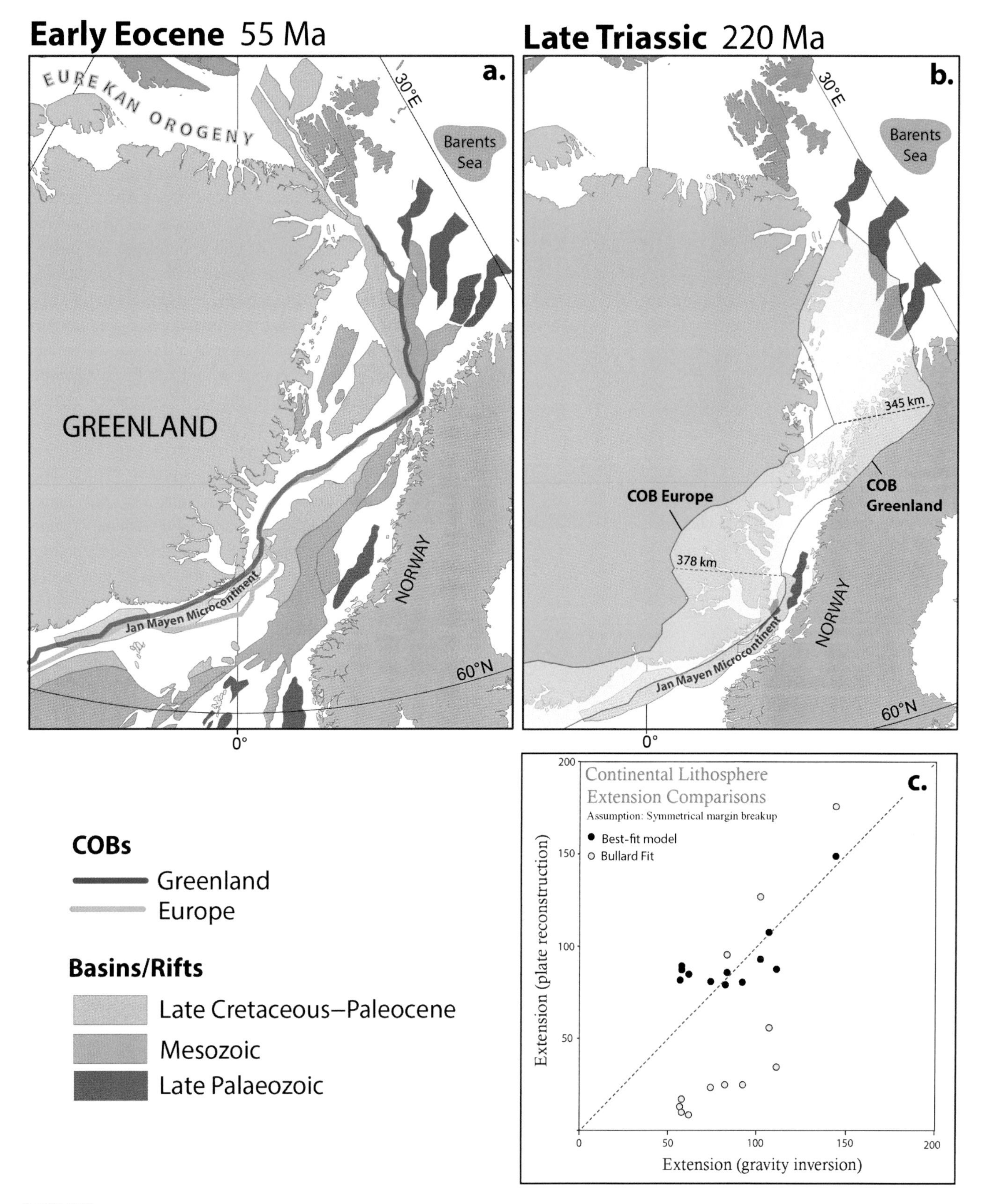

Fig. 2.14 Our plate model between Greenland and Norway (Europe) at 55 Ma (just prior to breakup and sea-floor spreading) and at around 220 Ma (a, b). The large COB (Continent–Ocean Boundary) overlap witnesses the total amount of pre-drift extension from then until breakup. We also show basins/rifts that developed from the Late Palaeozoic to the Paleocene. The Jan Mayen Microcontinent (now part of Eurasia) consisted of several distinct segments; the original size of these blocks at 220 Ma (b) was smaller than in present times because they are now composed of extended continental crust (18–20 km thick). (c) Comparing lithosphere extension derived from the plate model and gravity extension (crustal thickness) along 11 profiles between Greenland and Europe (Alvey et al., 2008). Symmetrical margin breakup is assumed and the values represent half of the continental overlaps in (b).

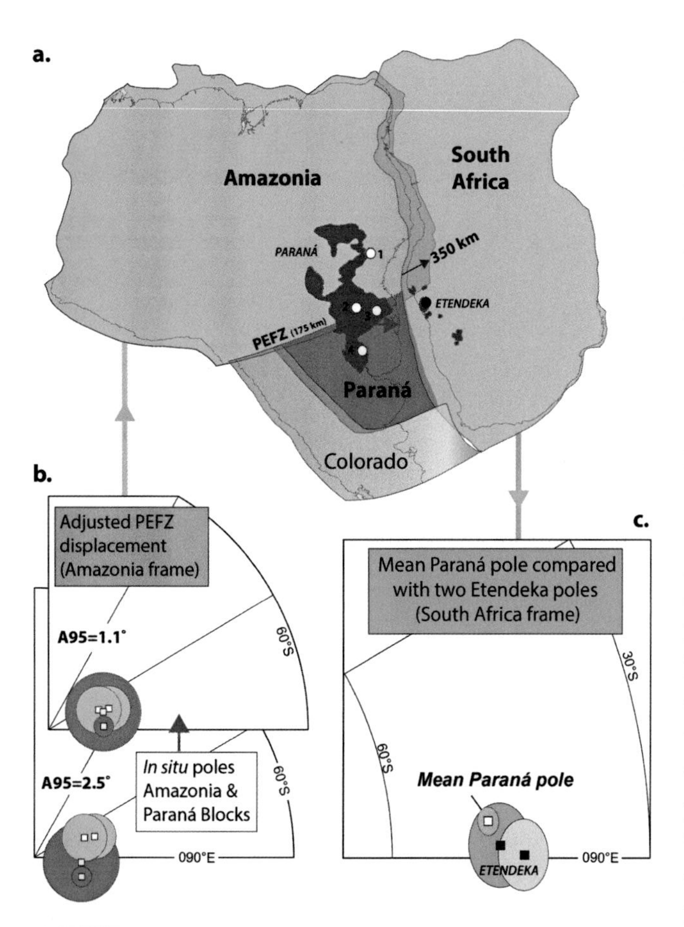

Fig. 2.15 (a) The South American intra-plate model where South America has been broken into several different blocks. Here we show the Amazonia, Paraná, and Colorado blocks and palaeomagnetic sampling sites across the PEFZ (Paraná–Etendeka Fracture Zone). South America is compared with the conjugate South Africa block in a pre-breakup (Jurassic) reconstruction. We also show the palaeomagnetic sampling location for the Etendeka lavas. The reconstruction includes 350 km of pre-breakup extension. (b) Four palaeomagnetic poles from the Paraná LIP; two south of and two north of the PEFZ. After adjusting for PEFZ displacement (Torsvik et al., 2009) the four poles fit perfectly (A95 = 1.1°). (c) Mean Paraná pole (N = 4) rotated to South Africa co-ordinates and compared with two recent 134 Ma Etendeka poles (Dodd et al., 2015; Owen-Smith et al., work in progress).

Palaeomagnetic data have also been used to test pre-drift South Atlantic fits and also intra-plate geometries in order to avoid unrealistically large continental overlaps (pre-drift extension) or gaps along the conjugate South Atlantic margins. Torsvik et al. (2009) divided South America into four main domains (Amazonia, Paraná, Colorado, and Patagonia). One of the most important boundaries (Fig. 2.15a) is between Amazonia and Paraná (Paraná–Etendeka Fault Zone), modelled as a transtensional boundary with an original lateral offset of about 175 km, and where dextral movements ceased at about 126 Ma. This boundary is important, not only to reduce the continental extended margin overlap on the Brazilian (Santos) margin to realistic numbers, but also because extension along this fault would made it easier, and a preferred location, for plume-related volcanism to occur (upside-down drainage) in order to explain the large extent of Paraná volcanism compared to that of Etendeka on the conjugate margin (Namibia). There is no consensus from either the surface geology or geophysical data on the exact localisation of intra-plate deformation in the South American continent, but the Paraná–Etendeka Fault Zone is recognised in the GOCE (Gravity field and Ocean Circulation Explorer) free air residual gravity field and in crustal and lithospheric thickness maps (Braitenberg, 2015; Assumpção et al., 2013; Chulick et al., 2013). High-quality palaeomagnetic data from the 134 Ma Paraná–Etendeka LIP came from both sides of the Paraná–Etendeka Fault Zone, and when the two Paraná Block mean poles are corrected for the Torsvik et al. (2009) model they become identical to the two Amazonian Block poles (Fig. 2.15b). Closing the South Atlantic (and rotating the mean Paraná pole to South African co-ordinates) shows that Paraná and two palaeomagnetic poles from Etendeka (Namibia) are in excellent agreement (Fig. 2.15c), thus demonstrating the power of palaeomagnetic data.

Calibrating Longitude: The Plume Generation Zone Method

LIPs (Appendix 1, Fig. 2.16a) are the result of catastrophic melting in the upper mantle, and although a globally agreed definition of a LIP does not exist, the original definition by Coffin and Eldholm (1994) stressed the importance of the large areal extent of predominantly mafic igneous rocks emplaced in an intra-plate setting, and characterised by igneous pulses of short duration.

By reconstructing the positions of LIPs over the past 300 Myr, it has become clear that most originated from plumes at the edges of Tuzo and Jason (Fig. 2.16b,c), the plume generation zones (PGZs; Burke et al., 2008). Fluxes of hot and buoyant material from the PGZs are related not only to the emplacement of LIPs, but also to about 85% of kimberlite intrusions (Fig. 2.16c; Torsvik et al., 2010), and many hotspot volcanoes (Fig. 2.2b) such as at Hawaii (Pacific Ocean), Réunion (Indian Ocean), and Iceland (North Atlantic Ocean). But there are anomalies from this pattern; for example, Yellowstone in western North America. In fact, several hotspots, one LIP (Columbia River Basalt), and most

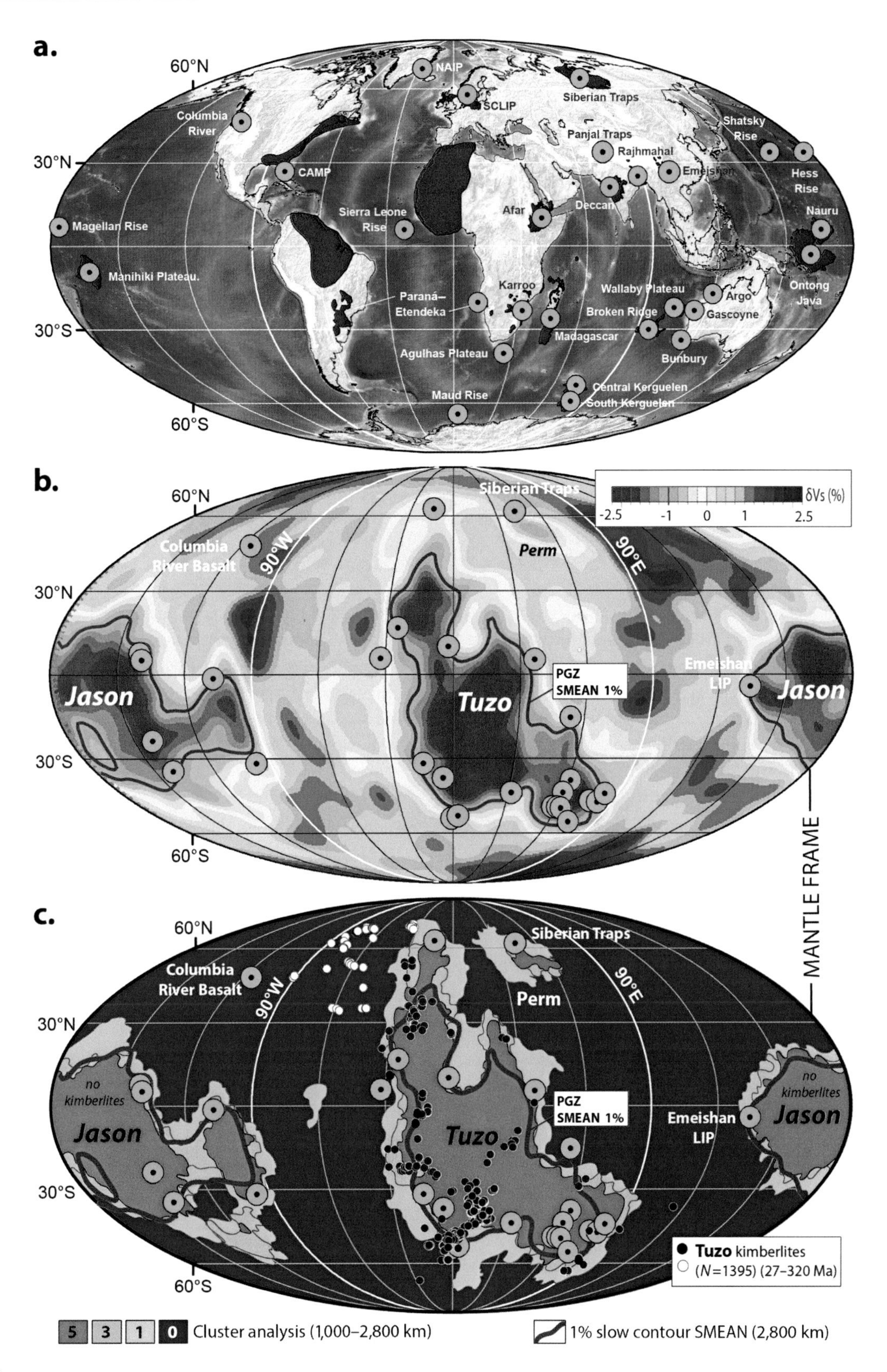

Fig. 2.16 (a) Large igneous provinces (LIPs, 15–297 Ma) and their estimated eruption centres. The Panjal Traps are allochthonous and thus associated with some reconstruction uncertainties in (b) and (c). The areal extent of the Central Atlantic Magmatic Province (CAMP) is liberal in the inclusion of all ~201 Ma basalts, sills and dykes. (b) The LIPs reconstructed and superimposed on the SMEAN tomographic

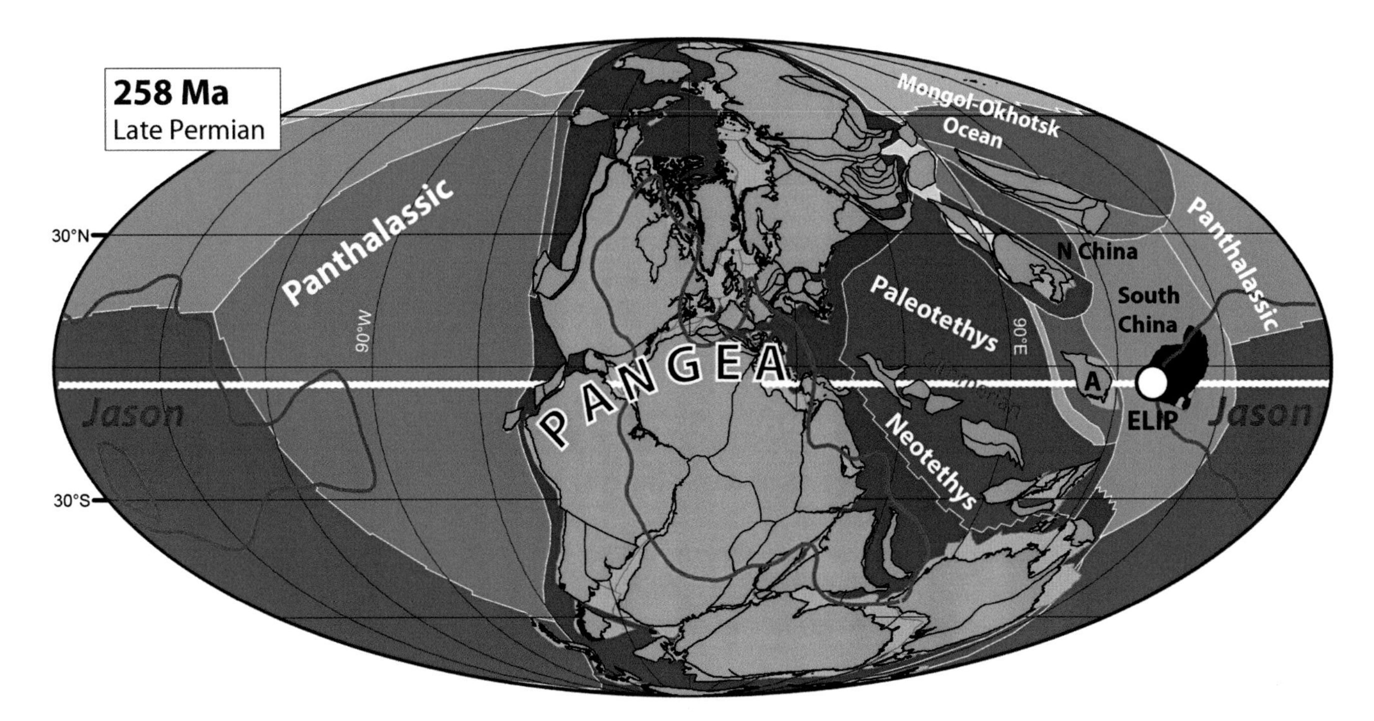

Fig. 2.17 Global palaeomagnetic plate reconstruction (Mollweide projection) at 258 Ma (Domeier & Torsvik, 2014) when Pangea was centred above Tuzo. Pangea formed in the Late Carboniferous but early breakup is witnessed already in the Early Permian by the opening of the Neotethys. The Cimmerian terranes leaving Pangea included parts of Iran, Turkey, Afghanistan, Tibet, Burma, Thailand, and Malaysia (Sibumasu). The white heavy line is the palaeolatitude determined palaeomagnetically for the ~258 Ma Emeishan LIP (ELIP) of South China. If the ELIP erupted above the plume generation zone (PGZ) it could have been located where the white line intersects a PGZ (here the 1% slow SMEAN contour in red), but the only likely position would be along the western edge of Jason. Net true polar wander at this time was zero (Fig. 2.23b).

Late Cretaceous/Tertiary kimberlites (white circles in Fig. 2.16c) from north-western America are anomalous.

The overall correlation of reconstructed eruption sites of LIPs and kimberlites (Fig. 2.16b,c), at least since about 300 Ma, when Pangea formed, indicates the long-term stability of Tuzo and Jason. That remarkable correlation between surface and mantle features provides an original way of reconstructing the longitudinal position of continents in which LIPs and kimberlites have been found. As an example, we can now reconstruct the longitude of South China in the Late Permian (Fig. 2.17). Although that was at a time when Pangea was amalgamated, the supercontinent did not include South China, which was thus without longitudinal constraints. In the Late Permian, South China, based

Fig. 2.16 (*cont.*) model (Becker & Boschi, 2002) at 2800 km depth (δV_s is the S-wave anomaly). LIPs are reconstructed using a moving hotspot frame or a TPW-corrected palaeomagnetic frame before 120 Ma. We also show the 1% slow contour in this model (SMEAN 1%) that we have used extensively as the proxy for the plume generation zones (Torsvik et al., 2006). The African and Pacific large low shear-wave velocity provinces (LLSVPs) are the Tuzo and Jason provinces; they are almost antipodal and centred nearly on the Equator. Columbia River Basalts and the Siberian Traps are not directly correlated with the edges of Tuzo and Jason, but the Siberian Traps overlie a smaller anomaly in the lowermost mantle beneath the Perm region (Lekic et al., 2012). The Emeishan LIP has been calibrated in longitude (Fig. 2.17) so that it falls above the western edge of Jason. (c) Reconstructed LIPs as in (b) but shown together with reconstructed kimberlites for the past 320 Myr and draped on a seismic voting-map in the lower mantle. Lekic et al. (2012) examined five global shear-wave tomographic models and produced a map that described whether a geographical location was above a seismically slower-than-average velocity region in the mantle below 1,000 km depth. Within contour 5, all five tomographic models show slower-than-average seismic velocities, whereas, for example, contour 3 outlines the area in which three of five models are in agreement. Contours 5–1 (only 5, 3, and 1 are shown here for clarity) define Tuzo and Jason (seismically slow regions) in addition to the smaller Perm anomaly. Contour 0 (blue) denotes faster regions in the lower mantle. The 1% slow SMEAN contour is shown for comparison. For the past 320 Myr, 80% of all reconstructed kimberlite locations (black dots) erupted near or over the Tuzo PGZ. The most 'anomalous' kimberlites (17%) are from Canada (white dots).

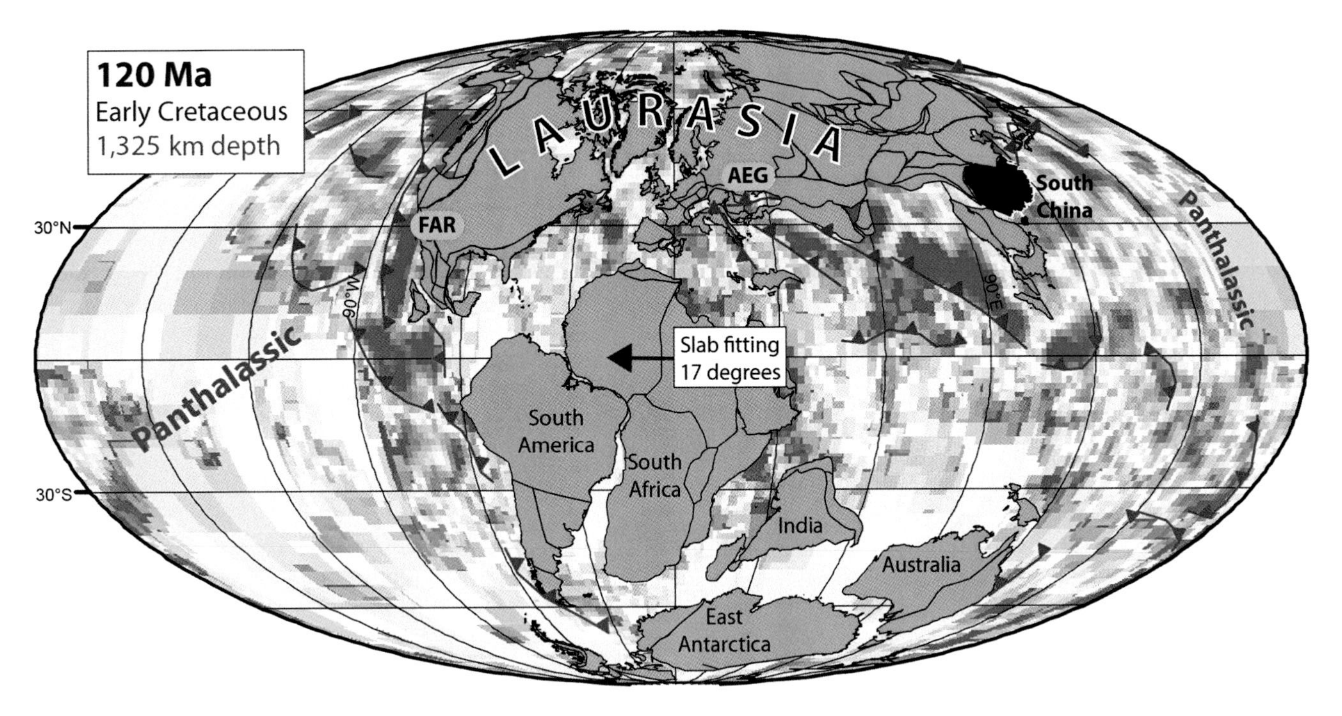

Fig. 2.18 Example of calibrating a 120 Ma reconstruction (based on palaeomagnetic data) in longitude (van der Meer et al., 2010). A reasonable fit was obtained between a 1,325 km tomographic depth-slice, Farallon and Aegean Tethys Ocean slabs when an absolute plate motion frame (Torsvik et al., 2010) was shifted 17° westwards. Interpreted subduction zones shown in red. AEG, Aegean slab; FAR, Farallon slab.

on palaeomagnetic and biostratigraphical data, was in tropical latitudes but the separation distance from Pangea is unknown. The 258 Ma Emeishan LIP is on the South China block, and palaeomagnetic results position it at latitudes around 4° S (white thick line in Fig. 2.17). If that LIP had erupted above a PGZ, there are several possible longitudinal locations where that white line of latitude crossed the PGZ at that time. Pangea covered two of those options (above Tuzo), leaving only the options related to Jason. The reconstruction in Fig. 2.17, with the Emeishan LIP above the western margin of Jason at ~134° E, is the only realistic alternative. Thus both the latitude and longitude (Appendix 1) of South China in the Late Permian can be established (Torsvik et al., 2008c).

Calibrating Longitude with Subducted Slabs (Tomography)

Palaeomagnetic reconstructions can also be calibrated in longitude using the positions of subducted material (slabs) which was subducted along convergent margins of the continents. Slabs can be observed in tomographic models at depths from the Earth's surface down to the core–mantle boundary. From such images, a first-order estimate of the amount of subducted material can be made, and used as a constraint on plate reconstructions. Van der Meer et al. (2010) pioneered this concept to interpret the depth and timing of subduction of a total of 28 remnants of slabs in the mantle. Through correlation of those slabs with their geological record (fossil arcs and orogens), they determined a sinking rate of 12 ± 3 mm/yr, which allows surprisingly consistent global correlations of palaeo-subduction zones at the surface with seismic tomography images at depth. Thus the longitudes of ancient active margins can be constrained as far back as the Permian by adjusting a global plate motion model based on palaeomagnetic data for improved fits with slab remnants in the tomographic model.

Figure 2.18 shows an example where a global palaeomagnetic reconstruction at 120 Ma was adjusted 17° westward in order to best fit the Farallon and Aegean Tethys slabs (van der Meer et al., 2010), i.e. the entire Earth was rotated westward around an Euler pole at the North Pole. As with all reconstruction methods, there are several possible sources of error in the interpretation of these slab-fitting exercises, including the unknown slab dip, lower mantle thickening of slabs, and tomographic imaging. Furthermore, subduction histories could have been more complex than often shown, due to back-arc spreading, and accretion and deformation at the plate edges.

The slab-fitting model of van der Meer et al. (2010) was the first of its kind and can be dubbed a global subduction reference frame. Since then the ages of several slabs have been reinterpreted, there has been revision of ocean closures (e.g. Mongol–Okhotsk) and also development of more complex subduction models, e.g. along the western margin of North America (Fig. 13.6).

The Rock Record

It is of pivotal importance that palaeomagnetic reconstructions (constraining latitude and rotation) comply with known geological and tectonic constraints (opening and closure of oceans, mountain building, and more). The rock record is therefore important, and convergent plate boundaries have the greatest diversity in terms of both igneous and metamorphic rocks (Fig. 2.19). Conversely, oceanic crust originally formed at divergent plate boundaries can occasionally be obducted onto a continent (ophiolite), and dating that piece of old oceanic crust can tell an important story about ancient sea-floor spreading around that continent.

The Distribution of Fossils

Faunal and floral provinces have been identified at various times and places for the past two hundred years. Modern provinces are chiefly caused by differences in climate and temperature, which are in turn linked broadly to latitude; but also by the ability or inability of the biota to get across physical barriers, chiefly land (in the case of marine life) and seas (for terrestrial plants and animals). The identification of various faunal and floral groupings useful for palaeogeographical studies is particularly important during the Palaeozoic (Fig. 2.20a), older than any ocean floor that has been preserved.

The faunal principles have not changed since first set out by Cocks and Fortey (1982). Before any animal can be used, both its age and its individual ecology must be assessed correctly. The distribution of those animals with a planktonic, pelagic, or nektonic (swimming) lifestyle is controlled by oceanic currents and temperature, and thus, although the deposits in which their fossils are found usually form part of one or more of the terranes that we recognise as separate, those occurrences are of no significance when assessing the closeness or individuality of such terranes (Fig. 2.20b, c). Because the great majority of such animals are dependent on temperature, their occurrence may only be generally correlated with the latitude in which the animals lived, since warmer or cooler ocean currents are often found covering varied latitudes (Cocks & Verniers, 2000). In the Lower Palaeozoic, those planktonic groups are best represented by graptolites, a minority of the trilobites such as the pelagic *Carolinites*, cephalopods, chitinozoa, acritarchs, and possibly conodonts (the detailed ecology of which is still poorly understood). As the dispersal of these animals was very often rapid, the quickly evolving members of these groups include most of the best fossils for global correlation. In contrast, there are those animals with a benthic lifestyle, which were and are confined to the sea floor for their adult life, such as brachiopods, most trilobites, bivalves, gastropods, and most ostracods, and which were also temperature dependent. These may be divided into two groups: the majority, which lived in shallower-water seas on the continental margins, and which were therefore both latitudinally related and also confined to particular terranes; and a smaller number that lived below the thermocline and were thus independent of palaeolatitude, and were distributed on the deeper parts of continental shelves and on the ocean floors. It is the former, larger, group of benthic animals upon which we rely most strongly to provide the faunal support for terrane reconstructions. Cladistic analysis has been used to support some of the faunal distributions, but there are not enough truly robust trees for palaeobiogeographical use, and there are few such data from the Lower Palaeozoic. Despite the fact that the adult brachiopods and most trilobites were confined to relatively small sites, their larval stage was planktonic for shorter or longer periods and thus the genera dispersed as time progressed.

As a working rule, although different animals have and had very different dispersal rates (McKerrow & Cocks, 1976), many faunas appear to be separable into recognisable provinces if the oceanic width is above 1,000 km. If two terranes are at the same latitude (Fig. 2.20b), then the composition of their benthos will be largely the same if the terranes are close to one another. However, if the terranes drift apart, then the larvae of the descendant species of the original benthos will not cross the intervening deeper ocean after a certain period, and thus the discriminating palaeontologist will be able to identify the two terranes as different and separate. Comparably, if two terranes at the same latitude but with different benthos drift towards each other, then the separate provinces recognised by the palaeontologist in the two terranes in the earlier period will gradually merge into a single faunal province in the later period. However, that proximity does not prove that the two terranes actually collided, which needs further evidence, such as ‘stitching’ granites across the suture zone.

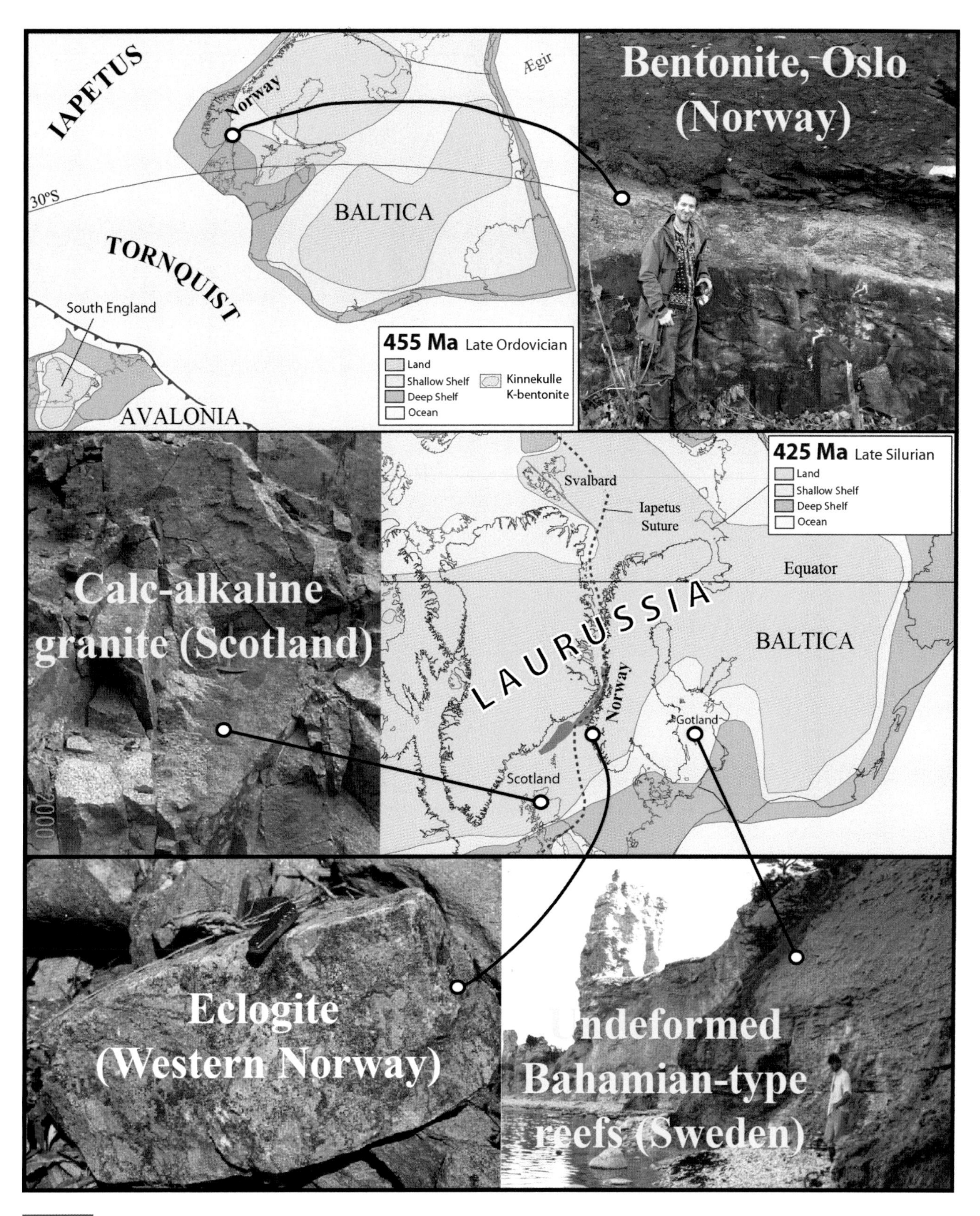

Fig. 2.19 Late Ordovician reconstruction where the Kinnekulle K-bentonite in Baltica is linked to an Avalonian magmatic arc (449–457 Ma), and ash fall-out was transported with westerlies all the way to St Petersburg, Russia. Photo to the left of the Late Ordovician reconstruction shows the bentonite in the Oslo region (grey–white layer), dated to around 457 Ma (Tucker & McKerrow, 1995; Svensen et al., 2014). The Late Silurian map (formation of Laurussia after collision of Avalonia and Baltica with Laurentia) is associated with three different pictures, one picture of calc-alkaline granites (425 Ma 'Newer' granite) linked to Iapetus subduction (Avalonian crust) beneath Scotland, a second of high-pressure metamorphic rocks in western Norway linked to subduction of Baltica continental crust beneath Greenland (and subsequent exhumation in the lower Devonian), and a third picture showing undeformed Bahamian-type reefs in the Caledonian foreland. The latter reflect the much better temperatures in southern Scandinavia at this time (subtropical to tropical).

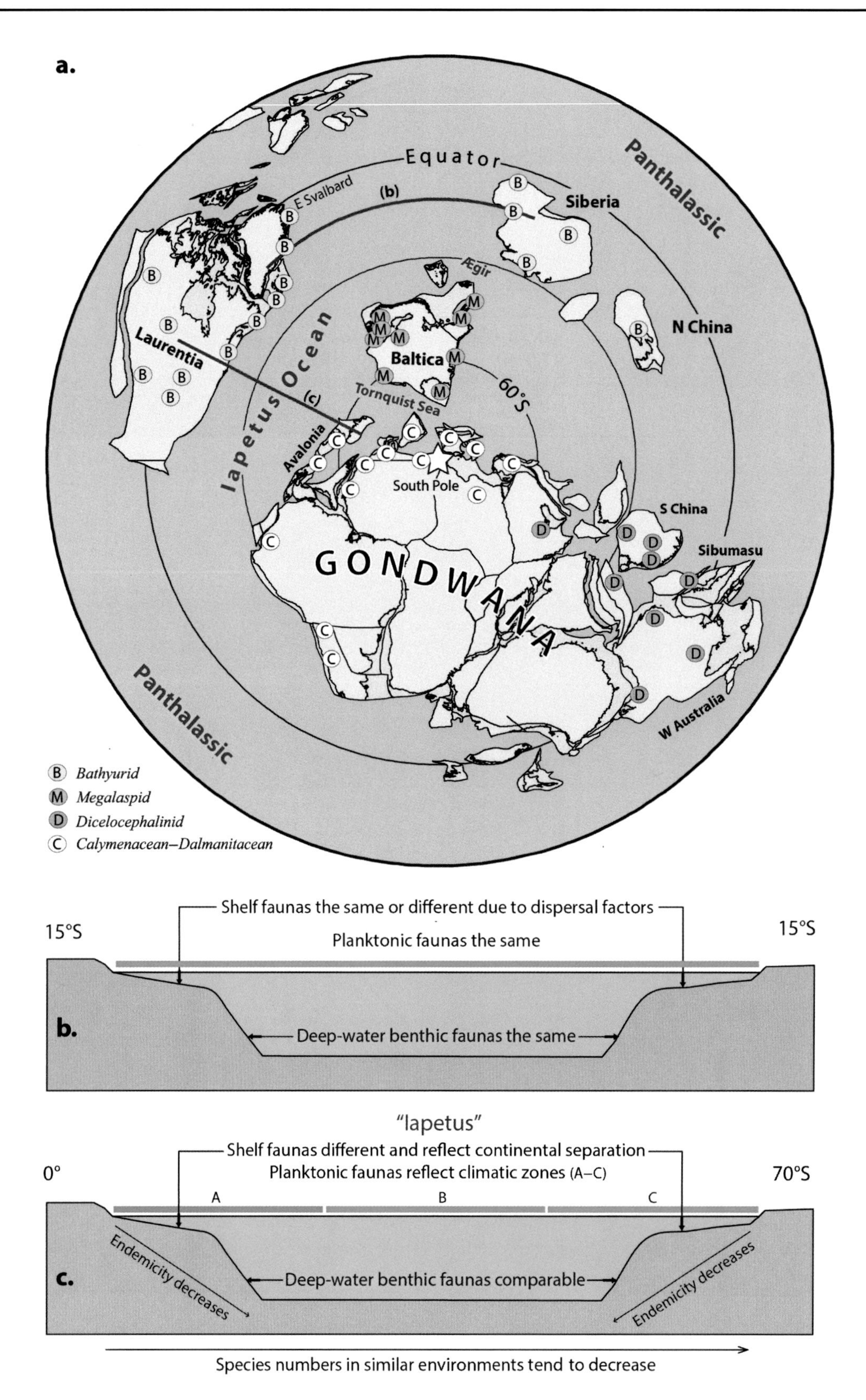

Fig. 2.20 (a) Early Ordovician reconstruction and the distribution of some key trilobite faunas (data from Torsvik et al., 1996 and Cocks & Torsvik, 2002). (b, c) Comparison of faunas found on opposite sides of an ocean at the same latitude or at the same longitude (after Cocks and Fortey, 1982). The red-line east–west profile (about 15° S) and the orthogonal north–south profile (0° to 70° S) in (a) are good analogues. Across the north–south Iapetus Ocean profile, the trilobite shelf faunas differ and reflect continental separation (Bathyurid as opposed to Calymenacean–Dalmanitacean provinces). Conversely, along a profile of constant latitude (arbitrarily shown as 15° S) the trilobite shelf faunas of Laurentia, Siberia, and North China are similar (Bathyurid).

In addition, some palaeocontinents, such as Gondwana, were so substantial that their margins covered many degrees of latitude, and thus the benthos at the northern and southern extremes of the continent can be very different, reflecting the changing climatic zones. However, between these two latitudinal extremes the faunas at intermediate latitude should show a gradation or cline, comparable with the cline seen today in the benthic molluscan faunas along the western seaboard of North America, which stretches from the tropics of Panama to the high latitudes of Alaska. That is allied to another principle: the lower the latitude, the larger the number of different species and genera that will be found; thus, the biodiversity is generally greater (in equivalent ecological situations) the nearer the Equator.

An integrated approach of palaeomagnetic and faunal analysis is applicable for the entire Phanerozoic, but works best for the Early Palaeozoic and, notably, the Early Ordovician (Fig. 2.20a). At that time, Gondwana stretched from the South Pole (Africa) to 20° north of the Equator (Australia and East Antarctica), Baltica occupied intermediate southerly latitudes, separated by the Tornquist Sea, whereas Laurentia straddled the Equator. Laurentia was separated from both Baltica and Gondwana by the Iapetus Ocean. The Iapetus Ocean (~5,000 km across the British sector) and the adjacent Tornquist Ocean (~1,100 km between southern Baltica and south-central Europe) were at their widest. This is probably why benthic trilobites from Laurentia (Bathyurid Province) and north-west Gondwana (Calymenacean–Dalmanitacean Province) are so markedly different from those of Baltica (Megalaspid Province). Bathyurid trilobites are also found in Siberia and North China but those continents were located at low latitudes (as was Laurentia), thus explaining the similarity in shelf fauna, although some authors had previously placed Siberia much closer to Laurentia in their reconstructions, partly because of the bathyurid fauna.

Sediment Distributions and Climate Patterns

Other semi-quantitative or qualitative methods, for example the distribution of latitude-sensitive and climate-dependent rock types, have also proved useful in deciphering old geographies, and they, like fossils, can also have the advantage of being derived from data gleaned from rocks that have suffered from much subsequent tectonic activity, in contrast to palaeomagnetism, when the original remanent magmatism has been overprinted and eradicated by later heating.

Carbonate buildups, often termed bioherms or reefs, are widespread throughout Phanerozoic rocks. However, they were largely laid down within tropical to subtropical latitudes (Fig. 2.21), and thus their distribution in ancient times can also be a useful check on the signals derived from palaeontological and other data. Comparably, evaporites are most usually found in belts on either side of the Equator (the subtropics), and so their presence questions the reality of reconstructions that show them at high latitudes. Conversely, coal deposits are commonly confined to the Equator (wet conditions, Fig. 2.21) or the northerly or southerly wet-belts at intermediate to high palaeolatitudes (Scotese & Barrett, 1990; Torsvik & Cocks, 2004; Boucot et al., 2013).

The recognition of glacial deposits has been important in placing continents since Wegener's (1915) original work. Much direct evidence of glaciers, such as eskers and drumlins, is rarely preserved in rocks older than the Pleistocene, but magnificent glacial striated pavements are known from various Precambrian glacial episodes, as well as those in North Africa caused by the latest Ordovician (Hirnantian) glaciation (Fig. 2.21), and in India and elsewhere from the lengthy but intermittent glaciations from the Middle Carboniferous to the Early Permian. In addition, glacial tillites are well represented in rocks of many ages, as well as dropstones deposited by icebergs. However, the latter must be treated with caution since, for example, the cold Labrador Current on the eastern side of Canada today can carry icebergs well down into temperate latitudes before they melt and drop their contained debris.

It is notable that all those sedimentary deposits mentioned above only reflect latitudinal bands, and since this is also the chief output from palaeomagnetic data, sedimentary distribution seldom plays a prime part in generating reconstructions of ancient palaeogeographies.

True Polar Wander

Motion of continents relative to the Earth's spin axis may be not only the motion of individual continents ('continental drift' as we have addressed so far), but also rotation of the entire solid Earth (crust and mantle) relative to its spin axis. That is termed true polar wander (TPW) and has been proposed as an explanation for such disparate phenomena as unusually rapid plate velocities, sea-level variations, climate changes, and geomagnetic reversal frequency.

TPW arises from the gradual redistribution of density heterogeneities within the mantle and corresponding changes in the planetary moment of inertia (Goldreich & Toomre, 1969; Steinberger & Torsvik, 2010). To establish the magnitude of TPW with confidence, the absolute velocity field and the plate geometry for both continental and oceanic lithosphere are required. That is difficult for pre-Cretaceous time and estimates of the relative magnitude of TPW must therefore rely on continental palaeomagnetic data (Fig. 2.23). In

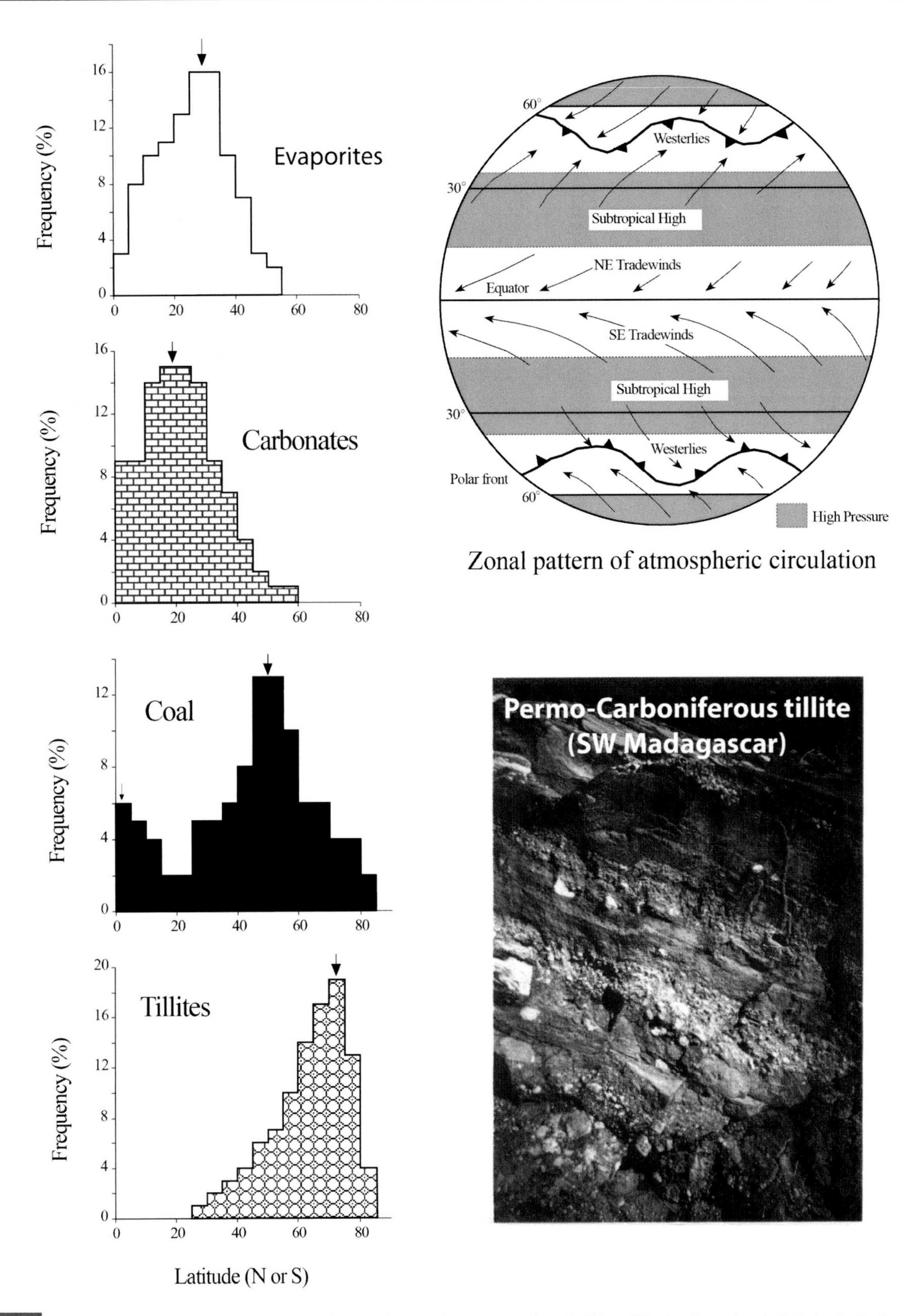

Fig. 2.21 Mesozoic and Cenozoic distributions of evaporites, carbonates, coal, and tillites. The horizontal axis is latitude for both hemispheres. Data in Scotese & Barrett (1990); pattern of zonal atmospheric circulation after Parrish (1982) and Scotese & Barrett (1990).

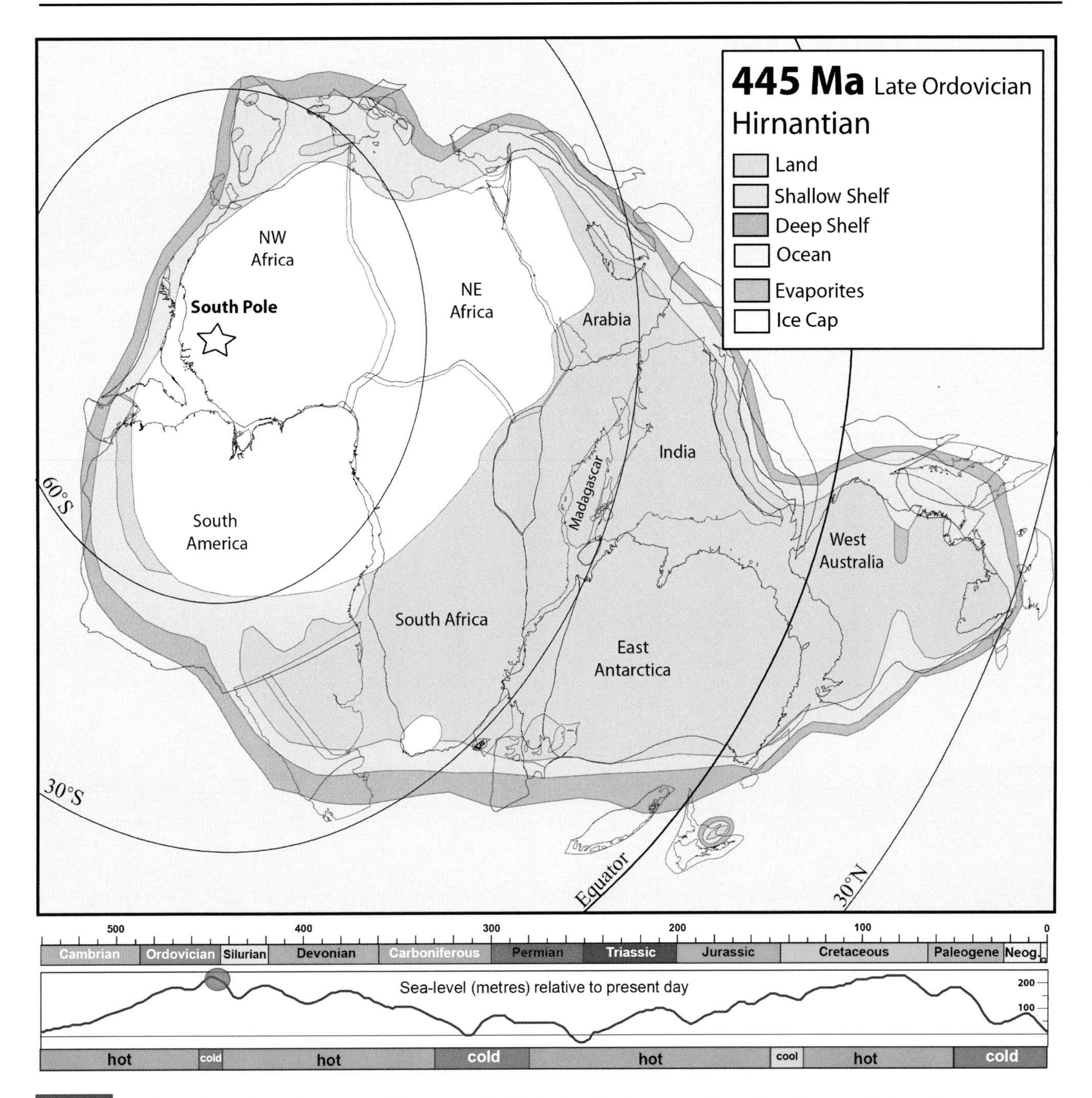

Fig. 2.22 Gondwana in the Latest Ordovician (Hirnantian: 445 Ma), showing the extent of the glacial ice cap. Bottom: Phanerozoic time scale (Endpaper), sea-level variations (Haq & Al-Qahtani, 2005; Haq & Shutter, 2008), and icehouse (cold) and greenhouse (hot) conditions.

spite of those difficulties, one can attempt to determine the magnitude of TPW by extracting the coherent (mean) rotation of all the continents around their common centre of mass in the palaeomagnetic reference frame of Steinberger and Torsvik (2008) (Fig. 2.23). Using that method, ten phases of slow and oscillatory TPW (around ~11° E and 191° E at the Equator) have been modelled for the entire Phanerozoic. Adding dense material in the upper mantle must be a prime candidate for causing TPW, and we expect that intermediate- to high-latitude subduction would preferentially induce TPW so that subduction zones shift towards the Equator, possibly with some time delay (Torsvik et al., 2014).

Attempts to link Earth's ancient LIPs and kimberlites (using palaeomagnetic data) to its interior require that TPW is accounted for. Because TPW is defined as the *rotation of the crust and mantle* (around a point at the Equator), TPW

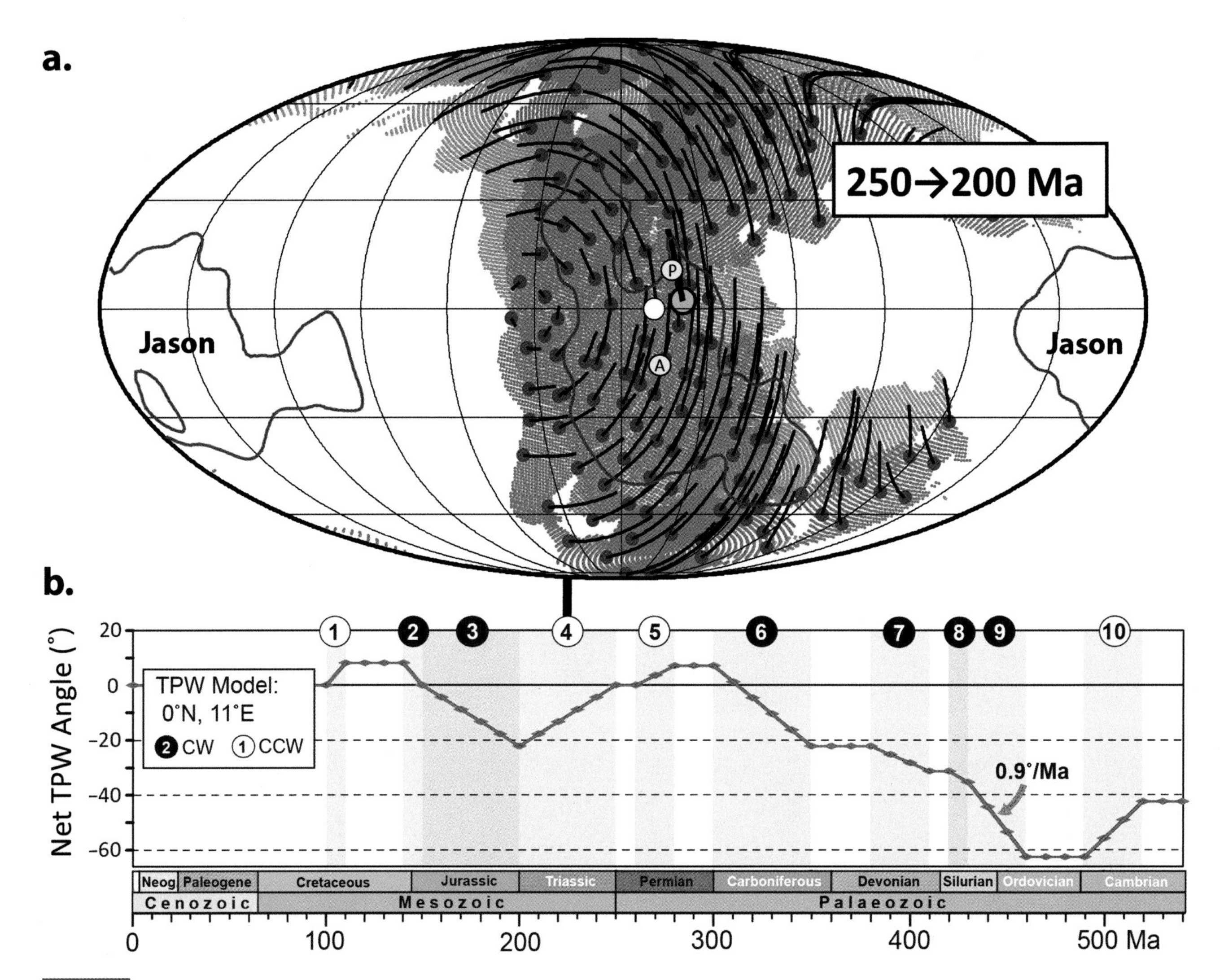

Fig. 2.23 (a) Motions of the continents reconstructed in the palaeomagnetic reference frame between 250 and 200 Ma. Total motions are shown as black lines, connected to blue dots (locations at the beginning of the time interval). Large green dot with thick black line indicates location and motion of the centre of mass of 'all' continents. Yellow dots (A and P) are the centre of mass or its antipode for Tuzo and the near antipodal Jason. The red line is the 1% slow contour in the SMEAN tomography model of Becker and Boschi (2002) at 2850 km depth. Open white circle is the preferred centre for true polar wander, TPW (0° N, 11° E). Blue and pink shading represent reconstruction at the start and at the end of the TPW episode (after Torsvik et al., 2012). (b) Ten identified Phanerozoic (before 100 Ma) phases of clockwise (CW) and counter-clockwise (CCW) rotations (TPW) around 0° N, 11° E (Torsvik et al., 2014).

correction may appear out of order since during a TPW event both the continent (with its embedded LIPs and kimberlites) and Tuzo and Jason must have rotated in the same manner (Fig. 2.24a). However, in such correlative exercises we compare surface reconstructions with fixed (present-day) features in the mantle (e.g. tomography) and thus the palaeomagnetic reconstruction *must* be corrected for TPW. Fortunately, the net cumulative effect of TPW through time (at least since the Late Palaeozoic) is at certain periods zero (e.g. Late Permian and Late Jurassic), or very small (Fig. 2.23b). Thus TPW-corrected or not corrected reconstructions can be quite similar, but that is not the case for the Early Palaeozoic.

In Fig. 2.23 TPW was estimated purely based on palaeomagnetic data; the axis of TPW rotation varies at certain intervals but averages to about 11° E (and 191° E) at the Equator (Fig. 2.24b) between 320 and 100 Ma, which was used as a constant value (Torsvik et al., 2012). TPW can also be calculated as the difference between a global moving hotspot reference frame and the palaeomagnetic one. Between 120 and 100 Ma calculated TPW is almost the same as that calculated from palaeomagnetic data alone, and, for the past 40 million years, the axis of TPW rotation is also close to that estimated from palaeomagnetic data. There is, however, a major deviation between 90 and 40 Ma (Late

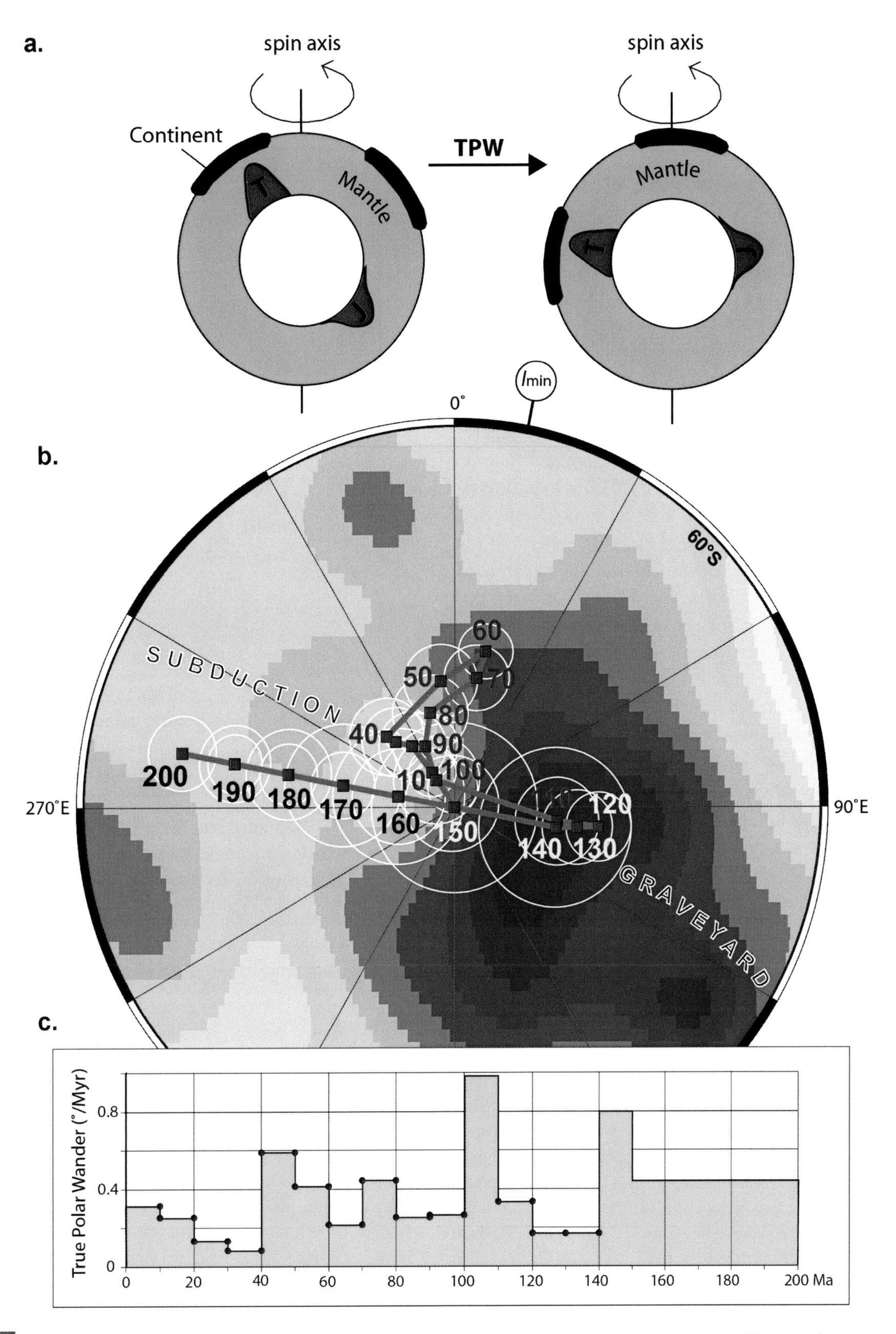

Fig. 2.24 (a) During a true polar wander (TPW) event both continents and the LLSVPs (Tuzo and Jason) will rotate the same amount (lower panel) around a rotation pole at the Equator. (b) TPW paths estimated from coherent rotation of all continents around an equatorial axis from only palaeomagnetic data (200–130 Ma) and from palaeomagnetic data (Torsvik et al., 2012) rotated into a GMHRF

Cretaceous to Middle Eocene), where TPW is almost orthogonal to that seen for most of the Mesozoic.

Putting It All Together

The past decade has seen the dawn of so-called *hybrid reference* frames, which combine different frames for different time periods. The first of these was based on a mantle reference (moving hotspot) frame for the past 100 Myr, and, before that, a reference frame derived from the African global APW path and making the assumption that Africa has not moved much in longitude (Torsvik et al., 2008b). Choosing Africa as a reference plate that has remained stationary (or quasi-stationary) with respect to longitude only works back to Pangea's assembly, because most relative plate circuits are reasonably well known from that time. In this book most reconstructions are based on a hybrid reference frame that uses the global moving hotspot reference frame (GMHRF) of Doubrovine et al. (2012) after 120 Ma and before that a palaeomagnetic reference frame corrected for TPW (Torsvik et al., 2012). The workflow to build this reference frame back to Pangea assembly is illustrated in Fig. 2.25, and the final outcome is a *Global Hybrid Mantle Reference Frame* that can link surface and deep Earth processes. The palaeomagnetic section of the reference frame was calibrated in longitude (known movement in longitude of Africa) by the GMHRF. The palaeomagnetic reconstruction is remarkably similar to the GMHRF reconstruction for Africa except for a shift in longitude of about 10° that is used to correct all palaeomagnetic reconstructions before 120 Ma. It is remarkable that two totally independent reference frames (and very different assumptions) produced similar reconstructions.

The method described above can potentially be carried back to the Middle Carboniferous at 320 Ma (e.g. Torsvik et al., 2008b), but here we only do this back to 250 Ma (Permian–Triassic boundary); before then we use a different

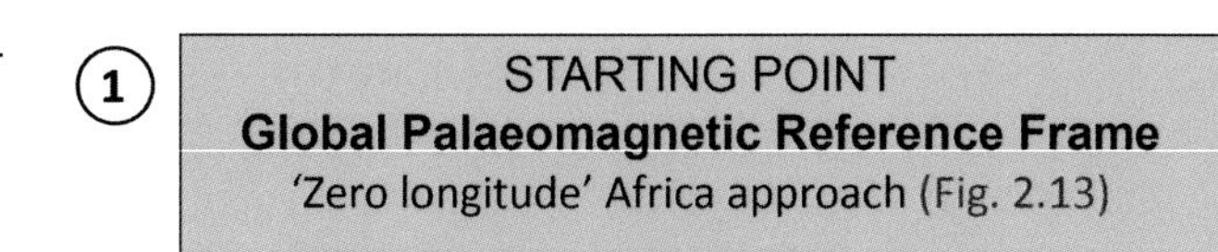

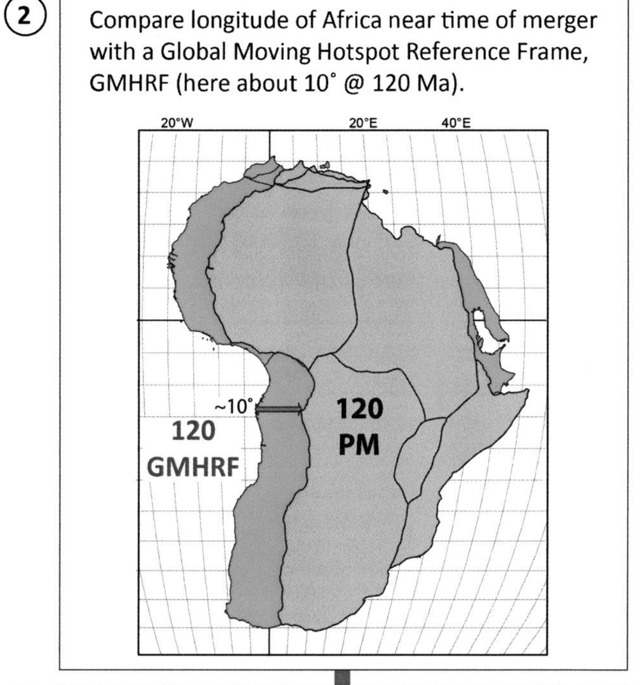

3 Adjust all longitudes for the palaeomagnetic frame at merger and for older times with the longitude difference in step (2) (constant longitude calibration)

4 **Correct for True Polar Wander**
(all palaeomagnetic Euler poles from merger time and using a TPW pole that is adjusted for the longitude change)

5 **Global Hybrid Mantle Reference Frame**

Fig. 2.25 How to build a global hybrid reference frame.

Fig. 2.24 (*cont.*) (Doubrovine et al., 2012) after 120 Ma. The two independent methods yield almost identical results between 120 and 100 Ma. Colour background map shows the variations of S-wave velocity (δV_s) near the core–mantle boundary at 2850 km depth (SMEAN model of Becker & Boschi, 2002). The TPW path is always confined above the high-velocity regions between Tuzo and Jason. The orientation of the minimum moment of inertia axis (I_{min}) of the Earth and therefore TPW is largely controlled by Tuzo and Jason (not shown in this figure), but is likely to change through time due to mass changes from subducted slabs and rising plume heads. In Torsvik et al. (2012) I_{min} was modelled at a constant equatorial location of 11° E (and 191° E) which closely matches the minimum moment of inertia axis for Tuzo and Jason combined (2.7° S, 11.9° E; Steinberger & Torsvik, 2010). I_{min} calculated from the GMHRF closely matches this estimate between 120 and 100 Ma and for the past 40 Myr. However, the TPW path (loop) between 90 and 40 Ma differs markedly. (c) TPW rates for the past 200 Myr. The fastest rates are observed between 100 and 110 Ma (~1°/Myr). A comparable peak is also observed from palaeomagnetic data alone (0.9°/Myr, Torsvik et al., 2012, 2014), and here by comparing a GMHRF and the palaeomagnetic data. The mystical 50 Myr TPW path between 90 and 40 Ma in (b) is not associated with notably higher-than-normal TPW rates (maximum 0.6°/Myr between 50 and 40 Ma). Note that there is only one total average for the whole 200–150 Ma interval, but the actual rates vary. Therefore reconstructions, for example in 10 Myr time intervals and corrected with a constant TPW rate, may result in artificial back-and-forth rotations of all the continents.

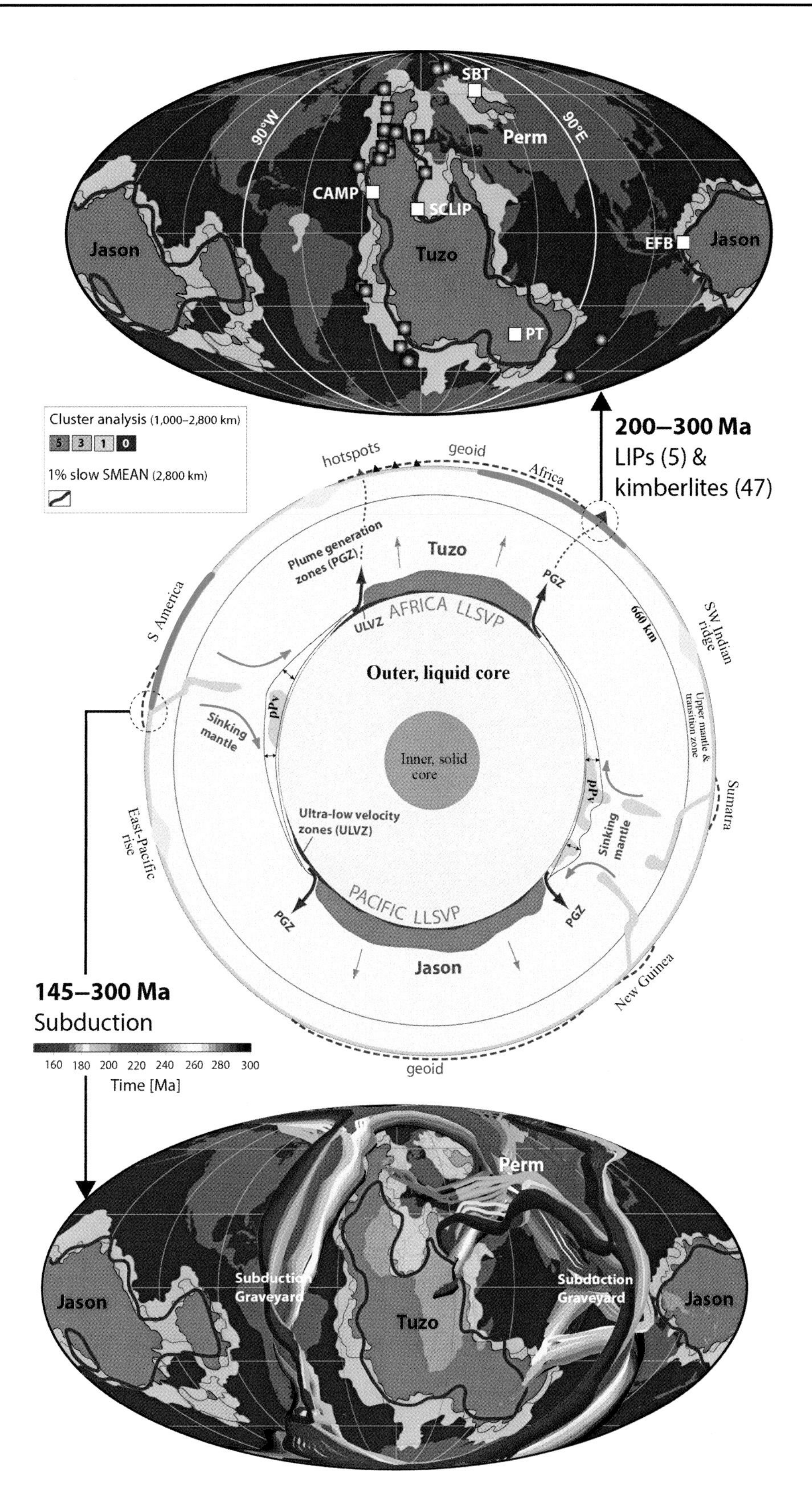

Fig. 2.26 Earth since Pangea: a planet where the geoid is dominated by a degree-2 mode, with elevated regions above large low shear-wave velocity provinces on the core–mantle boundary beneath Africa (Tuzo) and the Pacific (Jason). The edges of these deep mantle bodies (when projected radially to the Earth's surface) correlate with the reconstructed positions of large igneous provinces and kimberlites (top diagram showing LIPs and kimberlites erupted between 200 and 300 Ma) since Pangea formed at about 320 Ma. The geoid is largely a result of buoyant upwellings overlying Tuzo and Jason (thin red arrows), and subduction is largely confined between them (the subduction graveyards, lower diagram with subduction history between 145 and 300 Ma). EFB, Emeishan Flood Basalts; SBT, Siberian Traps; pPv, post-perovskite (bridgemanite).

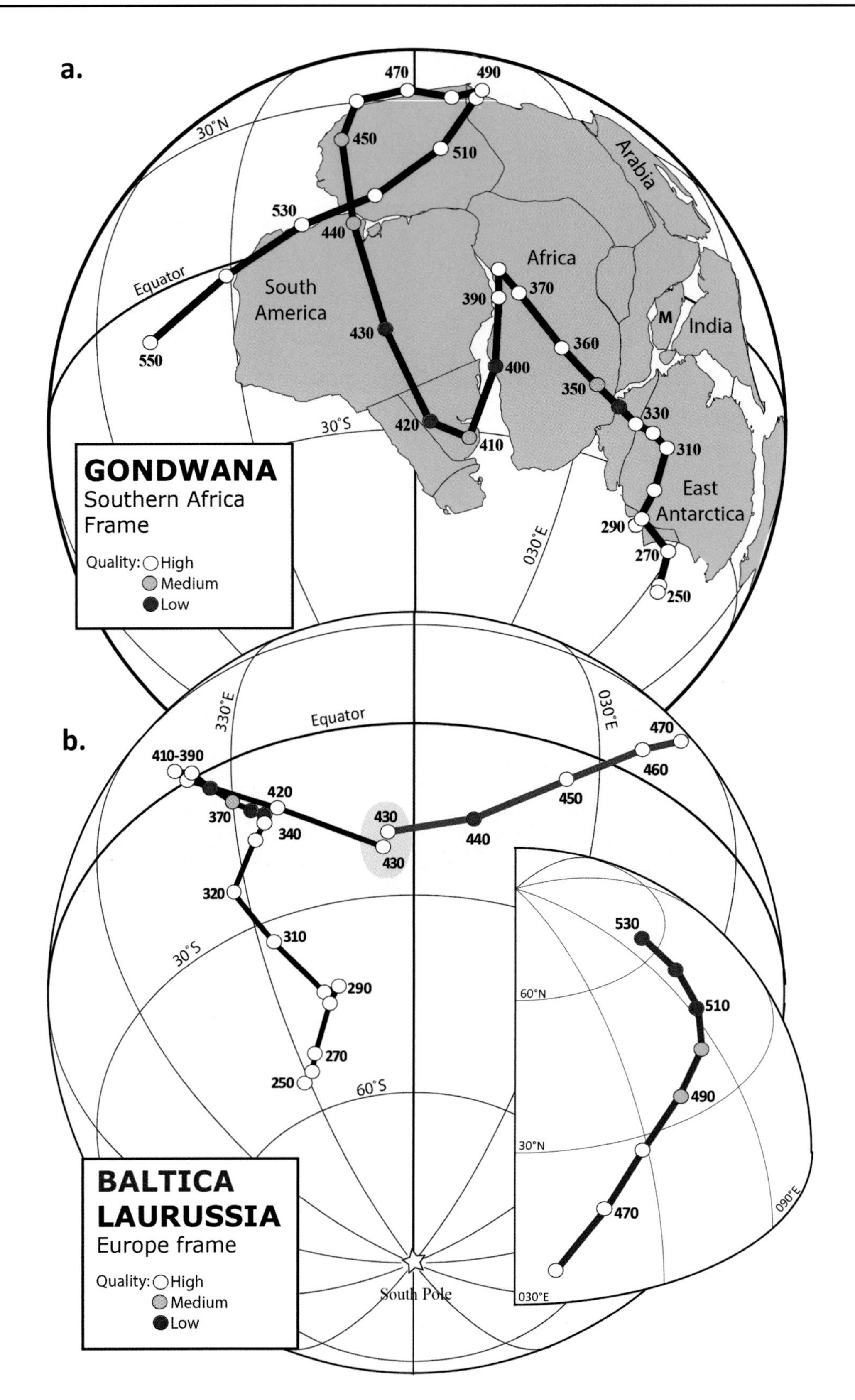

Fig. 2.27 (a) Phanerozoic smoothed spherical fitted apparent polar wander paths for Gondwana (550–250 Ma), Baltica (530–430 Ma), and Laurussia (Baltica/Stable Europe/Laurentia) from 430 Ma after correcting for the opening of the Labrador Sea (65–30 Ma) and the northeast Atlantic (55 Ma to present). Quality/reliability of the mean poles is classified as high, medium (only one pole) or low (purely interpolated).

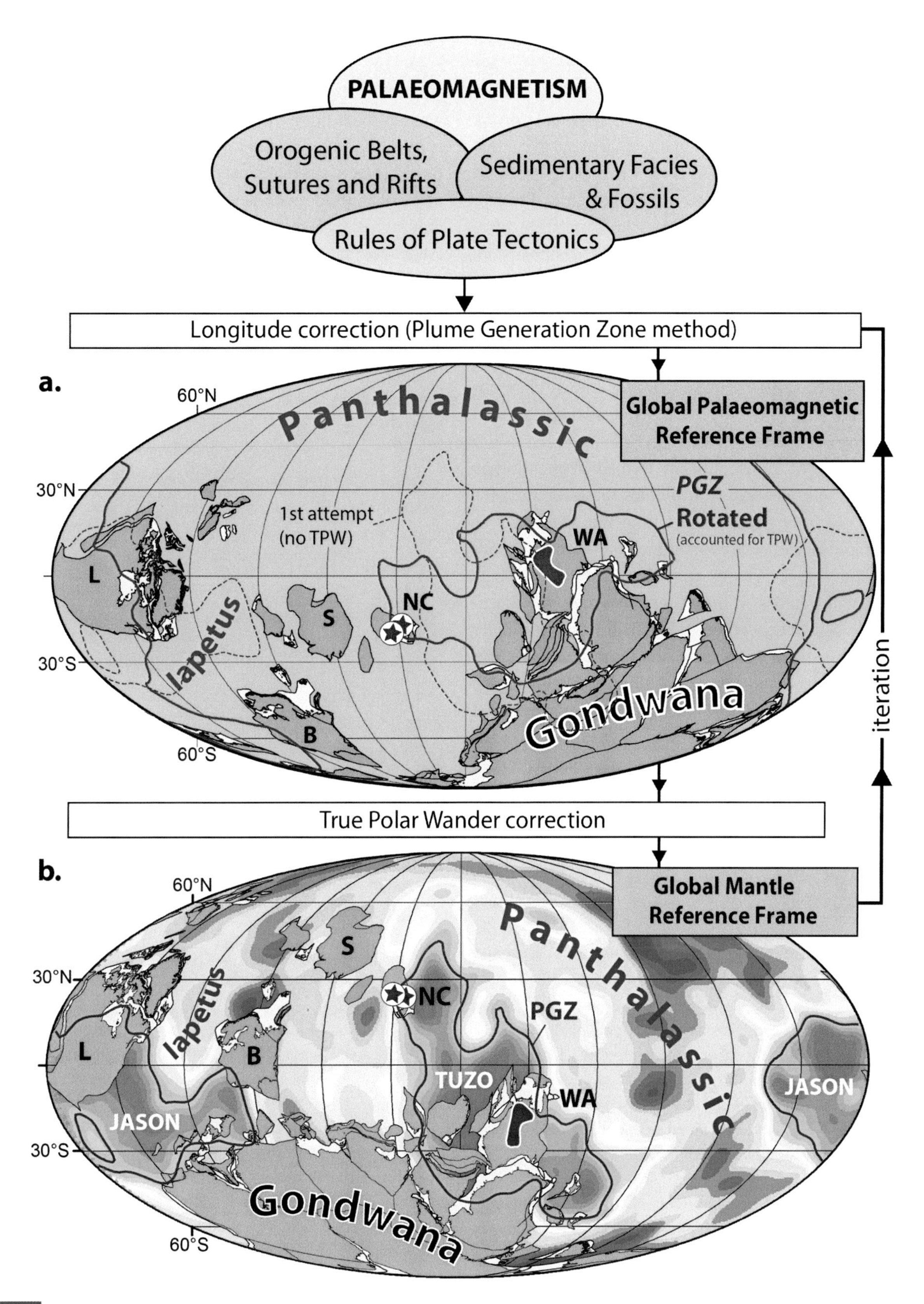

Fig. 2.28 Based on what we know about our planet for the past 320 Myr, notably the surface-to-core–mantle boundary correlation (Fig. 2.26), and our knowledge about how true polar wander (TPW) works, we use palaeomagnetic data (constraining latitude and rotation) in concert with geological information to first produce realistic global maps for times before Pangea. We then calibrate longitude with LIPs

technique based on the plume generation zone method (Fig. 2.28) and we merge the two reference frames together at this time. 'Absolute' reconstructions before Pangea are more uncertain because longitudes of continental blocks cannot be derived from palaeomagnetic data alone (although latitudes and azimuthal orientations can), and using Africa as a quasi-stationary reference plate (unlikely before Pangea) with respect to longitude is not tenable as relative plate circuits are unknown; except for times where continents are known to have come together (e.g. in the Silurian Caledonide Orogeny and the formation of Laurussia, when Laurentia collided with Baltica/Avalonia). Fortunately, Gondwana was mostly one coherent block during most of the Palaeozoic and Early Mesozoic (core-Gondwana with known plate circuits) except for the peri-Gondwanan terranes which rifted off at various times.

Since the Pangea supercontinent formed in the Middle Carboniferous at about 320 Ma, plumes that sourced LIPs and kimberlites have been derived from the edges of two stable thermochemical reservoirs at the core–mantle boundary. This is a degree-2 Earth with two major up-wellings (associated with plumes from the margins) and two down-wellings, the subduction graveyards which have essentially been confined between Tuzo and Jason for the past 320 Myr (Fig. 2.26). Taking a *uniformitarian approach* we can test whether it is possible to maintain this remarkable planet before Pangea; the alternative is a 'geodynamic dead end' because there are no other ways to build quantitative reconstructions. Palaeozoic plate reconstructions used here are based on APW paths for the major players such as Gondwana (Fig. 2.27a), Siberia (Fig. 9.11), Laurentia, Baltica (Fig. 2.11; combined to Laurussia after 430 Ma; Fig. 2.27b), and their later combinations into Pangea (Fig. 2.13). Euler poles were calculated from the APW paths; continents were then reconstructed in latitude and azimuthal orientation and subsequently calibrated in longitude using the plume generation zone method (Fig. 2.28) as explained in Torsvik et al. (2014). Ambiguity in the choice of plume generation zone was resolved through geological, palaeontological, and kinematic considerations; and, using a novel iterative approach for defining a palaeomagnetic reference frame corrected for true polar wander, we developed a model for absolute plate motion (global mantle reference frame) back to earliest Palaeozoic time (540 Ma).

Unless otherwise noted, we use a palaeomagnetic reference frame, the CEED frame, that is absolute in the sense that both latitude and longitude are determined with respect to the spin axis. In order to relate deep mantle processes to surface processes in the palaeomagnetic reference frame, we have also rotated the plume generation zones (PGZs) to account for TPW (red lines in Fig. 2.28a). In this elegant way, one can perceive how the surface distributions of LIPs (Appendix 1) and kimberlites relate to the deep mantle (Torsvik et al., 2014), at the same time plotting climatically sensitive sedimentary data that should not be corrected for TPW.

A full-plate model (e.g. Fig. 2.18a), unlike the continental rotation model described above, contains oceanic lithosphere (mostly synthetic before the Cretaceous) and requires an explicitly prescribed plate boundary network, which evolves independently, and with kinematic continuity, and in such a way that the collective system conforms to the basic principles of plate tectonics. Such a plate boundary model is now becoming available for Mesozoic to Cenozoic times (e.g. Seton et al., 2012; Shephard et al., 2014) and parts of the Palaeozoic (Domeier & Torsvik, 2014; Domeier, 2015), and is

Fig. 2.28 (*cont.*) and kimberlites and produce a global palaeomagnetic reconstruction (a). Our example is a 480 Ma reconstruction and kimberlites at this time are found in North China (NC) and ages vary between 485 and 474 Ma (Torsvik et al., 2014). Later in the Ordovician there are kimberlites from both Laurentia (L) and Siberia (S), and before 480 Ma kimberlites are also found in Gondwana. The map also shows the 511 Ma Kalkarindjii LIP in Western Australia (WA), which at that time was used for longitude calibration. For our initial model in (a), continental longitudes were defined so that kimberlites were located directly above a plume generation zone (PGZ), in this case Tuzo (stippled red line), ignoring any possible TPW (and plume advection). The next step from this idealised model was estimating TPW (as in Fig. 2.23a) and correcting the palaeogeographical reconstruction using the obtained TPW rotations. This produces a global mantle reconstruction (b). Because the TPW corrections would commonly degrade the LIP and kimberlite fits to the PGZs, the longitudes in the palaeomagnetic frame were then redefined to produce an optimal fit after the TPW correction. This produced the second approximation for the longitude-calibrated palaeomagnetic frame. The entire procedure of TPW analysis and longitude refinements was repeated six times in this case until no further improvement was observed in the TPW-corrected frame. The two reconstructions are therefore derived after the iteration procedure and the kimberlite fits are for both reconstructions. Note that in (a) the PGZ has been rotated according to the amount of TPW. In our global Palaeozoic model used in the book we did this procedure in 5 Myr steps from 540 to 250 Ma (Torsvik et al., 2014). B, Baltica.

used in this book whenever available. As stated above, practically all our reconstructions in this book use a palaeomagnetic reference frame, but our global maps from Triassic onwards (Figs. 11.1, 12.1, 12.3, 13.1, 13.2, 14.1, 15.1, and 15.2) are generated with GPlates using data supplied in Shephard et al. (2014). The reference frame there (termed EARTHBYTE mantle frame) *resembles* the global hybrid mantle frame of Torsvik et al. (2008b), some plate fits may also differ from the author's preference, and these figures mainly serve to portray the development of the oceanic lithosphere (Müller et al., 2008) in a vivid way.

GPlates rotation files and added information are available at http://www.earthdynamics.org/earthhistory/.

3 Tectonic Units of the Earth

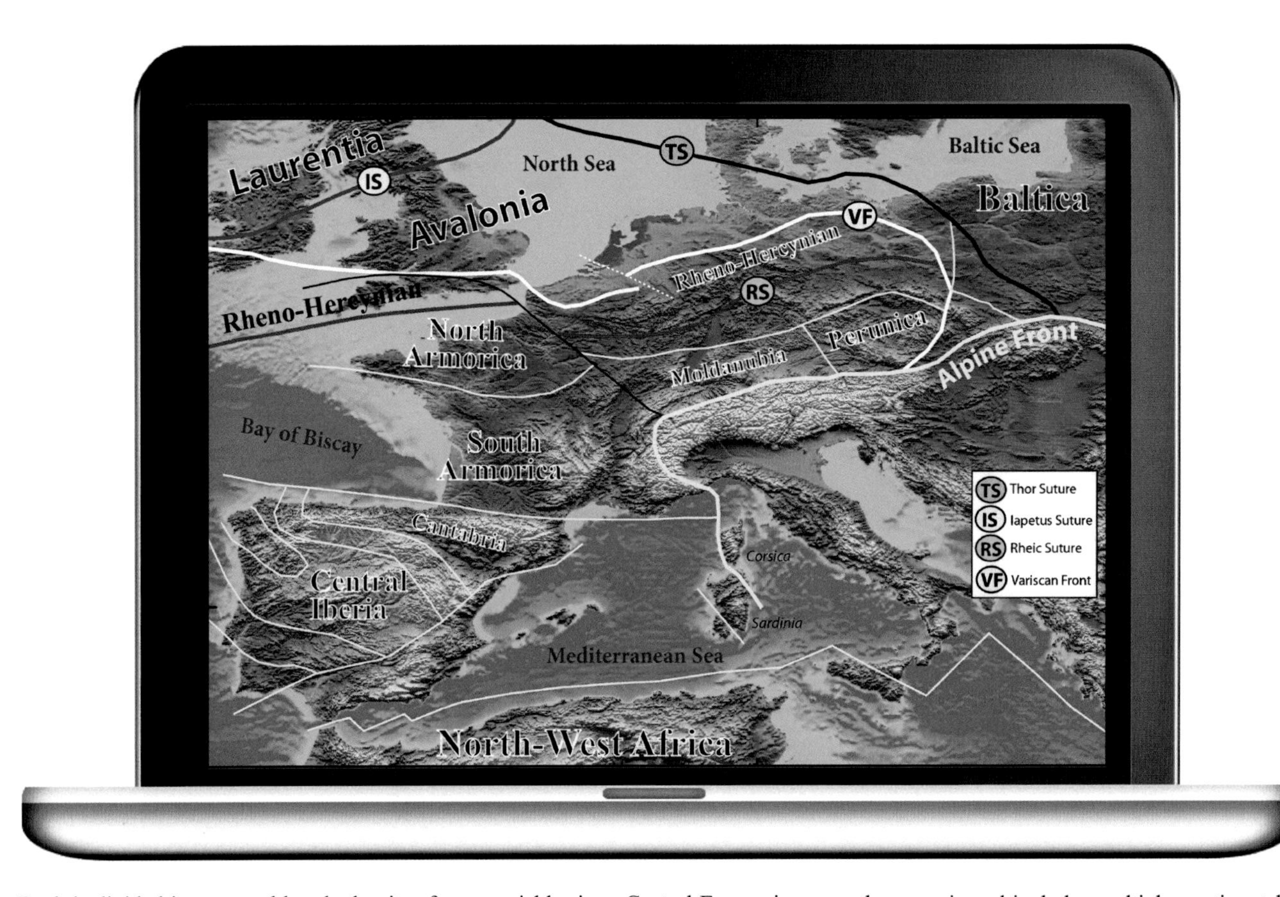

Earth is divided into several hundred units of very variable sizes. Central Europe is a complex mosaic and includes multiple continental units whose margins reflect the collisions between Baltica–Avalonia (along the Thor Suture in the latest Ordovician), Avalonia–Laurentia (along the Iapetus Suture in the Silurian), and the Armorican Terrane Complex (Iberia, France, and more, some collisions along the Late Palaeozoic Variscan Front). The yellow lines within the Iberian Peninsula reflect the different terranes within Unit 304 (below). Block boundaries, sutures, and major faults are draped on ETOPO1 bathymetry and topography.

This book is full of names of both modern geography and geological tectonic units. To identify and describe those units, they are divided into several hundred numbered units of very variable sizes (although many of the individual units also have several additional bewildering and often uncertain discrete but unnumbered tectonic fragments within them). We list the units in numerical order here, together with their places on the modern globe (Figs. 3.1 to 3.13), and with brief summaries of their geology, so that the reader can use this chapter as an illustrated index for the geological narrative in the rest of the book. The identity numbers largely follow the original Plates software numbers developed at the University of Texas Institute for Geophysics, but, since many units have been redefined and we have not used some others (including those for most of the oceans, apart from the Pacific area Fig. 3.13), there are many gaps in the numbering sequences used here.

North America

101, North American Craton.
102, Greenland.
103, Alaska North Slope (Arctic Alaska).
104, North-east Mexico.
105, Baja California.
108, Acadia. Western part of Avalonia.
109, Florida.
111 and 113, Mendeleev Ridge (111) and Northwind and Chukchi Blocks (113).
114, Lomonosov Ridge. Note also 136 and 140.
120 and 124, Sverdrup Basin, Axel Heiberg and North Ellesmere Island.
121 and 130, South-west Ellesmere (121) and Devon Island (130) ('Greenland Plate').
122 and 123, South-central (122) and Central (123) Ellesmere Island.
131 and 133, Framstredet Greenland Ridge (131) and Hovgaard (133).
134, Pearya.
136 and 140, Lomonosov Ridge. See 114.
142. Alpha Ridge of potential continental block.
153, Roberts Mountains Allochthon.
154, Belt–Purcell.
155, Farewell.
158, Cordillera Alaska.
159, Stikinia.
160, Ruby.
161, East Klamath and North Sierra.
162, Laurentia Parautochthon.
163, Caborca.
164, Cortes and Sierra Madre.
165, Wrangellia and Alexander.
170, Meguma.
171 and 172, Ganderia and Carolinia.

South America and Caribbean

201, Amazonia.
202, Paraná.
204, Chortis.
205, Yucatán.
215, Sierra Madre.
216, Mixteca and Oaxaquia.
217, Guerrero, South Mexico.
231, North-western South America.
232, Yucatán Basin.
233–236. Cuba.
237, Gonave microplate.
238 and 239, North and South Nicaraguan Rise.
240 and 241, Hispaniola Cordillera arc (240) and northern Hispaniola (241).
242, Puerto Rico.
243, Costa Rica.
244 and 245, Central (244) and Eastern (245) Panama.
246, Maracaibo Block West.
247, Accreted Andes Terranes.
248, Bonaire.
249, Aves Ridge.
250, Grenada Basin.
251, Lesser Antilles Arc.
252, Tobago Basin.
253, Accreted Antilles Prism (or Barbados Prism).
280, San Jorge Plate.
288 and 289, Falkland Plate, West Falklands, and East Falklands.
290, Colorado.
291, Patagonia.
296 and 298, Falklands Plateau and Maurice Ewing Bank.

Europe and Near East

302, Baltica.
303, Scotland and North-western Ireland.
304, Iberia.
305, Armorica.
306, Corsica and Sardinia.
307, Apulia.
309, West Svalbard.
311, East Svalbard.
315, Eastern Avalonia.
318, Hatton Bank–East Rockall Plateau.
319, Moesia.
322 and 333, Calabria and Central Apennines.
337, Tisia.
338, Rhodope.
340, Yermak Plateau.
346, Greece (including Adria).
347, Crete.
350, Timanian.
368, 369, 370, and 371, Jan Mayen Microcontinent.
373, Novaya Zemlya.
374, Bohemia.
375, Saxothuringia and Bruno-Silesia.
390, Urals.
391, Karakum.
392, South-eastern Peri-Baltica.
393, Caucasus–Mangyshlak.
394, Scythian–South Caspian.
395, Alpine Region.
397, Moldanubia.

Northern and Central Asia

401, Siberian Craton (Angara Continent and peri-Siberia).
405 and 415, New Siberian Islands and Anyui.
407, Chukotka.
408, Wrangel Island.
417, Kara.
420 and 421, Chersky and Omolon.
430, Altai–Salair, central Mongolian Basin.
431, West Sayan.
432, Tuva–Mongol.
433 and 434, Rudny Altai and Kobdin.
435, Eastern West Siberian Basin.
436, Barguzin.
437, Nadanhada–Sikhote–Alin.
438, Kamchatka.
440, Central Mongolia and Amuria.
441, Gobi Altai and Mandalovoo.
443, Kuriles.
450, Junggar.
451, Ala Shan.
452, Gurvansayhan.
453, Hutag Uul–Songliao.

454, Khanka–Jiamusu Bureya.
455, Nuhetdavaa.
456, Qaidam–Qilian.
457, Kunlun.
458, Kokchetav–Ishim.
459, North Tien Shan.
460, Chu-Ili.
461, Atashu–Zhamshi.
462, Chingiz–Tarbagatai.
463, North Balkhash.
464, Junggar–Balkhash.
465, Tourgai.
466, Karatau–Naryn.
467, South Tien Shan.
468, Western West Siberian Basin.
470, Stepnyak.
471 and 472, Selety and Boshchekul
480, Tarim.

India and the Middle East

501, India.
502, Sri Lanka.
503, Arabia.
504, Central Turkey and Taurides.
505, Alborz.
506, Afghan.
508, Sinai.
563, Greater Himalaya.
564, Lesser Himalaya.
570, Laxmi Ridge.
571, Murray Ridge.
572, Laccadives.
573, Chagos.
581, Pontides.
582, Sanand.
583, Lut.
584, Makran.
590, North Pamirs.
591, Central and South Pamirs.
592, Karakorum.

South-East Asia

600, Sulinheer.
601, North China.
602, South China (including South Qinling).
603 and 647, Sibumasu (northern (603) and southern (647) sectors).
604, Annamia; Indochina.
606, South Tibet (Lhasa).
607, West Burma.
614, Borneo; includes Kalimantan.
616, North Tibet (Qiantang).
617, East Sumatra and East Malaya.
624, Sakhalin.
625, Kitakami.
626, Western Hokkaido.
627 and 628, North-east and Central Honshu.
629, Kanto.
630, Maizuru.
631, Kurosegawa.
638, Sado Ridge.
651, Subawa–Flores.
667, 668, and 669, South-west (667), West (668), and East (669) Sulawesi.
670, Banggi–Sula.
673, Java.
675, Sumba.
674, 677, 678, and 691, West (674) and East (678) Philippines, Palawan (677), and Luzon (691).
681, Buru–Seram.
683, Wetar.
684, Timor.
686, South-west Sumatra.

Africa

701, South Africa.
702, Madagascar.
704, Seychelles.
705, Saya de Malha.
706, Oran Meseta.
707, Moroccan Meseta.
709, Somalia.
712, Lake Victoria Block.
714, North-West Africa.
715, North-East Africa.
775 to 777, Mauritius and Mauritia.

Australasia and Antarctica

801, West Australian Craton.
802, East Antarctica.
803, Antarctic Peninsula.
804, Marie Byrd Land.
805, Filchner Block.
806, North New Zealand.
807, South New Zealand and Chatham Rise.
808, Thurston Island.
809, Ellesworth–Whitmore Mountains.
827, New Hebrides
830, Solomon Islands
833, Central Lord Howe Rise. Includes New Caledonia Basin South.
835, Three Kings Ridge.
836, Louisiade Plateau.
841, Transantarctic Mountains.
842, Ross.
848, Eastern Australia.
850, Tasmania.
851 and 852, Western (851) and Eastern (852) South Tasman Rise.
854, Dronning Maud Land.
866, Chesterfield Plateau.
867, Gilbert Seamount.
868, Challenger Plateau.
869, North Lord Howe Rise and New Caledonia Basin.
878, Papuan Plateau.
883–886, North (883), Middle 1 (884), Middle 2 (885), and South (886) Dampier Ridge.
887, East Tasman Platform.
888, Eastern Plateau.
889, Mellish Rise.
890, Kenn Plateau.

Panthalassic–Pacific Oceanic Plates

901, Pacific.
902, Farallon.
903, Vancouver, Juan de Fuca (from 37 Ma).
908, Chazca.
907, Cache Creek.
911, Nazca.
918, Kula.
919, Phoenix, Catequil (from 120 Ma).
924, Cocos.
926, Izanagi.
982, Manihiki.
983, Hikurangi.

The Earth's lithosphere (crust and upper mantle) is broken up into tectonic plates. Today there are about a dozen large plates (e.g. Eurasia, North America, and Africa) and numerous minor plates whose numbers depend on how they are defined (Bird, 2003). Based on our own and previous work we have identified and separated 268 Phanerozoic tectonic units, which together make up all the Earth's surface. Each unit has an identity number, and is made up of polygons which can be manipulated by using the GPlates software (www.gplates.org).

The Earth's surface has changed dramatically during the Phanerozoic: for example, the European (plate identity number 3XX), Russian (4XX), Indian (5XX), and South-East Asian (6XX) units mostly belong to the single continent of Eurasia when plotted on simplified versions of today's tectonic plate boundaries (Fig. 3.1). In many cases, the individual units have been reshuffled several times during the Phanerozoic, sometimes by reactivating old plate margins and sometimes along fresh zones of weakness, a reshuffling often caused by the larger plates moving in different directions with respect to the other plates that surround them.

North America

The Palaeozoic of North America was reviewed by Cocks and Torsvik (2011). For more details on particular areas, including rocks of all ages, a valuable source is the many DNAG (Decade of North American Geology) volumes, which cover Canada and Mexico as well as the USA, published by the Geological Society of America between 1988 and 1998 (e.g. Hatcher et al., 1989; Plafker & Berg, 1994). An overview of the Arctic and nearby areas is in Spencer et al. (2011).

101, North American Craton. This continent has at its core the Precambrian craton of Laurentia, which mostly formed from about 2.0 to 1.8 Ga by the progressive amalgamation of six or seven large fragments of Archaean crust. Laurentia was an independent continent between the Proterozoic and the Middle Silurian, when it collided firstly with Avalonia–Baltica in the Silurian Caledonide Orogeny to form Laurussia, which in turn collided with Gondwana in the Carboniferous to form Pangea (Ziegler, 1989). Throughout the Phanerozoic numerous microcontinents and smaller terranes accreted to its margins, as reviewed for the Palaeozoic by Cocks and Torsvik (2011). North America regained its continental independence with the opening of the central Atlantic Ocean in the Early Jurassic and the North Atlantic in the Paleogene.

102, Greenland. An integral part of the Laurentia Craton, separated from it only by the Pleistocene erosional feature of the Nares Strait (Dawes, 2009). It is an Archaean and Proterozoic craton with Lower Palaeozoic shelf sediments on its northern margin and Devonian to Paleogene rocks on its western and eastern margins. The geology was summarised by Henriksen (2008), and the Caledonides on its eastern and northern margins by Higgins et al. (2008).

103, Alaska North Slope. The Arctic Alaska Terrane has Neoproterozoic to Jurassic strata. It is composite, and includes the Coldfoot, De Long Mountains, Endicott Mountains, Hammond, North Slope, and Slate Creek subterranes (Nelson & Colpron, 2007). It formed part of the Arctic Alaska–Chukotka Microcontinent from the Neoproterozoic to the Early Devonian, when it started its progressive accretion to the Laurentian Craton (Cocks & Torsvik, 2011). It rotated counter-clockwise in the Late Jurassic and Early Cretaceous, which opened the Amerasia Basin within the Arctic Ocean (Alvey et al., 2008). Since the Cretaceous, the area has formed the North Slope of Alaska on the margin of the North American continent.

104, North-east Mexico. Mesozoic to Recent strata, including some earlier Palaeozoic and Precambrian basement not considered separately here, on the north-eastern margin of the Gulf of Mexico, which moved as a single coherent unit after the breakup of Pangea in the Mesozoic.

105, Baja California. Proterozoic–Early Palaeozoic shelf strata and Mesozoic arc volcanics to the west of the Gulf of California, Mexico (Sedlock, 2003).

108, Acadia. An area stretching from Cape Cod, Massachusetts, north-westwards to eastern Newfoundland. A segment of north-western Gondwana until the Rheic Ocean opened at 490 Ma, when it became the western part of Avalonia, which was independent before its oblique docking with Baltica at the end of the Ordovician (about 443 Ma). Avalonia includes island arc rocks of various Lower Palaeozoic ages which were accreted during the Caledonian Orogeny, as well as the relatively flat-lying and unmetamorphosed Avalon Peninsula itself, which is mostly rocks of Cambrian and Early Ordovician ages in eastern Newfoundland.

109, Florida. Much of Florida state and the area to its north-east along the North American margin. Characteristic Cambrian trilobites in subsurface Florida indicate that it was part of Gondwana before the Carboniferous Alleghanian–Ouichita Orogeny, when it became part of Pangea until the Atlantic opening in the Jurassic. Included in the unit are the Carolina Slate Belt and various accreted Lower Palaeozoic island arc rocks in Alabama, which were originally peri-Gondwanan rather than peri-Laurentian. The southern part of Florida state now lies within the northern sector of the Bahamas carbonate platform.

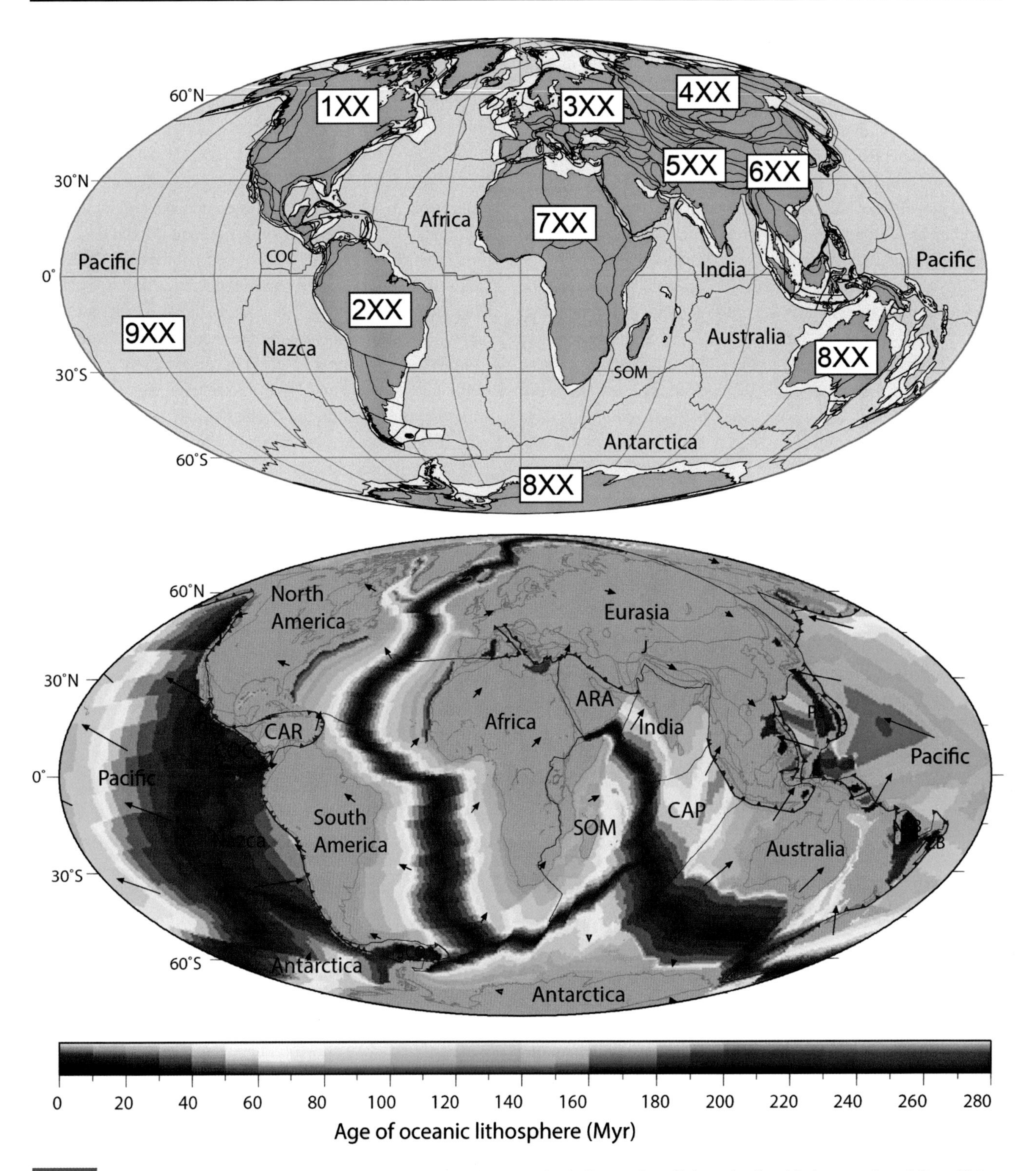

Fig. 3.1 (Top) The Earth divided into units used here. '7XX', for example, indicates that all the units listed below numbered from 701 to (potentially) 799 are within Africa and adjacent area. (Bottom) Continental lithosphere (grey) and oceanic crust with colours shaded to indicate the different ages of the ocean floors, with scale showing ages represented by the different colours in the map.

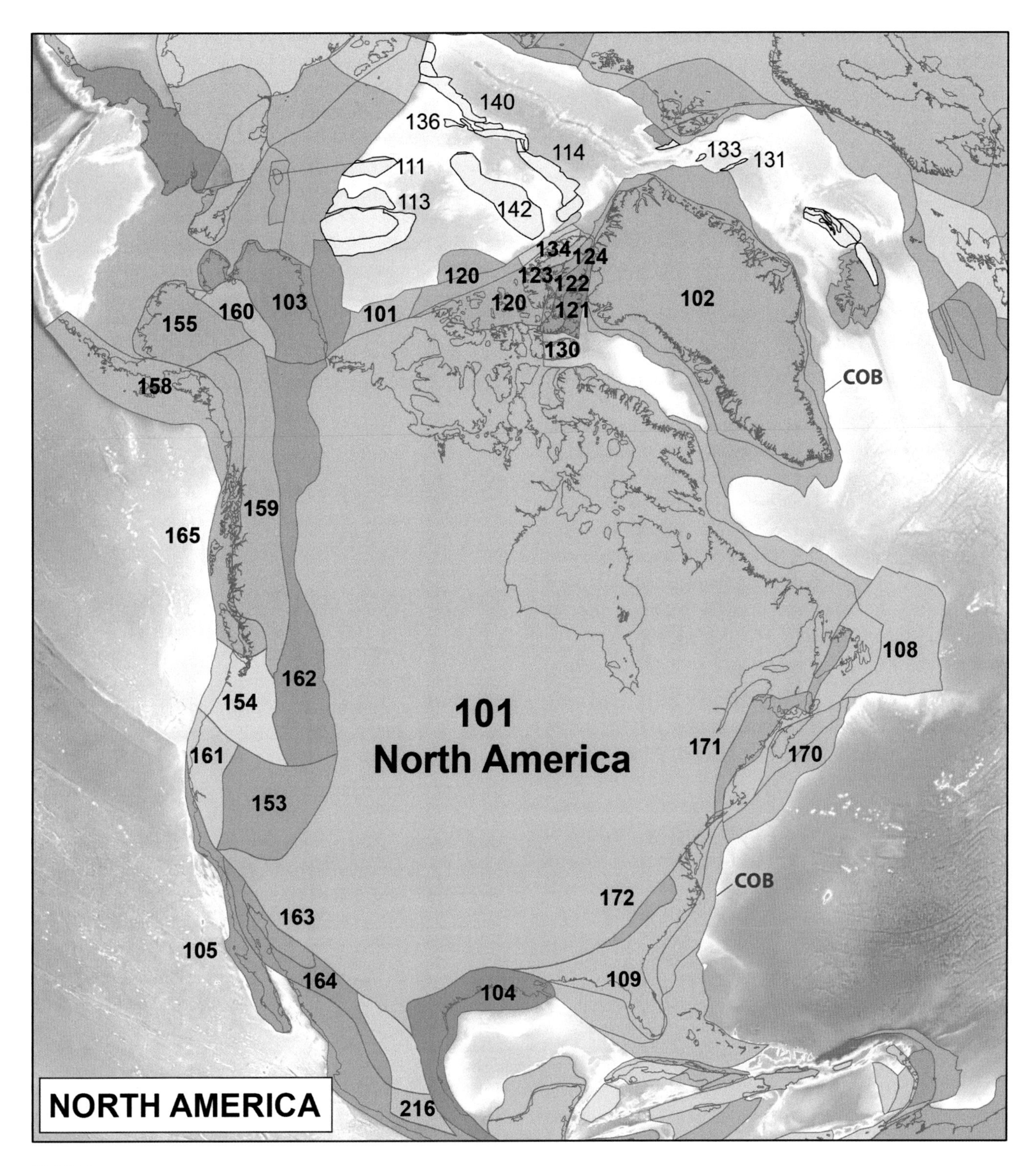

Fig. 3.2 North American units (plate identities 1XX). The Continent–Ocean Boundary transition zone (COB) is mainly defined from seismic, magnetic, and/or gravity data.

111 and 113, Mendeleev Ridge and Northwind and Chukchi Blocks. The wide continental shelf in the Arctic Ocean north of Siberia and Alaska is underlain by Palaeozoic rocks, few of which have been sampled; however, Northwind Ridge (in the Amerasia Basin at 76° N and 155° W) has yielded Cambrian to Permian fossils. The Palaeozoic of the other continental lithosphere blocks in the Arctic Ocean is speculative, but some are included

within the Arctic Alaska–Chukotka Microcontinent (Cocks & Torsvik, 2011).

114, 136 and 140, Lomonosov Ridge. A substantial elongate ridge under the Arctic Ocean which rifted off Eurasia in the Early Tertiary at about 55 Ma, opening the Eurasian Basin in the process. Its geology is largely known only through seismic surveys, but dredge sampling has uncovered possible Precambrian–Lower Palaeozoic sandstones that were metamorphosed (greenschist facies) at around 470 Ma (Marcussen et al., 2015).

120 and 124, Sverdrup Basin and Axel Heiberg, and North Ellesmere Island. A section of Arctic Canada in which the Sverdrup Basin (Unit 120) curves from Prince Patrick Island eastwards through Axel Heiberg Island to include most of Ellesmere Island (Spencer et al., 2011). The basin is about 1300 km long and up to 400 km wide, and was formed by rifting over an Early Palaeozoic fold belt and the rise of adjacent Arctic areas in the Visean, and sagged substantially from the Carboniferous to the Cretaceous, and more slowly up until the present day, with many hydrocarbons deposited in the Triassic and Jurassic. The area has been united with Pearya (Unit 134) from the Devonian onwards.

121 and 130, South-west Ellesmere Island (121) and Devon Island (130). A north-westward extension of the 'Greenland Plate' (Unit 102) within Arctic Canada, with Precambrian rocks unconformably overlain by Palaeozoic sediments in places. Not affected by the Tertiary Eurekan Orogeny. The Greenland Plate is shaded green in Fig. 3.3.

122 and 123, South-central and Central Ellesmere Island. Extending from the central sectors of Ellesmere Island in the east, and westward to include the northern part of Bathurst Island and the southern part of Melville Island. Both areas are largely Lower Palaeozoic rocks and were affected by the Tertiary Eurekan Orogeny, in contrast to Unit 121.

131 and 133, Framstredet Greenland Ridge (131) and Hovgaard (133). The Hovgaard Ridge (now located north-east of Greenland) is a fracture zone (Eldholm & Myhre, 1977), warping of oceanic crust (Engen et al., 2008), or a continental sliver (Myhre et al., 1982) which was detached from the West Barents Sea Hornsund margin during the Early Oligocene at 33.3 Ma (Chron 13). The origin of the Greenland Ridge is enigmatic, but it might be a microcontinent which was torn off the West Barents Sea Senja margin.

134, Pearya. The northern sectors of Ellesmere and Axel Heiberg islands and adjacent ocean, Arctic Canada. Middle Proterozoic metamorphics are overlain by Late Proterozoic to Ordovician metasediments and arc volcanics are intruded by unmetamorphosed 480–460 Ma Middle Ordovician plutons after the local M'Clintock Orogeny (Trettin, 1998). An originally independent terrane, which probably accreted to Laurentia (Units 120 and 124) as part of the Middle Silurian Caledonide Orogeny.

136 and 140, Lomonosov Ridge. See above under Unit 114.

142, Alpha Ridge. Of potential continental origin (C. Gaina, pers. comm. 2016).

153, Roberts Mountains Allochthon. In northern California, composite fault-bounded packets of Middle Cambrian to Early Carboniferous oceanic and island arc rocks, with Ordovician to Devonian benthic faunas indicating that they were then peri-Laurentian. The arcs are imbricated into a tectonic wedge which was thrust onto the Laurentian Craton in the Late Devonian to Early Carboniferous Antler Orogeny (Wright & Wyld, 2006).

154, Belt–Purcell. To the north of the Roberts Mountain and east Klamath areas, a substantial part of Oregon and Washington State consists of the Belt–Purcell terranes, whose rocks are mostly post-Palaeozoic magmatic arcs (Cascade).

155, Farewell. A composite terrane in western Alaska, including the Mystic, Dillinger, and Nixon Fork subterranes. Neoproterozoic basement above which are Palaeozoic fragments representing a microcontinental platform which accreted to the North American Craton in the Late Jurassic. Many of the Palaeozoic faunas are of Siberian affinity (Blodgett & Stanley, 2008).

158, Cordillera Alaska. Post-Palaeozoic terranes and island arcs, now in Alaska, which have been accreted to north-western North America since the end of the Palaeozoic.

159, Stikinia. This terrane in central British Columbia includes Devonian carbonates and Carboniferous volcanic rocks and sediments: there are also Lower Permian deeper-water clastics and Lower and Upper Permian platform limestones, some resting on tholeiitic metabasalts (Gabrielse & Yorath, 1992). It accreted to North America in the Jurassic, after the closure of the Slide Mountain Ocean (Nelson & Colpron, 2007).

160, Ruby. Grouped with the Innoko Terrane here, the Ruby Terrane in central Alaska consists of possibly Precambrian

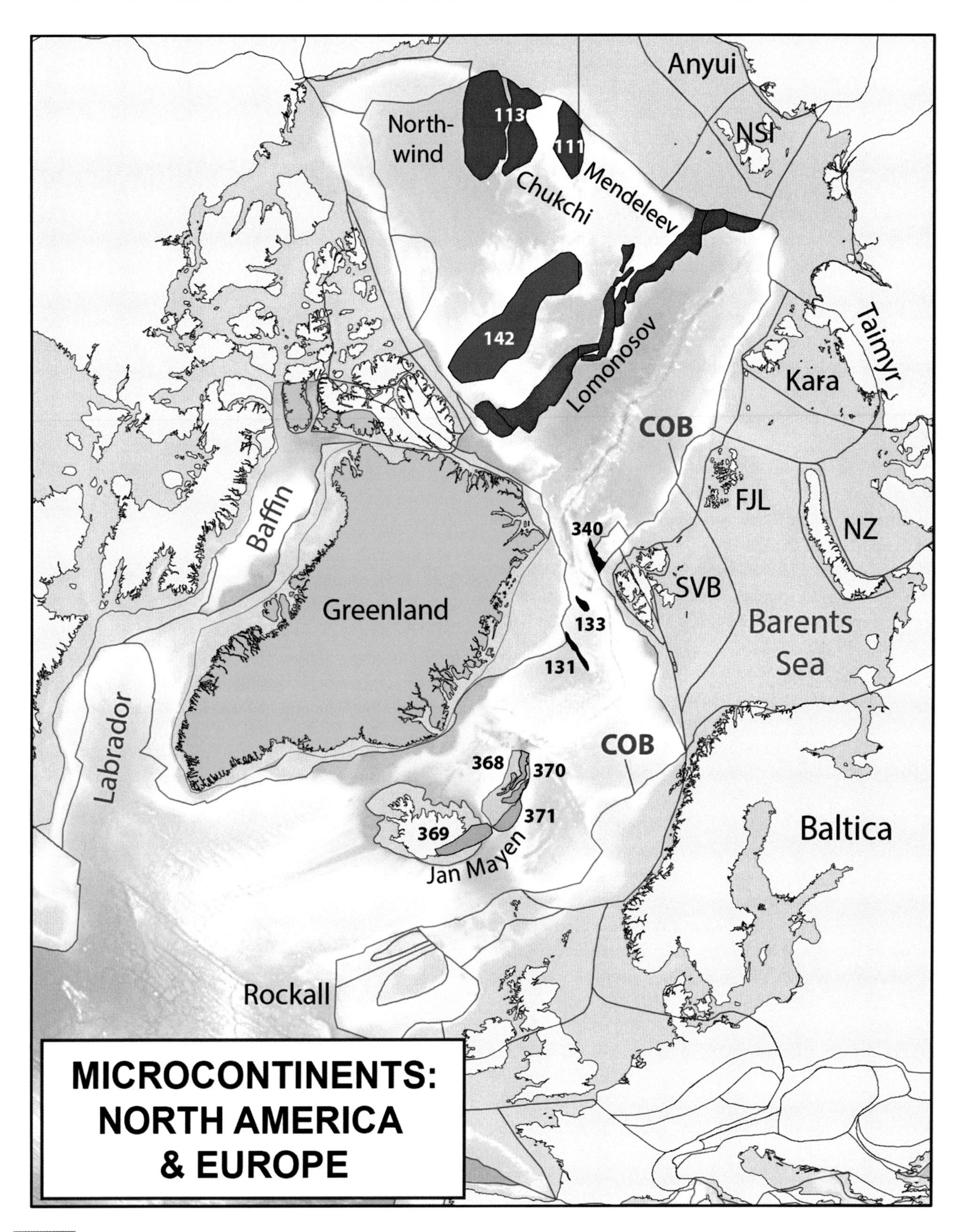

Fig. 3.3 Microcontinental units in the northern Atlantic and Arctic areas (plate identities 1XX; high Arctic) and Europe (plate identities 3XX; north-east Atlantic). The 'Greenland Plate', which behaved as a semi-coherent plate during the Cenozoic opening of the Labrador Sea and Baffin Bay, is shaded green (Greenland, south-west Ellesmere and Devon Island). FJL, Franz Josef Land; NZ, Novaya Zemlya; NSI, North Siberian Islands; SVB, Svalbard. The Continent–Ocean Boundary transition zone (COB) is mainly defined from seismic, magnetic, and/or gravity data.

and certainly Palaeozoic metamorphic sedimentary and volcanic rocks. Middle Ordovician and Middle Devonian conodonts are known, but its spatial affinities are not, although the rocks were probably laid down in ocean basins and island arcs off Laurentia (Blodgett & Stanley, 2008).

161, East Klamath and North Sierra. Occupying much of California and extending eastwards into Nevada, this group of terranes also includes the Yreka, Trinity, North Fork, Fort Jones, Forest Mountain, and Black Rock terrane units, some of which are themselves composite. They all consist of Palaeozoic rocks ranging from Cambrian to Permian, many of them metamorphosed. Most represent island arcs, whose fossils, e.g. the Ordovician and Devonian brachiopods from the Klamath Mountains (Potter et al., 1990), indicate that they were peri-Laurentian.

162, Laurentia Parautochthon. An elongate and substantial strip adjacent to the west of the Laurentia Craton (Unit 101) and stretching from Mexico to Alaska. The Parautochthon makes up much of the Western Cordillera of North America, and includes over 60 terranes with very varied rocks and ages, many of which are grouped together in our reconstructions; however, only its eastern sector is included within Unit 162.

163, Caborca. Within Sonora Province, north-west Mexico, the Caborca Terrane was originally part of the Laurentian Craton, but separated from it in the Carboniferous Ouachita Orogeny and then thrust back onto the craton in the Permian–Triassic Sonora Orogeny (Dickinson & Lawton, 2001).

164, Cortes (or Cortez) and Sierra Madre. Varied Ordovician to Permian sediments in the Cortes Terrane, Mexico, were overstepped by Late Jurassic and Cretaceous magmatic arc rocks, which accreted to North America in the Tertiary Laramide Orogeny. To the east of Cortes, the Sierra Madre Terrane is a metamorphic basement of uncertain age overlain by Silurian sediments, themselves unconformably overlain by Triassic and later rocks (Keppie, 2004).

165, Wrangellia and Alexander. A composite terrane area stretching from Alaska to British Columbia, including the Wrangellia Terrane (not to be confused with Unit 408, Wrangel Island, north of Siberia), Peninsular, and Chilliwack terranes, sometimes grouped with the Alexander Terrane to be termed the Insular Superterrane. The oldest rocks in Wrangellia in Alaska and Yukon are Late Carboniferous and Permian island arcs, but there are Late Devonian arcs in British Columbia. The Alexander Terrane includes the Admiralty (Ordovician to Permian) and Craig (Ordovician to Triassic) subterranes, and has yielded various unmetamorphosed Palaeozoic fossils. The Alexander and Wrangellia terranes were stitched by a Late Carboniferous pluton. There are also voluminous Late Triassic ocean basalts and the two terranes were not accreted to North America until the Late Jurassic or Cretaceous (Colpron & Nelson, 2009).

170, Meguma. The south-western two-thirds of Nova Scotia, Canada, separated from the rest of the province by the Cobequid–Chedabucto Fault Zone, both ends of which are in the Atlantic Ocean. That fault defines the northern boundary of the Meguma Zone, whose Cambrian successions are comparable to those in North Wales (Waldron et al., 2013), then part of Avalonia (Unit 315), but whose Ordovician to Devonian rocks are different. Like Avalonia, the Meguma Terrane was part of Gondwana until the end of the Cambrian, after which its drift history as an independent terrane is controversial until its union with Avalonia, which may not have occurred until the Late Carboniferous Alleghanian Orogeny.

171 and 172, Ganderia and Carolinia. Ganderia, Avalonia (Unit 108), Meguma (Unit 170), and Carolinia (Unit 172) were all peri-Gondwanan in the Late Neoproterozoic and Early Palaeozoic. Ganderia and Carolinia probably separated from Avalonia during the Early Palaeozoic and later Silurian–Devonian subduction of Avalonia beneath Ganderia occurred after Ganderia had accreted to Laurentia.

South America and Caribbean

South America was part of Gondwana in the Early Palaeozoic and then Pangea from the Carboniferous onwards until the opening of the South Atlantic Ocean in the Jurassic, since when it has been a separate continent. Its Palaeozoic history and geography were reviewed by Torsvik and Cocks (2013), the geology of Chile by Moreno and Gibbons (2007), and the extensive sedimentary basins in South America, many of which contain substantial hydrocarbon deposits, by Tankard et al. (1995). The Mesozoic to Recent geological history of the Caribbean region was summarised by James et al. (2009) and Boschman et al. (2014).

201, Amazonia. A large unit which includes the northern two-thirds of South America apart from its north-western margin (Units 231, 246, and 247). Two Archaean and Paleoproterozoic cratons (Amazonia and São Francisco) united to become part of the main Gondwana Craton in the Neoproterozoic. Much of the craton is overlain by Phanerozoic sediments, particularly in the Solimões and Amazonas Basins in the centre of the unit, which were caused by sag warping during the Palaeozoic. That continued until the intrusion of the Central Atlantic Magmatic Province (CAMP) LIP over nearly half of the north-central area of Amazonia at the end of the Triassic at 201 Ma, and the consequent opening of the Atlantic Ocean.

202, Paraná. Most of this large area in east-central South America is underlain by the large Archaean and Paleoproterozoic Rio de La Plata Craton (which is known through boreholes to be more extensive than its present outcrop indicates). That craton became part of Gondwana in the Neoproterozoic, and remained so until the intrusion of the Central Atlantic Magmatic Province LIP at 201 Ma. The Neoproterozoic is overlain in many places by Phanerozoic sediments, particularly in the extensive Paraná Basin, which include Late Ordovician to Middle Silurian and Late Devonian to Early Permian glaciogenic rocks.

204, Chortis. In southern Mexico, Belize, Honduras, and Guatemala, Palaeozoic outcrops are very sporadic and much masked by Mesozoic to Recent rocks, and may represent several originally independent terranes, including the Maya Terrane of some authors; nevertheless they are grouped as the Chortis Terrane here. There are Proterozoic basement fragments, a Late Silurian (418 Ma) granite in Belize, and Late Carboniferous and Permian clastics unconformably overlain by Jurassic rocks (Dickinson & Lawton, 2001). The Tarahumara Terrane in northern Mexico, which lies just south of the Carboniferous and Permian Ouachita Front, is included, and consists of poorly dated basinal facies which were deformed and metamorphosed in the Permian before being unconformably overlain by Late Jurassic rocks (Keppie, 2004), but it is not shown separately here.

205, Yucatán. The Yucatán Peninsula in southern Mexico has a basement with Lower Palaeozoic volcano–sedimentary rocks and granitoids, which may represent either a separate peri-Gondwanan terrane or an integral marginal sector of Gondwana. There are Neoproterozoic metamorphic rocks overlain by Late Carboniferous to Middle Permian shallow-water clastics, carbonates, and metavolcanic rocks in the Yucatán Peninsula, which was displaced southwards in the Jurassic during the formation of the Gulf of Mexico (Dickinson & Lawton, 2001).

215, Sierra Madre. The Sierra Madre and Cortez (Unit 164) terranes are adjacent to each other to the south of the west end of the Ouachita Front (which forms the southern boundary of Unit 101 – Laurentia) and are grouped here. Late Ordovician sediments unconformably overlie Devonian, Carboniferous, and Permian psammitic and pelitic rocks in the Cortez Terrane, which are overstepped by Late Jurassic and Cretaceous magmatic arc rocks which became part of North America in the Tertiary Laramide Orogeny. The Sierra Madre Terrane (including the Tampico Block) has a metamorphosed basement of uncertain age unconformably overlain by Silurian sediments themselves unconformably overlain by Triassic rocks (Keppie, 2004).

216, Mixteca and Oaxaquia. Two terranes, parts of the Middle American Oaxaquia Microcontinent (Keppie, 2004). Both have Precambrian basements overlain by Palaeozoic rocks which fringed the Amazonia sector of Gondwana (Unit 201) from the Late Neoproterozoic until the Carboniferous. Their Palaeozoic margins are masked by Mesozoic and later rocks. Deformation associated with the Laurentian–Gondwanan collision started in the Late Carboniferous. The Mixteca Terrane has Ordovician granites and associated rifted tholeiites (Keppie et al., 2008).

217, Guerrero. A substantial part of southern Mexico, which may have been associated with the Chortis Terrane (Unit 204) in the Cretaceous; however, it is largely made up of Cenozoic arc rocks which accreted to North America relatively recently.

231, North-western South America. A large area, including the Maracaibo Block (Boschmann et al., 2014), but to the east of the smaller Western Maracaibo Block (Unit 246). It includes the Mérida Andes, from which Late Ordovician to Devonian Gondwanan benthic marine faunas are known, as well as some Caribbean areas, including Trinidad. To the north, there are Early Cretaceous volcanic arc rocks which were parts of the Great Arc of the Caribbean. There was Late Jurassic to Middle Cretaceous (Coniacean) rifting from South America, but Late Cretaceous (Campanian) to Miocene subduction from Venezuela to Trinidad drove upper crustal nappes cratonwards to become reunited with the South American continent, of which it today forms part of the margin.

232, Yucatán Basin. A mainly ocean basin bounded by the Cayman Trough to the south, the Yucatán Peninsula (Unit 205) to the west and Cuba to the north. Its western sector is formed by Paleocene to Eocene oceanic crust, and the eastern sector, including the Cayman Rise, is volcanic arc material probably resting directly upon pre-Cenozoic ocean crust.

233–236, Cuba. Cuba has four units, each of which is a different thrust slice. To the west a fold-thrust belt rests upon Neoproterozoic basement overlain by Paleogene clastic successions. There are Late Jurassic (150 Ma) ophiolites and a variety of igneous rocks ranging from 160 to 50 Ma (Jurassic to Eocene), probably representing formation of oceanic lithosphere, as well as a central range made up of a 3 km thick Cretaceous volcanic–sedimentary and plutonic complex dated from 132 to 90 Ma, and forming part of the Great Arc of the Caribbean (Boschman et al., 2014).

237, Gonave. A block of Mesozoic and later continental lithosphere in the western part of Hispaniola (largely Haiti) and extending westwards to Jamaica, where it merges with the North Nicaraguan Rise (Unit 239). The microcontinent is bounded to the north by the Oriente Fault and to the south by the Walton and Enriquillo–Plantain Garden Faults.

238 and 239, North Nicaraguan and South Nicaraguan Rise. The island of Jamaica forms the northern part of the North Nicaraguan Rise, and consists of a basement of Cretaceous and Paleogene volcanic and plutonic arc-related complexes, some metamorphosed as blueschists, and was originally part of the Cretaceous Great Arc of the Caribbean. There are varied Tertiary to Recent carbonates and other sediments.

240 and 241, Hispaniola. The Hispaniola Cordillera Arc (Unit 240) is divided from Northern Hispaniola (Unit 241) by the Septentrional Fault, and is a complex tectonic melange overlain by ophiolitic, plutonic, and island arc rocks formed from 116 Ma onwards. There are unmetamorphosed Late Cretaceous rocks in a forearc basin overlain by Late Eocene to Recent sediments, faulted against oceanic flood basalts. Northern Hispaniola (Unit 241) is a subducted part of the North American continental margin, including metasedimentary rocks, eclogites, and ophiolites, overlain by Lower Cretaceous to Middle Eocene arc rocks.

242, Puerto Rico. Part of the Great Arc of the Caribbean, which was an active volcanic arc until its Late Eocene collision with the Bahamas Platform. Cretaceous to Eocene island arc rocks are underlain by a Late Jurassic to Middle Cretaceous serpentinite-matrix melange, and overlain by Eocene to Pliocene sedimentary rocks.

243, Costa Rica. Lying between Central Panama (Unit 244) and the Chortis and Maya Terranes (Unit 204), and whose eastern half includes the Suna Terrane of some authors. There are volcanics and other ultramafic igneous rocks, and it probably also formed part of the Lower Cretaceous Great Arc of the Caribbean, which developed on the ocean floor and was partially thrust over and accreted to the Chortis block (Unit 204) in the Late Cretaceous.

244 and 245, Panama. Central Panama (Unit 244) and Eastern Panama (Unit 245). The Panama Block lies to the north of the North Andean Block and has andesites, dacites, diorites, and granodiorites of Cretaceous and later ages. Units 243, 244, and 245 have together been termed the Panama–Chocó Block. To the north is the relatively small 'Panama Deformed Belt', not differentiated as a separate unit here.

246, Maracaibo Block West. A relatively small unit lying west of the North-western South America Block (Unit 231) is the western part of the Maracaibo Block. It is currently moving slowly (6 mm/yr) away from South America.

247, Accreted Andes Terranes. Northernmost Peru, and parts of Ecuador, Colombia, and north-western Venezuela make up a complex tectonic region which represents more than four terranes which accreted to each other during Mesozoic and later times (Kennan & Pindell, 2009).

248, Bonaire. An area largely in the Caribbean, including some of the Leeward Antilles (Aruba, Curaçao, and Bonaire), but also including rocks in South America north of the east–west-trending Oca Fault in eastern Colombia and western Venezuela. Mostly consisting of Cretaceous nappes, between which are basins filled with Oligocene to Recent sediments, but its palaeogeography is poorly unravelled.

249, Aves Ridge. A remnant island arc, west of the Grenada Basin, which was active from the Late Cretaceous to the Paleocene, and which was also previously a part of the Great Arc of the Caribbean. Including parts of the Leeward Antilles (Las Aves, Los Roques, La Orchilla, La Blanquilla, Los Hermanos, and Los Testigos).

250, Grenada Basin. An oceanic back-arc basin, the eastern extended forearc of the Aves Ridge (Unit 249) in the Paleocene to Middle Eocene, which extends between the Aves Ridge to its west and the Lesser Antilles Arc (Unit 251) to its east.

251, Lesser Antilles Arc. Between the Grenada and Tobago basins, an arc formed from Early Cretaceous (Albian) times onwards, and volcanic arc activity still continues there. To its east lies the accreted Antilles Prism (Unit 253), which includes Barbados.

252, Tobago Basin. Largely Caribbean ocean floor to the east of the Lesser Antilles Arc and west of the Barbados Prism. Cretaceous oceanic island arc rocks and pluton (110–103 Ma) in the south, with unconformable Pliocene to Recent sediments, Cretaceous (128 Ma) meta-igneous and greenschist facies in the north.

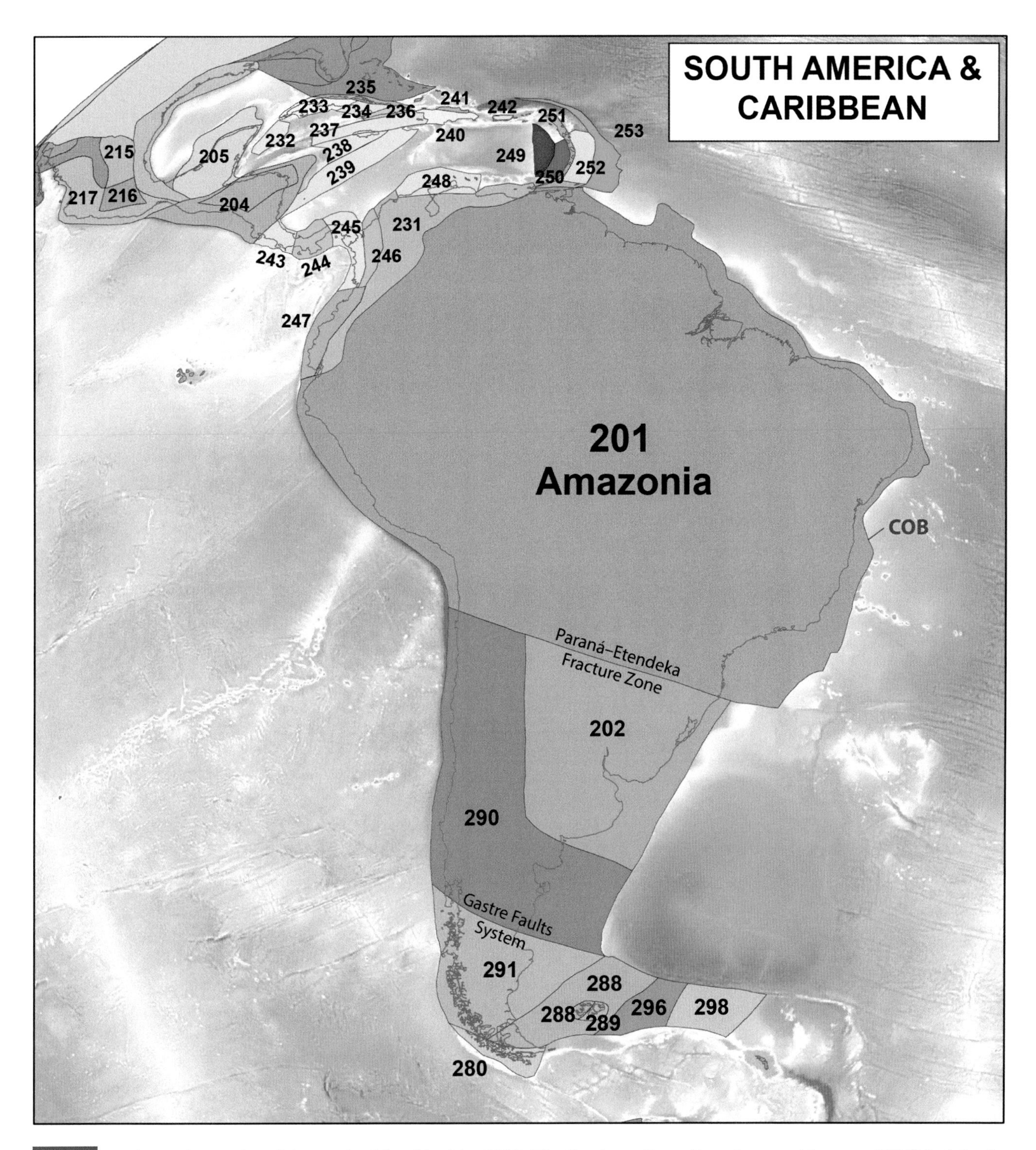

Fig. 3.4 South American and Caribbean units (plate identities 2XX). The Continent–Ocean Boundary transition zone (COB) is defined mainly from seismic, magnetic, and/or gravity data.

253, Accreted Antilles Prism. The most easterly part of the Caribbean Ocean crust, sometimes termed the Barbados Prism. It lies east of the Tobago Basin (Unit 252), and is bounded to its east by the Lesser Antilles Trench.

280, San Jorge Plate. Independent ocean-floor plate off southwestern South America. Whilst geophysical evidence points to continental lithosphere there, no rocks have yet been recovered from the area, and so the age range of the plate is unknown.

288, Falkland Plate and West Falklands, and 289, East Falklands. The western part of the Falkland Islands, and the South Atlantic Ocean surrounding them, extending westwards to abut with Patagonia (Unit 291) and which includes the southernmost tip of South America. The East Falklands are separated from the West Falklands by the substantial Falkland Sound strike-slip fault. Within the Falkland Islands there is a Neoproterozoic basement unconformably overlain by possibly Silurian and certainly Early Devonian sediments, with brachiopods of the cooler-water Malvinokaffric Province. Before the Middle Jurassic, this terrane group was situated to the south-east of South Africa (Unit 701).

290, Colorado. A unit that includes the northern sector of geographical Patagonia, as well as the northern two-thirds of Chile and the southernmost part of Peru. Its south-eastern third is a Precambrian craton, but its western sector consists of several terranes making up the Patagonian Andes (Moreno & Gibbons, 2007). Those terranes include volcanic arcs ranging from Silurian to Recent, with many volcanoes, some of which are still active, all intruded by a variety of Palaeozoic to Recent batholiths.

291, Patagonia. The southern area of South America (apart from the westward tip of Unit 288, West Falklands). In the north, adjoining Unit 290 (which includes the northern sector of geographical Patagonia), there are Ordovician (475 Ma) granitoids intruded into lightly metamorphosed Cambrian and Early Ordovician sediments which unconformably overlie a Neoproterozoic basement.

296 and 298, Falklands Plateau (296) and Maurice Ewing Bank (298). Two entirely submerged adjacent areas projecting into the South Atlantic Ocean to the east of Unit 288 (Falkland Plate). Whilst geophysical evidence points to continental lithosphere, no rocks have yet been recovered from the two areas, and so their age ranges are unknown.

European and Near East

The geology of central Europe was summarised in the two volumes edited by McCann (2008) and western and Central Europe in the annotated maps of Ziegler (1990). Europe is divided between the central North Sea and the Black Sea by the Tornquist Suture Zone, and is dominated by Baltica (Unit 302), but south-west of the Tornquist Zone are many separate areas, some included in the 'Armorican Terrane Assemblage', which were fragments of the Gondwana margin prior to the Variscan Orogeny. Europe has also been united with various sectors of Asia along the Ural Mountains since before the end of the Permian. Europe's western and northern margins have been passive since then (apart from the failed Triassic and Jurassic north–south rifting in the North Sea), and there have been continuing smaller tectonic readjustments in the Ural area; but its southern margin has been active since the start of the Alpine Orogeny at about 100 Ma. Some of the Asian units have slightly longer entries here since we have not included them in our published reviews.

302, Baltica. Bounded to the east by the Ural Mountains (Unit 390), to the south-west by the Tornquist Zone, and to the north-west and north by the North Sea and Arctic Ocean. Baltica has at its core an Archaean and Proterozoic metamorphic craton overlain by Lower Palaeozoic sedimentary rocks, and was an independent continent from the Neoproterozoic until its union firstly with Avalonia (Units 108 and 315) in the end Ordovician at 443 Ma and subsequently with Laurentia (Unit 101) in the Silurian Caledonide Orogeny to form a sector of Laurussia. The north of Baltica includes the Timanian area (Unit 350), largely today in north-east Russia, which merged with Baltica in the Timanide Orogeny just before the Cambrian at about 550 Ma. The north-east of Baltica abuts the Kara Plate (Unit 417) and Novaya Zemlya (Unit 373).

303, Scotland and North-western Ireland. Essentially united during the Palaeozoic, the two countries were integral parts of north-western Laurentia until the merger with Avalonia–Baltica in the Silurian Caledonide Orogeny at about 420 Ma. The unit is Archaean and Proterozoic Laurentian cratonic fragments in the north of Scotland, with Neoproterozoic and Lower Palaeozoic island arcs to the north of the Iapetus Suture, which were accreted as the Iapetus Ocean closed. Later rocks include Devonian granites and continental Old Red Sandstone sediments, substantial Carboniferous marine deposits and Permian–Triassic New Red Sandstone continental deposits, as well as the Tertiary Igneous Province LIP (from 62 Ma) in north-west Scotland. The area stayed in Europe after the Mesozoic opening of the Atlantic Ocean to its north-west.

304, Iberia. The unit includes the Iberian Peninsula and the Atlantic Ocean to its west as far as the Continent–Ocean Boundary. The main pre-Mesozoic Iberian Massif is exposed in the western half of the Iberian Peninsula, and bounded to the north by the Pyrenees Mountains, and was tectonised during the Variscan and pre-Variscan orogenies. It is a complex and composite agglomeration of terranes (Gibbons & Moreno, 2002), all including Proterozoic to Mesozoic rocks. They are, from north to south, Cantabria, West Asturian–Leone (WALZ), Galicia–Tras os Montes (GMZ), Central Iberia (CI), Ossa Morena (OMZ), and South Portuguese (SP). All were originally parts of Gondwana, or island arcs adjacent to it. The zones are separated today by the NW–SE-trending South Portuguese and Badajoz–Cordoba Shear Zones of Variscan age, but an Early Devonian (410 to 400 Ma) tectonic event probably occurred before the main Variscan Orogeny at 380 Ma and later (Arenas et al., 2014). A trough with much east–west strike-slip movement developed in the Cretaceous to the north of Iberia (the Ebro Massif) and Corsica–Sardinia and to the south of the Massif Central of France (Unit 305, Armorica). That trough filled and became inverted progressively during the Tertiary to form the Pyrenean Mountains.

305, Armorica. Southern France consists largely of the Palaeozoic Massif Central (including the Maures Massif), exposed sporadically for over 400 km from the Montagne Noire of south-west France to the north, with boreholes showing extension eastwards under the Mesozoic and Tertiary Aquitaine Basin. Much of the Palaeozoic is also overlain by the Mesozoic Paris Basin. There are Cambrian and Ordovician clastics, Devonian limestones, and a thick syntectonic Visean–Namurian Carboniferous sequence, mainly turbidites. Much of the massif is allochthonous; but the main part is a Late Precambrian and Lower Palaeozoic metamorphosed sequence intruded by Cambrian–Ordovician (540–460 Ma) diorites and granites. Early Devonian (385–380 Ma) nappes, Late Devonian and Early Carboniferous metamorphism peaking from 350 to 340 Ma and granite plutons emplaced between 330 and 305 Ma were all phases of the Variscan Orogeny (Franke et al., 2016). Synorogenic flysch and olistostromes were emplaced in the Visean of the Montaigne Noire over a Devonian platform sequence.

The northern sector is the Late Precambrian and Lower Palaeozoic Armorican Massif of Brittany and Normandy (Shelley & Bossière, 2000), much of which was deformed in the Neoproterozoic to Early Cambrian Cadomian Orogeny from 645 to 540 Ma, followed by sporadic Late Cambrian volcanism and Ordovician and Silurian tectonism, and the main Variscan Orogeny in the Late Devonian and Carboniferous. The margin is less deformed and includes fossiliferous Late Ordovician sediments unconformably overlain by Early Devonian reef limestones below a Carboniferous basin sequence, although in Brittany the Lower Palaeozoic marine rocks extend up to the Famennian. Upper Palaeozoic granitoids are dated at 340 Ma, 300 Ma, and 290 Ma. Because the progressive development of Lower Palaeozoic terrane units and their geometries are so poorly constrained, Armorica is only shown as a single area before the Devonian.

306, Corsica and Sardinia. Sardinia did not leave the Iberian Peninsula until as recently as 30 Ma (Oligocene), and has a relatively unmetamorphosed External Zone in the south-west and two more deformed areas in the centre and north-east. The south-west has Cambrian and Ordovician benthic faunas typical of the higher-latitude Mediterranean Province. The north-eastern sector contains Variscide nappes, which include Cambrian metasandstones covered firstly by rhyolites and then Ordovician trangressive deposits, gneisses, and igneous intrusions (Helbing & Tiepolo, 2005). Corsica has no rocks older than Devonian.

307, Apulia. Southern Italy apart from Calabria and most of Sicily. Oceanic Ordovician and Silurian peri-Gondwanan basins with graptolitic shales are followed by a variety of later Palaeozoic and Mesozoic rocks, some shelf and others basinal.

309 and 311, West and East Svalbard. Svalbard was part of Laurentia prior to the Silurian Caledonide Orogeny. West Svalbard has Neoproterozoic and Lower Palaeozoic rocks, which are tectonised and overlain unconformably by undeformed Devonian Old Red Sandstone and later Upper Palaeozoic marine sediments, as well as Mesozoic and Tertiary rocks. In contrast, the Lower Palaeozoic rocks in East Svalbard are relatively undeformed and are platform carbonates with tropical benthic faunas typical of the Laurentian Craton, of which it was a part; thus the two halves of Svalbard were some way apart prior to the Caledonide Orogeny.

315, Eastern Avalonia. Southern Ireland, England, Belgium, the Netherlands, Denmark, and a small part of northern Germany. Unified in the Palaeozoic with Unit 108 as the Avalonia Microcontinent until the two halves were separated by the Jurassic Atlantic Ocean opening. Avalonia was an integral marginal sector of Gondwana until the Rheic Ocean opened at 490 Ma, but was an independent microcontinent only in the Ordovician before its amalgamation with Baltica at 443 Ma. Proterozoic basement (the Midlands Microcraton of England) is unconformably overlain by Lower Palaeozoic and Upper Palaeozoic to Tertiary sediments, mostly marine apart from the Devonian and Permian–Triassic Old Red and New Red Sandstones.

318, Hatton Bank–East Rockall Plateau. A submerged area of continental crust of uncertain ages, apart from the small island of Rockall, north-west of Scotland, which is a 60 Ma (Paleocene) granite.

319, Moesia. Including parts of Bulgaria, Serbia, and southern Romania. Extensive Mesozoic and Tertiary deposits in the southern Carpathian Mountains overlying Cambrian and Ordovician sediments in boreholes. Mediterranean Province Ordovician brachiopods occur in eastern Serbia (Krstić et al.,

1999), and thus the area was probably originally peri-Gondwanan (Yanev et al., 2006).

322, Calabria and 333, Central Apennines. Calabria (which includes eastern Sicily) left the Iberian Peninsula at about 30 Ma. Its Ordovician and Silurian sediments are deeper-water graptolitic shales reflecting a peri-Gondwanan positioning then. The Ordovician and Silurian sediments in the Apennines are also deeper-water graptolitic shales and also indicating a peri-Gondwanan positioning, but they were not provably close to Iberia in the Palaeozoic.

337, Tisia. The Tisia Terrane (also known as the Tisza Mega-Unit in Hungary, Croatia, Serbia, and Romania) forms the basement of parts of the Pannonian Basin. Tisia is a Variscan orogenic collage which accreted to southern Europe during the Carboniferous and Permian, but subsequently broke away in the Jurassic (Bathonian), and the present boundaries of Tisia are governed by Alpine (Miocene) structures (Szederkényi et al., 2012).

338, Rhodope. Extends over parts of southern Bulgaria and north-eastern Greece in the northernmost Aegean region. Dominated by a metamorphic basement comprising pre-Alpine and Alpine units of continental and oceanic affinities. Late Cretaceous to Early Miocene granitoids intrude the basement and are covered by Late Cretaceous/Early Cenozoic–Neogene volcanic and sedimentary sequences (Bonev, 2006).

340, Yermak Plateau. A sliver extending north of western Svalbard (Riefstahl et al., 2013). Crustal nature and age largely unknown, but assumed to be stretched continental crust strongly affected by alkaline magmatism at around 51 Ma.

346, Greece. Including Adria, the northern part of the Adriatic from north-east Italy to the western parts of the former Yugoslavia, and which has been a north-westerly-pushing indenter from the Late Cretaceous (Maastrichtian) to the present. No rocks older than Carboniferous are known.

347, Crete. Lies in the forearc of the Hellenic subduction zone, currently consuming the remnants of Tethyan sea floor, which is subducting northward beneath Crete. Most of the geology was assembled by thrust imbrication during Oligocene times and the oldest rocks (marine carbonates) are probably Permian in age.

350, Timanian. A substantial part of north-western Europe, which became part of Baltica (Unit 302) in the Late Neoproterozoic to Early Cambrian Timanian Orogeny (Gee & Pease, 2005). Largely Neoproterozoic rocks with a few unconformable Palaeozoic rocks. Included is the Barents Sea, a flooded area of Caledonian metamorphosed basement overlain by unmetamorphosed Early Devonian to Recent marine and non-marine sediments (Smelror et al., 2009). In the north is Franz Josef Land, an archipelago with tectonically disturbed Lower Palaeozoic rocks of uncertain palaeogeographical affinity, unconformably overlain by unmetamorphosed Carboniferous, Permian, and Mesozoic rocks.

368–371, Jan Mayen Microcontinent. Divided into several distinct blocks (Units 368, 370, and 371; Gaina et al., 2009; Peron-Pinvidic et al., 2012), with Unit (369) extending beneath Iceland (Torsvik et al., 2015). These blocks were originally marginal to East Greenland and probably contain Precambrian basement, Caledonian nappes, and younger rocks. Unit 369 detached in the Early Eocene from 53 to 47 Ma, and by the Oligocene (27 Ma) all parts of the Jan Mayen Microcontinent had become parts of Eurasia.

373, Novaya Zemlya. Originally an integral part of Baltica and a northern sector of the Urals (Unit 390), but displaced westwards in an arcuate structure by Baltica's accretion with north-western Siberia in the Triassic (Buiter & Torsvik, 2007). The Lower Palaeozoic platform carbonates with typically Baltic Province benthic faunas there are relatively undeformed.

374, Bohemia. Sometimes termed Perunica, and occupying most of the Czech Republic as well as the Sudetes of south-west Poland. An integral part of Gondwana until the Palaeotethys Ocean opened in the Early Devonian. The central Prague Basin (known as Barrandium) has unmetamorphosed Cambrian to Devonian sediments with many benthic fossils, indicating the cooler-water low-diversity Mediterranean Province in the Lower Ordovician, but reflecting decreasing latitudes until the Devonian. There are Ordovician granites in the Bohemian Massif as well as many Variscan volcanics and igneous intrusions (McCann, 2008).

375, Saxothuringia and Bruno-Silesia. Saxothuringia is an elongate area to the south of the Rhenian–Hercynian, and includes a Silurian to Early Devonian island arc which was highly metamorphosed in the Variscan Orogeny (Franke et al., 2016). Cadomian (750–540 Ma) basement overlain by thin Lower Palaeozoic sediments intruded by 490–485 Ma granites.

390, Urals. The Ural Mountains form the eastern boundary of Europe, and are a varied complex with six zones (Scarrow et al., 2002); from west to east the Pre-Uralian and Central Uralian zones, marginal sectors of Baltica in the Palaeozoic

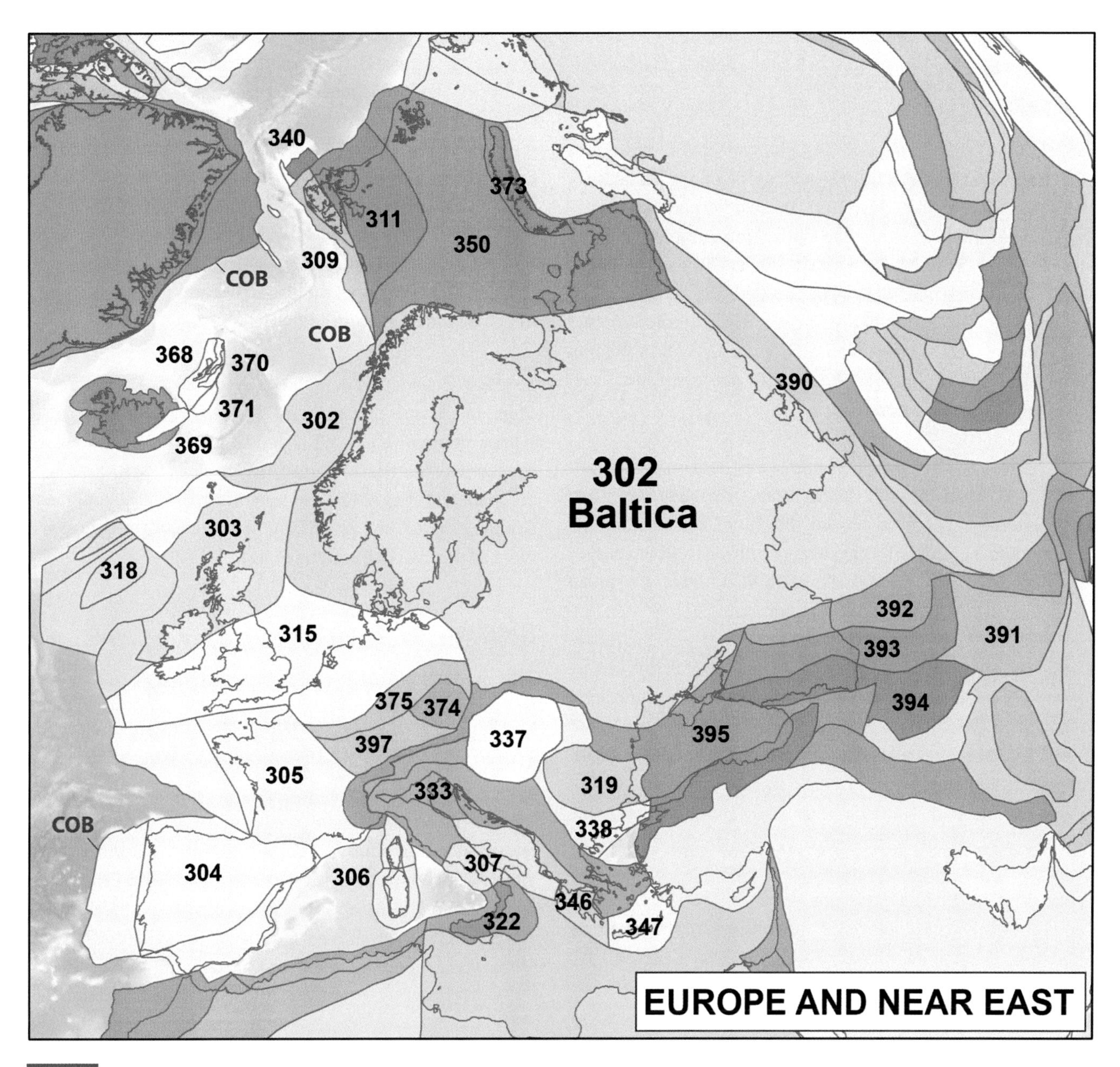

Fig. 3.5 European and Near East units (plate identities 3XX). The Continent–Ocean Boundary transition zone (COB) is mainly defined from seismic, magnetic, and/or gravity data.

(Unit 302), and the West Uralian, Tagil–Magnitogorsk, East Uralian, and Trans-Uralian zones. These are amalgamated island arcs which originally lay in the ocean to the east of Baltica. The straight alignment of the Ural Mountains is due to strike-slip faulting during the Late Carboniferous and Early Permian Uralian Orogeny, when Baltica and Kazakhstania were united in the formation of Pangea.

391, Karakum. A composite terrane, largely within Turkmenistan and extending south-east into Pakistan. Its core includes remains of a Precambrian continent of some size, including the Amudarya Block in the east sector (Daukeev et al., 2002). Palaeozoic and earlier rocks are obscured by Mesozoic to Recent tectonised rocks in the Kizil Kum Desert, including the North Pamir–Mashhad Arc (Natalin & Sengor, 2005), apart from Karakum's boundary with the Caucasus–Mangyshlak Terrane (Unit 393) to the north-west. The Precambrian basement on the Tajikistan–Uzbekistan border is exposed in the Baisun and Garm massifs, and overlain by deformed Ordovician to Devonian clastic rocks and

Carboniferous volcanics (Biske & Seltmann, 2010). The unit probably accreted to Kazakhstania in the Middle Carboniferous (Windley et al., 2007).

392, South-eastern peri-Baltica. An elongate area south-east of the Baltica Craton (Unit 302) and north of the Caucasus–Mangyshlak Terrane (Unit 393). Its eastern part lies south-west of the southern part of the Urals (Unit 390), and it spans the Caspian Sea. Poorly known Lower Palaeozoic rocks underlie thick Upper Palaeozoic and Mesozoic strata. The south-eastern extension between Units 391 (Karakorum) and 497 (South Tien Shan) is arbitrarily included within the unit here: the geology of that tectonically deformed and fragmented region is poorly known.

393, Caucasus–Mangyshlak. The unit lies both west and east of the Caspian Sea in Kazakhstan, Turkmenistan, and Uzbekistan. It has been termed the Turan Block; although the modern Turan Platform extends northwards to the east of the Aral Sea. There are no rocks older than middle Devonian near the eastern side of the Caspian.

394, Scythian–South Caspian. The Caspian Sea is divided by two WNW–ESE-trending tectonic zones. The northerly zone, with the Karpinsky Swell to the west and the Central Ust Yurt Fault to the east (Natal'in & Şengör, 2005), separates the Scythian block from the Terek Basin to the west of the Caspian and the peri-Caspian Depression from the Mangyshlak–Turan block to the east of the Caspian. That basin has more than 25 km of sediments above a basement consisting of Middle Devonian oceanic floor (Zonenshain et al., 1990). The southerly zone runs through the Great Caucasus Mountains of Azerbaijan, eastwards beneath the Caspian (including the Apsheron Sill), and through the Kopet Dagh area of Turkmenistan (Egan et al., 2009). There it is divided into the northern Greater Caucasus Terrane and the southern Lesser Caucasus Terrane in the west, and a South Caspian Terrane in the centre. Both have marine Tournaisian and Lower Visean sediments, but later rocks are terrestrial, apart from Svanetia in the Great Caucasus, where deeper-water marine shelf sediments extend to the top of the Carboniferous (Daukeev et al., 2002). The South Caspian sector has only Mesozoic rocks. There was much Middle and Late Devonian volcanism in the Greater Caucasus Terrane; however, the Lesser Caucasus Terrane did not leave Gondwana until the Permian opening of the Neotethys Ocean (Ruban et al., 2007). There are terrestrial strata in the North Caucasus Mountains, with Carboniferous and Permian floras of the Euramerian Province. To the south of the Caucasus Terrane (Scythian Terrane) is a Jurassic volcanic arc.

395, Alpine Region. The Alps extend from the south-eastern coast of France and curve north-eastwards to the north of Units 333 and 337, and are adjacent to the Baltica Craton (Unit 302) to its north-east. Highly tectonised Tertiary to Recent nappes of the Alpine Orogeny include varied Precambrian to Tertiary rocks, including Late Ordovician and Silurian deeper-water sediments in the Carnic Alps, Austria. Fragments of Cambrian island arcs in both Western and Eastern Alps, and Ordovician volcanics and granitoids, all indicate that the area was at the active north-western sector of the Gondwanan margin in the Palaeozoic.

397, Moldanubia. An elongate area to the south of Saxothuringia and Bruno-Silesia (Unit 377) and to the north of the Alpine region (Unit 395), including the Vosges area of France and the Black Forest of Germany. Its structures are complex and it may include several small Palaeozoic microcontinents (McCann, 2008), but it is attached to Armorica on most of our reconstructions (e.g. Fig. 8.10).

Northern and Central Asia

This vast region includes Siberia, Mongolia, Kazakhstan, and much of the northern and eastern sectors of China. Although the Ural Mountains form a natural boundary to its west, the Arctic Ocean to its north, and the Pacific Ocean to its east, the southern boundaries with India and the Middle East (Unit 500 onwards) and South-East Asia (Unit 600 onwards) are arbitrary. Those parts in the former Soviet Union were reviewed by Zonenshain et al. (1990) and Şengör and Natal'in (1996), and the Palaeozoic geography of Siberia and its adjacent terranes by Cocks and Torsvik (2007). South of the large Siberian Craton and to the north of North China and Tarim lies an extensive fold belt, variably termed the Altaids or the Central Asian Orogenic Belt (CAOB), which includes many terranes, mostly composite, which were accreted to each other during the assembly of Pangea in the Late Palaeozoic and Triassic (Figs. 3.8 and 8.9). Since we have not yet reviewed much of this region, many of the unit entries here are longer than average for the rest of this chapter.

401, Siberian Craton. Also named the Angara Continent, with a central craton surrounded by many peri-Siberian rocks, many obscured by thick Mesozoic to Recent sedimentary basins, although the western (Unit 468) and the eastern (Unit 435) West Siberian Basin are separate units here. The Verkhoyansk and Kolyma fold belts to the east are included. Archaean and Proterozoic rocks in the Anabar Massif in the north-centre and the Aldan Shield in the south of the craton are overlain by thick unmetamorphosed Neoproterozoic (locally termed Riphean and Vendian) and Palaeozoic sediments. To the south there is a substantial indenter largely filled by the Barguzin granite (Unit 436), but between Barguzin and the main craton lies Patom, which we include within Unit 401. Patom is entirely heavily folded Neoproterozoic rocks, unlike those of the same age on the main craton. Siberia was intruded by the Siberian Traps Large Igneous Province at the end of the Permian at 251 Ma.

405, New Siberian Islands and 415, Anyui. The New Siberian Islands are an archipelago mostly of Mesozoic to Recent rocks in the Arctic Ocean. However, Kotelny Island has a largely complete Cambrian to Permian succession; Belkov Island has Devonian to Carboniferous sediments overlain by lavas of the Permian Siberian Traps; Bennett Island has thick Cambrian and Ordovician sediments; and Bolshoi Lyakhov Island has Neoproterozoic metavolcanic rocks with Permian and Mesozoic fold belts. Anyui lies to the east of the New Siberian Islands and is also a part of the Arctic Ocean underlain by continental lithosphere, although it is poorly known.

407, Chukotka. The north-eastermost part of political Siberia, but it was (with Units 408, 160, and others in North America) part of the Arctic Alaska–Chukotka Microcontinent prior to the Devonian. The basement is Mesoproterozoic (1.9–1.6 Ga) and is overlain by Ordovician to Lower Carboniferous sediments.

408, Wrangel Island. An area of continental lithosphere in the Arctic Ocean, including Wrangel Island, where Late Silurian and Upper Palaeozoic sediments and volcanics unconformably overlie the Late Proterozoic Wrangel Complex. It was also part of the Arctic Alaska–Chukotka Microcontinent prior to the Devonian. Not to be confused with Wrangellia (Unit 162), in the North American Cordillera.

417, Kara. Includes the northern Taimyr Peninsula of Russia, as well as October Revolution Island and others in the Severnaya Zemlya archipelago within the Arctic Ocean. It has Cambrian to Devonian marine sediments, and was an independent microcontinent during much of the Lower Palaeozoic, but merged with Baltica in the Early Silurian and with Siberia during the Permian.

420, Chersky and 421, Omolon. Alternatively known as the Kolyma–Omolon Terrane, this unit lies to the east of the Verkhoyansk Mountains of north-east Siberia, and is underlain by the north-western sector of the North American Plate today. Under Mesozoic and later rocks, there are several Palaeozoic and possibly Precambrian massifs, of which the Omulevka Massif has a continuous Lower Ordovician to Middle Devonian (Givetian) sedimentary sequence with marine benthic faunas.

430, Central Mongolia and Altai–Salair. The Altai Mountains extend for over 1000 km from Kazakhstan, through southern Russia and north-western China into Mongolia, and thus the name 'Altai' recurs in the separate Units 430, 433, and 441 here. Altai–Salair includes several terranes, many composite, including Tomsk, Salair, Western Altai, Eastern Altai, Batenov, and Kuznetsk–Alatau (Cocks & Torsvik, 2007). Many are Lower Palaeozoic volcanic island arcs and associated accretionary complexes.

431, West Sayan. Lower Cambrian marine clastic rocks overlain by Middle and Upper Cambrian subaerial volcanics. Island arc and adjacent Ordovician to Silurian accretionary complexes, although there are also shelly Lower Silurian (Llandovery) shelf deposits. Accreted to the main Siberian Craton (Unit 401) in the Silurian, after which it was part of a passive margin.

432, Tuva–Mongol. A terrane assemblage, including the 'Eastern Sayan' of some authors, in west-central Mongolia, with 16 separate terranes identified by Badarch et al. (2002), many with island arcs. Dominated by sedimentary melanges, including oceanic deposits, and with many nappes, from Archaean to Ordovician in age. Tuva–Mongol collided with West Sayan in the Middle Ordovician (Sennikov, 2003), and both were accreted to the main Siberian Craton in the Late Ordovician.

433, Rudny Altai and 434, Kobdin. Rudny Altai is a composite and highly metamorphosed terrane with Vendian to Lower Cambrian ophiolites and Cambrian to Lower Carboniferous island arc rocks with shallow-water carbonates, turbidites, and olisthostromes, as well as Middle Devonian and Lower Carboniferous calc-alkali volcanics. Kobdin is Cambrian and Lower Ordovician turbidites unconformably overlain by Upper Ordovician andesites, clastics and limestones. Middle to Upper Devonian clastics and cherts in the south-west of the unit may represent an accretionary complex or forearc basin fill. The two terranes are stitched by Late Carboniferous granites.

435 and 468, West Siberian Basin. The large area to the east of the Ural Mountains and west of the Siberian Craton (Unit 401) is divided into Eastern (Unit 435) and Western (Unit 468) sectors. Thick Tertiary to Recent deposits obscure poorly known Precambrian and Palaeozoic rocks. Unit 435 includes Ob–Saisan–Surgut, where boreholes reveal that an Upper

Devonian to Lower Carboniferous accretionary complex is unconformably overlain by Middle and Upper Carboniferous terrestrial deposits intruded by Carboniferous and Permian granites. The whole area became an integral sector of the Siberian Craton (Unit 401) before the end of the Carboniferous.

436, Barguzin. Proterozoic metamorphic rocks and ophiolites underlie Vendian to Cambrian sediments (Zonenshain et al., 1990). Barguzin collided with the Siberian Craton passive margin in the Vendian, although shortening of the Patom Thrust Belt in the Early Devonian implied continued tectonic separation throughout the Lower Palaeozoic. Most of the unit is the vast Late Devonian Barguzin granodiorite, possibly connected with the contemporary rifting of the Viljuy Basin aulacogen within the main Siberian Craton (Unit 401).

437, Nadanhada–Sikhote–Alin. An area to the east of Khanka–Jiamusu–Bureya (Unit 454) and bordered to the east by the Pacific Ocean. Some remnants of Middle Palaeozoic to Early Cretaceous accretionary terranes and forearc turbidites, but mainly an agglomeration of Mesozoic to Recent volcanic arcs and accretionary prisms which became progressively welded to Khanka–Jiamusu–Bureya from the Late Cretaceous onwards.

438, Kamchatka. In the north-east of political Siberia, between the Pacific Ocean to its east and the Sea of Okhotsk to its west, this unit is a Tertiary to Recent island arc on the ocean floor, and is still volcanically active.

440, Central Mongolia. A terrane assemblage with 14 units, mostly island arcs and accretionary prisms (Badarch et al., 2002), lying east of the Tuva–Mongol Terrane Assemblage (Unit 432). Nine units have Precambrian cores, and Central Mongolia became united from the Cambrian to the Devonian, and then formed part of Amuria before becoming a unified continent with North China (Unit 601) during the Permian along the Solonker Suture.

441, Gobi Altai and Mandalovoo. These are two separate terranes, largely within Mongolia, which merged before the Devonian and are thus grouped here (Badarch et al., 2002). Both consist of Ordovician and later Palaeozoic rocks, including serpentinites and gabbros, volcanics and clastics. In the Mandalovoo area, Silurian (Ludlow) clastics have the *Tuvaella* brachiopod fauna, indicating that the terrane was then peri-Siberian rather than peri-Gondwanan.

443, Kuriles. A volcanic island arc, running from the southern end of Kamchatka (Unit 438), on the north-western margin of the Pacific Ocean, active from the Neogene to the present day. In front of the arc are growing accretionary wedges fringing the north of the deep subduction-related Pacific Ocean trenches. To the west of the Kuriles is the Sea of Okhotsk, which is probably an accreted Cretaceous oceanic plateau.

450, Junggar. The Junggar Basin lies between the Altai Mountains to its north and the Tien Shan Mountains to its south, and is mostly Mesozoic to Recent, although the rims include Palaeozoic outcrops. The unit is an aggregation of Cambrian to Permian volcanic arcs, accretionary prisms and segments of obducted ocean floor, all intruded by later granites (Xiao et al., 2009b), which was welded to Tarim (Unit 480) in the Devonian and Carboniferous (Charvet et al., 2007). There are Ordovician (Darriwilian onwards) deeper-water shelf sediments with some volcanics; and Devonian (Lochkovian to Famennian) volcanics, including ignimbrites, with Lower Devonian shallower-water sediments and Frasnian and Famennian deeper-water sediments. The Permian is terrestrial, with interspersed volcanics, apart from a shallow-marine incursion during the Late Kungurian (Daukeev et al., 2002). Tarim and Junggar amalgamated either before the latest Carboniferous (Zhou et al., 2001) or near the end of the Permian (Xiao et al., 2008).

451, Ala Shan. This composite terrane spans the China–Mongolia border, and Xiao et al. (2009a) included it within the North China continent: it is made up of the Beishan and Liuyan Terranes, the Dunhuang and Alax Massifs, and the Altun Faulted Block (Zhou & Dean, 1996). The sinistral Altyn Tagh Fault between Tarim and the Dunhuang sector of Ala Shan to its north and Kunlun and Qaidam–Qilian to its south, has a displacement of more than 400 km. Alax includes Archaean and Proterozoic, and Cambrian to Middle Ordovician rocks (Wang et al., 2007). In Mongolia the unit includes the Atasbogd, Hashaat, and Tsagaan Terranes of Badarch et al. (2002). There are also Devonian and Carboniferous Dananhu volcanic island arc rocks surrounding the largely Cenozoic Tuha (or Turfan) Basin. Collision between South Tien Shan, Junggar, and Tarim occurred progressively during the Latest Carboniferous to Early Permian, between 300 and 280 Ma (Zhang et al., 2008). There is a Late Devonian (about 370 Ma) stitching pluton between Ala Shan and Qilian (Unit 456) (Xiao et al., 2009a).

452, Gurvansayhan. A broad belt within south-central Mongolia which is combined with the Edren Terrane of Badarch et al. (2002), which is in a similar structural position, and has two or more island arcs with a Cambrian ophiolite, Ordovician to Silurian greenschists, Upper Silurian to Lower Devonian radiolarian cherts and tholeiitic pillow basalts, and Middle Devonian to Lower Carboniferous volcanoclastics.

There are many imbricate thrust sheets, and the unit accreted to some of its adjacent terranes in the Late Carboniferous, although it did not accrete to Ala Shan to its south until the Permian (Jian et al., 2008). There are Middle Carboniferous island arc andesites (336 Ma) and granites (321 Ma) (Batkhishig et al., 2010). The Edren Terrane is a metamorphosed Devonian island arc, with clastics interbedded with volcanic rocks succeeded by Carboniferous sedimentary rocks, all intruded by Permian alkaline granites. Gurvansayhan accreted to Ala Shan in the Late Permian at about 260 Ma and to the Gobi Altai–Mandalovoo Terranes of peri-Siberia at about the same time.

453, Hutag Uul–Songliao. A triangular area straddling the Chinese, Russian, and Mongolian borders and adjoining Khanka–Jiamusu–Bureya (Unit 454), with which it was linked for much of the Palaeozoic, becoming part of the Amuria continent. A tectonised amalgamation of many Precambrian and later units, but most of the earlier rocks are obscured by Mesozoic and later volcanics and sediments. It is bounded to its south by the Solanker Suture Zone, which closed obliquely from its west end in the Early Permian to its eastern end by the end of the Permian, thus amalgamating Hutag Uul–Songliao with North China (Jian et al., 2008). The Songliao Massif is a Mesozoic sedimentary basin, but boreholes have yielded weakly metamorphosed Palaeozoic sediments and granites with Proterozoic zircons. Wu et al. (2011) dated 282 granites from the terrane, all Mesozoic and younger, apart from six Cambrian–Ordovician and seven Carboniferous.

454, Khanka–Jiamusu–Bureya. To the east of Unit 453 are the Bureya and Khanka areas, between which is the Jiamusu Massif. The Khanka Massif and Jiamusu Massif have Late Cambrian (about 500 Ma) high-grade metamorphism, and the same Neoproterozoic zircon signatures and have been united since then (Zhou et al., 2010), although they are largely composed of Lower Palaeozoic and Permian granitoids, and the Triassic Heilongjiang metamorphic complex. The unit became a sector of Amuria in the Early Devonian. There are marine Early Cambrian and Ordovician rocks, and Jiamusu and Bureya both include similar Devonian and Lower Carboniferous continental rift-related volcanics and sedimentary rocks. In Russia, there is a Palaeozoic and Mesozoic cover of clastics, carbonates, and volcanics overlying the Neoproterozoic, intruded by a Middle Ordovician (471 Ma, Dapingian) syenite (Zonenshain et al., 1990). There are Early Permian volcanics and interbedded terrestrial rocks with marine horizons with Middle and Late Permian brachiopods in the South Primorye area, Russia. Khanka–Jiamusu–Bureya might have been originally derived from either Siberia or Gondwana, but does not appear to have been near either North China or South China in the Lower Palaeozoic.

455, Nuhetdavaa. A composite terrane with several island arcs combining the Mongolian Nuhetdavaa Terrane of Badarch et al. (2002) and its eastwards extension as the Hinggan Terrane and the Turan Block within China. Neoproterozoic metamorphics are overlain by Palaeozoic sedimentary and volcanic rocks and are all intruded by Ordovician to Permian granites (Wu et al., 2011).

456, Qaidam–Qilian. Composite terrane to the south-east of Tarim (Unit 480) and south of Ala Shan (Unit 451). Both Qaidam and Qilian include a variety of independent Palaeozoic units, but we combine the two units. Included are the Cambrian to Early Ordovician Hexi Corridor (a continental slope); the North Qilian belt of Middle Cambrian and Ordovician ophiolites, intra-continental volcanics, olisthostromes, and sediments with a North China Province trilobite fauna (Zhou & Dean, 1996); the Qilian–Laji Block (a Proterozoic cratonic fragment with overlying Palaeozoic to Mesozoic sedimentary rocks); the Qaidam Massif, with Cambrian and Early Ordovician carbonates; and the Qimantag belt, with Lower Palaeozoic clastic and volcanic rocks. The terrane accreted to North China (Unit 601) in the Devonian. There are also Lower Silurian (428 Ma) granites and Carboniferous shallow marine rocks with brachiopods (Chen et al., 2003).

457, Kunlun. Within this composite unit in China are the Songpan–Ganze Belt and the Qamdo–Simao Terrane (Metcalfe, 2006). The latter accreted to peri-Gondwana (Qiangtang) along the Lancangjiang Suture in the Early Triassic (Chen et al., 2010). There was arc-ophiolite obduction in the west Kunlun Range (on the northern periphery of the Tibet Plateau) onto the Tarim continent (Xiao et al., 2002), and the arc- or ophiolite-derived turbidites contain Late Ordovician to Early Silurian and Late Devonian to Early Carboniferous radiolarians. The Kunlun Belt is the remnants of Carboniferous to Triassic island arcs superimposed on Ordovician (490–450 Ma) arcs into which Middle Devonian (389–384 Ma) batholiths were intruded (de Jong et al., 2006). In north-east Kunlun, only Ordovician fossils are securely dated (Zhou & Dean, 1996). Further east there is Proterozoic basement which is similar to the basement of the Qaidam block, and thus Kunlun may be an accretionary wedge on the southern margin of Qaidam (Metcalfe, 2006). The Dzhetym Range in the Tien Shan Mountains has Lower Cambrian ignimbrites overlying Vendian rocks succeeded by Middle Cambrian deeper-water shelf sediments and Upper Cambrian shallower-water marine sediments. The

Carboniferous has Upper Visean to Kazimovian shallow-water marine sediments unconformably on Middle Silurian rocks, succeeded by Carboniferous (Gzhelian) deeper-water sandstones (Daukeev et al., 2002).

458, Kokchetav–Ishim. The Ulutau and Kalmykkol–Kokchetav massifs were united before the Middle Cambrian at 510 Ma and are substantial Precambrian metamorphic cratons, unconformably overlain by Late Neoproterozoic (Vendian) to Middle Ordovician sediments and island arc volcanics, including the Ishim arc (Dobretsov et al., 2006). The Devonian in Ishim has Lochkovian to Pragian volcanics unconformably succeeded by Givetian and Frasnian terrestrial sediments, with the Famennian absent; but elsewhere there are terrestrial rocks with minor volcanics from the Lochkovian to the Early Famennian, succeeded conformably by younger Famennian shallow-water marine carbonates (Daukeev et al., 2002). In the west, there are Middle Carboniferous to Middle Permian non-marine basins with substantial evaporites.

459, North Tien Shan. A composite unit, with Precambrian metamorphic complexes and Cambrian to Middle Ordovician (Darriwilian) island arc andesites and basalts and interbedded sediments, some terrigenous (Degtyarev & Ryazantsev, 2007). The boundary with Chu-Ili (Unit 460) is the Zhalair–Naiman Fault Zone, which closed at about 440 Ma: there are latest Ordovician to Early Silurian stitching granites (Popov et al., 2009). Eifelian and later sediments lie unconformably on Ordovician rocks, whilst in Northern Kyrgyzstan there is a mixture of Lochkovian to End Devonian sediments and volcanics (Daukeev et al., 2002). The eastern sector merged with South Tien Shan in the Carboniferous (Zhou et al., 2001), followed by latest Carboniferous molasse deposits and Lower Permian Type A granites. The Late Ordovician is overlain by Middle to Upper Devonian acid volcanics and Lower Carboniferous red beds, Upper Carboniferous marine clastics and Permian volcanics (Bazhenov et al., 2003), and there are Middle Carboniferous to Middle Permian non-marine basins. The eastern sector (in north-west China) includes the Devonian and Carboniferous North Tien Shan–Bogdo Shan volcanic arc (Pirajno et al., 2008).

460, Chu-Ili. There are Neoproterozoic outcrops, with 2.8 Ga Archaean zircons, and an Early to Late Ordovician accretionary wedge in front of the North Tien Shan Microcontinent (Unit 459), with which Chu-Ili merged in the earliest Silurian at about 440 Ma. Bounded to the south-east by the Zhalair–Naiman strike-slip fault complex, a subduction–accretion complex with Middle Cambrian to Middle Ordovician (Darriwilian) oceanic deposits. There are also Middle Cambrian to Late Ordovician island arc deposits, including ophiolites and cherts within latest Ordovician nappes, as well as abundant diverse shallow-marine faunas, especially brachiopods (Popov & Cocks, 2017). There was a passive margin at the south-western border from the Late Cambrian to the Middle Ordovician, before it and North Tien Shan became accreted (Popov et al., 2009). There are terrestrial sediments with volcanics from the Lochkovian to the top of the Devonian, apart from in the north-west, where Early Famennian shallow-water marine sediments with shelly faunas are succeeded by Late Famennian deeper-water shelf clastics (Daukeev et al., 2002).

461, Atashu–Zhamshi. There are substantial Late Neoproterozoic (Vendian) rocks, and the shallower-marine clastic and carbonate slope rise Cambrian succession is largely complete; but that was followed by Darriwilian and Sandbian volcanics and Katian and Hirnantian deeper-water shelf sediments with radiolarians. In the north there are extensive Devonian volcanics, with much palaeomagnetic data (Levashova et al., 2009), with volcanism continuing to the Permian in the north-east. In the east there are Lochkovian to Frasnian volcanic rocks, succeeded by Famennian terrestrial rocks and shallow-water marine sediments. Even further east, to the west of Lake Balkhash, there is a relatively small Yili Terrane in north-west Xinjiang (Zhu et al., 2009), where Precambrian metamorphosed basement is unconformably overlain by Cambrian to Ordovician carbonates and clastics and Silurian turbidites. Yili also includes Devonian continental rocks and a Late Devonian and Carboniferous (361–313 Ma) volcanic arc and arc-derived granitoids, and is bounded to its north-east by the North Tien Shan Fault and a Late Carboniferous (325 Ma) ophiolite, and to its south by a Permian ophiolite included within South Tien Shan (Unit 467).

462, Chingiz–Tarbagatai. A composite terrane which has three units with fragmentary Precambrian cores (Zonenshain et al., 1990). An island arc developed directly upon Ordovician oceanic crust at the southern margin; in contrast to the Late Neoproterozoic to Silurian island arc in the north-west, which developed on basement of both oceanic and continental fragments (Windley et al., 2007). Between the two arcs is a Middle Cambrian to Late Ordovician accretionary wedge, where substantial Tremadocian to Sandbian volcanics are succeeded by shallower-water Katian and Hirnantian sedimentary rocks. There are also Lower to Middle Devonian island arcs with widespread volcanics in the south-west. In the Chingiz area, Late Pragian to Frasnian terrestrial rocks with volcanics lie unconformably on the Silurian, above which is a Famennian shallow-water marine sequence; but in the north-west there are Lochkovian to Frasnian terrestrial rocks with volcanics unconformably on Llandovery rocks succeeded by a Famennian shallow-water marine sequence. Permian terrestrial sediments are interbedded with tuffs, ignimbrites and other volcanic rocks (Daukeev et al., 2002).

463, North Balkhash. Largely an Ordovician to Devonian accretionary wedge (Windley et al., 2007). The Middle Ordovician has turbidites with fragments of Early Ordovician ophiolites, all within an island arc which also includes Middle and Late Ordovician andesites and basalts, and shallower-marine sediments with brachiopods (Daukeev et al., 2002). Most of the Devonian is shallow marine sediments; but there are also earlier Devonian continental slope rocks, and in the Frasnian there was terrestrial deposition. In the south-east there was shallow marine deposition up to the end of the Givetian, with both shallow marine and terrestrial deposits in the Frasnian, and then terrestrial deposits only during the Famennian. The Permian consists of terrestrial rocks with many volcanics (Daukeev et al., 2002).

464, Junggar–Balkhash. The area in the east of the Kazakh Orogen is a Late Devonian to Carboniferous accretionary sequence (Şengör & Natal'in, 1996). Devonian sections contain shallow marine limestones and clastic sediments throughout the entire Devonian, as well as Eifelian basalts; and there are also deeper-water sediments with more abundant lavas between the Eifelian and the Famennian (Daukeev et al., 2002). The Junggar–Balkash Fold Belt lies largely to the west of the Junggar Terrane area and to the north of the Atasu–Zhamshi Terrane, but the limits of that fold belt do not coincide exactly with the margin of Unit 464.

465, Tourgai. Although geographically in the north-western Kazakh Orogen, Tourgai was peri-Baltic (Hawkins et al., 2016). A Devonian volcanic arc was deposited on oceanic crust, and there are Middle Devonian to Middle Carboniferous sediments in the Tourgai Basin, and substantial Carboniferous volcanics and plutons, some now important ore deposits, In north-east Tourgai there are Lochkovian to Lower Eifelian basic volcanics, sporadically interspersed with clastics from the Upper Eifelian to the Middle Frasnian, but volcanics are absent from the largely carbonate rocks of Upper Frasnian and Famennian age (Daukeev et al., 2002). The Permian consists of terrestrial sediments.

466, Karatau–Naryn. A complex unit, whose boundaries lie under Mesozoic to Recent rocks (Popov et al., 2009). Included is the southern half of the Ishim–Naryn Rift Zone, which has Early Cambrian to Ordovician (Darriwilian) clastic sediments, some terrigenous (Degtyarev & Ryazantsev, 2007). In the south, earlier accretionary sediments are followed by Devonian and Carboniferous volcanics; and Neoproterozoic to Carboniferous passive margin sediments. There was an Andean-type magmatic arc in the south-west from the Middle Ordovician (Late Darriwilian) onwards, and 466–438 Ma granites, and substantial Early Devonian and Middle Carboniferous to Permian volcanism, as well as terrestrial sedimentation. Volcanoes were also active in the north from the Devonian to Late Carboniferous with interbeds of Lower to Middle Carboniferous marine carbonates. To the south-west is a Devonian and Carboniferous island arc, where Late Ordovician granites stitch thrusts and folds which deform Late Riphean to Darriwilian clastics. The core has Precambrian to Early Ordovician and Middle Devonian to Permian rocks (Allen et al., 2001), and there are Middle Cambrian to Early Ordovician seamounts, and Cambrian shallow-water marine limestones followed by black shales and shallower-water carbonates. Karatau–Naryn merged with North Tien Shan (Unit 459) in the Late Ordovician (Popov & Cocks, 2017). There are Lochkovian to Eifelian lavas (Daukeev et al., 2002). There was a Middle Devonian to Late Carboniferous carbonate platform which changed from shallow-marine reef and sand shoals in the Late Devonian to deeper-water ramps and skeletal mounds in the Tournaisian and Early Visean and to skeletal mounds and sand-shoal-rimmed margins in the Middle Visean to Bashkirian (Cook et al., 2002). In the south-east, there was Early Carboniferous southward thrusting of arc rocks over the carbonate platform, followed by laterite and bauxite deposition in the Bashkirian and Early Moscovian at 320–315 Ma, with overthrusting and flysch deposition in the south, and olistostromes and further nappes in the Late Moscovian at about 310 Ma (Belousov, 2007).

467, South Tien Shan. An elongate composite unit stretching from the eastern end of Karatau–Naryn and through Tajikistan and Kyrgyzstan to north-west China, which includes the Middle Cambrian to Early Ordovician Narat Terrane of Zhou & Dean (1996). In the Silurian of the Pamir Mountains there are both deep-water black shales and shallow-water marine carbonates. Further east, there are deeper-water Llandovery sediments, as well as contemporary shallow-water marine sediments, but all the Wenlock to Pridoli rocks are shallow-water marine sediments, largely carbonates (Daukeev et al., 2002). The Tien Shan Mountains in the north of Tarim are a series of nappes formed after the closure of the Turkestan Ocean between Tarim and Kazakhstania, which started in the Middle Carboniferous (Burtman, 2008). The lower nappes were originally parts of the passive northern margin of Tarim, which was subducted. The upper nappes formed as

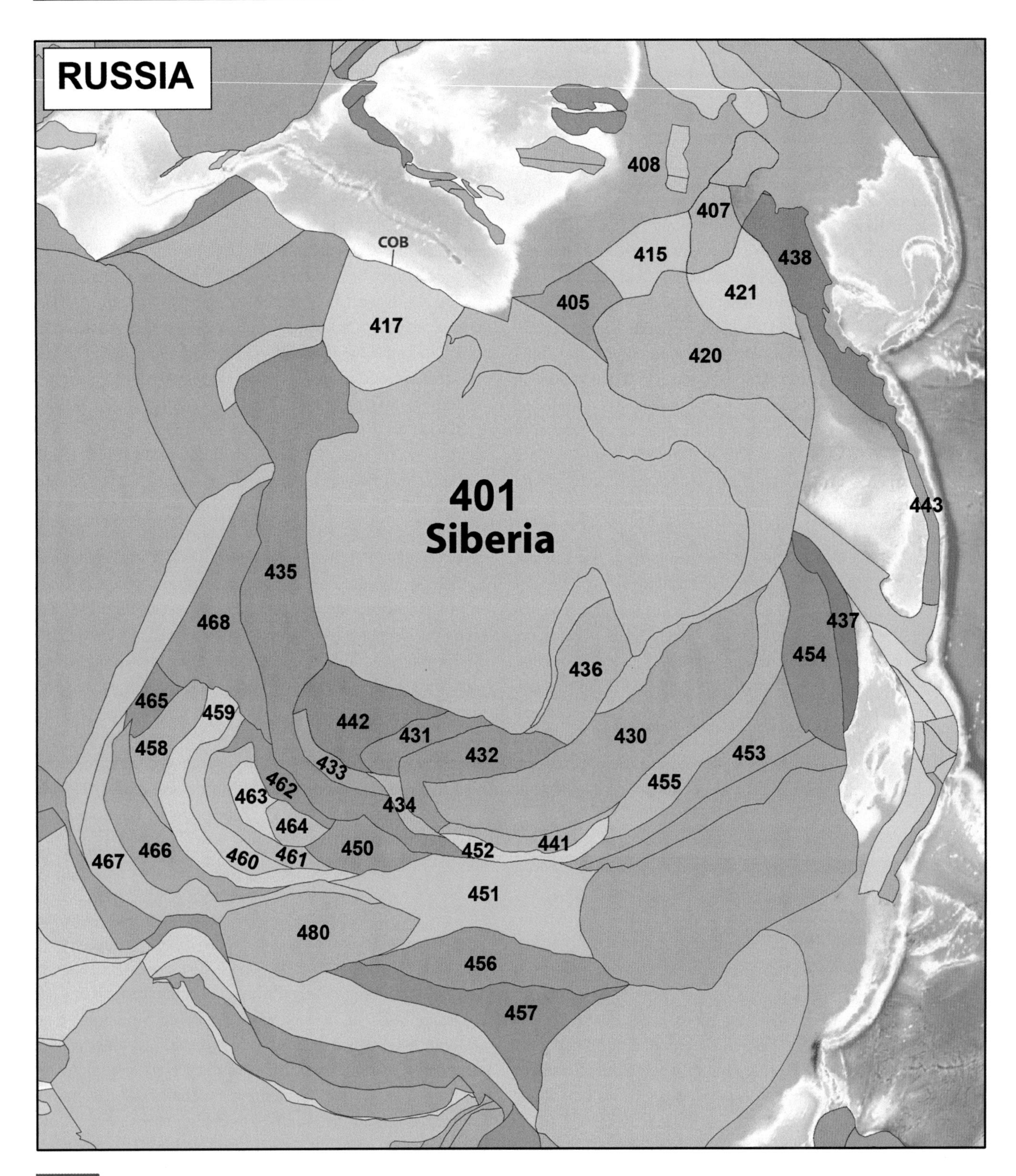

Fig. 3.6 Russian and adjacent units (plate identities 4XX). The Continent–Ocean Boundary transition zone (COB) in the Eurasian Basin is mainly defined from seismic, magnetic, and/or gravity data.

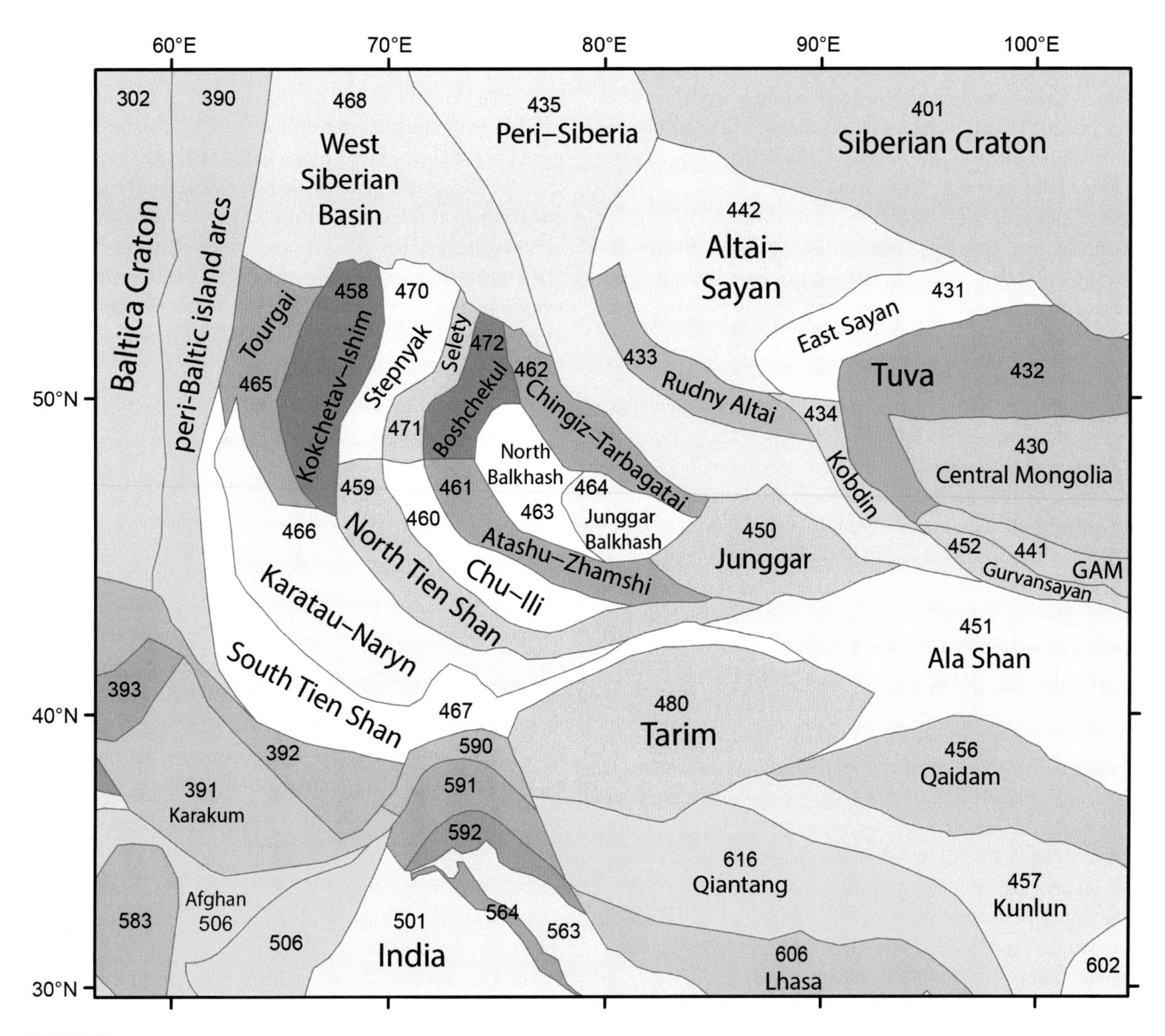

Fig. 3.7 Enlargement of south-eastern and central Asia (including much of the Central Asian Orogenic Belt and the Kazakh Orogen) and northern Indian units.

an accretionary prism at the margin of Kazakhstania and were obducted over Tarim. As well as the Central Tien Shan arc, the terrane also includes an accretionary complex which borders the passive margin of northern Tarim to its south (Xiao et al., 2009b).

468, Western West Siberian Basin. See above under Unit 435.

470, Stepnyak. There was an accretionary wedge in a back-arc basin and the Stepnyak volcanic island arc, which evolved from the Early Ordovician (Tremadocian: 485 Ma) to the Early Devonian (490–400 Ma) on a Cambrian basement of both oceanic and continental crust (Windley et al., 2007). There was a collision between it and the Kokchetav Microcontinent (Unit 458) in the Middle Ordovician from 480 to 460 Ma (Floian to Sandbian), and Late Ordovician granites were intruded between 460 and 440 Ma (Dobretsov et al., 2006): much of the unit is Ordovician and Devonian granitoids (Kheraskova et al., 2003, Fig. 8). In the north, Tremadocian volcanics unconformably overlie the Middle Cambrian, and are followed by both shallow- and deep-water Floian to Katian sediments, above which there is an unconformity below the Lower Silurian (Daukeev et al., 2002).

471, Selety and 472, Boshchekul. Selety has Early Cambrian to Tremadocian volcanics overlain by or interbedded with sediments, some of which are terrigenous and formed the eastern margin of Unit 472, the Boshchekul Microcontinent, with which it merged in the latest Ordovician (Popov & Cocks, 2017). In Boshchekul there are two Precambrian cores unconformably overlain by Lochkovian shallow-water marine sediments and Givetian to Early Famennian terrestrial sediments with volcanics, and Middle Famennian shallow-water marine sediments and Late Famennian deeper-water shelf facies; however, elsewhere shallow-water marine sedimentation persisted from the Lochkovian to the Famennian (Daukeev et al., 2002).

480, Tarim. The Tarim Microcontinent is mostly within China and includes most of the Taklimakan Desert, with the Tien Shan Mountains in the north and the Kunlun Mountains in the south. The Precambrian block lies to the south, with a marginal basin fringing the northern margin (Zhou & Dean, 1996). The centre is covered by Cenozoic sediments, although boreholes have penetrated to an old craton with Archaean to Lower Cambrian rocks (Xiao et al., 2008), over which lie dominantly carbonate Cambrian and Ordovician sequences (Daukeev et al., 2002). Latest Silurian to earliest Carboniferous rocks are absent. There are extensive Permian volcanics (280–270 Ma) (Pirajno et al., 2008). Late Devonian (Famennian) to Upper Carboniferous (Bashkirian) carbonate and clastic sediments on the continental slope are bordered to the north by deeper-water calcareous and siliceous sediments, followed by flysch, prior to the accretion of Tarim to the active Kyrgyzstan margin of Kazakhstania from the Middle Carboniferous after 310 Ma (Windley et al., 2007). The southern margin was active, with an Early to Middle Ordovician Andean-type continental margin and progressive accretion of island arcs (de Jong et al., 2006), including in the south-west the North Kunlun Terrane (Xiao et al., 2002), with a Triassic (214 Ma) granite stitching Tarim and Kunlun. The south-eastern boundary is a substantial strike-slip fault dividing Tarim from the Qaidam, Kunlun, and Ala Shan terranes, and contains fragments of 509 to 487 Ma Upper Cambrian to Lower Ordovician eclogites, and 487 Ma Tremadocian gneisses.

India and the Middle East

The region includes the south-western part of Asia and also the Indian subcontinent. Much of it originally lay within the Gondwana continent and later Pangea (Torsvik & Cocks, 2009, 2013).

501, India. A subcontinent which was an integral sector of Gondwana until the Indian Ocean opening in the Late Cretaceous. Dominated by its Archaean to Neoproterozoic craton, which is unconformably overlain by a few Lower Cambrian marine rocks in places, themselves unconformably overlain by Permian non-marine rocks which yielded the first record of the *Glossopteris* Flora. In the Salt Range of Pakistan there are substantial glaciogenic deposits of latest Carboniferous (Gzhelian) to Early Permian (Asselian) ages. At India's northern margin lies the Himalayan region (Units 563 and 564), which was within the margin of the old Gondwana and where the rocks are more varied. Much of north-west India is covered by Cretaceous–Tertiary (65 Ma) lavas of the Deccan Traps LIP.

502, Sri Lanka. Formerly named Ceylon, and an integral sector of Gondwana. Largely Precambrian rocks form a craton which was probably originally a sector of the Indian Craton.

503, Arabia. An extensive area which was an integral sector of Gondwana until the Red Sea opened in the Tertiary. Neoproterozoic basement is overlain by Lower and Middle Palaeozoic rocks in many places, particularly Saudi Arabia and south-western Iran, which include higher-latitude Mediterranean Province invertebrate faunas in the Ordovician. Peri-Tethyan rocks, largely carbonates, were deposited in several basins from the Carboniferous onwards which have been folded from the Mesozoic to the present.

504, Central Turkey and Taurides. An integral sector of peri-Gondwana, although it changed position and orientation with respect to core Gondwana during the Neoproterozoic to Ordovician (Ghienne et al., 2010). It includes Cyprus, and has a wide variety of Phanerozoic igneous and sedimentary rocks of most ages. The Taurides have Lower Palaeozoic rocks with high-latitude Mediterranean Province faunas. To the north lie the Pontides (Unit 581). From the Mesozoic onwards, both units have formed parts of the Anatolian Plate, which includes all of Turkey and adjacent areas in southern Greece, Syria, and Iraq.

505, Alborz. Originally an integral marginal sector of Gondwana, now in Iran. Its Palaeozoic brachiopod faunas were the same as those of Libya, Afghanistan, and Pakistan until at least the Devonian. The Neotethys Ocean between it and Gondwana opened to its south in the Permian. On the eastern side of the Caspian, the Kopet Dagh area forms the northern part of the terrane, and there is Late Palaeozoic deformation in the Alborz Mountains (Gaetani et al., 2009). In Kopet Dagh, Early Silurian graptolitic shales and limestones are overlain by thick Devonian limestones.

506, Afghan. Like Alborz, the terranes in Afghanistan combined in this unit were also originally integral sectors of the Gondwana margin, with the same Devonian and Carboniferous brachiopod faunas; and the Neotethys Ocean between them and Gondwana also opened to their south in the Permian. The Palaeozoic cores of Lut, Alborz, and the Afghan terranes are separated from each other by Mesozoic and later accretionary belts, including peri-Tethyan marine faunas.

508, Sinai. A relatively small area now forming the Sinai Peninsula of Egypt and its northwards extension to Israel, originally an integral part of the Gondwana core with Neoproterozoic and Lower Palaeozoic rocks including fossiliferous Cambrian limestones, and tectonically isolated only by the Early Paleogene opening of the Red Sea between Africa and Arabia and the Neogene development of the Dead Sea transform fault separating Sinai from the Arabian Plate (Unit 503).

563, Greater Himalaya. Tethyan Himalayan Microcontinent which defines the leading edge of the Greater Indian Basin ('Greater India'). Deformed sedimentary rocks range from Neoproterozoic to Eocene in age, and Early Palaeozoic successions are well preserved. Palaeomagnetic data from Ordovician red beds (Torsvik et al., 2009) and geological data demonstrate that the Tethyan Himalaya was located near the Indian craton during Early Ordovician times, and then formed a continuous margin. From palaeomagnetic data, van Hinsbergen et al. (2012) concluded that Greater India underwent extension and ocean basin formation between 118 and 68 Ma (Middle to latest Cretaceous). The Tibetan Himalaya drifted more than 2000 km northward relative to cratonic India (Unit 501), followed by convergence of a similar magnitude after collision with Lhasa (Unit 606), the leading edge of Southern Asia, at around 50 Ma (Eocene).

564, Lesser Himalaya. Defines the northernmost part of the Indian Plate and has a lower sequence of Paleoproterozoic rocks interpreted as a passive margin or volcanic arc, overlain by Late Precambrian to Early Palaeozoic carbonates and quartzites (Kohn et al., 2014).

570, Laxmi Ridge. Thinned continental crust beneath the Laxmi Ridge (Collier et al., 2009) is probably of Neoproterozoic age (Torsvik et al., 2013), and was originally juxtaposed to the Seychelles and the Malani province in north-west India. The Laxmi Ridge and Seychelles were detached from India (Gop Rift) in the Late Cretaceous. Sea-floor spreading started between the Laxmi Ridge and the Seychelles at 62 Ma (Paleocene).

571, Murray Ridge. Like the Laxmi Ridge (Unit 570), this unit could be a small thinned continental fragment originally located offshore of north-west India (Calves et al., 2011).

572, Laccadives. See Unit 777.

573, Chagos. See Unit 777.

581, Pontides. Although now part of Turkey to the north of Central Turkey (Unit 504), the Early Palaeozoic faunas reveal that the Pontides was then separate from the other sectors of Gondwana. The Pontides has been considered as several terranes, including an Istanbul Terrane, but we treat it as a single unit. The Lower Ordovician (Tremadocian) is unconformable on possibly Precambrian basement gneiss, and there is no known Cambrian. The Ordovician faunas in the Pontides were quite different from those in the Taurides (Unit 504), but similar to those in Avalonia (Dean et al., 2000), indicating that the Pontides were then further to the west, and possibly at higher latitudes, than the Taurides. Late Carboniferous flysch marks the onset of 'Variscan' deformation and is unconformably overlain by Lower Triassic continental clastics. The derived zircons in that flysch suggest that the Pontides might have been near Bohemia (Unit 374) in the Late Palaeozoic, but the terrane was translated to its present position by pre-Cretaceous strike-slip faulting, and from the Mesozoic onwards has formed a sector of the Anatolian Plate.

582, Sanand. Sometimes termed the Sandaj–Sirjan Terrane, now in Iran. An elongate strip of presumably Gondwanan metamorphic Lower Palaeozoic basement overlain by Carboniferous and later shelf sediments adjacent to the north-east margin of Arabia. Between Sanand and Arabia (Unit 503) is a Permian ophiolite, marking their separation as part of the Neotethys Ocean opening.

583, Lut. Also now in Iran, Lut was separate from Gondwana in the Cambrian, since it has extensive andesite and trondhjemite volcanics near its western margin, at a time when the adjacent northern margin of Gondwana was passive. The Palaeozoic core of the unit may represent three different Lower Palaeozoic smaller terranes.

584, Makran. A Mesozoic to Recent assemblage of accretionary units adjacent to the Himalayan Orogen, situated today in southern Iran and Pakistan between the western Himalaya and the Indian Ocean.

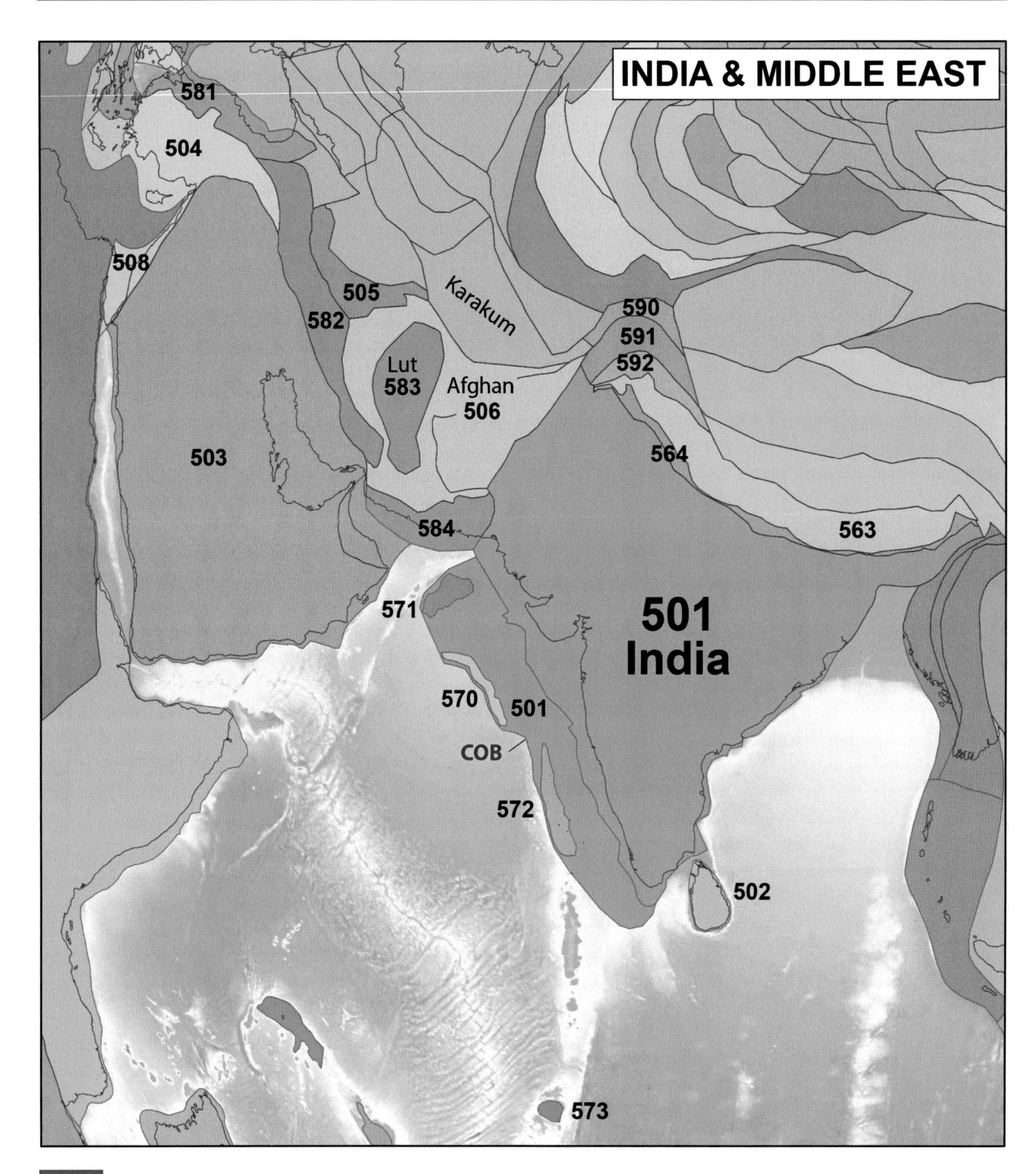

Fig. 3.8 India and Middle East units (plate identities 5XX). The Continent Ocean Boundary transition zone (COB) is mainly defined from seismic, magnetic, and/or gravity data.

590, North Pamirs. The Pamir Mountains in Tajikistan, Afghanistan, India, and western China are three separate units, Northern, Central, and Southern (Zanchi et al., 2000). The northern Pamirs were part of one of the many composite 'Kazakh' terranes, but whether Karatau–Naryn, South Tien Shan, or Kunlun is arbitrary: we identify it as a separate unit west of Kunlun (Stampfli & Borel, 2004). The Lower Palaeozoic margin with Gondwana is at the Wanch–Ak

Baktail Suture Zone between the Central and North Pamirs. Within the North Pamirs, there is Precambrian basement in a southerly nappe and Palaeozoic oceanic and island arc complexes within a northerly nappe. The Cambrian is largely black shales; the Silurian shallow-water marine sediments; Carboniferous volcanic rocks are overlain by shallow-water marine sediments (Daukeev et al., 2002). The Pamirs were all much deformed in the Himalayan Orogeny. However, the Carboniferous and Permian deformation in the North Pamirs is absent from the South and Central Pamirs (Zonenshain et al., 1990), which confirms their separation. There is no unambiguous Palaeozoic terrane boundary between the North Pamirs and South Tien Shan (Unit 467); however, we do not combine them because such a united terrane would run both north and south of Tarim, which is unrealistic. It is difficult to draw a boundary between the North Pamirs and Kunlun (Unit 457): the former may be a western extension of the latter (Zhou Zhiyi pers. comm., 2012).

591, Central and South Pamirs. The South and Central Pamirs can be grouped with the Karakorum Terrane (Unit 592), since the latter was stitched by lavas to the southern sector of the Pamirs to its north by Middle Devonian time. The fossils have a Gondwanan signature, such as the trilobites from the Upper Tremadocian (the oldest Palaeozoic rocks in South Pamir), which include temperate forms such as *Vietnamia* and *Birmanites* (Fortey & Cocks, 2003). The Karakorum Terrane, South and Central Pamir, and the Afghan Terrane left Gondwana in the Early Permian opening of the Neotethys Ocean (Stampfli & Borel, 2004).

592, Karakorum. Karakorum lies largely in the north-west of Pakistan, and has Ordovician to Cretaceous sediments (Gaetani, 1997). Above a pre-Ordovician crystalline massif lie sediments containing Early to Middle Ordovician acritarchs and chitinozoans (Quintavalle et al., 2000) and Early Ordovician conodonts from the western Karakorum and the adjacent Hindu Kush. A second cycle appears to be of Late Devonian to Early Permian ages, and a third cycle ran from the Early Permian to after the end of the Jurassic.

South-East Asia

600, Sulinheer. A relatively narrow strip at the northern margin of North China, consisting of Silurian and Devonian metamorphic rocks, including ophiolites, intruded by gabbros and granodiorites, all overlain unconformably by Carboniferous and Permian clastics and limestones. It was an integral sector of North China from the end of the Devonian onwards, and forms the southern margin of the Solonker Suture, which closed progressively from the Late Carboniferous to the Middle Permian.

601, North China. Includes much of eastern and northern China as well as the Korean Peninsula, and is often termed Sino-Korea. Its centre is an amalgamation of three Archaean cratons which were united during the Proterozoic. North China has a good Cambrian to Middle Ordovician marine sequence, but there is a substantial unconformity between the Late Ordovician and the Upper Carboniferous, indicating that it was a land area then. The northern margin is the Solonker Suture, formed during the accretion of Hutag Uul–Songliao (Unit 453) progressively between the Late Carboniferous and the Middle Permian. Its southern margin amalgamated with South China (Unit 602) along the Qinling–Dabie Suture in the Late Jurassic or Early Jurassic. Qaidam–Qilian (Unit 456) accreted to North China's western margin in the Devonian.

602, South China. A stable craton (including some Archaean 2.8 Ga basement) since its Proterozoic amalgamation at about 1 Ga. Largely undeformed Palaeozoic shelf clastics are found on the craton, with many depositional breaks within a relatively complete sequence which has yielded diverse Cambrian to Devonian benthic faunas, succeeded by later Palaeozoic terrestrial sediments with characteristic Cathaysian floras. Its south-eastern Cathaysia sector (including Hainan Island) was displaced southward by strike-slip faulting in the Mesozoic. South China amalgamated along its southern boundary with Annamia along the Ailoshan and Song Ma sutures in the Triassic (Cai & Zhang, 2009). At its north is the Qinling–Dabie orogenic belt, which includes the small South Qinling Terrane, which separated from South China in the Late Ordovician at about 465 Ma and was reunited with it at about the end of the Permian.

603, and 647, Sibumasu. Eastern Burma and northern and central Thailand formed the northern sector (Unit 603) of the elongate Sibumasu Terrane, which stretched from Sumatra, Indonesia, up to eastern Burma (Metcalfe, 2006; Ridd et al., 2011). The southern sector (Unit 647) runs from western Thailand through the Malay Peninsula to Sumatra, and was united with the northern sector (Unit 603) throughout the Phanerozoic. Sibumasu was an integral sector of Gondwana until the Neotethys Ocean opening in the Permian, and has Neoproterozoic basement rocks, and Palaeozoic clastic, carbonate, and volcanic rocks, with many local unconformities: most were tectonically disturbed during the intrusion of substantial Permian–Triassic granites. Its Late Ordovician (Sandbian) faunas indicate that it was close to South China (Unit 602) (Fortey & Cocks, 2003).

604, Annamia. Otherwise termed Indochina, Annamia is mostly made up of the Indochina Peninsula, but also includes adjacent parts of China and Thailand (Sone & Metcalfe, 2008). Its core is a Paleoproterozoic craton, which has Palaeozoic intrusions, and above it there is a substantial unconformity under Late Palaeozoic continental red beds, themselves intruded by Late Triassic to Cretaceous granites. Annamia was possibly integrated with South China as a unified continent from the Neoproterozoic to the Middle Devonian (Cocks & Torsvik, 2013). There are also Carboniferous and Permian marine carbonates, originally parts of oceanic seamounts, in the Thai sector (Ridd et al., 2011).

606, South Tibet (Lhasa). Lhasa is the more southerly of the two chief Tibetan terranes, which were integral parts of the Gondwana Superterrane until the opening of the Neotethys Ocean during the Permian. Lhasa and its northern neighbour Qiantang (Unit 616) were apparently united during the Palaeozoic (Metcalfe, 2006), but they separated in the Late Triassic, and subsequently became reunited in the Early Cretaceous.

607, West Burma. To the west of the Sibumasu Terrane (Unit 603) there are largely Triassic to Tertiary rocks forming a separate terrane unit, which subsequently drifted northwards from the Antarctica sector of Gondwana.

614, and 617, East Malaysia and Borneo. Includes eastern Sumatra and Kalimantan. Bordered by Sibumasu (Unit 647) to its west by the Raub–Bentong Suture, which includes ophiolites, and with another ophiolite belt in the east. Largely Cretaceous and Tertiary deformed flysch belts with olistostromes, and including Oligocene (26 Ma) granites. Rapidly changing Cretaceous to Recent palaeogeography (Hall & Holloway, 1998).

616, North Tibet (Qiantang). Qiantang is the more northerly of the two chief Tibetan terranes, which were integral parts of Gondwana until the opening of the Neotethys Ocean during the Permian. Qiantang and its southern neighbour Lhasa (Unit 606) were apparently united during the Palaeozoic (Metcalfe, 2006), but they separated in the Late Triassic, and subsequently became reunited in the Early Cretaceous.

624, Sakhalin. Located along the western margin of the Okhotsk Sea and between mainland Russia and the Pacific Ocean. Early Palaeozoic to Recent sediments and Mesozoic accretionary complex rocks affected by several phases of Cretaceous–Cenozoic deformation.

625, Kitakami. Eastern part of Hokkaido Island, Japan, to the south of Sakhalin (Unit 624). Cambrian to Middle Ordovician metamorphosed basement, probably representing an accretionary prism and volcanic island arc, intruded by ophiolites and 466 Ma trondhjemites. Silurian to Devonian island arc volcanics and sediments unconformably overlain by Lower Carboniferous (Tournaisian) shales and Visean limestones. There was fresh arc activity in the Permian.

626, Western Hokkaido. Hokkaido, the northernmost main island in Japan, is divided into several parts and the western region is part of the Kurile Arc. A Jurassic accretionary complex intruded by Cretaceous granite is unconformably overlain by Neogene to Holocene volcanic rocks and sediments.

627 and 628, North-east and Central Honshu. Part of the Japan Arc but also includes Silurian to Early Cretaceous marine sediments, a Jurassic accretionary complex and Early Cretaceous felsic plutonic rocks. Carboniferous and Permian fossils have been linked with shallow marine fossils in South China.

629, Kanto. Part of the Japanese arc and includes Mesozoic high-pressure metamorphic rocks, Jurassic to Cenozoic accretionary complexes, and ash-bearing Quaternary sediments.

630, Maizuru. Part of Honshu Island, Japan. Boreal Realm Carboniferous brachiopods (in contrast to the mixed Boreal and Tethyan realm faunas of the other Japanese terranes) suggest that this terrane represented an isolated oceanic seamount then, in contrast to the others, which were marginal to the South China continent (Unit 602).

631, Kurosegawa. Part of Kyushu Island, Japan. Lower Palaeozoic metamorphic rocks above which are Late Silurian to Devonian volcanics and clastic rocks representing an island arc. They are overlain by Permian arc rocks. The terrane is mostly composed of Late Jurassic to Early Cretaceous nappes.

638, Sado Ridge. Located west of the Japan Islands (Japanese Sea) and probably consisting of Palaeozoic and Mesozoic basement rocks overlain by volcanics and Miocene tuffaceous sediments.

647, Sibumasu (southern sector). See Unit 603.

651, Subawa–Flores. Part of the Lesser Sunda Islands, a volcanic arc formed by subduction along the Sunda Trench

667, 668, and 669, South-west (667), West (668), and East (669) Sulawesi. Located at the junction of three major tectonic plates (Eurasia–Pacific–Australia) and has undergone a complex history of Late Cretaceous–Recent subduction and collision, and Miocene and younger extension and faulting.

670, Banggi–Sula. Located in eastern Indonesia between Sulawesi and the Banda Sea in a complex zone of convergence between the Pacific, Philippine Sea, Indo-Australian, and Eurasian plates. Includes the Banggai–Sula Microcontinent considered to have split off from Australia in the Mesozoic and collided with East Sulawesi some time between the Middle and Late Miocene. Rocks are mostly Carboniferous or older in age.

673, 675, and 686, 673 (Java), 675 (Sumba), and 686, South-west Sumatra. Volcanic island arc formed by northward subduction of the Indo-Australian Plate and belonging to the Sunda or Java Trench, which extends from Burma in the north-west to Sumba Island in the south-east. West Java includes Mesozoic rocks accreted to the Sundaland core. Sumba Island contains volcanic, plutonic, and volcanoclastic rocks, which record arc volcanism extending from Late Cretaceous to Oligocene time. Sumba is an exotic terrane originating from north-west Australia or Sundaland.

674, 677, 678, and 691, West (674) and East (678) Philippines, Palawan (677), and Luzon (691). A complex collage in the Philippines of metamorphic terranes, magmatic arcs, ophiolitic complexes, sedimentary basins, and continental blocks of Eurasian affinity. Continental microblocks of Late Permian to Jurassic ages are the North Palawan Block, located in northern Palawan (677), southern Mindoro (674), and the Romblon Island Group and Buruanga Peninsula in Panay Island (674). The rest of the Philippines is termed the Philippine Mobile Belt, which is a product of the interaction between the Eurasian and Philippine Sea plates. The Philippine islands are bordered by the Manila Trench to the west and the Philippine Trench to the east, both active subduction zones.

681, Buru–Seram. These islands form part of the Maluku Islands of Indonesia. Buru is located in the outer Banda Arc, has Late Palaeozoic and younger rocks and is considered by many as a microcontinent derived from Australia. Seram has Late Palaeozoic to Miocene rocks that underwent extreme extension (mantle exhumation) at 5–6 Ma (Miocene–Pliocene) as a result of slab roll-back into the Banda Embayment (Pownall et al., 2013).

683, Wetar. Part of the Lesser Sunda islands, which formed as a result of subduction of the Indo-Australian Plate beneath the Sunda–Banda Arc since the Miocene.

684, Timor. Located in the non-volcanic Outer Banda Arc and part of the Australian continental margin. Probably underlain by Australian Precambrian basement, but the oldest exposed sediments are Permian. Rifted off the north-west Australian margin during the Jurassic but only emerged as an island in the Pliocene after colliding with the fore-arc of the Banda volcanic islands (Hall, 2012).

Africa and the Western Indian Ocean

The Palaeozoic geography of Africa, which then formed the central part of Gondwana and subsequently Pangea, was reviewed by Torsvik & Cocks (2009, 2013). It became today's independent continent when the Atlantic Ocean opened from the Late Jurassic onwards (Torsvik et al., 2009), and the Indian Ocean from the Late Cretaceous (Torsvik et al., 2013).

701, South Africa. This area, much larger than today's Republic of South Africa, has two Archaean and Paleoproterozoic cratons (Congo and Kalahari), which are overlain in many places by Phanerozoic sediments, chiefly in South Africa. The latter include the metamorphosed Late Neoproterozoic Klipheuwel Group, the marine Ordovician to Early Devonian Table Mountain Group, and the chiefly deltaic Bokkeveld Group; the latter yielding Early Devonian brachiopods of the cooler-water Malvinokaffric Province. Over a larger area, in South Africa and to its north, the Carboniferous to Jurassic Karoo Supergroup consists of interbedded marine, freshwater, continental and glaciogenic sediments with the Permian *Glossopteris* Floras, as well as spectacular fossil reptiles. The southern sector was affected by the Cambrian Cape Fold Belt tectonics, which ended with the intrusion of Middle Cambrian granites in South Africa and Namibia. The unit became part of Gondwana in the Neoproterozoic, and remained so until the intrusion of the Central Atlantic Magmatic Province LIP in the end-Triassic at 201 Ma, with the consequent opening of the South Atlantic Ocean. The Karoo LIP was intruded during the Jurassic at 183 Ma.

702, Madagascar. The island of Madagascar was originally part of the cratonic Archaean and Paleoproterozoic Late Victoria Block of Africa (Unit 712), which is overlain in many places by Phanerozoic sediments. It became part of Gondwana in the Neoproterozoic, and remained so until the

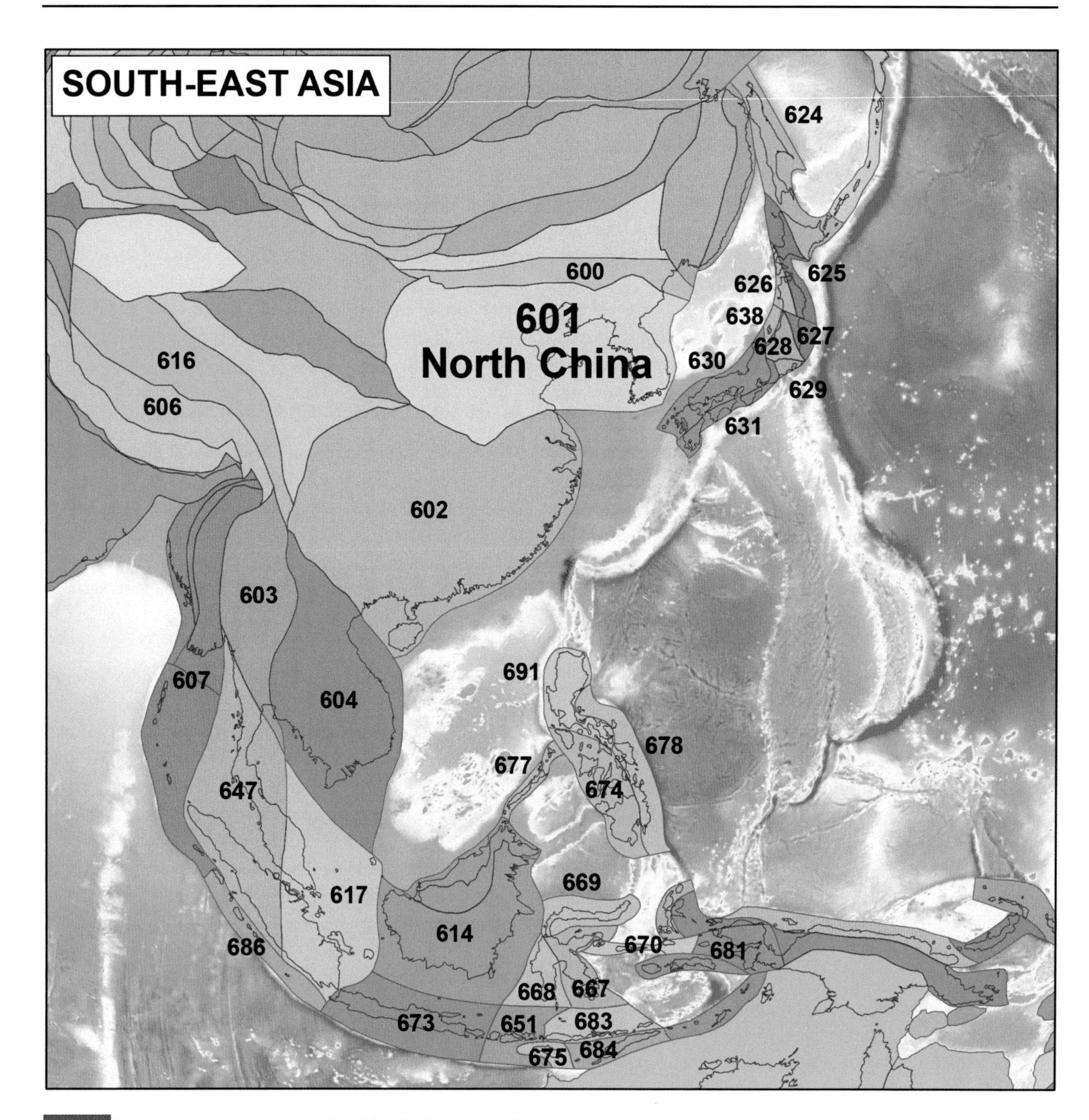

Fig. 3.9 South-East Asia units (plate identities 6XX). The Continent–Ocean Boundary transition zone (COB) is mainly defined from seismic, magnetic, and/or gravity data.

opening of the Indian Ocean in Mesozoic (Jurassic) times, after which it remained attached to India until the Late Cretaceous before finally becoming an independent island.

704, Seychelles. The Seychelles is a continental fragment that separated from India (Laxmi Ridge, Unit 570) in 63–62 Ma (Paleocene), soon after the peak Deccan magmatism at 65 Ma. The Seychelles islands mostly consist of undeformed granites and granodiorites (Ashwal et al., 2002; Tucker et al., 2001) of Neoproterozoic age (700–800 Ma), but two of the islands (Silhouette and North Island) include a younger plutonic–volcanic complex (63 Ma: Paleocene) that is contemporaneous with the final stages of the Deccan Traps volcanism (Owen-Smith et al., 2013). The Seychelles, Madagascar, and India were together from the Neoproterozoic to the Late Cretaceous. A Late Cretaceous LIP event (91–84 Ma) blanketed most of Madagascar and parts of SW India, and the Mascarene Basin opened shortly thereafter, separating India and the Seychelles from Madagascar. Soon after the separation from India at 62 Ma, sea-floor spreading terminated in the Mascarene Basin, and at 56 Ma (Paleocene–Eocene boundary time) the Seychelles became a permanent part of the African (Somalian) Plate (Torsvik et al., 2013).

705, Saya de Malha. See Unit 777.

706, Oran Meseta. An area of north-west Africa, including part of eastern Morocco (sometimes termed the Eastern Morocco Meseta), northern Algeria, and western Tunisia, separated from the Moroccan Meseta (Unit 707) by the Middle Atlas Mountains (Michard et al., 2008) and from North-West Africa (Unit 714) by the Eastern and Sahara Atlas Mountains. Varied Palaeozoic sediments were affected by Palaeozoic to Early Jurassic (often wrongly termed 'Variscan') rifting. There was basin marine and non-marine sedimentation along a passive margin from the Triassic to the Late Cretaceous.

707, Moroccan Meseta. The north-western sector of Morocco, including the Rif (sometimes termed the Coastal and Central Mesetas). An arcuate accreted nappe series, best exposed in the High Atlas Mountains, with a variety of rocks, including Triassic and latter accretionary prisms, which became progressively welded onto the north-western sector of the African Craton during the Mesozoic and Tertiary (Moratti & Chalouan, 2006). Much of the area was covered by the Central Atlantic Magmatic Province (CAMP) LIP basalts at the end of the Triassic at 201 Ma.

709, Somalia. This large area of eastern Africa stretches southwards from the southern Red Sea, and includes the Horn of Africa (Ethiopia and Somalia), eastern Kenya and Tanzania, and Mozambique. There is no Archaean or Paleoproterozoic craton, but there is some Mesoproterozoic basement in Mozambique. The unit formed land during much of the Phanerozoic. There are extensive Late Tertiary flood basalts, particularly in Ethiopia and mostly extruded since the Oligocene at about 30 Ma.

712, Lake Victoria Block. An Archaean and Paleoproterozoic craton, possibly connected with the Congo Craton of South Africa (Unit 701), and overlain in many places by Phanerozoic sediments. Lake deposition of the Karoo Supergroup started in the Late Carboniferous and continued on into the Early Jurassic. The unit became part of Gondwana in the Neoproterozoic, and remained so until the opening of the Indian Ocean. The southern sector of the African Rift Valley is included.

714, North-West Africa. The Archaean and Paleoproterozoic West African Craton, united in the end-Proterozoic Pan-African Orogeny, became part of Gondwana in the Late Neoproterozoic. The craton is overlain in many places by Phanerozoic sediments, although the area was largely land until the intrusion of the end-Triassic Central Atlantic Magmatic Province (CAMP) LIP at 201 Ma with the consequent opening of the Atlantic Ocean. The CAMP basalts cover more than half the unit's area.

715, North-East Africa. This very large unit has no Archaean rocks and only a small Paleoproterozoic craton, but the Neoproterozoic basement is extensive and had become an integral part of Gondwana in the Pan-African Orogeny before the beginning of the Cambrian. The southern two-thirds of the unit was land for nearly all of the Palaeozoic, but the northern third has unconformable Cambrian and later sediments, including extensive Late Ordovician (Hirnantian) glaciogenic rocks and overlying Silurian shales rich in hydrocarbons, particularly in Libya (Ghienne et al., 2007). The unit includes a substantial area of continental lithosphere which underlies the south-eastern Mediterranean Sea to the north of Libya and Egypt.

775 to 777, Mauritius and Mauritia. The island of Mauritius (0–8.9 Myr old) is part of the Southern Mascarene Plateau and linked to the volcanic chain that has developed above the Réunion mantle plume over the past 65.5 Myr (Paleogene and later). Paleoproterozoic and Neoproterozoic (840–660 Ma) zircons have been found in Mauritian beach basalts, and Torsvik et al. (2013) argued that the old zircons were assimilated from ancient fragments of continental lithosphere beneath Mauritius. Mauritius and the Mascarene Plateau may therefore overlie a Precambrian microcontinent named Mauritia, which possibly included parts of the Cargados Carajos (776), Nazareth (775), Saya de Malha (705), Laccadives (572), and Chagos (573) units. Mauritia was separated from Madagascar and fragmented into a ribbon-like configuration by a series of mid-ocean ridge jumps during the opening of the Mascarene ocean basin from 83.5 to 61 Ma (Late Cretaceous to Paleocene).

Australasia and Antarctica

Australia and Antarctica were integral parts of Gondwana throughout its history (Torsvik & Cocks, 2013); their margins were reviewed by Vaughan et al. (2005), and the evolution of Antarctica by Torsvik et al. (2008a). Today the Antarctic Plate is neighboured by six different tectonic plates and is almost entirely surrounded by spreading ridges.

801, West Australian Craton. The large western sector of Australia with two Proterozoic originally separate cratons,

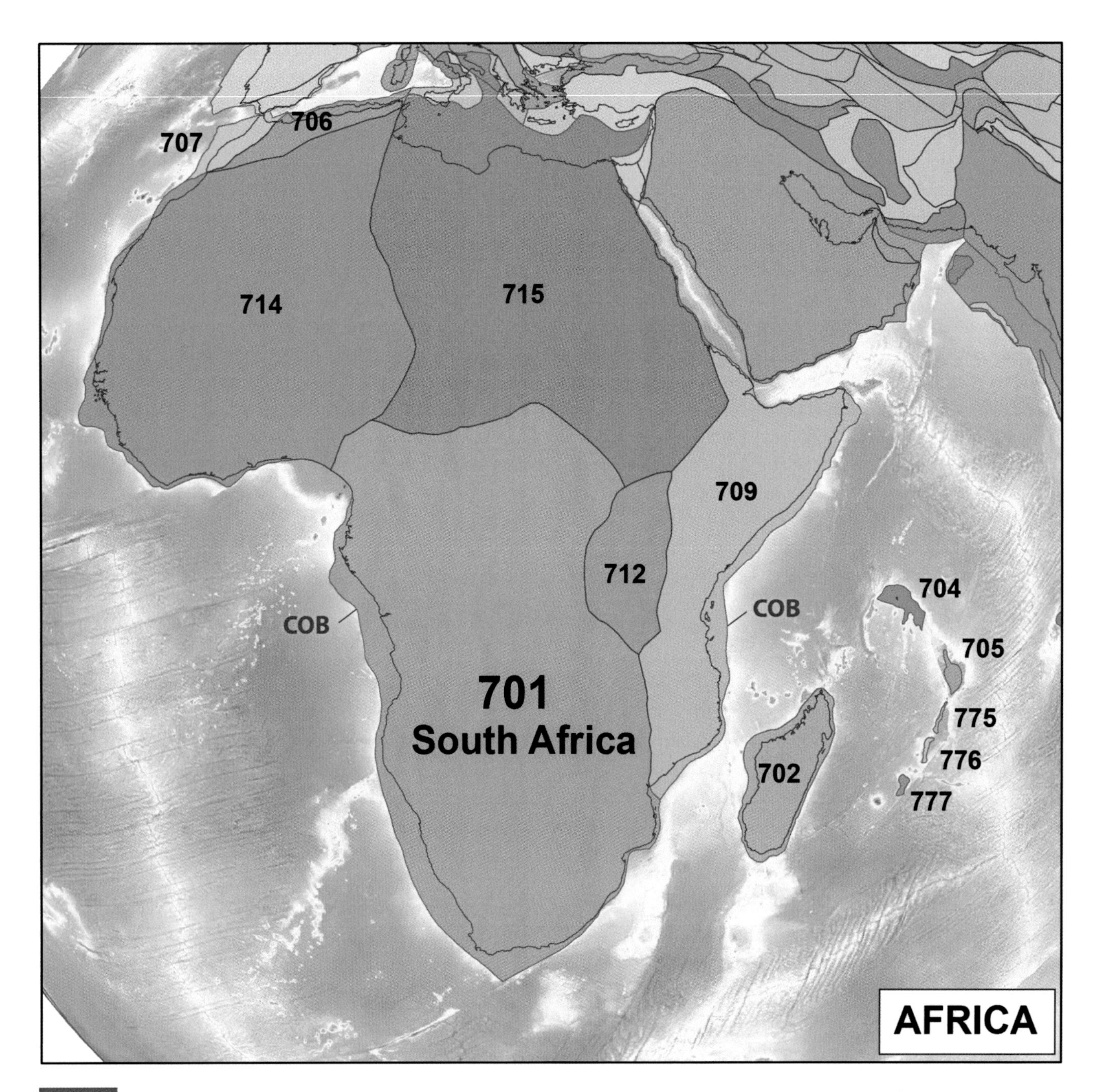

Fig. 3.10 Africa units (plate identities 7XX). The Continent–Ocean Boundary transition zone (COB) is mainly defined from seismic, magnetic, and/or gravity data.

the North Australian Craton and the Pilbara–Yilgarn Craton, which together formed a substantial sector of Gondwana. The Kalkarindji LIP flood basalts were intruded into the northern part of the West Australian Craton during the Cambrian at 512 Ma (Glass & Phillips, 2006). Although land for much of the Phanerozoic, there were marine basins over the margin at various times, particularly when the Larapintine Sea across the centre of the unit from north to south divided the continent into two land areas during the Ordovician.

802, East Antarctica. The eastern two-thirds of the continent, and much of the adjacent Weddell Sea, comprises Archaean and Proterozoic to Early Cambrian terranes (the Lützow, Holm Bay, and Northern Victoria Land Terrane Assemblage) which amalgamated during Precambrian and Early Cambrian times to form a major sector of Gondwana. After tropical marine Cambrian limestone deposition, East Antarctica drifted southward, so that by the Late Carboniferous it lay over the South Pole, which resulted in the widespread

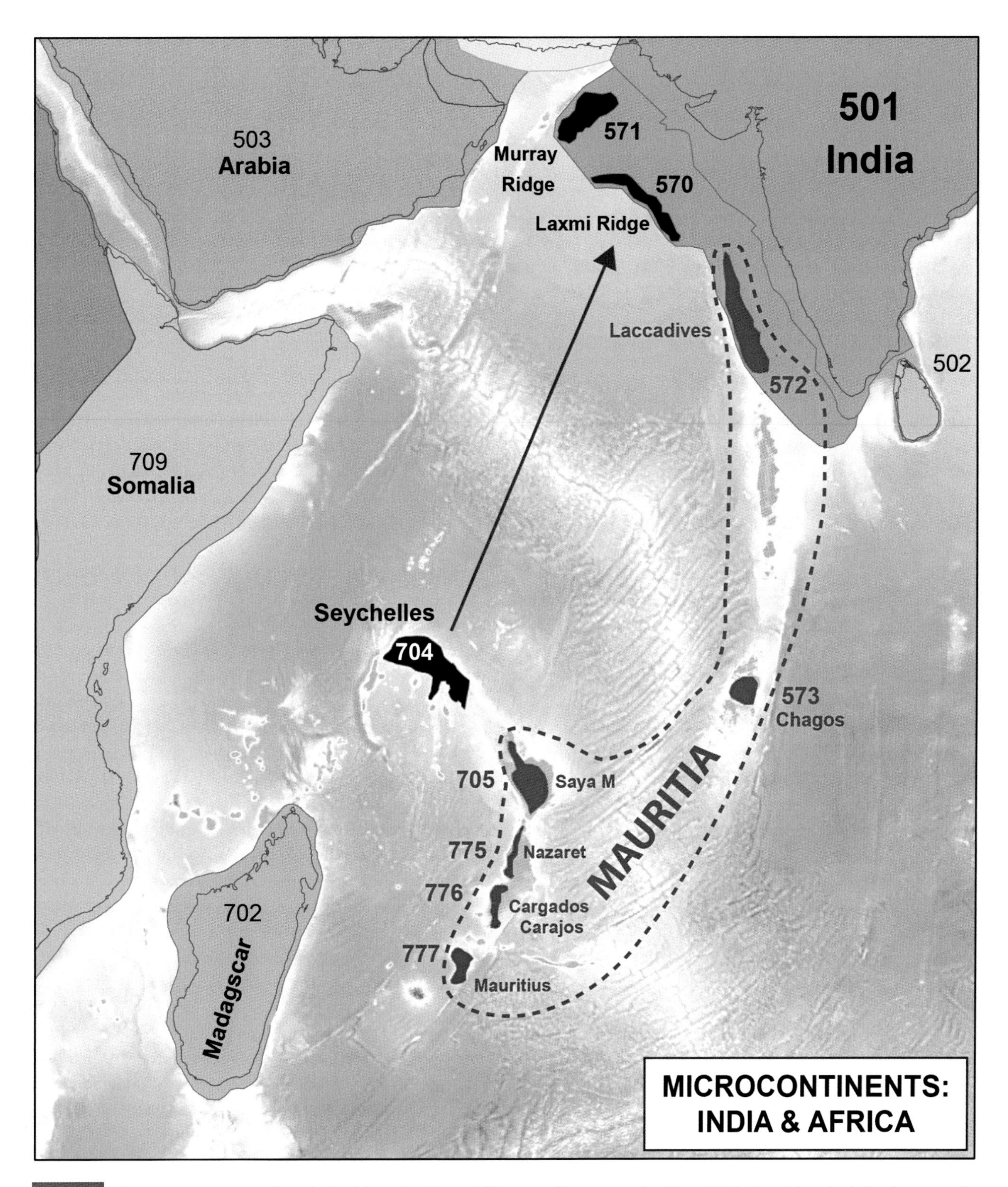

Fig. 3.11 Microcontinental units from India (plate identities 5XX) and Africa (plate identities 7XX). Dark blue shaded units were all part of the Mauritia Microcontinent before the Late Cretaceous (Torsvik et al., 2013), and the Seychelles was conjugate to the Lazmi Ridge until the Early Paleogene. The Continent–Ocean Boundary transition zone (COB) is mainly defined from seismic, magnetic, and/or gravity data.

deposition of thick glacial deposits and it also included the provincially diagnostic *Glossopteris* Flora then.

803, Antarctic Peninsula. An amalgamation of three fault-bounded terranes, with complex geology and many floras and faunas, which together made up a Palaeozoic to Cenozoic continental arc system formed above an eastward-dipping Palaeo-Pacific subduction zone. By the end of the Permian it was close to Patagonia (Unit 291). The Antarctic Peninsula moved away from Eastern Antarctica in the Jurassic and Early Cretaceous between 175 and 140 Ma in a slow clockwise motion which opened the Weddell Sea, followed by convergence and partial subduction of the Weddell Sea.

804, Marie Byrd Land. This is a substantial area on the margin of West Antarctica between Thurston Island (Unit 808) and the Ross Shelf (Unit 842). Scattered exposures beneath the ice cap show that the western part is folded Lower Palaeozoic sandstones and turbidites intruded by the large Late Devonian–Early Carboniferous Ford Granodiorite and smaller Cretaceous granitoids. The eastern part is a Lower Palaeozoic metamorphic basement with Permian and Cretaceous igneous intrusions.

805, Filchner Block. This is another substantial area in West Antarctica just west of the Transantarctic Mountains and abutting Dronning Maud Land. Mostly unknown submerged continental crust, with its northern sector a younger extended continental margin. Onshore widely separated exposures within the ice cap indicate that it includes one or more Proterozoic cratonic fragments, which were probably originally attached to South Africa (Unit 701).

806, North New Zealand. The North Island of New Zealand, as well as the north-western part of South Island (to the west of the major strike-slip Alpine Fault) is a composite unit that is mostly several Mesozoic to Recent volcanic island arc systems. However, the Takaka Terrane, mainly in north-western South Island, includes the remnants of one or more Palaeozoic arc systems which have yielded low-latitude Cambrian brachiopod faunas, a cooler-water *Hirnantia* Fauna in the latest Ordovician Hirnantian glacial interval, and Early Devonian shelly faunas in the Reefton Complex, as well as many Ordovician and Silurian graptolites from the surrounding deeper-water shales.

807, South New Zealand and Chatham Rise. A major arcuate strike-slip fault (the Alpine Fault) through South Island of New Zealand separates most of it from northern New Zealand (Unit 806). The rocks to the east of that fault represent at least seven formerly independent volcanic arcs, largely of Jurassic to Late Cretaceous ages. New Zealand was subsequently a passive margin from the Late Cretaceous to the Paleocene, but the margin has been active again since the earliest Miocene. Adjacent to the eastern side of southern New Zealand, and included within Unit 807, is a substantial area, now under the Pacific Ocean, beneath which is continental lithosphere, and which is divided into the Chatham Rise to the north and the Campbell Plateau to the south.

808, Thurston Island. On the western Antarctic margin between the Antarctic Peninsula (Unit 803) and Marie Byrd Land (Unit 804). Sporadic exposures document Carboniferous to Late Cretaceous volcanic arc magmatism, but there are Late Carboniferous (about 300 Ma) gneisses and gabbros, as well as Triassic diorites. Palaeomagnetic data suggest that it was emplaced into its present position by about 300 km of dextral movement and some clockwise rotation during the Cretaceous between 130 and 110 Ma, and there was silicic volcanism prior to the cessation of subduction in the Late Cretaceous between 100 and 90 Ma, and also Miocene post-subduction alkaline basalts.

809, Ellsworth–Whitmore Mountains. These mountains are a displaced portion of the old Gondwanan cratonic margin. There is a relatively complete Middle Cambrian to Permian succession, commencing with Cambrian tropical carbonates, overlain by non-marine sediments and with Early Devonian sandstones yielding cooler-water Malvinokaffric Province brachiopod faunas. There are also Permian–Carboniferous glacial diamictites. Although now part of West Antarctica, the unit may have been located in the Natal Embayment of South Africa (Unit 701) in the Cambrian, when it underwent intra-continental extension. There was intrusion of Middle Jurassic granites, and 90° counter-clockwise rotation of this unit, both related to the breakup of Pangea in the Jurassic at about 175 Ma.

827, New Hebrides. Active island arc at the western margin of the North Fiji Basin, thought to have originated as part of an earlier east–west trending arc system to the north under which the Pacific Plate was subducting (Johnson et al., 1993).

830, Solomon Islands. Complex collage of crustal units or terranes which have formed and accreted within an intra-oceanic environment since Cretaceous times. Cretaceous basaltic basement sequences are divided into a plume-related Ontong Java Plateau Terrane, a 'normal' ocean ridge-related

South Solomon mid-ocean ridge basalt (MORB) terrane, and a hybrid 'Makira Terrane' which has both MORB and plume/plateau affinities (Petterson et al., 1999).

833, Central Lord Howe Rise. Includes New Caledonia Basin South. A microcontinent which rifted away from East Australia during the Cretaceous from 90 to 64 Ma. The Lord Howe Rise may represent the offshore continuation of Palaeozoic–Mesozoic orogens and basins of eastern Australia, and the Central Lord Howe Rise (Unit 833) is one of 13 tectonic pieces (Gaina et al., 1998). Other pieces in the Tasman Sea and south-western Pacific puzzle (see below) include Units 850 (Tasmania), 851 and 852 (Western and Eastern South Tasman Rise), 866 (Chesterfield Plateau), 867 (Gilbert Seamount), 868 (Challenger Plateau), 869 (North Lord Howe Rise), 883–886 (Dampier Ridge segments), and 887 (East Tasman Platform).

835, Three Kings Ridge. Early Miocene (22–19 Ma) volcanic arc offshore from Northern New Zealand (Unit 806) (Mortimer et al., 1998).

836, Louisiade Plateau. One of several possible continental slivers of unknown age which rifted off north-east Australia during the opening of the Coral Sea in the Paleocene and Eocene between 63 and 52 Ma (Gaina et al., 1999). Other possible microcontinental blocks include Units 878 (Papuan Plateau), 888 (Eastern Plateau), 889 (Mellish Rise), and 890 (Kenn Plateau).

841, Transantarctic Mountains. An orogenic belt stretching across Antarctica at the margin of East Antarctica (Unit 802) which represents much Palaeozoic tectonic activity from the Middle Cambrian onwards. Includes the Shackleton, Pensacola, and Horlick Mountains, and the Beardmore Glacier, which have Lower Palaeozoic rocks and fossils. There is a Middle Palaeozoic erosion surface which separates gently tilted Devonian to Triassic rocks and Jurassic continental tholeiites from an underlying Proterozoic to Early Palaeozoic deformed orogenic belt, the Ross Orogen (Unit 842), and there are substantial Carboniferous and Early Permian glaciogenic deposits. The mountains have undergone episodic uplift since the Early Cretaceous. The Transantarctic Mountains divide Antarctica into two geological provinces: cratonic East Antarctica (Unit 802) and the collage of tectonic blocks that make up West Antarctica.

842, Ross. The Ross Sea is probably underlain by continental lithosphere, and appears to represent a separate tectonic unit of extended continental crust between Marie Byrd Land (Unit 804) and East Antarctica (Unit 802). Relative extension between East and West Antarctica probably commenced in the Late Cretaceous–Early Cenozoic.

848, Eastern Australia. The Tasman Orogenic Belt or Tasmanides (itself divided into a northern Thompson Fold Belt and a southern Lachlan Fold Belt) represents the union of many terranes which accreted to today's eastern margin of the West Australian Craton (Unit 801) progressively during the Palaeozoic at the eastern end of Gondwana (Glen, 2005). There are the Late Neoproterozoic to earliest Ordovician Delamerian Cycle, the Middle Ordovician to Early Silurian Benambran Cycle, the Middle Silurian to Middle Devonian Tabberabberan Cycle, and the Middle Devonian to Carboniferous Kanimblan Cycle, each of which resulted in substantial additions to the margins of Gondwana. This belt is now collectively nearly half of Australia.

850, Tasmania. The western half of Tasmania has a metamorphosed Neoproterozoic basement unconformably overlain by Lower Palaeozoic rocks, both shallow marine and terrestrial. The eastern half has Lower Palaeozoic flysch and shallower-marine rocks, and the two halves were united in the Early to Middle Devonian Tabberabberan Orogeny, resulting in Late Devonian granite intrusions from 380 to 360 Ma. Permian and Mesozoic sedimentary rocks are intruded by extensive Jurassic dolerites.

851, Western (851) and Eastern (852) South Tasman Rise. See Unit 833 and Gaina et al. (1998).

854, Dronning Maud Land. Originally an independent Neoproterozoic microcontinent, probably near South Africa, it now forms a substantial integral sector within western Antarctica. A Proterozoic basement is unconformably overlain by Lower Cambrian sediments in the west and Middle to Upper Cambrian sediments in the east.

866, Chesterfield Plateau, 867, Gilbert Seamount, and 869, North Lord Howe Rise and New Caledonia Basin. See Unit 833 and Gaina et al. (1998). Individual parts of the south-western Pacific Ocean and the Tasman Sea.

868, Challenger Plateau. Schists, granitoids, and greywackes are exposed in small islands on the Challenger Plateau to the north-west of New Zealand, indicating that the region is underlain by continental crust, sometimes termed 'Zealandia', which includes North New Zealand (Unit 806). See Unit 833 and Gaina et al. (1998).

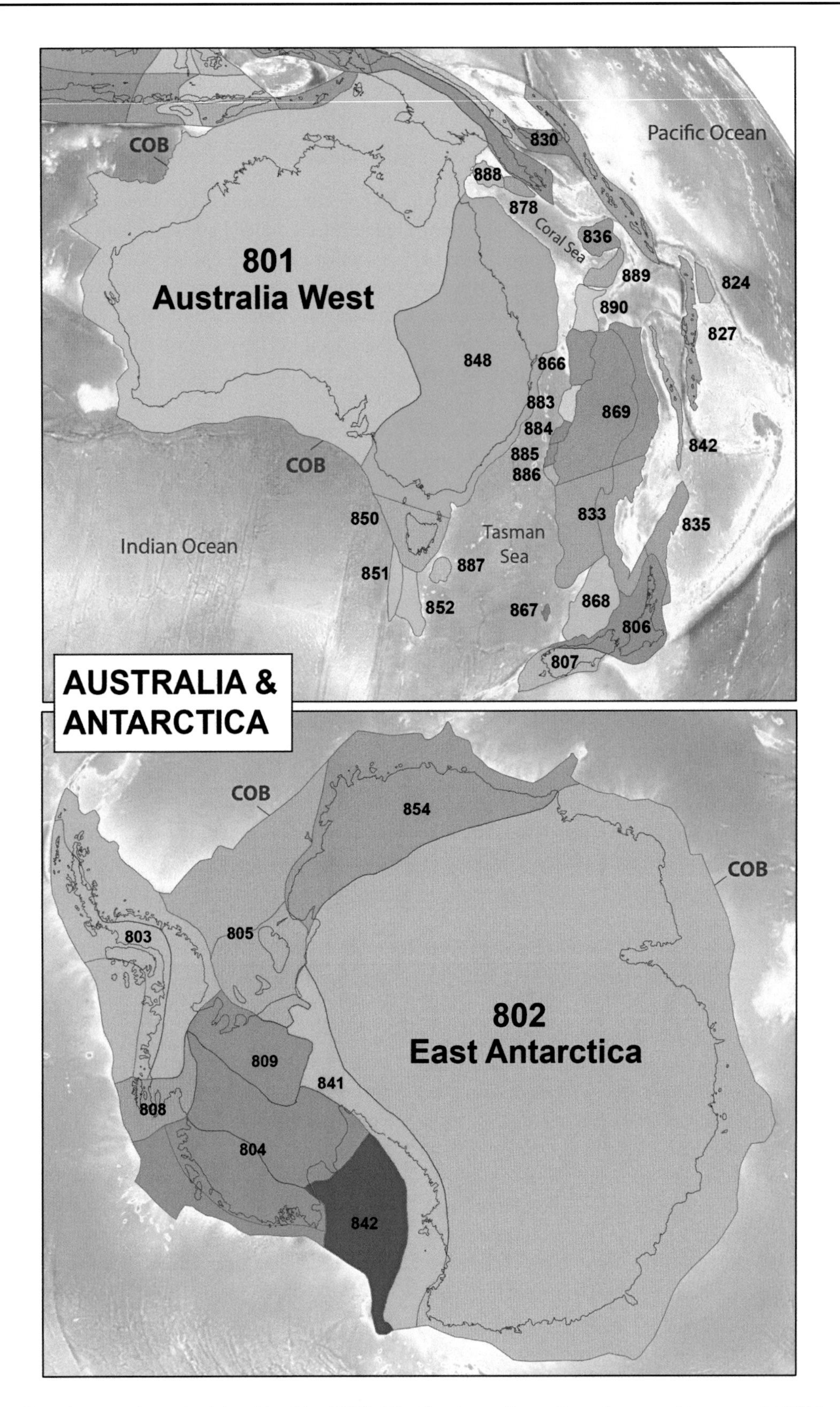

Fig. 3.12 Australia and Antarctica unite (plate identities 8XX). The Continent–Ocean Boundary transition zone (COB) is mainly defined from seismic, magnetic, and/or gravity data.

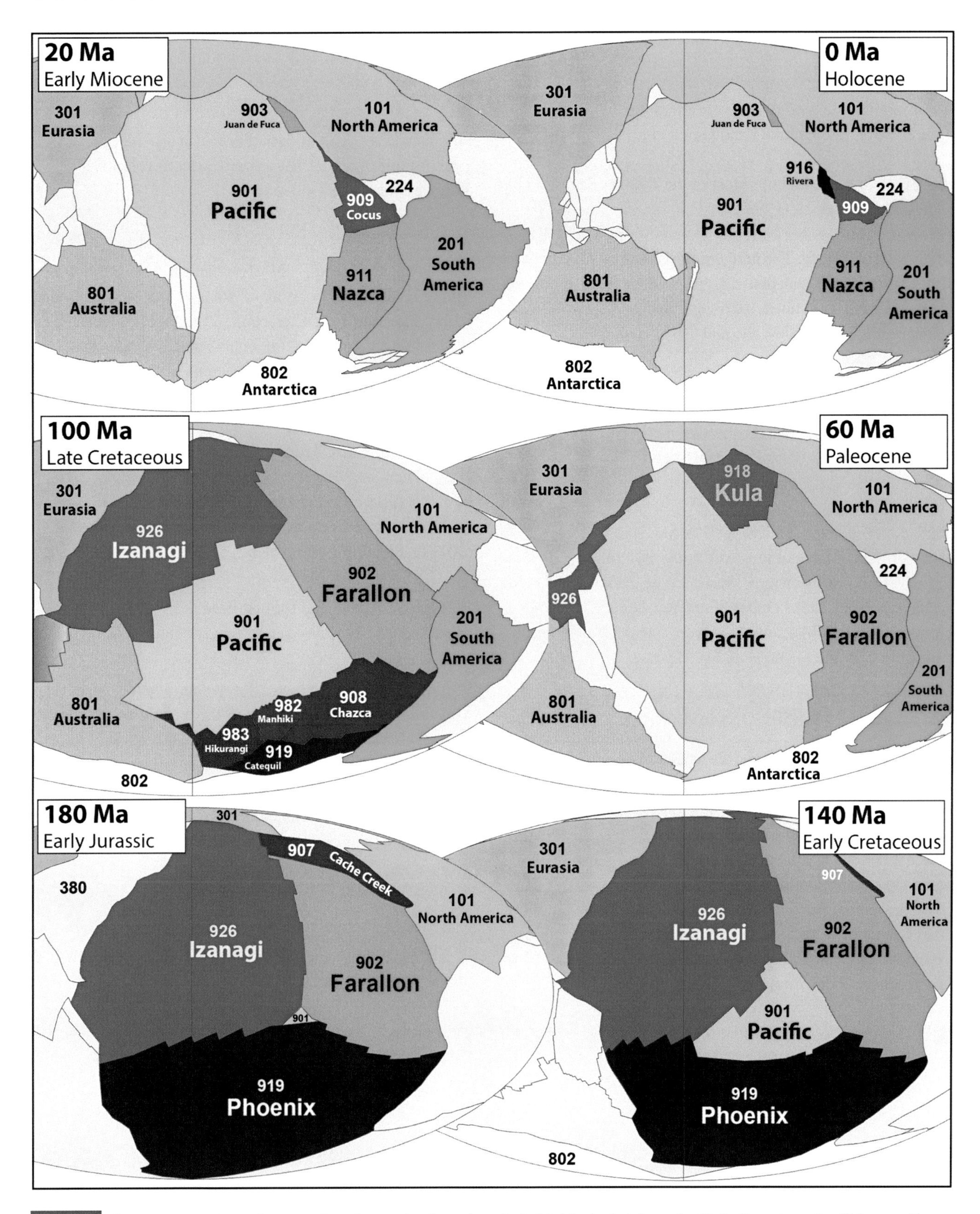

Fig. 3.13 Plate reconstructions of some selected oceanic plates (mostly in bluish shades) from the Early Jurassic to the Holocene. For 'directional reference' we have also highlighted Eurasia, North and South America, Australia, and the Antarctic plates in some of the reconstruction frames. The Caribbean Plate is Unit 224 (yellow shading).

878, Papuan Plateau. See Unit 836 and Gaina et al. (1999).

883–886, North (883), Mid1 (884), Mid2 (885), and South (886) Dampier Ridge. Various parts of the south-western Pacific Ocean floor. See Unit 833 and Gaina et al. (1998).

887, East Tasman Platform. Most of the Tasman Sea. See Unit 833 and Gaina et al. (1998).

888, Eastern Plateau. The northernmost marginal plateau of the NE Australian continental margin, underlain by deformed, fault-bounded and tilted basement blocks of possible Late Cretaceous age. See Unit 836 and Gaina et al. (1999).

889, Mellish Rise. Continental sliver believed to have separated from the Australian continental margin as a result of extension between the Australian, Lord Howe Rise, and Louisiade Plateau crustal elements between 62 and 52 Ma. See Unit 836 and Gaina et al. (1999).

890, Kenn Plateau. The Kenn Plateau is a large piece of submerged continental crust off north-eastern Australia that rifted from north-eastern Australia at about 63–52 Ma (Paleocene–Eocene). One of several thinned continental fragments east of Australia that were once part of Australia. See Unit 836 and Gaina et al. (1999).

Panthalassic–Pacific Oceanic Plates

All the units described above are static today, but the predominantly oceanic plates, when shown in our Mesozoic–Cenozoic reconstructions (e.g. Fig. 11.1) are dynamic polygons (generated through GPlates) that vary in size though time. We list below only the oceans in the Panthalassic and subsequent Pacific region, which extended at times to cover nearly half the globe. Some units have no existing oceanic crust today, and have thus been generated almost entirely from synthetic modelling. The region is illustrated in Fig. 3.13 by six reconstructions from Early Jurassic (180 Ma) to today.

901, Pacific. The largest active oceanic plate on Earth, and probably formed at about 190 Ma.

902, Farallon. Largely synthetic, probably formed in the Palaeozoic and was split into Nazca (911) and Cocos (924) at around 23 Ma.

903, Vancouver, Juan de Fuca. Formed at around 52 Ma (Early Eocene) and from 37 Ma (latest Eocene) referred to as the Juan de Fuca Plate (active today).

907, Cache Creek. An ocean plate within the Panthalassic, which opened during the Triassic off the north-western former Laurentian margin of Pangea but was subducted before the end of the Cretaceous. Although the oceanic part of the unit is synthetic, the continental lithosphere on both east and west margins are different rocks still preserved within the Western Cordillera of North America.

908, Chazca; 919, Catequil; 982, Manihiki; 983, Hikurangi. All four plates are largely synthetic and formed at around 120 Ma (Early Cretaceous) from the Phoenix Plate (919). At 85 Ma (Late Cretaceous) they became part of the Pacific (901) or Farallon (902) plates.

911, Nazca. Active and formed from the Farallon Plate (902) at around 23 Ma (Paleogene–Neogene boundary time).

918, Kula. Largely synthetic, formed at around 83 Ma (or 79 Ma), and later became part of the Pacific Plate (901) at around 40 Ma.

919, Phoenix. Synthetic, perhaps formed in the Palaeozoic which split into the Chazca (908), Catequil (919), Manihiki (982), and Hikurangi (983) plates at around 120 Ma (Early Cretaceous).

924, Cocos. Active and formed from the Farallon Plate (902) at around 23 Ma (Oligocene–Miocene boundary time).

926, Izanagi. Synthetic plate that perhaps formed in the Palaeozoic and vanished at around 55 Ma (Eocene).

4 Earth's Origins and the Precambrian

The most popular explanation for the origin of the Moon, the 'Giant Impact Hypothesis' or the 'Moon-Forming Event', is that it resulted from debris after collision of Earth with a Mars-sized proto-planet, probably about 100 million years after the formation of the Solar System. The Moon was very close to the Earth when it formed and has since travelled away from the Earth. Credit: Steve A. Munsinger/Science Source.

There is no accurate date for the formation of the Universe; however, an approximation of over thirteen billion years (13.8 Gyr) appears likely. From that time onwards, the Universe has continuously expanded, with the creation of galaxies and stars within them. Our own galaxy, the Milky Way, was formed at around 10 Ga, and according to the nebular hypothesis, our own star, the Sun, formed at the centre of a proto-planetary disk at 4.57 Ga. Planetismals collided and ultimately acquired sufficient mass to form Earth and the other planets.

The Precambrian Earth. Until the twentieth century, 'real' fossils with hard parts had only been found in Cambrian and later rocks, and all rocks formed earlier than that were thus termed 'Precambrian', a term still widely used today. Before and during the nineteenth century there was no good means of knowing the age of the Earth, and estimates varied from the six thousand years calculated from the addition of biblical dates by Bishop Ussher in the seventeenth century to the few hundred million years suggested by Charles Lyell in the nineteenth century. However, it is now known that the oldest rocks on Earth so far radiometrically measured are about four billion years old, and the individual minerals within some of them have been dated to be from as far back as 4.4 Ga (Wilde et al., 2001). Thus there are enormous geographical and chronological age ranges within Precambrian rocks, whose description and discussion are largely outside the scope of this book. The rocks older than about 2.5 Ga are termed the Archean, and those dated between the Archean and the base of the Cambrian at 543 Ma the Proterozoic, which is divided into the Paleoproterozoic, Mesoproterozoic, and Neoproterozoic successively (see Endpaper). The age unrepresented by rocks between the Earth's origin and the Archean is termed the Hadean.

The Origin of the Earth and the Moon. Because of the restless tectonic evolution of our planet, the rocks originally formed near the times of the origin of the Earth have long since been subducted and recycled within the mantle. However, from lunar and other planetary studies, scientists assume that most, if not all, the heavenly bodies in the Solar System were formed within a relatively short time of each other. Therefore the age of the Sun, planets, and asteroids should be close to the age of primitive meteorites, and from that the age of the Earth's formation is now agreed to have probably been at 4.57 Ga (Connelly et al., 2012). The most popular explanation for the origin of the Moon, which is known as the 'Giant Impact Hypothesis' or the 'Moon-Forming Event', is that it resulted from debris after collision of Earth with a Mars-sized proto-planet, probably about 100 million years after the formation of the Solar System. It is interesting to note that George Darwin (the second son of the pioneer biologist and geologist Charles Darwin) had suggested as early as 1898 that the Earth and Moon had been one body in the deep past and that the Moon later travelled away from the Earth.

Early Earth. We here refer to 'Early Earth' as roughly the first 1.5 billion years of Earth's evolution, before the onset of plate tectonics, and it includes the Hadean and parts of the Archean Eons. Nearly all the traces of Early Earth, including a massive bombardment history, have been demolished by younger plate tectonic processes, but clues recorded and preserved in some of the oldest rocks, combined with numerical modelling and the cratered surface of the Moon and other terrestrial planets, yield important insights into the Early Earth. How continental crust formed and evolved through time is highly controversial, but during the Early Earth new continental crust must have been dominantly mafic, had a lower silica content, and was probably much thinner (<20 km), and some argue that the net crustal growth rate may have been fairly constant (Fig. 4.1).

Early Earth was certainly very different from the planet we know today; the mantle was clearly hotter than the modern mantle, mantle plumes of very hot magma were abundant, and the lithosphere was thinner and more buoyant. This very different and hot geodynamic regime, facilitated by high heat production due to radioactive decay, core differentiation, and the Moon-forming event, would have limited subduction of tectonic slabs (e.g. O'Neill et al., 2007), or perhaps resulted in flatter subduction. Vertical tectonic processes appear to have been dominant over horizontal ones, and the mode of mantle convection was probably one of stagnant lid convection, as proposed for other terrestrial planets in the Solar System where there is no evidence for plate tectonics.

After the Moon-forming event, it is likely that a fast spinning Earth (a day was only a few hours) retained enough heat to melt large parts if not all of the mantle for tens of millions of years, giving rise to what is generally termed a magma ocean. The two large low shear-wave velocity provinces (LLSVPs) in the lowermost mantle beneath Africa and the Pacific Ocean, named Tuzo and Jason (Fig. 2.13) have been argued by some to have been stable for a very long time (Torsvik et al., 2014), and peridotitic materials enriched in iron and perovskite provide a suitable fit to their seismic properties. Magmatic segregations of Fe-rich peridotitic or komatiitic materials could have formed during early magma ocean crystallisation or shortly afterwards, and thus an Early Earth origin for these LLSVP deep mantle structures is feasible but controversial.

Supercontinents. Pangea is the only Phanerozoic supercontinent, and is therefore known in considerable detail (Fig. 4.2). Wegener (1912) first proposed that all the continents once formed a single supercontinent surrounded by a vast marine area (the Panthalassa Ocean). His Pangea reconstruction (originally named Urkontinent) was based on the similarity between coastlines on opposite margins of the Atlantic, the distribution of old glacial deposits, and identical Permian and Carboniferous plant and animal fossils from a number of continents now separated by oceans. There are many similarities between Wegener's and modern reconstructions of Pangea, but the most important difference is that we are now able to position Pangea in its ancient latitude and longitude (Fig. 4.2).

The reconstruction of supercontinents relies on palaeomagnetism in concert with the alignment of orogenic features, matching of conjugate margins, geochronology, detrital zircon geochronology, large igneous provinces (LIPs), mafic dyke swarms, and more. Even though Precambrian palaeomagnetic data are often sparse and geometric relationships can be ambiguous, at least two Precambrian supercontinents, Rodinia (Fig. 4.3) and Nuna (Columbia), are now recognised by most workers, although variable continental make-ups have been proposed for them. From those three supercontinents one may argue for an assembly periodicity of 750 Myr (Meert, 2012) based on formation times at 1.9–1.8 Ga (for Nuna), 1.1–1.0 Ga (for Rodinia) and 0.32 Ga (for Pangea). Dispersals of Nuna, Rodinia (e.g. Li et al., 2008), and Pangea (Buiter & Torsvik, 2014; Torsvik & Cocks, 2013) all appear to have been contemporaneous with massive LIP volcanism, which perhaps assisted their break-ups, but LIPs are also found well away from supercontinent dispersal time, and not all promoted plate boundary changes and breakup.

Supercontinents have also been linked to sea-level changes, but that can only be demonstrated for Pangea. If mantle flow patterns have remained fairly stable, then patterns of dynamic topography at the Earth's surface should also have remained stable. Tuzo (Fig. 2.2b) was located beneath the heart of Pangea (Fig. 10.1) at around 250 Ma, and the concentration of continents above this region of

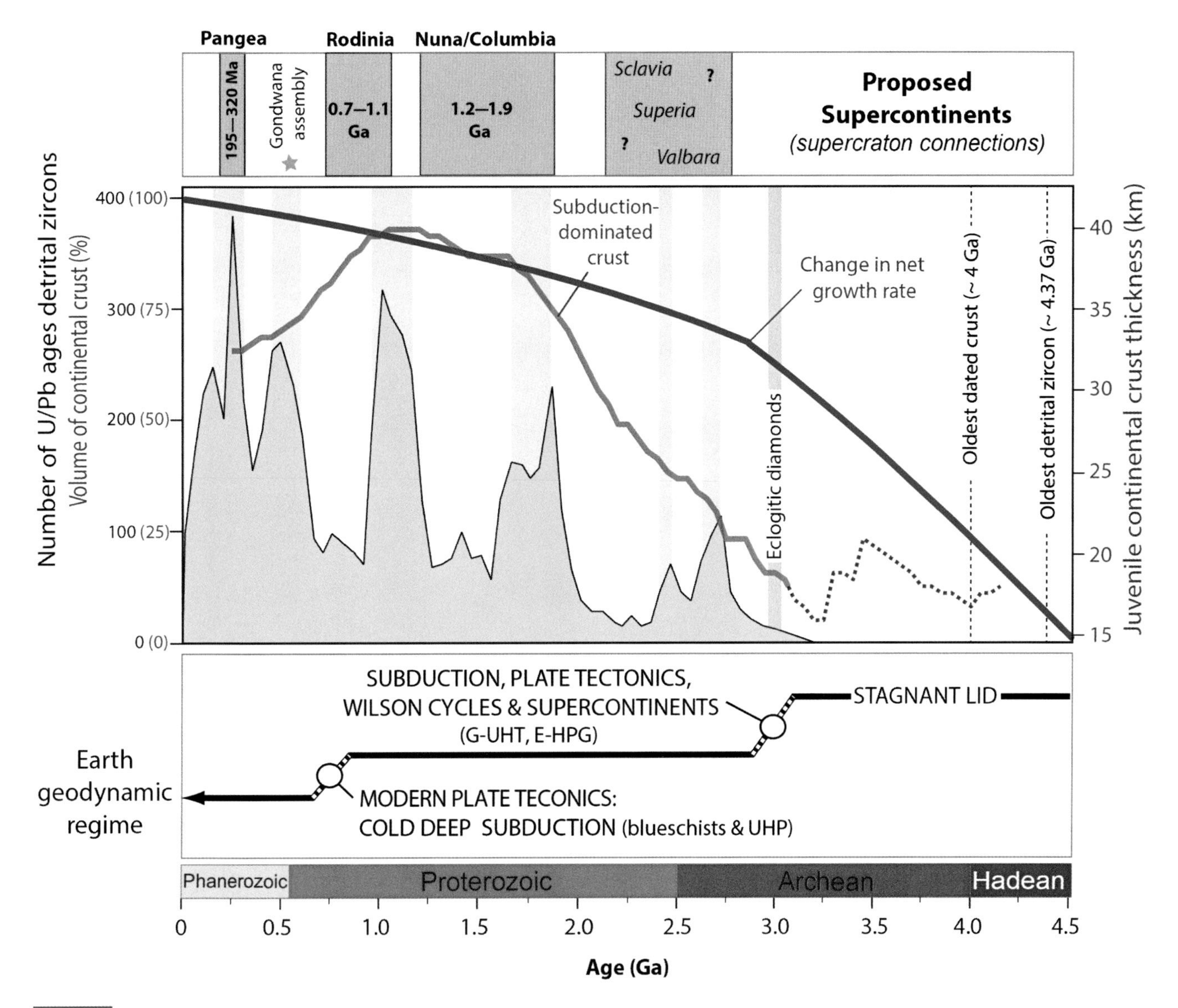

Fig. 4.1. Supercontinent (supercraton/superterrane) time lines compared with the spectra (number) of U/Pb detrital zircon crystallisation ages (Hawkesworth et al., 2010), estimated cumulative continental crust growth (blue thick line), and the thickness of the continental crust (thick and stippled red lines) through time (Dhuime et al., 2012, 2015). The oldest detrital zircon found on Earth is 4.37 billion years old (Wilde et al., 2001), whilst the oldest dated crust is about 4 billion years old (Bowring & Williams, 1999; Mojzsis et al., 2014). Eclogite diamonds started to dominate from 3 Ga (Shirey & Richardson, 2011). The onset of steep subduction and plate tectonics probably occurred at around 3 Ga but modern-day-style plate tectonics with deep cold subduction attendant on ultra-high-pressure (UHP) metamorphism probably did not occur until the Neoproterozoic. G-UHT, granulite–ultra-high-temperature metamorphism; E-HPG, eclogite–high-pressure granulite metamorphism.

stable high dynamic topography (upwelling) should have contributed to low sea levels (Conrad et al., 2014). Indeed, Phanerozoic sea level was at an all-time low near the Permian–Triassic boundary (Fig. 16.2). Earth was in an ice-free greenhouse condition at that time, and the subsequent dispersal of Pangea with continents moving towards regions of negative dynamic topography led to a global eustatic sea-level rise of the order of 50–100 metres since the Early Jurassic. That magnitude is comparable with that of other important mechanisms for sea-level change, such as crust production rates and glaciations.

Supercontinent Attractors. There are some striking similarities between many Pangea (Fig. 4.2), Rodinia (Fig. 4.3), and Nuna reconstructions; for example, the Baltica–Laurentia–Siberia and East Gondwana elements (India–Australia–East Antarctica–Madagascar) form groups that largely resemble each other in all supercontinents (Fig. 4.4). The term 'strange attractor' was introduced by Meert (2014) for

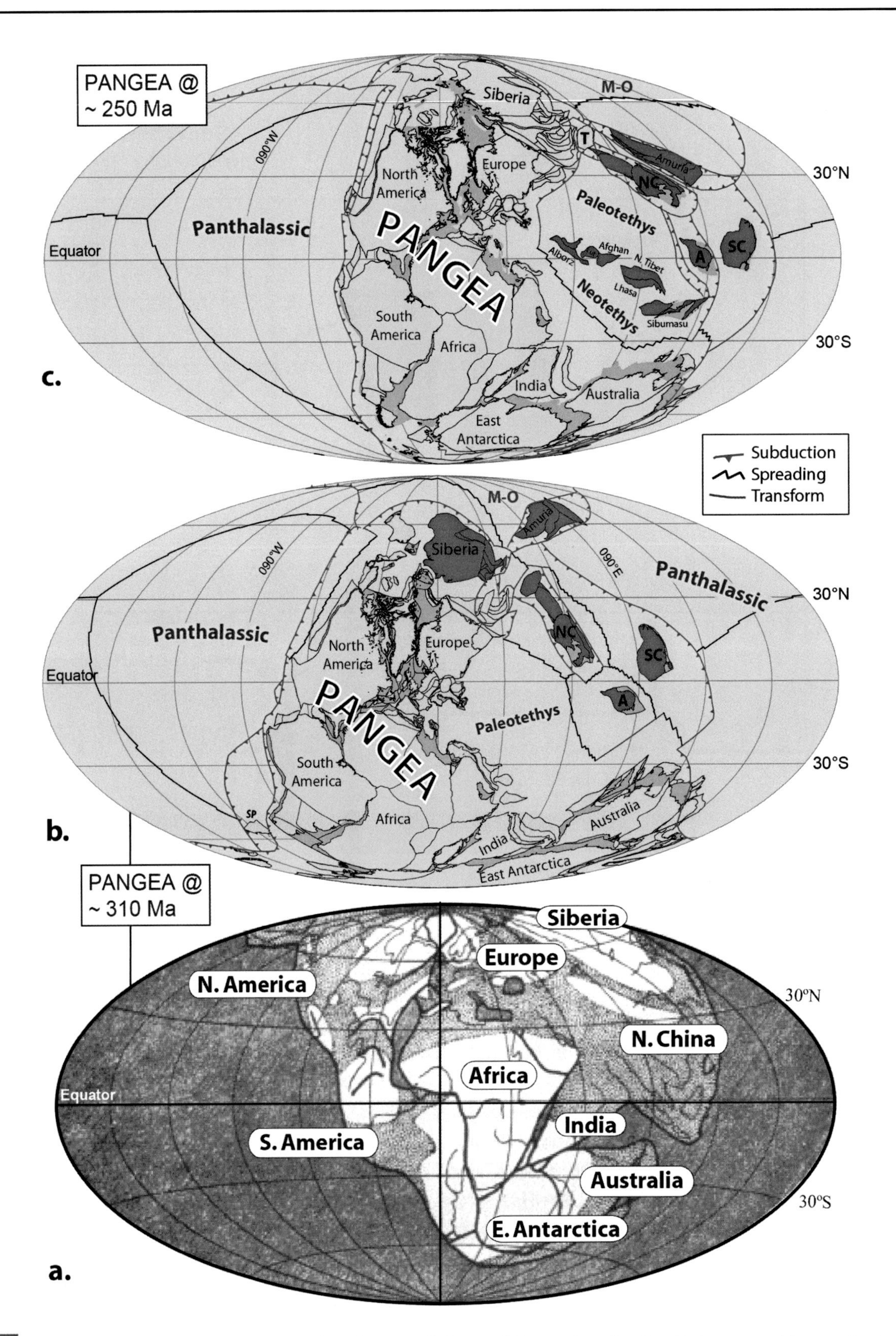

Fig. 4.2. (a) Wegener's (1912) relative reconstruction of Pangea (Africa and Europe kept fixed), with ice caps (white areas), at around 310 Ma (Late Carboniferous), and (b, c) two modern reconstructions of Pangea at around 310 and 250 Ma (Permian–Carboniferous boundary). (b) and (c) are reconstructed to Pangea's latitude and longitude with palaeomagnetic data using the 'zero longitude' approximation for Africa (see Chapters 2, 9, and 10). Not all the continents and terranes were part of Pangea at any given time and those shaded red were either separate from Pangea or only loosely connected with it in Late Carboniferous to Early Mesozoic times.

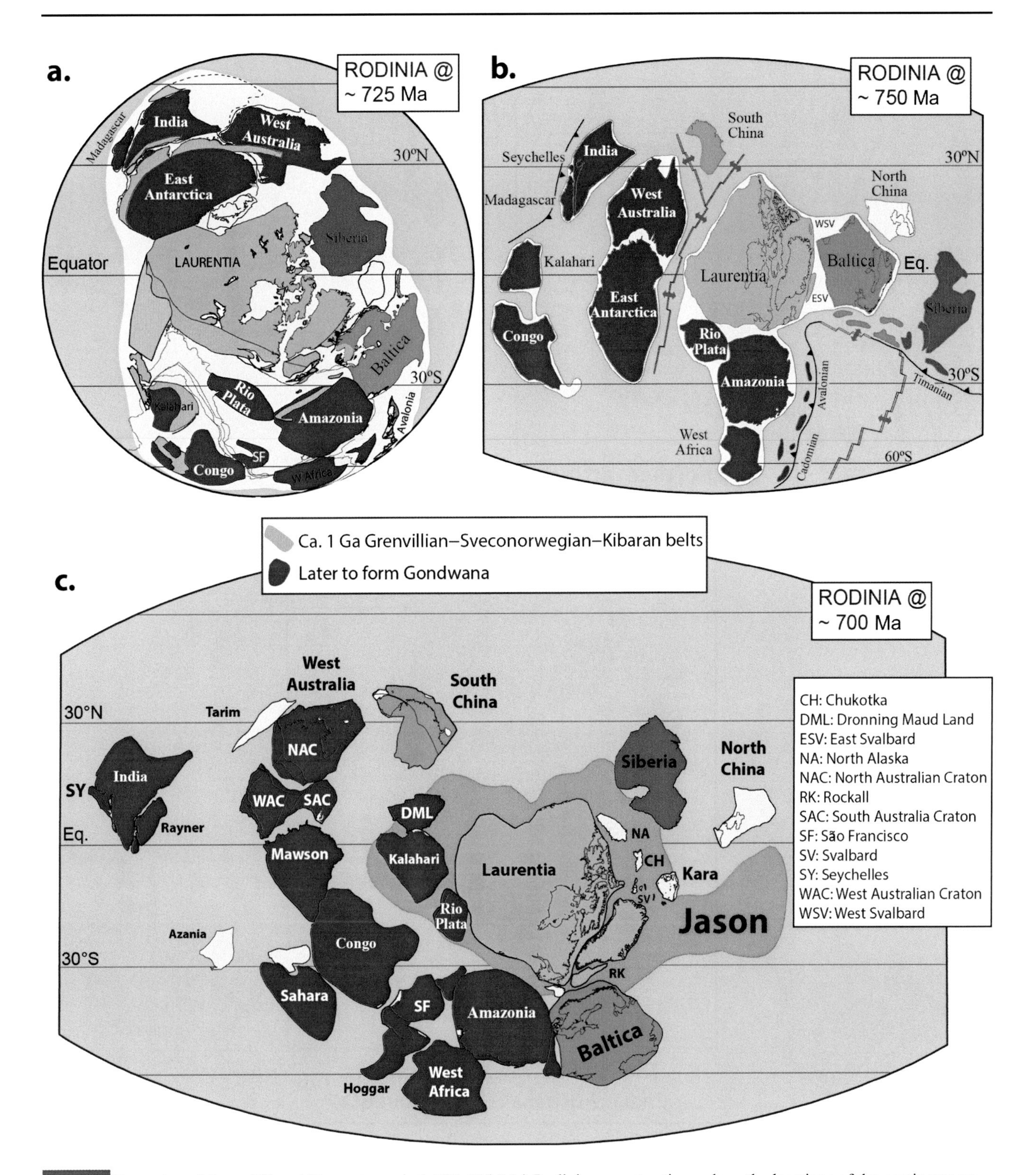

Fig. 4.3. Examples of three different Neoproterozoic (~700–750 Ma) Rodinia reconstructions where the locations of the continents are taken from (a) Dalziel (1997), (b) Torsvik (2003), and (c) Li et al. (2008). Laurentia (including North America, Greenland, Scotland, and Rockall) forms the core continent in all varieties and is centred at the Equator. Continents or cratons shaded in red later amalgamated to form Gondwana at around 550 Ma. Pangea dispersal was associated with many LIPs fed by plumes above the African (Tuzo) LLSVP in the deepest mantle (Fig. 2.16). If Tuzo and its Pacific antipode (Jason) were stable as far back as Rodinian times we show the present size of Jason (the plume generation zone, 1% slow contour in the SMEAN tomographic model).

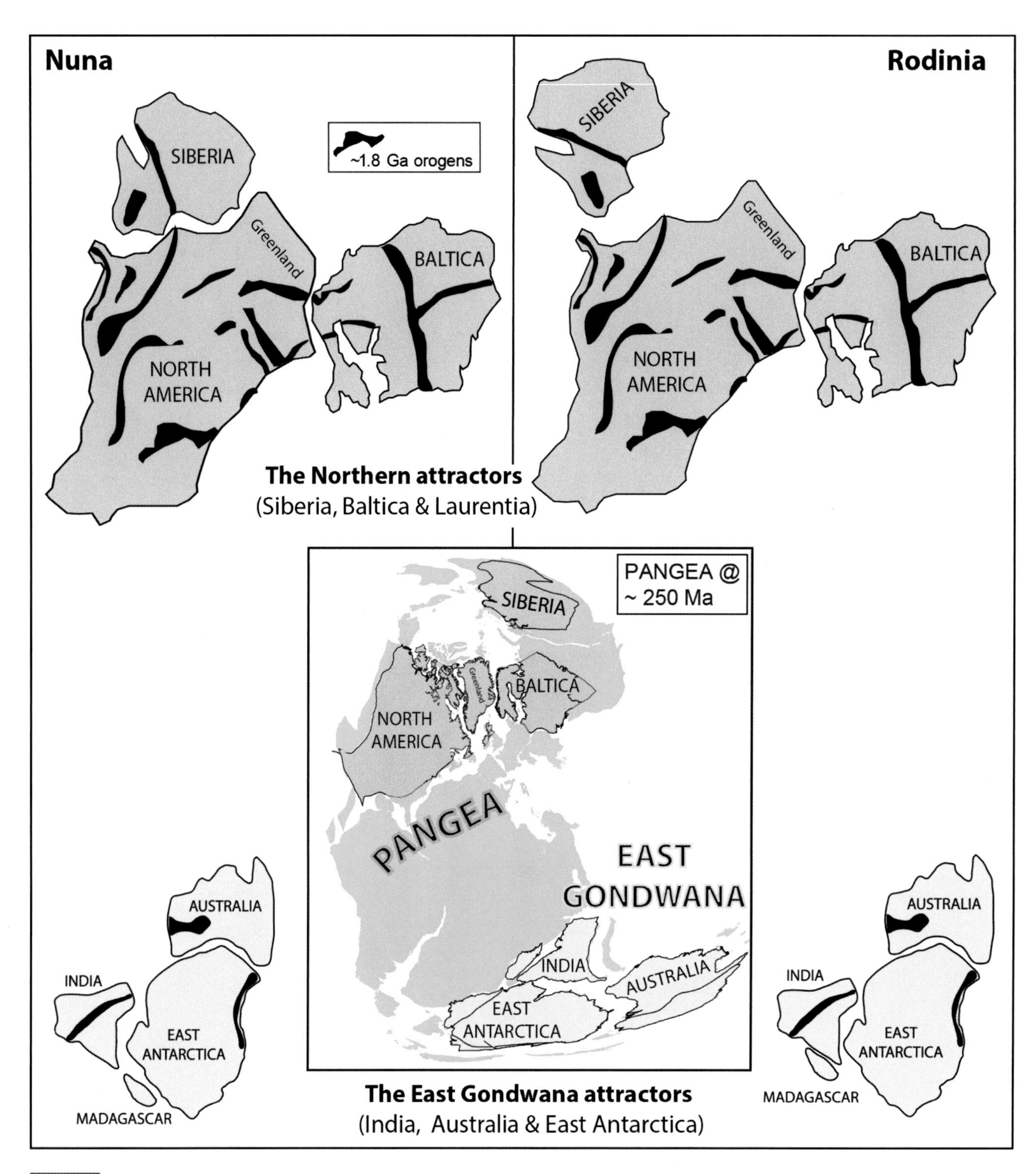

Fig. 4.4. The Northern and East Gondwana attractors identified by Meert (2012). The examples show the relatively similar grouping of North America–Baltica–Siberia within the Nuna (Zhao et al., 2004), Rodinia (Li et al., 2008), and Pangea (Fig. 4.2c) reconstructions. Note that the India–East Antarctica–Australia connections are essentially similar for all the three supercontinents, and also for the Gondwana Superterrane, which formed at around 550 Ma.

landmasses that formed a coherent geometry in all three supercontinents. An extreme form of vertical-dominated tectonics or Wilson Cycle tectonics, with continents assembling and dispersing in more or less the same position for billions of years, is very unlikely, and a probable reason for the 'strange attractor' is that our visions of old supercontinents are moulded by the known connections between the Gondwana Superterrane and Laurussia, which amalgamated to form the relatively recent supercontinent of Pangea in the Late Carboniferous. A very different Rodinia map by Torsvik (2003) is nearly free of the strange continental attractors (Fig. 4.3b), with Baltica geographically inverted (though next to Greenland), Siberia is shown to the east of Baltica and India conjugate to West Australia (Torsvik et al., 2001), rather than East Antarctica as in all other supercontinent reconstructions from Nuna to Pangea.

Zircon Record and Supercontinents. Zircon forms the archive of continental crust (Hawkesworth et al., 2010; Roberts & Spencer, 2014) and many authors have noted that crystallisation age peaks (Fig. 4.1) correspond with periods when the Earth was in a supercontinent regime. Those peaks are commonly interpreted as the result of enhanced preservation associated with supercontinent assembly or episodic growth (O'Neil et al., 2007; Parman, 2015) of continental crust (e.g. massive magmatism linked to deep-seated mantle plumes). Crustal 'preservation or growth' peaks appear correlated with Nuna, Rodinia, and Pangea, but there is also a notable peak at around 500 Ma, which has been linked to the Gondwana Superterrane, which constituted about 64% of all today's land areas. Before the assembly of Nuna, a pronounced zircon peak is also seen at around 2.7 Ga and that could have been linked to various postulated supercraton connections such as Vaalbara, Superia, and Sclavia (Evans, 2013).

Onset of Plate Tectonics. The bulk composition of continental crust is similar to that of rocks produced in subduction settings, and therefore the time of the onset of convergent plate interaction, plate tectonics, and ultimately supercontinents, are leading questions in Earth history. When and how modern plate tectonics (with subduction) started is debated: the Archean remains the most popular time frame for the onset, but this question critically depends on how plate tectonics is defined. The prevalence of eclogitic diamond compositions at around 3 Ga (which are interpreted as demonstrating the onset of formation of eclogite through subduction and its capture via continental collision) suggests a major change in the geodynamic regime at that time (Shirey & Richardson, 2011) and perhaps the onset of plate tectonics. Rb/Sr ratios in juvenile continental crust also increased around that time (Dhuime et al., 2015), suggesting that the newly formed crust became more silica-rich, and therefore probably also thicker, reaching maximum thickness (~40 km) at Rodinia assembly time (Fig. 4.1). A gradually cooling mantle and the onset of cold, deep and steep subduction (blueschist and ultra-high-pressure metamorphic conditions), comparable with the present day, first occurred in Neoproterozoic times (e.g. Stern, 2008). Ultra-high-pressure terranes represent subduction of continental crust to depths of 100 km or more as a result of continent–continent collision and later exhumation. Western Norway is a prime example of a large ultra-high-pressure terrane shaped through the Late Silurian collision of Baltica with Laurentia (Greenland: Figs. 2.19, 7.5, and 7.6), and which was subsequently exhumed relatively rapidly over 20 million years, during the Early Devonian (e.g. Andersen et al., 1991).

The Earth's Early Atmosphere. A critical milestone occurred at about 2.2 Ga, termed the Great Oxidation Event, when the Earth's atmosphere changed from a reducing one (consisting largely of carbon dioxide, nitrogen, water vapour, and inert gases, with subsidiary amounts of hydrogen, methane, and ammonia) to an oxidising atmosphere (the ancestor of the air of today). That change was brought about chiefly by the cumulatively increasing quantities and activities of a very varied and slowly evolving range of primitive biota discussed below, much of it too simple to be allocated to either the animal or the plant kingdoms, but chiefly cyanobacteria. Those organisms absorbed the carbon gases and released free oxygen. Although the Great Oxidation Event involved a sizeable increase of free oxygen in the atmosphere and hydrosphere, the increase continued steadily after that, and accelerated again in the Late Proterozoic, at about 600 Ma (Lyons et al., 2014; Andersen et al., 2015).

The Origin of Life. It is uncertain exactly when life started, but it appears to have occurred only once, at approximately 3.7 Ga, when the oldest DNA molecules evolved from a complex inorganic soup. Even Darwin suggested in 1859 that 'probably all the organic beings which have ever lived on this Earth have descended from some one primordial form, into which life was first breathed', and no reputable scientist has subsequently documented multiple creations. DNA enables the unique mechanism through which cells are repetitively reproduced through interlinked spirals (thus defining what we call life) in all the many organisms from that time to the present day. There is unresolved debate as to whether the first organisms were photosynthetic cyanobacteria, which lived near the sea surface, chemosynthetic organisms (e.g. chemolithotrophs), which lived around deep-sea vents, or anoxygenic photosynthesisers, which could live closer to the surface: all three categories still live today (Taylor et al., 2009). Those most primitive creatures, which include bacteria, are classified within the Prokaryotes, which differ from Eukaryotes in their lack of a nucleus. Because the first organisms were largely soft-bodied, their early record as actual fossils is tantalisingly incomplete; however, the steady increases in the overall proportions of carbon isotopes seen in rocks from about 3.5 Ga onwards appear to indicate that biological systems were in place by that time, which was over a billion years before the Great Oxidation Event.

Laminated structures known as stromatolites have been described from Archean rocks as old as 3.4 Ga in Australia and South Africa; however, the oldest stromatolites appear to have been inorganic in origin. In contrast, most of the stromatolites deposited after about 3.0 Ga are certainly of organic origin, and were the earliest known rock-builders to have made a key contribution to the amount of free oxygen in the atmosphere. Although the range and diversity of those complex organisms peaked in the Middle Mesoproterozoic and later Precambrian times, stromatolites are still to be found living today at a variety of sites, most famously as somewhat amorphous mounds in Shark Bay on the west coast of Australia. The oldest Eukaryotes are dated at about 2.0 Ga, and the earliest metazoans at about 1.1 Ga.

Away from the seas, there is some evidence for the existence of terrestrial life from isotope compositions in rocks as early as 2.7 Ga, also before the Great Oxidation Event, which would probably have been in the form of microbial ecosystems driven by oxygen-generating photosynthetic cyanobacteria, although there is no direct evidence of those organisms from actual fossil remains until the Late Proterozoic at about 1.1 Ga (Wellman & Strother, 2015). As in arid regions today, microbial mats and biological crusts consisting of communities of organisms were probably important in soil formation, and Proterozoic primitive soils (palaeosols) are known (Taylor et al., 2009).

During the later Neoproterozoic, more distinctive and larger animals are found as identifiable fossils. Chief of those is the *Ediacara* Fauna, named after Ediacara in the Flinders Range of South Australia where it was first noticed, and representatives of the *Ediacara* Fauna have now been found around the world in more than 20 areas; however, most of the soft-bodied species in that fauna can only loosely be placed within the 'modern' phyla found from Cambrian times onwards. Although variable numbers of different animals are known from the different localities, no recognisable faunal provinces of potential use in assessing global palaeogeography are apparent from analysis of the known distribution and generic composition of the *Ediacara* Fauna (McCall, 2006).

Snowball Earth. An Earth covered by ice from pole to pole has been suggested for parts of the Precambrian (e.g. Kirschvink, 1992; Hoffman et al., 1998), perhaps lasting for some millions of years and associated with chilling global mean temperatures as low as −50 °C. Two Snowball Earth intervals have been postulated for the past billion years, one at around 710 Ma (Sturtian glaciation) and the last ending at about 635 Ma (Marinoan). Both the existence of and reasons for Snowball Earth conditions are debated, but reduced atmospheric greenhouse gases, essentially carbon dioxide (CO_2) and methane (CH_4), to near present-day levels through tectonically mediated rock weathering is one explanation. Arguments for Snowball Earth conditions include their global distribution, association of glacial deposits with carbonate successions, and a few palaeomagnetic studies (e.g. from Australia) suggesting glacial deposition close to the ancient Equator.

Our Unique Planet. Scientists and others have long searched for planets similar to ours, but so far in vain. The Earth is unique in many ways. The most obvious is the presence of life here, and that is linked with the abnormal abundance of free surface water. A magnetic field protects us from cosmic radiation, and modulates atmospheric escape, and the Earth's magnetic field may have affected climate as well as evolution. A magnetic field is not unique to Earth, but plate tectonics have not been identified on any other terrestrial planets. The distributions of natural resources such as hydrocarbons, gas, coal, metal deposits, and industrial minerals are directly linked to plate tectonics. Biodiversity is well tied to the distribution of continents, and human origins themselves may be linked to plate tectonics. Human activity, with unrestrained burning of natural resources (mainly coal, oil, and gas), which originated from plate tectonic processes, may ultimately prove catastrophic for our planet.

5 Cambrian

Two specimens each of the ptychopariid trilobite *Modocia typicalis*, about 18 mm long, and the smaller agnostid *Ptychagnostus* sp. from the Middle Cambrian Marjum Formation, Utah, USA. Credit: James L. Amos/Science Source.

The Cambrian is famous for the incoming of hard-shelled fossils for the first time, and marks the start of the Phanerozoic. The system was named from the Cambrian Mountains in Wales by Adam Sedgwick (in Sedgwick & Murchison, 1835) where he had found the oldest rocks with the distinctive trilobites preserved. Since then, Cambrian rocks have been recognised in many places around the world, and the base of the Cambrian is now formally defined in Newfoundland, Canada, and has been dated at 541 Ma. It is now known that other smaller shelly fossils preceded the trilobites, and are found in sediments older than those preserved in Wales, and trilobites are first known only from rocks which are dated from some time into the Early Cambrian at about 535 Ma. The global distribution of the continents and oceans (but not the margins of land and sea) near the beginning of the Cambrian at 540 Ma, in the Middle Cambrian at 510 Ma, and in the Late Cambrian at 490 Ma, is shown in Fig. 5.1. Within the Cambrian, which was a long period lasting for some 54 Myr, there are four series, the basal Terreneuvian (a name derived from the French word for Newfoundland), succeeded by two which have yet to be formally named, and then the youngest series, the Furongian, whose base is defined in Hunan Province, South China.

Tectonics and Igneous Activity

Early Cambrian Orogenesis. The orogenies which united many old cratons to form Gondwana had begun in the Late Neoproterozoic at about 570 Ma, but they continued on into the Cambrian, and were largely over before Middle Cambrian times. Chief among them was the Pan-African Orogeny, whose tectonic activity surrounded much of the African continent (Meert, 2003; van Hinsbergen et al., 2011). Other contemporary orogenies were the Cadomian Orogeny in Europe (see below), the Timanide Orogeny of the northern Baltica area and north-western Siberia, the Kuungan Orogeny between India on the one hand and eastern Antarctica and western Australia on the other, the East African Orogeny between Africa and India and Arabia, and the Pampean Orogeny between southern South America and southern Africa (Fig. 5.2a). Unravelling of the former intraplate rotation shows substantial overlap between the South Australian Craton and the orogenic belts to its east, which must have developed through continental growth largely after 550 Ma (Fergusson & Henderson, 2015). Gondwana was by far the largest Cambrian continent (Torsvik & Cocks, 2013).

The Cadomian Orogenic belt (not shown in Fig. 5.2a) was first defined in the Armorican Massif of north-western France in the early nineteenth century, and consists of a variety of shear zone and fault-bounded microcontinental, volcanic arc, accretionary prisms, and margin basin units, all of which were the products of an active subduction zone outboard of Gondwana. The Cadomian affected a large area of today's Europe, certainly including the Iberian Peninsula and much of western, central, and southern Europe, and it possibly extended laterally to include areas now in North America, such as the Carolinia and Meguma terranes, parts of West Avalonia (Unit 170; Figs. 5.3 and 6.2a), and even south-eastwards as far as Turkey (McCann, 2008). However, post-Cambrian tectonism, including the Caledonide, Variscan, and Alpine orogenies, severely affected much of the Cadomian area, which is now broken up into many disconnected sectors and outcrops.

Oceans. Like today, the Cambrian oceans covered most of the tectonic plates that make up the Earth's crust; however, the convergent, divergent, and transform margins of those units are so poorly constrained for that age that we do not attempt to show them on the global maps (Fig. 5.1). We do, however, show some interpreted plate boundaries on regional reconstructions (e.g. Figs. 5.3 and 5.6), including the Late Cambrian (500 Ma) plate patterns within the Iapetus (Domeier, 2015). The immense Panthalassic Ocean occupied and dominated most of the northern hemisphere, as it continued to do throughout the Palaeozoic.

Another major ocean was the Iapetus, which lay between Laurentia to its west, Gondwana to its south, Baltica to the south-east, and Siberia to the north-east. Other significant oceans included Ægir (between Baltica and Siberia) and the Ran Ocean (Fig. 5.3) between Baltica and Gondwana. The Ran had developed near the start of the Cambrian and merged with the subsequent Rheic Ocean to become a sector of the latter during the Ordovician. The Iapetus had opened at some time in the Late Neoproterozoic, following rifting beneath north-west Gondwana, and widened progressively during the Cambrian (Fig. 5.3) to reach its maximum width by about 500 Ma, in the Late Cambrian. There was active convergence at its then southern margin at that time, with subduction north of the north-western Gondwanan margin, parts of which included the areas that were subsequently to become the Avalonia (including Meguma), Ganderia, and Carolinia microcontinents. The northern Iapetus margin was fringed with a string of peri-Laurentian terranes (including Cuyania, now in South America; Lushs Bight and Dashwoods, now in North America; and Clew Bay–Highland Border and Midland Valley–South Mayo, now in north-western Europe) which were becoming separated from the main Laurentian Craton by an active northward subduction zone. There was also probable sea-floor spreading in the western sector of the ocean (Fig. 5.3), which caused the

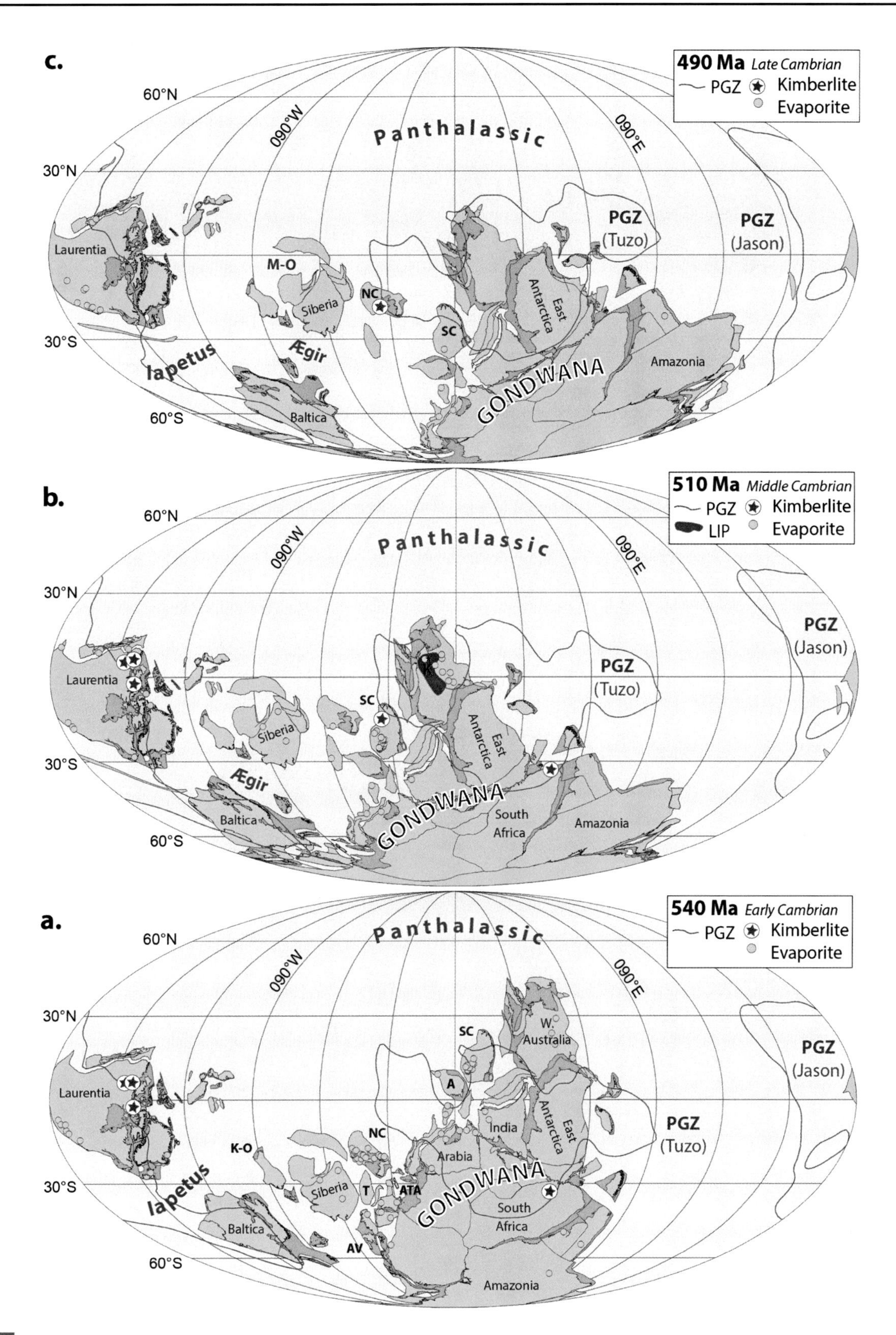

Fig. 5.1. Outline Earth geography at (a) Early Cambrian at 540 Ma (not very robust), (b) the Middle Cambrian at 510 Ma, and (c) the Late Cambrian at 490 Ma, including the postulated plate boundaries, outlines of the major crustal units, and the more substantial oceans. A, Annamia; ATA, Armorican Terrane Assemblage; AV, Avalonia; K (in (b)), Kalkarindji Large Igneous Province; K-O, Kolyma–Omolon; M-O, Mongol–Okhotsk Ocean; NC, North China; PGZ, plume generation zone; SC, South China; T, Tarim.

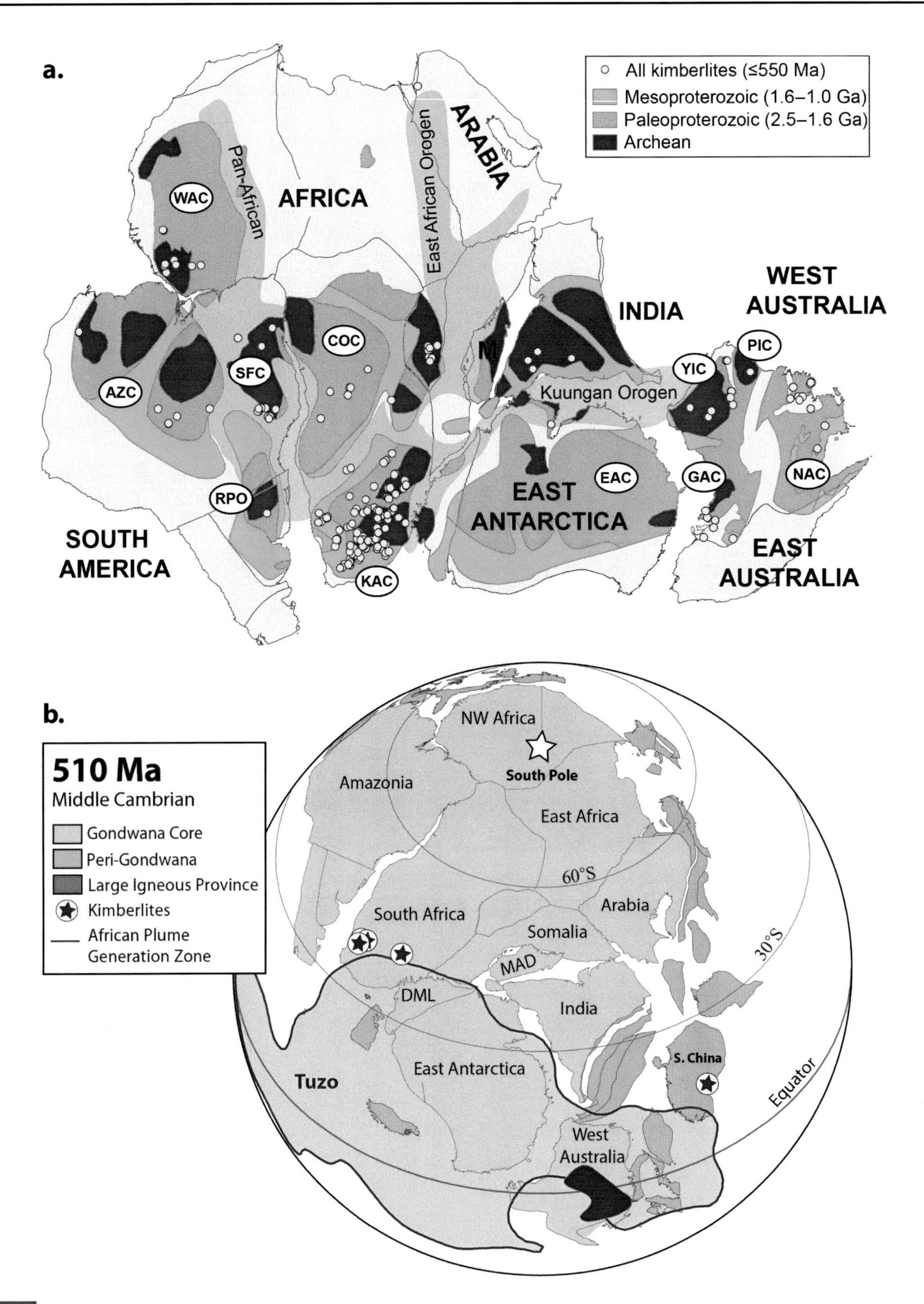

Fig. 5.2 (a) The Precambrian cratons and the Phanerozoic distribution of kimberlites in the Gondwana area plotted on an end-Palaeozoic base map, demonstrating the many Late Precambrian and Early Cambrian orogenic belts (in blue) that welded the cratons into the united

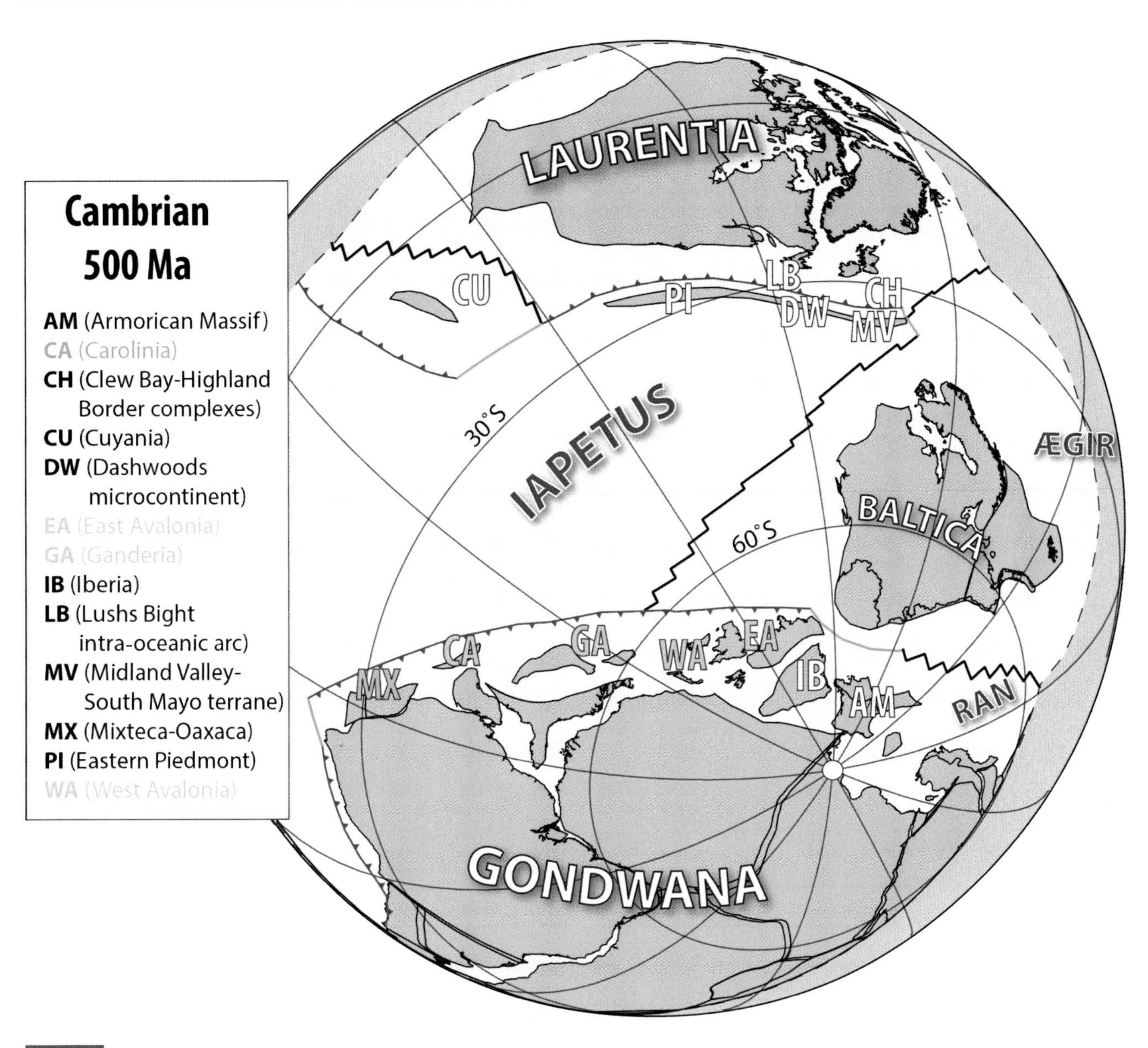

Fig. 5.3. Late Middle Cambrian reconstruction at 500 Ma. Solid blue lines are subduction zones with teeth on the upper plate, black lines are spreading centres, and green lines are transform plate margins. The dashed boundary to the grey areas in the reconstruction margins is an arbitrary perimeter (Domeier, 2015) marking the outer limit of the Iapetus and Ran domains.

Cuyania Terrane (often termed the Precordillera Terrane and now in Argentina) to move away from the rest of the string and make its lonely way across to its eventual position in the south-western South American sector of Gondwana, which it reached in the Late Ordovician at around 455 Ma (Domeier, 2015).

The oceanic domains that undoubtedly existed outside today's Atlantic area in the Cambrian are poorly known

Fig. 5.2 (*cont.*) very large continent. The Cadomian Orogen (not shown) lay in northern Africa and extended to southern Europe. AZC, Amazonian Craton; COC, Congo Craton; EAC, East Antarctic Craton; GAC, Gawker Craton; KAC, Kalahari Craton: M, Madagascar; NAC, North Australia Craton; PIC, Pilbara Craton; RPO, Rio de La Plata Craton; SFC, São Francisco Craton; WAC, West Africa Craton; YIC, Yilgara Craton. From Torsvik & Cocks (2013). (b) New palaeomagnetic reconstruction of core Gondwana at 510 Ma (orthogonal projection), demonstrating that Gondwana stretched from over the South Pole in North-West Africa to north of the Equator in West Australia, and that kimberlites in South Africa and South China and the substantial 511 Ma Kalkarindji Large Igneous Province in Australia erupted near-vertically above the margins of the Tuzo Plume Generation Zone.

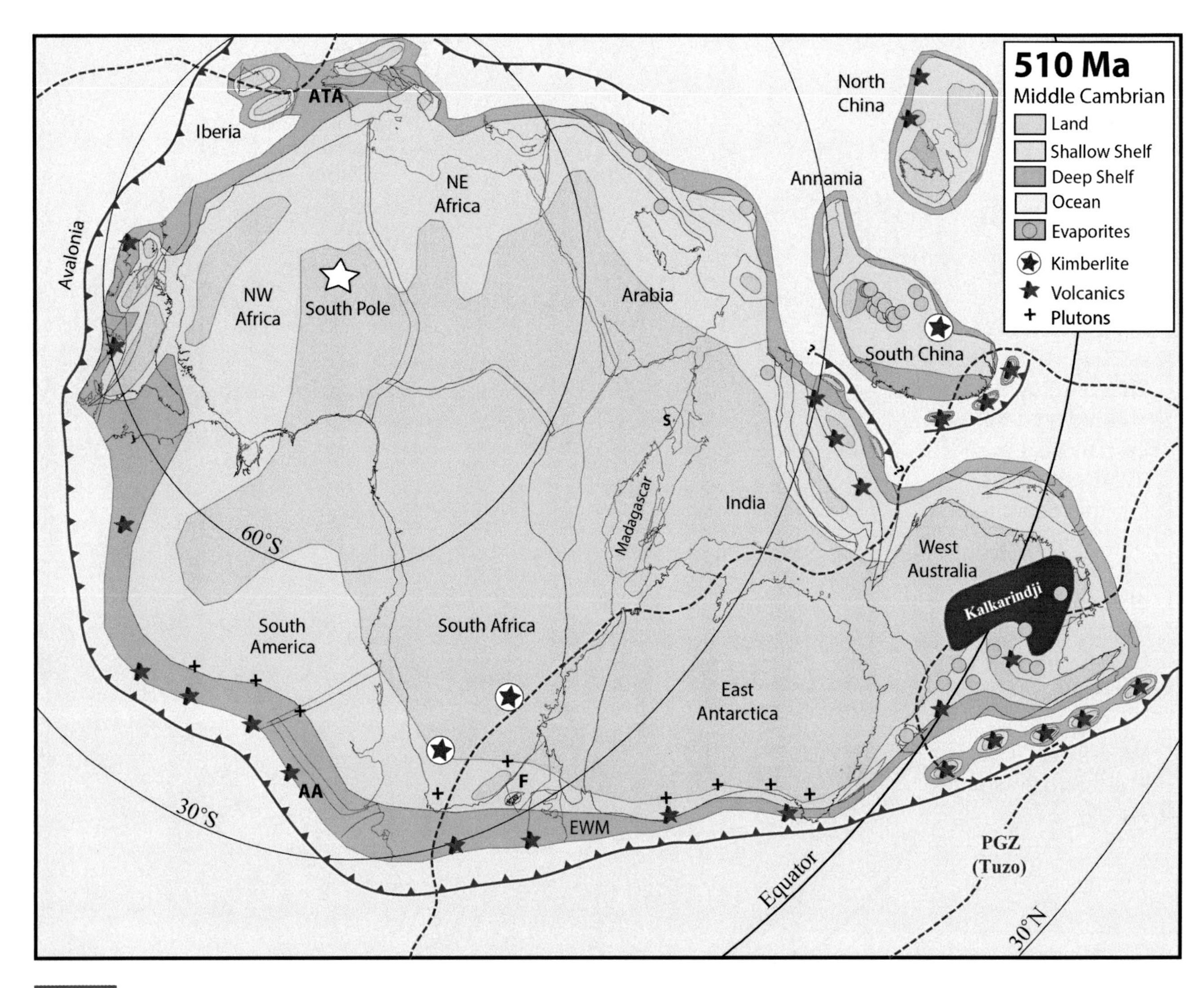

Fig. 5.4. Gondwana and adjacent areas at 510 Ma (Middle Cambrian), showing the pattern of lands and seas. Dotted red lines are the plume generation zones (PGZ), and pink dots are evaporites. Solid blue lines are subduction zones, with teeth on the upper plate. The strings of island arcs shown are diagrammatic, since the individual extent of each unit within the arcs and the extent to which each unit was above sea level are uncertain. The Kalkarindji LIP in Australia erupted at 511 Ma. Middle Cambrian kimberlites occur in southern Africa and Dahongshan, South China. AA, Arequipa–Antofalla Block; ATA, Armorican Terrane Assemblage; EWM, Ellsworth–Whitmore Mountains; F, Falkland Islands; S, Seychelles.

and have been interpreted in many ways by different authors. For example, although we show in Fig. 5.5 an oceanic area between Gondwana and Siberia occupied only by the North China, South China, and Tarim continents in the Middle Cambrian at 510 Ma, that area must also have hosted the several microcontinents with Precambrian cores which today form different units within the Kazakh terrane assemblage (Fig. 3.7), as described in the following chapters.

Gondwana. Gondwana was an enormous continent, more than three times the size of any other in the Early Palaeozoic, around 100×10^6 km^2 or about 64% of all land areas today. It included Africa, South America, India, Arabia, Antarctica, and Australasia, as well as many smaller units at its margins, such as Florida, the Taurides of Turkey, and various parts of central Asia and China, notably the Tibetan terranes, and Sibumasu. What subsequently became the independent Avalonia Terrane remained an integral part of the Gondwana Craton during the Cambrian, probably outboard of Gondwana's North-West Africa and Amazonia sectors. However, although rifting between Avalonia and Gondwana started near the end of the Cambrian, analysis of the stratigraphy and faunas on both sides indicates that detachment did not occur until the Early Ordovician (Cocks & Fortey, 2009). To the east of Avalonia, along the North African part of the

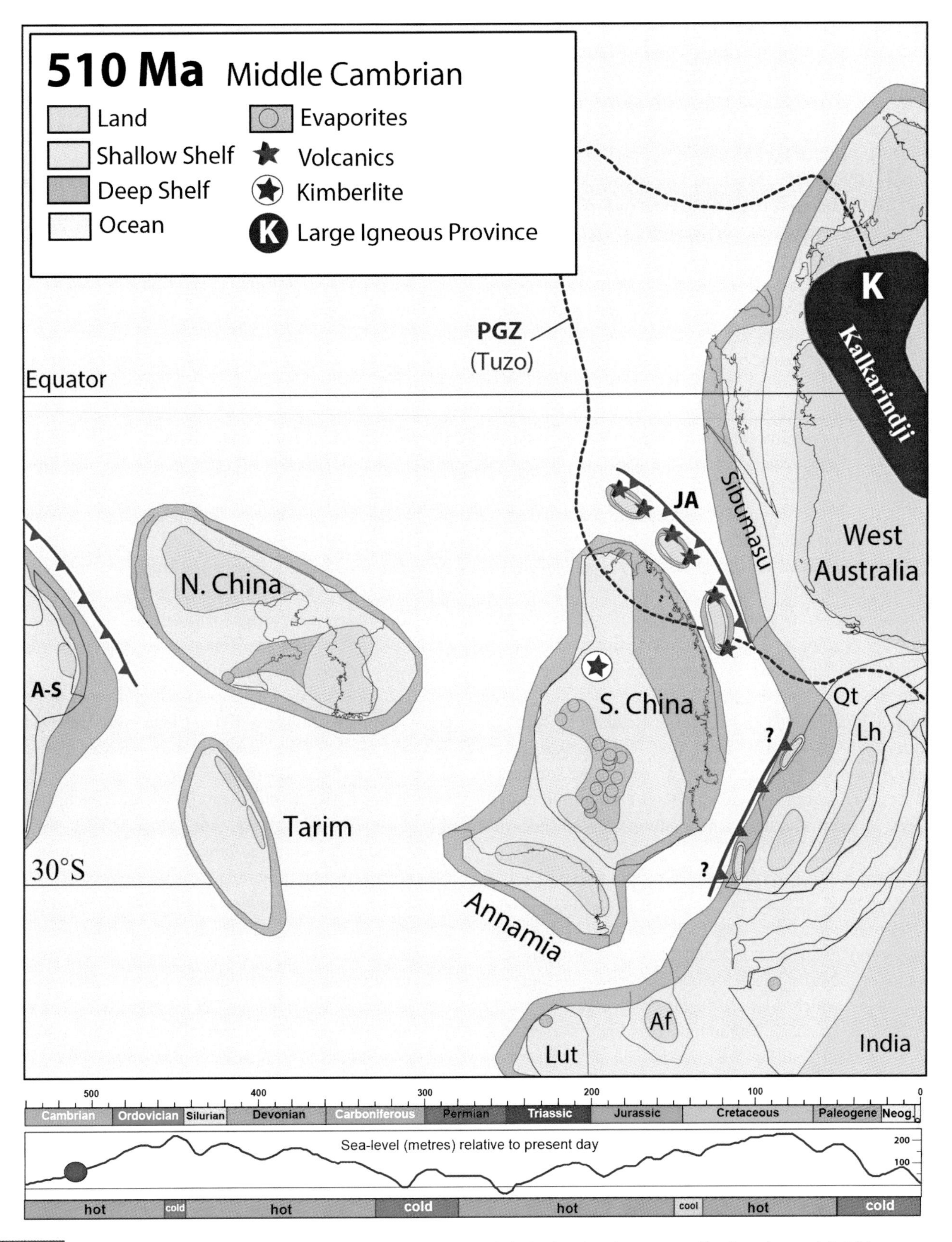

Fig. 5.5. Eastern Asia terranes and continents at 510 Ma (Middle Cambrian), showing the pattern of lands and seas. Af, Afghan Terrane; A-S, the Altai–Sayan sector of Siberia; JA, Japanese arcs; Lh, Lhasa Terrane; Qt, Qiantang Terrane. Dotted red line is the African (Tuzo) plume generation zone (PGZ). The kimberlite is at Dahongshan, South China. Solid blue lines are subduction zones, with teeth on the upper plate. Bottom: Phanerozoic time scale and sea-level variations (Haq & Al-Qahtani, 2005; Haq & Shutter, 2008) and icehouse (cold) and greenhouse (hot) conditions.

Gondwanan margin and still then an integral part of the latter, there lay the Armorican Terrane Assemblage (ATA, often more simply termed Armorica, although it was not a single united terrane), whose tectonically complex and varied parts today make up most of France, the Iberian Peninsula, parts of Italy (Sardinia, Adria, and Apulia), and probably Bohemia and Moldanubia. However, all the sectors in the ATA remained as part of the Gondwana Craton until at least the Late Ordovician, and many did not leave until later than that (Fig. 8.9).

Continuing from parts of the Late Neoproterozoic and earliest Cambrian orogenic belts, there was a continuous subduction zone along the very extensive active margin, which apparently extended nearly all the way around Gondwana from North-West Africa southwards round South America, South Africa, East Antarctica, and to the eastern edge of West Australia (Fig. 5.4). There were also smaller and shorter subduction zones to the north of Arabia and probably north of India. At 511 Ma the Kalkarindji Large Igneous Province (formerly called the Antrim Plateau Basalts) was intruded into a very substantial area ($>2 \times 10^6$ km^2) of Gondwana in north-western Australia (Glass & Phillips, 2006). In eastern Australia, the Delamerian Orogeny was variably active, with ages ranging from Late Neoproterozoic in some areas, including Tasmania, but not starting in the Thomson Orogen to the north until the Middle Cambrian after 514 Ma, but its direct causes are not yet completely understood (Fergusson & Henderson, 2015). In South America, a small ocean (the Punoviscana Ocean) closed in the Early Cambrian, thus causing the substantial and elongate Arequipa–Antofalla Block (AA on Fig. 5.4) to its west to unite with the main Gondwana Craton (Escayola et al., 2011).

Eastern Asia. Much of eastern Asia was affected by Late Neoproterozoic (Ediacaran) to Early Cambrian tectonic activity, including the Early (539 Ma) and Middle (526 Ma) Cambrian volcanics in South China, and there are remnants of a Cambrian island arc in rocks within the Song Ma Suture Zone, which itself accreted to South China before the Devonian (Fig. 5.5). Within modern China, the Junggar Terrane (Unit 450 in Fig. 3.8) was formed by the accretion of Cambrian and later island arcs (Xiao et al., 2009b), and the Qaidam–Qilian Terrane includes Middle Cambrian ophiolites which were probably formed near the south-western margin of the North China continent. Lower Cambrian ignimbrites occur in the east of Kunlun (Unit 457: Daukeev et al., 2002). Between the areas of the former North and South China continents today is the Qinling Orogenic Belt, which includes the Songshugou ultramafic massif, which has been dated as Middle Cambrian (510 Ma); although it is uncertain exactly where that massif lay during the Cambrian (Cocks & Torsvik, 2013).

Not shown on Fig. 5.5 are the Hutag Uul–Songliao Terrane of Mongolia (Unit 453), in which there are Late Cambrian to Ordovician (498–461 Ma) near-trench plutons as well as juvenile arc crust (Jian et al., 2008); and the nearby Khanka–Jiamusu–Bureya Terrane (Unit 454), into which Early to Middle Cambrian (525–515 Ma) granitoids were intruded. Both units were substantial, but their detailed situations in the Cambrian are poorly constrained, although the two were close to each other and probably accreted to the Central Mongolia Assemblage by the Late Silurian, subsequently to become parts of the Amuria continent (Fig. 7.7).

Rocks originally at the margins of both Gondwana and North China, but now in the Himalayas, have yielded zircons indicating a tectonic event at 500 Ma, but it is not agreed where North China was situated in the Cambrian. McKenzie et al. (2011) placed North China as an integral part of Gondwana bordering today's north-eastern India; however, we consider it was a separate and independent continent and place it at some distance away from Gondwana (Fig. 5.5). Annamia and South China were combined into a single continent which was also situated off north-central Gondwana, perhaps near Afghanistan; however, the western (Thai) part of the Annamia continent is made up of younger rocks which did not exist in the Cambrian (Ridd et al., 2011). Whether the eastward-moving strike-slip faulting that transported the united Annamia–South China from Afghanistan to Australia (~3000 km) commenced in the Cambrian or at some time in the Ordovician is unconstrained; we show it from our Early Ordovician (480 Ma) reconstruction (Fig. 6.4) onwards. At some time before 520 Ma, the Cathaysian margin of south-eastern South China changed from passive to active, and the first parts of the oceanic volcanic island arcs which are today identified as the Japanese terranes (JA in Fig. 5.5) were formed offshore from there (Isozaki et al., 2010).

Laurentia. The Early Cambrian was not marked by any very substantial tectonic events in Laurentia, which straddled the Equator some distance from Gondwana throughout the Cambrian (Fig. 5.6). A passive Laurentian margin existed in northern Canada continuously from the Late Neoproterozoic to the Silurian (Dewing et al., 2004), which extended eastwards to include North Greenland (Bradley, 2008). The western margin was also passive from Mexico to eastern Alaska, although much of it was separated as a large parautochthon, shown on our Silurian and later maps (e.g. Fig. 7.5), until its eventual Mesozoic or Tertiary reunification with the main Laurentian Craton. In the southern Cordillera, the Okanagan High, which bordered the western margin of

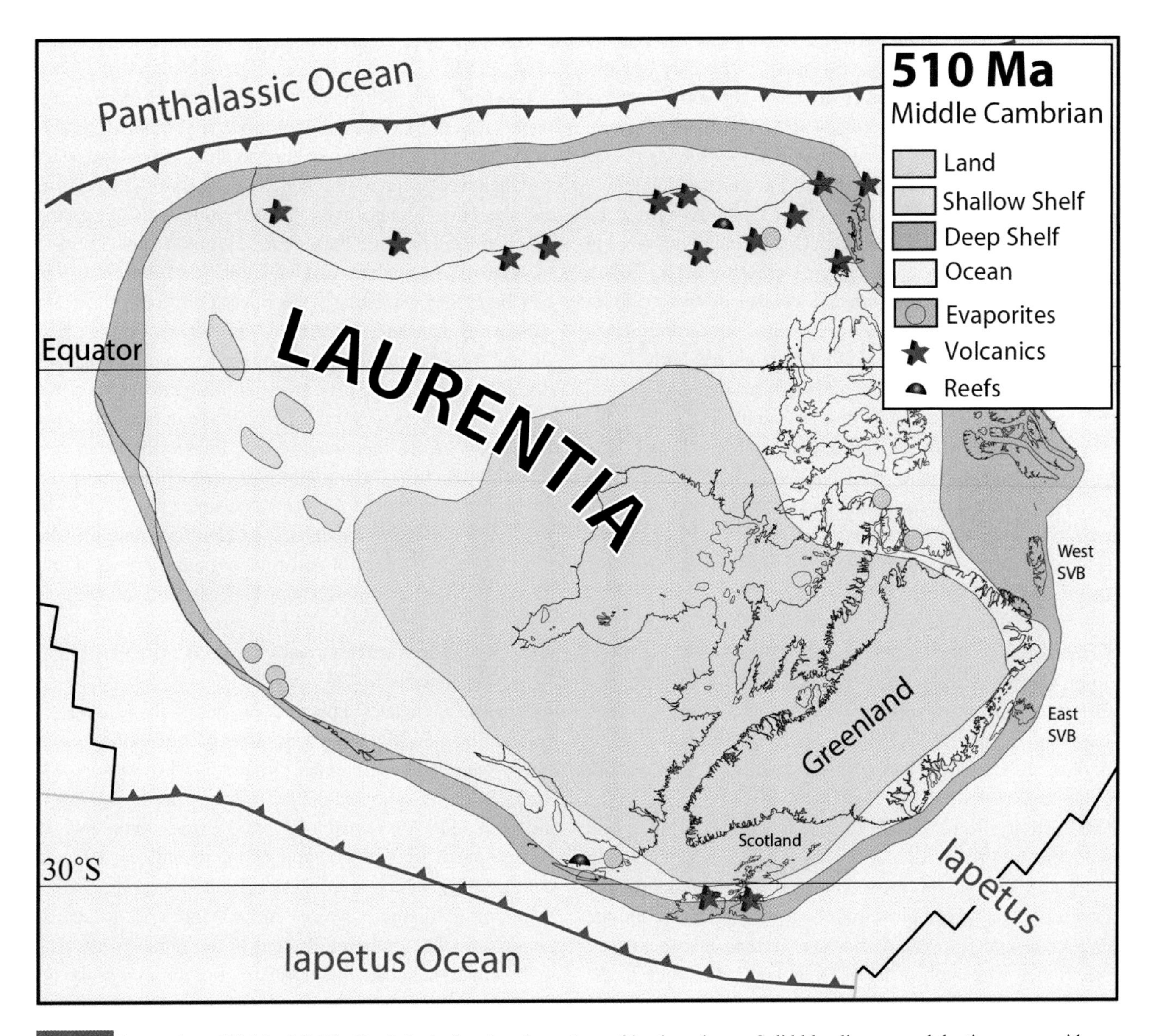

Fig. 5.6. Laurentia at 510 Ma (Middle Cambrian), showing the pattern of lands and seas. Solid blue lines are subduction zones with teeth on the upper plate, black lines are spreading centres and transform faults, green lines are transform plate margins. No kimberlites at this time. SVB, Svalbard.

the craton, separated the relatively flat miogeocline from the shelf edge of the Panthalassic Ocean (Colpron & Nelson, 2006). However, it is uncertain whether or not there was land above sea level in that area and there are few coarse clastic rocks: we do not show any land there on Fig. 5.6. There was also some calc-alkaline igneous activity in both the southern and northern areas of the Cordillera. Those eruptions were submarine and often violent, with associated diatreme breccias with clasts cemented by ferroan carbonates in places (Goodfellow et al., 1995).

To judge by its diverse but undoubtedly Laurentian benthic marine faunas, the relatively small Cuyania (or Precordillera) Terrane of Argentina certainly formed some part of peri-Laurentia in the Cambrian, but, as shown in Fig. 5.3, a spreading centre must have developed between the two at some time in the later Cambrian, with the result that Cuyania followed an individual route until its eventual Late Ordovician situation near the margin of South America (Domeier, 2015) (Fig. 6.2a).

In the Newfoundland area of Laurentia, the Humber Seaway between the Laurentian Craton and the Dashwoods, Lushs Bight, and other terranes continued its steady opening (Fig. 5.3) until, near the end of the Cambrian at 495 Ma (Furongian), subduction started at the western margin of the

Dashwoods area (shown later in Fig. 6.5), and the seaway started to close (Waldron & van Staal, 2001). That culminated in the Latest Cambrian (Furongian) Taconic 1 event at about 495 Ma (van Staal et al., 2009). The first of several island arcs in the Notre Dame Arc within the Dunnage Zone of western Newfoundland also became accreted to Laurentia.

Siberia. In Siberia, there are rocks of Cambrian ages in the fold belts surrounding the Siberian Craton, including those in Mongolia (Astashkin et al., 1995; Dobretsov et al., 2003). There had been very substantial volcanism from 550 to 520 Ma in the Altai–Sayan fold belts which started in the latest Neoproterozoic, and continued on through the Early Cambrian (locally termed the Tommotian, Atdabanian, Botomian, and Toyonian Stages) and Middle Cambrian (Amgan and Mayan Stages), with the thickest arc volcanic material in the Altai–Sayan terranes of Tuva, West Sayan, and Salair (see Fig. 7.7 for their locations), and thick basalts and tuffs in the Altai and adjacent mountains. In the Late Cambrian (Fig. 5.7) that volcanism was restricted to selected areas (the Salair and Kuznetz Alatau terranes). Each group that has presented palaeogeographical reconstructions for the Lower Cambrian (e.g. Zonenshain et al., 1990; Dobretsov et al. 1995; Şengör & Natal'in, 1996) has come up with a very different model, particularly for the Altai–Sayan terranes, and most of these models were themselves composite. However, all of these authors agreed that there were accretionary complexes, ophiolites, island arcs, and other indicators of extreme tectonic activity from Vendian to Middle Cambrian times, and we make no pretence that that part of peri-Siberia shown on our Late Cambrian reconstruction (Fig. 5.7) is anything more than a possibility, compiled from both previous work and the data known to us. We are, however, certain that the Altai–Sayan area was north of the Siberian continental area in the Cambrian since, firstly, the Siberian Craton was certainly inverted by comparison with the present day, and, secondly, the Altai–Sayan area was accreted to the craton before the Devonian, when the craton was still inverted (Fig. 9.11).

Although there are no provably pre-Silurian rocks to confirm it, the Mongol–Okhotsk Ocean between the Central Mongolian Terrane collage and the Siberian Craton probably opened slowly in the Cambrian (Cocks & Torsvik, 2007): however, the distance between the two does not appear to have been very great (Kravchinsky et al., 2001), and we show the two land areas as adjacent at their then eastern ends in Fig. 5.7. Thick Lower Cambrian ocean volcanics with some interbedded Lower to early Middle Cambrian (Amgan Stage) fossiliferous clastics and carbonates lie unconformably under the Silurian in the centre and west of Tuva: only in the north-east of Tuva (in the Kidrik River sections) is there a full Cambrian sequence which is unconformably overlain by the Upper Ordovician (Astashkin et al., 1995).

Central Asia. Şengör and Natal'in (1996) reconstructed their Lower Palaeozoic Altaid terranes as former constituents of two enormous island arcs, termed the Kipchak Arc, which they believed to have stretched between Baltica and Siberia, and the Tuva–Mongol Arc, situated north-west of the main Siberian Craton in the Palaeozoic. However, analysis of the biological relationships and endemicity of the Ordovician benthic faunas, particularly trilobites, from four of the more substantial 'Kipchak Arc' areas (Altai–Sayan, Chingiz, Chu-Ili, and Tien Shan, all largely now in Kazakhstan) indicates that at least the latter three of the four must have formed parts of the complex peri-Gondwanan collage during the Lower Palaeozoic rather than having any links with either peri-Siberia or Baltica. Only Altai–Sayan (which was more than a single terrane in the Lower Palaeozoic: Fig. 7.7) shows substantial Siberian elements in its benthic fauna, and thus probably did form part of peri-Siberia in those times (Fortey & Cocks, 2003). Therefore the Kipchak Arc, although an elegant concept, probably did not exist and is not mentioned again here. The Kazakh terranes are further discussed in the Ordovician chapter below.

Baltica. In the latest Precambrian, during the Ediacaran at around 560–550 Ma, today's northern part of Baltica was an active margin (the Timanian Orogeny), whose suture with the main Proterozoic and earlier Baltica Craton is shown on the later (510 Ma) map (Fig. 5.8). That event enlarged Baltica much by the accretion of microcontinental blocks, the largest of which were Timan–Pechora and the Barents Sea (Unit 350), and Novaya Zemlya (Unit 373), More contentious are the Cambrian placings of the Kara Terrane (Unit 417, which includes the northern Taimyr Peninsula and Severnaya Zemlya, both now in Arctic Russia), and the existence or otherwise of island arcs in one or more of the oceans surrounding Baltica. We show Kara near Baltica (Fig. 5.8) and there are some palaeomagnetic data that support its latitudinal positioning, but they are not very robust. During most of the Cambrian, there was land which would have been identified then as two continents, the larger Sarmatia, over today's south-east of the Precambrian craton, and the smaller Fennoscandia, which surrounded much of today's Baltic Sea area.

In the Uralian sector of Baltica (Fig. 5.8), Zonenshain et al. (1990) showed substantial Late Cambrian thrusting and folding offshore of the Urals, which was caused by the collision of the Baltic margin with 'island arcs and some microcontinents'. That is supported by the sub-Ordovician angular unconformities overlying Lower Cambrian rocks west of the Urals, but the detailed ages and causes of those

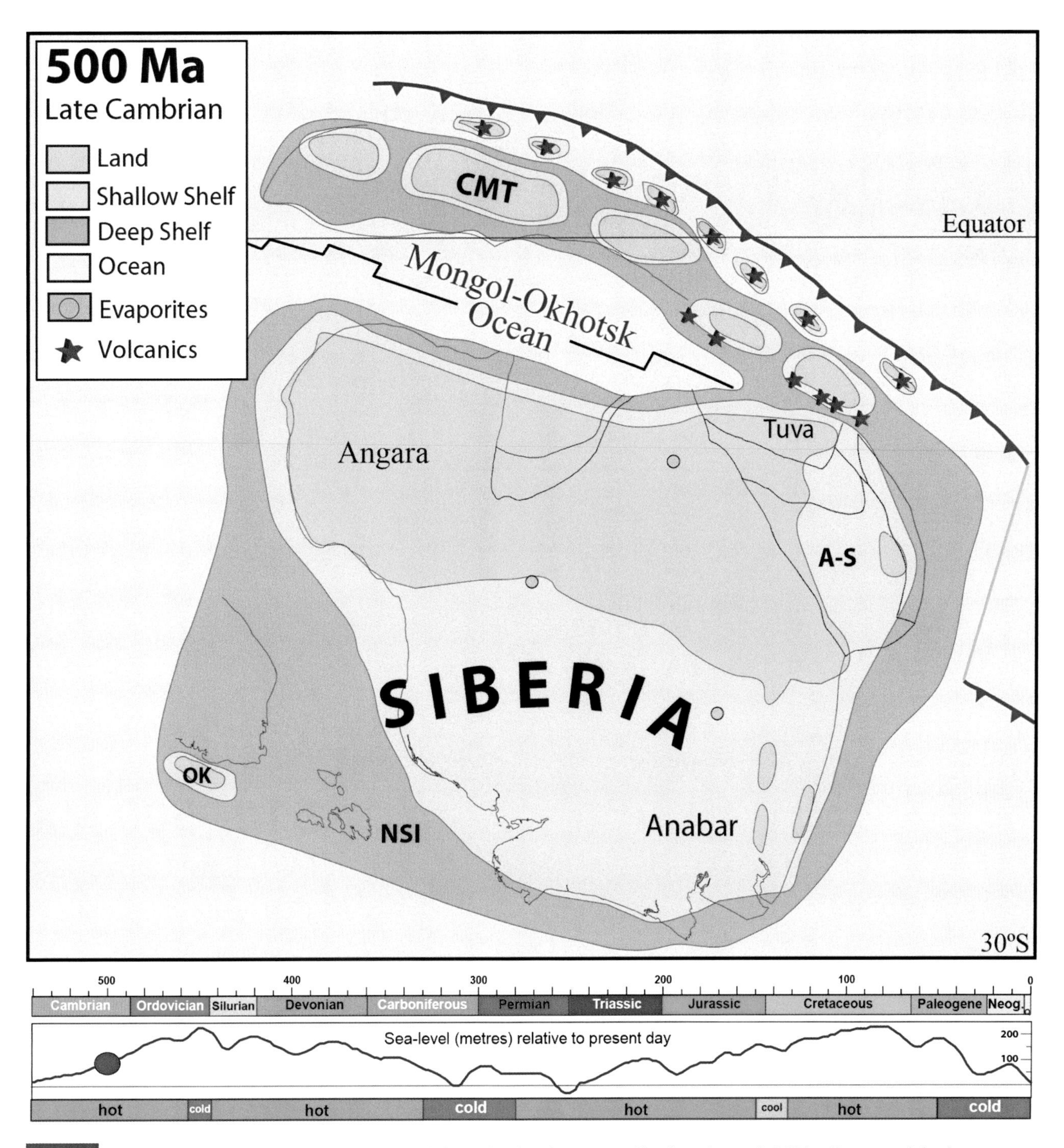

Fig. 5.7. Siberia and peri-Siberia at 500 Ma (Late Cambrian), showing the pattern of lands and seas. Solid blue lines are subduction zones with teeth on the upper plate, black lines are spreading centres and transform faults, and green lines are transform plate margins. No known kimberlites. A-S, Altai–Sayan; CMT, Central Mongolian Terranes; NSI, New Siberian Islands; OK, Okhotsk Massif. Bottom: Phanerozoic time scale and sea-level variations (Haq & Al-Qahtani, 2005; Haq & Shutter, 2008) and icehouse (cold) and greenhouse (hot) conditions.

orogenic events remain somewhat obscure. Due to Baltica's substantial rotation in the Ordovician, the terrane that lay opposite the Uralian margin of Baltica in the Cambrian was probably the Armorican sector of the gigantic Gondwanan continent (Fig. 5.8), rather than any part of the complex collage that makes up the Altaids of Central Asia today.

In contrast to the relatively stable conditions within the centre of Baltica during most of the Early Palaeozoic, at all

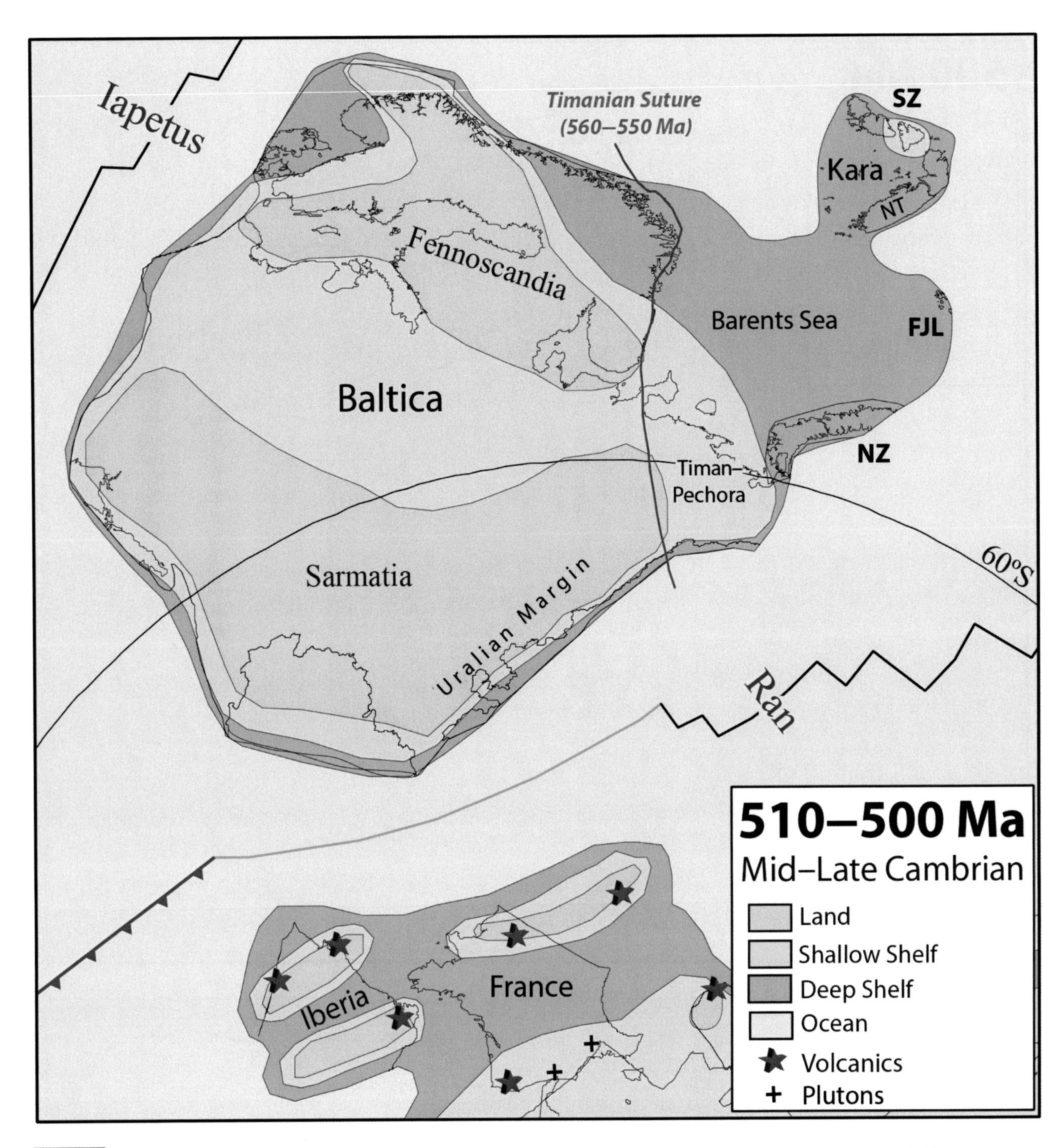

Fig. 5.8. Baltica and surrounding areas at 510–500 Ma (Middle Cambrian), showing the pattern of lands and seas, including the continental areas of Fennoscandia and Sarmatia. Solid blue line is subduction zone with teeth on the upper plate, black lines are spreading centres, and green line is transform plate margins. The Late Precambrian (*c.* 560–550 Ma) was a period of active accretion, the Timanian Orogeny, in which various microcontinental blocks in the Timan–Pechora, northern Ural, and Novaya Zemlya areas of north-west Russia were united with Baltica to form a much expanded continental area. FJL, Franz Josef Land; NT, North Taimyr; NZ, Novaya Zemlya; SZ, Severnaya Zemlya.

of its margins there was often violent tectonic activity at one time or another during the same long period. Thus on our maps (Figs. 5.8 and 6.11) there are some areas of both shallow and deep shelf seas shown that are outside the currently preserved margins of Baltica, and apparently even overlying ocean floors. However, the Lower Palaeozoic crust that must have supported those shelf areas has been lost, either due to subsequent subduction or possibly displaced laterally to other areas, which we have not been able to identify and restore to their original positions (Cocks & Torsvik, 2005).

Because of Baltica's rotation, which started in Cambrian times but peaked in the Early Ordovician (Fig. 2.10), there was progressive and substantial strike-slip movement in the Ran Ocean between Baltica and Gondwana. The Ran initially developed as an arm of the Iapetus Ocean (Hartz & Torsvik, 2002). In our reconstructions for the later Palaeozoic, the Ran is shown as united with the Rheic Ocean, but the latter name cannot properly be used prior to the Early Ordovician, since the Rheic only came into existence after Avalonia had separated from Gondwana at 490 Ma (Figs. 6.1 and 6.2).

Facies and Faunas

'The Cambrian Explosion'. Although trilobites are usually thought of as the typical Cambrian fossils, the earliest trilobites are only known from rocks more than 6 million years younger than the basal Cambrian boundary. In the first half of the Cambrian there was a very substantial radiation in the animal kingdom, many of the species of which were of quite small sizes (and are thus termed 'Small Shelly Fossils'), into a variety of metazoan groups, many of whose direct descendants are still alive today, although other groups subsequently became extinct. Although that radiation is often termed 'The Cambrian Explosion', the name is hardly appropriate for an event that took as long as it did (about 15 million years); nevertheless, it was still an exceptional period of evolutionary expansion which justifies its publicity. As new animals arose, they needed habitats in which to live, and thus often ousted the existing inhabitants, which in turn either became extinct or displaced others from their ecological niches. The individual distributions of animals around the globe therefore changed immensely during the Cambrian, as well as forming new benthic and planktonic communities. The massive Early Cambrian faunal radiations also caused a fundamental change in the prevalent carbonate-dominated sediments on the shallow seas bordering or upon the various continental cratons, with the characteristic Late Proterozoic and earliest Cambrian soft substrates sealed by microbial mats gradually giving way to more 'modern' sediments, which are usually much reworked by bioturbation (Dornbos & Bottjer, 2000). Evidence for those global radiations is best seen in Laurentia, partly because of its equatorial position (Fig. 5.1), which naturally maximised the biodiversity, and partly because of the large amount of research published on North American fossil faunas during the past two centuries.

Until recently, most palaeontologists have thought that there was the Cambrian Explosion in the first half of the Cambrian, followed by less dynamic radiation in the second half of the era and then by the Great Ordovician Biodiversification Event in the Middle and much of the Late Ordovician (Webby et al., 2004). However, as more and more exceptionally preserved fossils have come to light both from earlier strata (such as the Chengjiang fauna from China) and also from later periods, that sequence is now known to be less correct, since radiations in many phyla are now known to have proceeded without pause, although at variable rates, throughout all of the Cambrian and also through the Ordovician until the end-Ordovician extinction. For example, the Early Ordovician (Tremadocian) Fezouata Biota in southern Morocco includes many Cambrian holdovers (such as the anomalocaridid arthropods also known from the Middle Cambrian Burgess Shale of Canada) along with a surprising number of newly evolved crown group taxa which are the ancestors of better-known later Ordovician forms (Van Roy et al., 2015).

Gondwana. Gondwana was by far the largest continent in the Cambrian, but its craton was variably overlain by shallow shelf seas, which varied rapidly, due partly to eustatic changes in sea level (Fig. 5.5) but more substantially to local tectonics in the different regions. There were numerous separate land areas around and upon the Gondwana Craton, although the main continental land area stretched uninterrupted from the Cambrian South Pole, which lay under North Africa, to north of the Equator in the Australasian area (Fig. 5.4). Rocks laid down in the shallow shelf seas along the Gondwanan margins and over the craton at successive times in the Cambrian have yielded many trilobites (Álvaro et al., 2003), and the abundant benthic polymerid trilobites have been divided into three groups on a global basis, termed the Redlichiid, Olenellid, and Bigotinid realms. All the lower-latitude parts of Gondwana were within the Redlichiid Realm, and the bigotinids were in the higher latitudes, whilst the Olenellid Realm was centred on Laurentia, which was also almost entirely tropical (Fig. 5.1). The intermediate-latitude Australian Redlichiid Realm trilobites were more diverse than those in Europe, and varied between the various shallow marine basins there throughout the Cambrian (e.g. Shergold, 1991) (Fig. 5.4). In contrast, the sponge-like archaeocyathids were more cosmopolitan, and their genera are broadly comparable between Australasia and Laurentia, as were the brachiopods. However, since the latter were mostly inarticulated, and it is well known that their larval stages lived longer than those of the articulated brachiopods (which became more dominant from the Ordovician onwards), the Cambrian brachiopods demonstrated less provincial differentiation than the trilobites (Cocks, 2011).

Central and Eastern Asia. The Cambrian to Early Ordovician brachiopod faunas from the Chu-Ili and Karatau–Naryn terranes (Units 460 and 466) of Kazakhstan were very similar to some in South China (Holmer et al., 2001); however, those similarities probably reflect the comparable low palaeolatitudes of the two terrane groups (Fig. 5.1) and the lengthy larval stages of most inarticulated brachiopods, rather than the close proximity of those terranes in the Cambrian. The presence of the same species of trilobites such as *Neoanomocarella*, *Parablackwelderia*, *Sudanomocarina*, *Fuchouia*, and *Redlichia* in both the Himalaya and South China (N.C. Hughes pers. comm., 2011) supports the positioning of South China relatively close to Gondwana in the Cambrian, as shown here.

South China moved northward from a palaeolatitude ranging from 15° to 35° S in the Middle Cambrian (Fig. 5.5) to straddle the Equator by the Late Ordovician (Fig. 6.4) and was entirely north of the Equator by the Late Silurian (Fig. 7.1), but Gondwana moved much less during that period, and thus there must have been considerable strike-slip movement in the oceans between those two continents. Consequently, South China moved from being offboard from India to near the Sibumasu and Australian sectors of Gondwana during the Cambrian, a movement reflected in both the changing facies (with a progressive increase in limestones as the latitudes decreased) and also the increasing faunal diversity with time in South China. However, as can be seen by analysis of the conodonts and the trilobites (McKenzie et al., 2011), North China (including Korea), South China, and the Himalaya were all in a largely similar faunal province in the Cambrian, and thus North China (and probably Tarim, Fig. 5.5) was not too distant from both the united South China–Annamia and also Gondwana (Cocks & Torsvik, 2013).

Laurentia. The shelf margin of the Laurentian Craton has preserved within it one of the most famous exceptionally well preserved fossil deposits (Lagerstätten) anywhere, the Burgess Shale of British Columbia, Canada, which has yielded a large variety of Middle Cambrian animals with soft parts which are rarely found as fossils. Much of the craton was flooded at different times within the Cambrian (Fig. 5.6), and relatively small sea-level changes caused numerous transgressions and regressions which brought newly evolved offshore trilobite species inshore, where their diversification has been used to define successive stratigraphic divisions termed 'biomeres'. The biomeres have been used in the past as correlation tools; although, since the transgressions were not all instantaneous, some of the alleged 'correlations' have subsequently been proved to be diachronous.

The Early Cambrian *Olenellus* trilobite Fauna was well established over much of the Laurentian Craton; for example, near the passive margin in Ellesmere Island (Dewing et al., 2004), but since the characteristic genus also occurs in the adjacent but then quite separate Arctic Alaska–Chukotka Microcontinent, that fauna is not so unequivocally Laurentian as is often thought. A shallow- to deep-water sequence of Late Cambrian trilobite communities at the craton margin is preserved in limestone boulders contained as sedimentary particles within off-shelf sediments of the Cow Head Group in Newfoundland. There is another shallow- to deep-water sequence at the western margin of the craton in Nevada, and, although the basinal trilobites there had been thought of as having 'Asian' affinities, they have subsequently been deemed cosmopolitan, as is more usually the case with deeper-water biota (Cocks & Fortey, 1982; Fig. 2.20). However, the echinoderms reflect how distinct and diverse the low-latitude Laurentian faunas were by comparison with those from Gondwana, Baltica, and other areas at higher palaeolatitudes (Lefebvre & Fatka, 2003). That endorses the conclusion, after assessment of a variety of benthic faunas and supported by analysis of the palaeomagnetism, that the width of the Iapetus Ocean was very substantial by the beginning of the Cambrian, perhaps as much as 6,500 km between Laurentia and West Gondwana (Amazonia) and about 2,000 km between Laurentia (Greenland–Scotland) and Baltica (Fig. 5.1a).

The Late Middle Cambrian trilobites from the Farewell Terrane, today in Alaska (Unit 155), originated in a cool-water outer shelf setting, and strongly resemble Siberian faunas of the same age (and also have some similarity to faunas in Baltica) but they have nothing in common with those in Laurentia. That reinforces the suggestion that Farewell was at some distance from Laurentia at the time, and it did not accrete to North America until the Jurassic (Cocks & Torsvik, 2011). However, its Cambrian position is poorly constrained, although we tentatively show it as near the Arctic Alaska–Chukotka Microcontinent in the Silurian (Fig. 7.5) and the Late Carboniferous (Fig. 9.6).

Siberia. Siberia was also at some distance from Laurentia, but at relatively low tropical latitudes (Fig. 5.1). Deposition continued with scarcely a break between latest Neoproterozoic (locally termed Riphean and Vendian) sediments and the earliest Cambrian, forming rocks which are still flat-lying and unmetamorphosed today. The concomitant warmth encouraged biological speciation, and the Cambrian of the Siberian Craton is famous for the variety and preservation of its fossils, in particular the Small Shelly Fossils of uncertain biological affinity near the base of the Cambrian. Another striking feature is the number of archaeocyathan reefs upon and around the craton, which are chiefly of Early Cambrian (Atdabanian) age; for example, those which are well developed in the Anabar Massif (Fig. 5.7). The areas of submerged shelf were very extensive, particularly over today's northern part of the craton; for example, there was no known large land area in the Anabar region (Keller & Predtechensky, 1968). Nevertheless, there seems to have been a substantial Cambrian continental landmass in the then north of Siberia, and the adjacent Altai–Sayan and Mongolian peri-Siberian terranes would have had substantial areas of mountain and highland near their tectonically active margins.

It is difficult to correlate the successive shallower-water Siberian Platform trilobite faunal zones with those from elsewhere, and a distinctive Siberian Province was defined by Shergold (1988), which includes the New Siberian Islands (Fig. 5.7) in the Arctic Ocean. There were certainly many endemic trilobites on the inner shelves of Siberia throughout the Cambrian; however, on the deeper shelves there were some genera, for example *Kootenia*, *Erbiella*, *Paradoxides*, and *Hebediscus*, which are related to those from western Gondwana, including Morocco and southern Britain; and there are eight deeper-water trilobite genera which occur in both Siberia and Gondwana in the Early Cambrian and 27 in the Middle Cambrian (Álvaro et al., 2003). The reasons for the Siberian shallower-water endemicity are varied, but are partly due to some taxonomic artefacts (the erection of genera which truly also occur in other terranes but have been identified by other names there), partly because of the relative geographical isolation of Siberia, and partly because Siberia and some other large continents (including Baltica) were covered by relatively widespread reduced oxygen levels in the sea-water ocean floors at various times, which supported olenellid trilobites but not much else. However, by the Late Cambrian there were more faunal links between Siberia and other areas, and trilobites and brachiopods from Severnaya Zemlya, then in the Kara Terrane near Baltica (Fig. 5.8), provided correlation between Siberia, Kara, and Baltica (Rushton et al., 2002).

There are two broad facies divisions in the Cambrian of the Siberian Platform. In today's south-east, the sediments consist largely of dolomites with interbedded anhydrites, and only occasional limestone beds with sparse and endemic trilobites and algal biostromes, and also rocks of terrigenous origin. In contrast, in the north-west, there are bioherms of stromatolites in the Early Cambrian, as well as open-marine limestones and marls throughout the Cambrian, with occasional bituminous shales of deeper-water origin (Pegel, 2000). The total Cambrian rocks thicknesses in the two extensive areas are about the same, at between 1,500 and 2,000 m.

In the Mongolian sector of Siberia, to the then north of the Angara Massif (in the Kasagt–Khairkan Ridge and South Khubsugul Lake areas), the Lower Cambrian is largely complete, with many archaeocyathans and endemic Siberian Small Shelly Fossils in the Tommotian, Atdabanian, and Botomian all well represented. The deposition of those carbonates persisted until the early Middle Cambrian (Amgan Stage), when there were a few interbedded acid volcanics. Most of the Middle and Upper Cambrian rocks are thick and poorly fossiliferous molasse deposits unconformably overlain by Permian rocks (Astashkin et al., 1995). Various parts of the Central Mongolian Terrane Assemblage have typically Siberian faunas, and were therefore not far from the main Siberian continent. The Middle and Upper Cambrian shelly faunas of Siberia, in particular the brachiopods, have more in common with those of North America (Laurentia) than Baltica or Gondwana, but that was probably due just as much to their comparable equatorial palaeolatitudes as to their possible proximity. However, the benthic communities are not as diverse and the fossiliferous beds are not as abundant in Siberia as in the comparable low palaeolatitudes of Laurentia.

Baltica. The Cambrian sediments of Baltoscandia lie unconformably upon the metamorphosed Precambrian rocks of the craton and thus represent a marine transgression, which had started in the Late Ediacaran, onto the peneplained older basement of the Baltic Shield (Cocks & Torsvik, 2005). Judging by the thinness of the Middle Cambrian deposits of Norway, Sweden, and the East Baltic, and the extensive lateral extent of many of the facies, as well as the general lack of coarse clastic rocks, the centre of the continent must have been relatively low in topography (and hence sediment supply) for most of the period. Much of the craton was submerged under shelf seas for long parts of the 54 Myr of the Cambrian, and reflects steadily rising global eustatic sea levels (Fig. 5.8). As a consequence, the Late Cambrian olenid trilobite fauna, which is abundant in Sweden and elsewhere, represents a fauna living in niches under relatively deep water on the shelf, and the same animals are also found in comparable conditions in other continents such as Laurentia and Siberia. It seems probable that the unusually widespread distributions of the olenid faunas were due more to global low oxygenation in the sea water, rather than that the trilobites lived at any great water depths on the Baltica Craton and elsewhere, not least since there is no evidence of any local tectonic activity which would have led to the formation of deeper-water basins there. Baltica was in greater proximity to nearby continents across the narrower Ran Ocean (Fig. 5.8) during this period in comparison with the much wider oceanic separations in the Early Ordovician (which was when there was far greater endemicity within the benthic shelly fauna), and thus the Cambrian distribution of the larvae of the olenids would have been facilitated.

The limited distribution and occurrences of Cambrian articulated brachiopods can also be explained by the relative isolation of Baltica as well as its relatively high palaeolatitude in the southern hemisphere (Fig. 5.1). Only *Oligomys* from the Middle Cambrian and *Orusia* from the Upper Cambrian are known from the well-collected sections in Sweden, and in Novaya Zemlya only *Diraphora* from the uppermost Middle Cambrian and *Billingsella*, *Ocnerorthis*, and *Huenellina* from the Upper Cambrian are recorded. There are no articulated brachiopods at all in the equally well-known St Petersburg and Estonian localities until rocks of Early Ordovician (Floian) age, a lack which may have been due to the lack of original suitable ecological niches in the available biofacies or possibly to the post-mortem diagenetic destruction of the calcitic brachiopod shells by the acidic fluids within the widespread Alum Shales there.

In the centre of Baltica, the two major land areas of Fennoscandia and Sarmatia overlay the Fennoscandian Shield and the Sarmatian Shield (Fig. 5.8). The adjacent shallow-water marine facies include conglomerates and glauconitic sandstones, offshore of which the Middle Cambrian Andrarum Limestone can be traced in an arcuate belt from the west coast of Norway, north-eastwards within the Lower Allochthon of the Scandinavian Caledonides (which consists only of materials derived from Baltica), and curving southwards to the Baltic Sea. The trilobites and brachiopods of that limestone are relatively cosmopolitan, and some of the genera are found as far away as Australia, which suggests that some of the oceans neighbouring Baltica were not nearly as wide as they became by the end of the Cambrian. The deeper shelf facies in Sweden to the south-east of the Andrarum belt consists mainly of the distinctive Alum Shales.

6 Ordovician

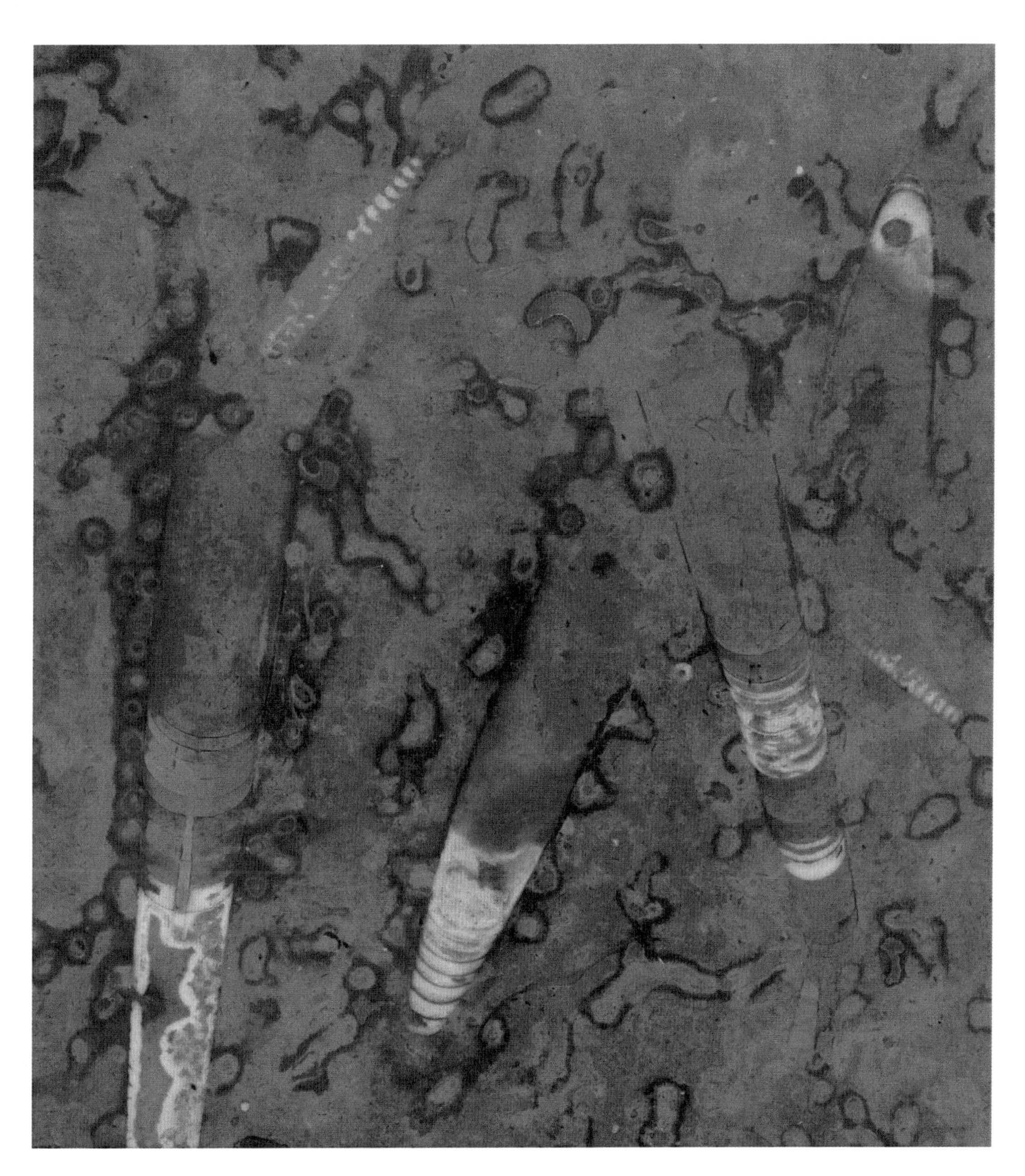

The dead shells of straight cephalopod molluscs abound in the deeper-water Orthoceras Limestone from southern Sweden. The remains of burrows, perhaps dug by arthropods, are picked out by brown secondary deposits.

The Ordovician is noted for the very large amounts of tectonic activity and volcanism that occurred in many regions; and also because of the oceanic distances separating many of the major continents which caused the distinctive faunal provinces that are found in the marine benthos of the continental shelves. There was also exceptional biological radiation, the Great Ordovician Biodiversification Event.

Because Adam Sedgwick and Roderick Murchison had disagreed so vehemently as to where the top of Sedgwick's Cambrian and the base of Murchison's Silurian should be taken, it was not until after the deaths of both men that Charles Lapworth suggested (1879) that the rocks of overlapping ages previously in contention should be named the Ordovician System, a compromise that eventually became universally accepted. Because the originally disputed rocks are largely in Wales, Lapworth named his system after an old Welsh native tribe, the Ordovices. The global stratotype of the Cambrian–Ordovician boundary is now formally defined at the shelf edge of the Laurentian Craton in the Cow Head Peninsula of western Newfoundland, Canada, and is estimated at 487 Ma (Landing et al., 2015). The global distributions of the continents and oceans at four periods within the system, at 480 Ma (Tremadocian), 470 Ma (Floian), 460 Ma (Darriwilian), and 450 Ma (Katian) are shown in Figs. 6.1 and 6.2. The Ordovician is formally divided into seven stages, in ascending order, the Tremadocian, Floian, Dapingian, Darriwilian, Sandbian, Katian, and Hirnantian, which together total up to 55 million years, about the same length of time as the preceding Cambrian. However, many authors still use the successive traditional stage terms of Tremadoc, Arenig, Llanvirn, Llandeilo, Caradoc, and Ashgill, which were originally defined in England and Wales, even for rocks well away from the British Isles.

Tectonics and Igneous Activity

Oceans. The Panthalassic Ocean still dominated more than half the globe. The Iapetus Ocean between Laurentia, on the one side, and Baltica and Gondwana on the other, had reached its maximum width near the end of the Cambrian and closed progressively throughout the Ordovician. Also near the Cambrian–Ordovician boundary time (at about 490 Ma) there was the initial rifting and subsequent widening of the Rheic Ocean between Gondwana and the Avalonia terranes (Fig. 6.2a). By the Middle Ordovician, the Rheic Ocean was of comparable size to Iapetus (Fig. 6.2b), and had merged with the Ran Ocean to its east, and that enlargement continued into the Silurian (Fig. 7.2). The Tornquist Sea (Fig. 6.2a) was essentially an arm of the Iapetus and separated Baltica from the Avalonia and Armorican sectors of Gondwana (Cocks & Fortey, 1982). On the other side of Baltica, between it and Siberia, lay the Ægir Ocean (Hartz & Torsvik, 2002). Further to the east, in the large area bounded by the continents of Siberia, Gondwana, and Baltica, there is no agreed oceanic name, although the name Paleoasian Ocean has been used by some authors; however, we do not recognise it until the Carboniferous (Fig. 9.1). The relatively small Mongol–Okhotsk Ocean lay between Siberia and the Central Mongolian Terrane area (Fig. 6.7).

Gondwana. Further eastwards from the original site of Avalonia in Gondwana, a chain of tonalites and granodiorites was intruded within the Central Iberian Zone of Spain during the Floian at about 478 Ma, indicating activity on that northwestern margin of Gondwana. Near the end of the Ordovician, at about 446 Ma, that activity extended as far eastwards as southern Turkey (Fig. 6.9), where metagranites are known from just north of the Tauride Mountains, and on the opposite Gondwanan margin a string of peri-Gondwanan island arcs developed in Australia (Cocks & Torsvik, 2013). However, between those two active areas the northern Gondwanan margin was largely passive and remained so until the Triassic (Torsvik & Cocks, 2009). That was apart from the final phase of the Bhimphelian Orogeny in the Himalayas, which had peaked before the end of the Cambrian at about 490 Ma, but whose effects rumbled on into the Ordovician in the form of a granite emplaced as late as the Dapingian at 470 Ma.

In South America, there was an elongate Ordovician marine basin on the site of the modern Andes which was floored by continental crust and extended westwards to the Famatina Massif (Fig. 6.3), which had accreted to Gondwana in the Late Neoproterozoic to Early Cambrian. There was also substantial plutonism, ranging from ultramafic intrusive rocks to calc-alkaline granitoids, as well as much associated metamorphism in the adjacent mobile belt in Argentina from the Early Tremadocian at about 490 Ma up to the Late Ordovician at about 450 Ma, all termed the Famatinian Orogeny (Dahlquist et al., 2013). However, deformation had started earlier in that general region at about 495 Ma, before the end of the Cambrian. Metagabbros there are dated from 474 to 452 Ma (Floian to Katian), and their chemical signatures indicate that they were enriched mid-ocean ridge basalts (MORBs) intruded into a back-arc basin. There were contemporary Early Ordovician gabbros and calc-alkaline volcanics in the Famatina area, accompanied by extensive volcanoclastic turbidites from the Dapingian to the Sandbian. Inboard, extensive Famatinian granites were intruded in the proto-Andean area from 486 to 468 Ma (Tremadocian to Dapingian) times (Hervé et al., 2013). In Central America, metagranitoids which include inherited zircons from the

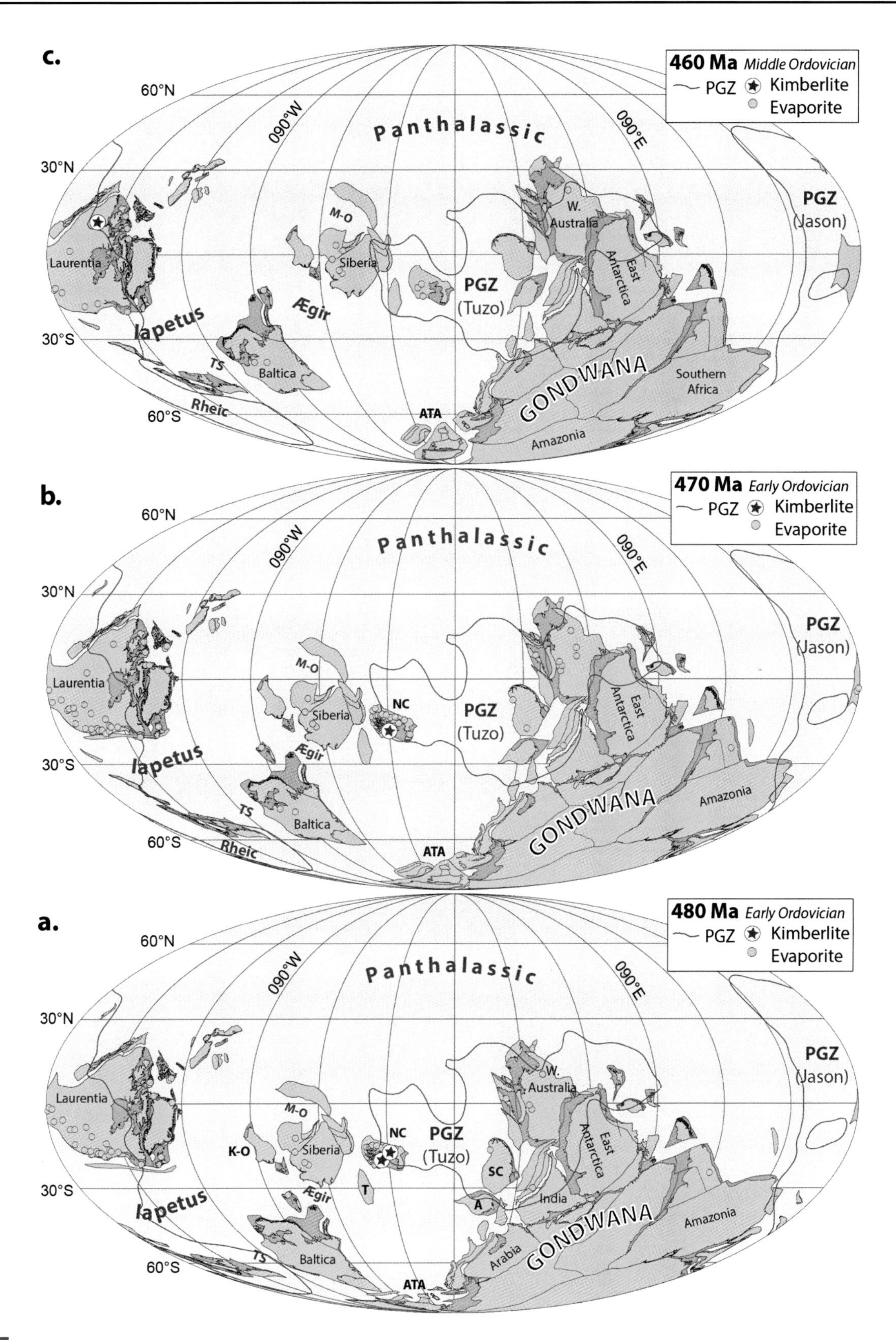

Fig. 6.1 Outline Earth geography in the Early and Middle Ordovician at (a) 480 Ma (Tremadocian), (b) 470 Ma (Dapingian), and (c) 460 Ma (Sandbian), including outlines of the major crustal units, and more substantial oceans. Red lines are plume generation zones (PGZ). Kimberlites are only known in North China and Laurentia. A, Annamia; ATA, Armorican Terrane Assemblage; K-O, Kolyma–Omolon; NC, North China; M-O, Mongol–Okhotsk Ocean; SC, South China; T, Tarim; TS, Tornquist Ocean.

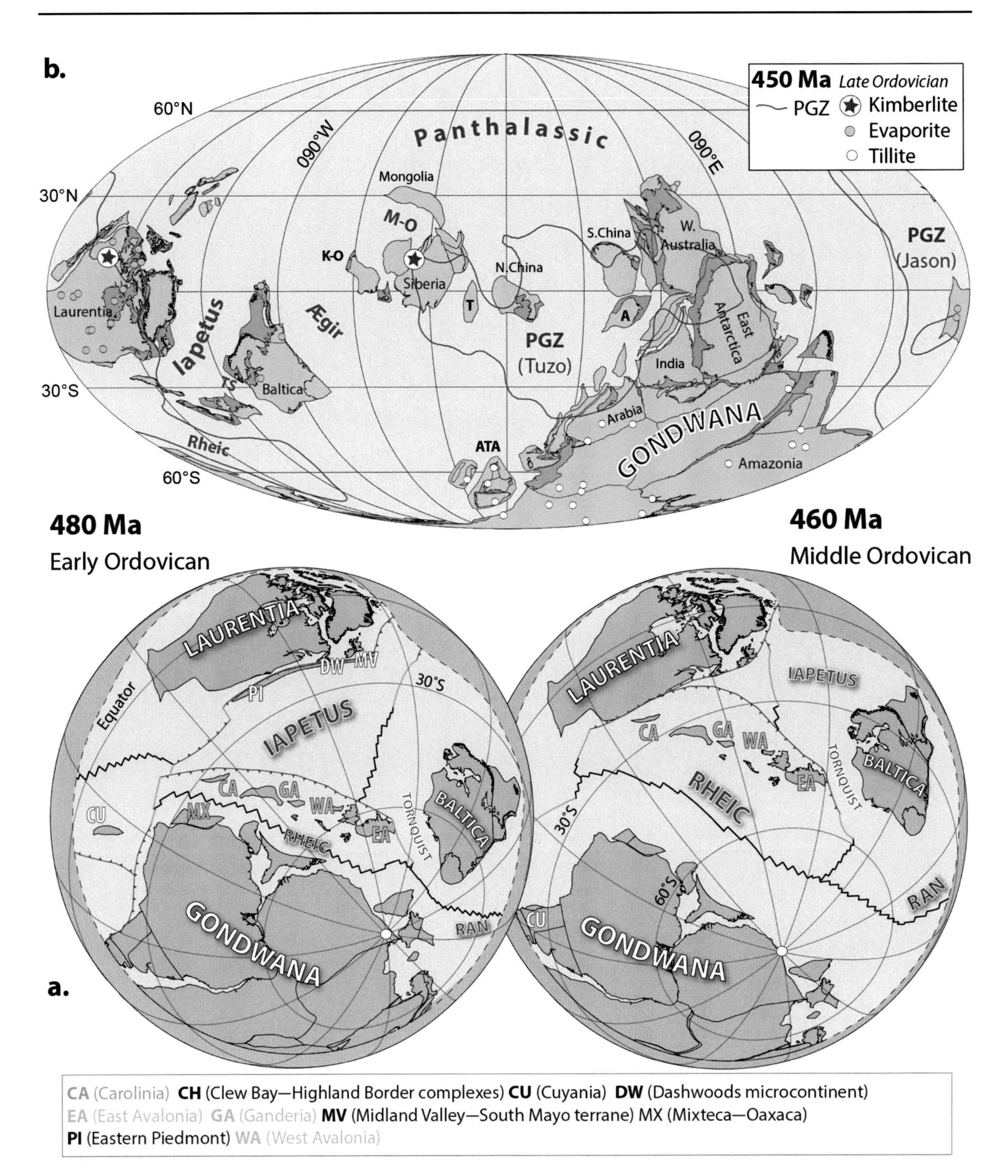

Fig. 6.2 (a) Early Ordovician (480 Ma: Tremadocian) and Middle Ordovician (460 Ma: Darriwilian) reconstructions. Solid blue lines are subduction zones with teeth on the upper plate, black lines are spreading centres, and green lines are transform plate margins. The dashed boundary to the grey areas in the reconstruction margins is an arbitrary perimeter (Domeier, 2015) marking the outer limit of the Iapetus and Rheic/Ran domains. (b) Outline of global Earth geography in Late Ordovician at 450 Ma (Katian), including the outlines of the major crustal units, and more substantial oceans. Kimberlites are known in Laurentia and Siberia. A, Annamia; ATA, Armorican Terrane Assembly; K-O, Kolyma–Omolon; M-O, Mongol–Okhotsk Ocean; PGZ, plume generation zones; T, Tarim; TS, Tornquist Ocean.

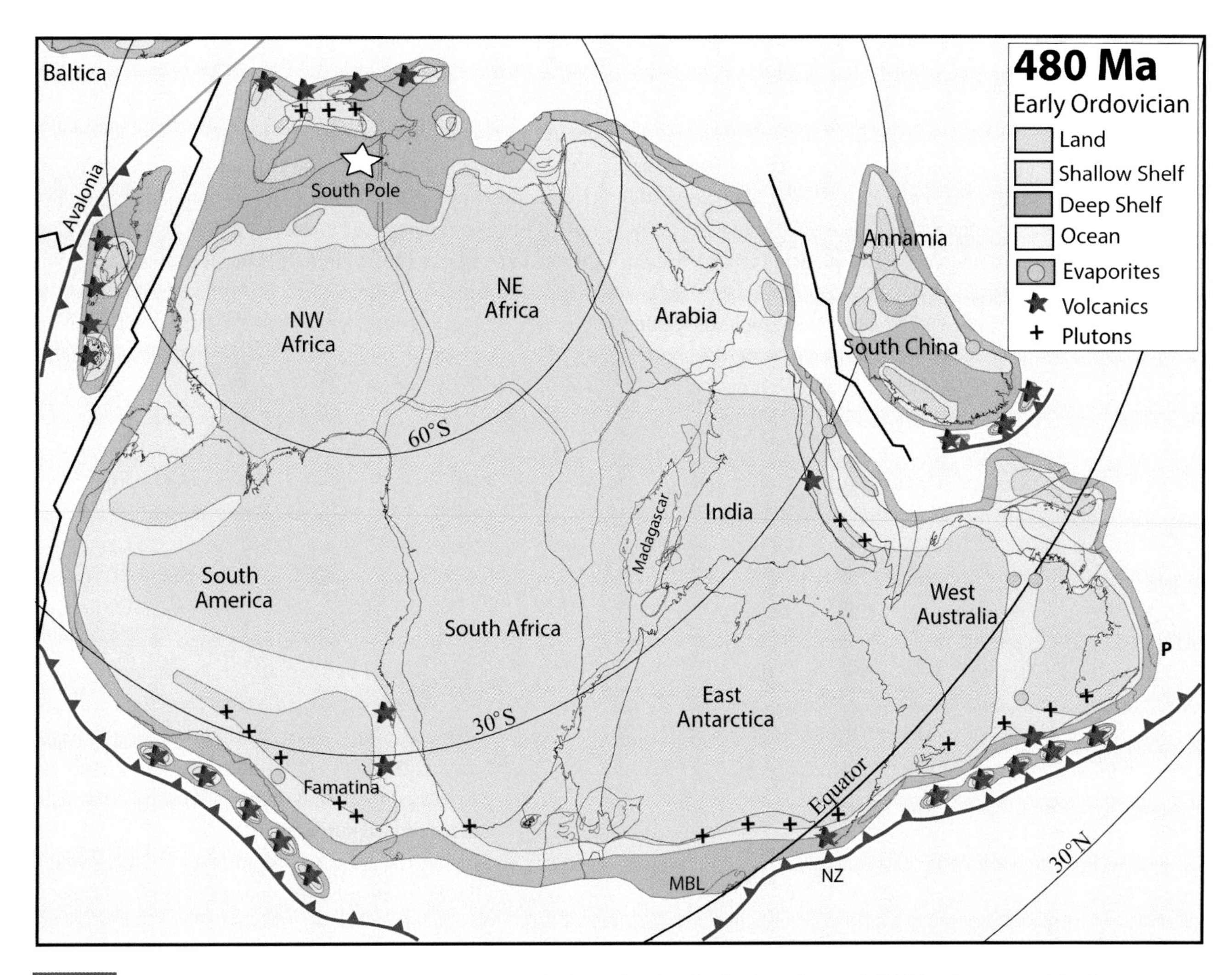

Fig. 6.3. Gondwana and adjacent areas at 480 Ma (Tremadocian), showing the lands and seas. Solid blue lines are subduction zones with teeth on the upper plate, black lines are spreading centres and transform faults, and green line is transform plate margin. The strings of island arcs shown are diagrammatic, since the size of each unit within the arcs and the extent to which each was above sea level are uncertain. MBL, Marie Byrd Land; NZ, New Zealand; P, Papua New Guinea.

underlying Proterozoic basement in southern Mexico, part of the Mixteca–Oaxaca Microcontinent (Fig. 6.2a), are dated at about 452 Ma (Katian). There are also earliest Silurian (442 Ma: Rhuddanian) dykes intruding low-grade metamorphic rocks in the same region; all of which are interpreted as the final episode of a prolonged 480–440 Ma series of events associated with the progressive widening of the Rheic Ocean from the Gondwana Craton in that region (Keppie et al., 2012).

Gondwana's eastern margin remained active throughout the Ordovician, where the Benambran Orogeny within the fold belt of eastern Australia represented an extensional event and the further accretion of oceanic island arcs to the Gondwana Craton, which continued on into the Early Silurian (Glen, 2005). Early Ordovician (473–463 Ma: Floian to Dapingian) granites, now tectonised, were also intruded in Australia (Fergusson & Henderson, 2015). The Ordovician and earliest Silurian Macquarie Arc of New South Wales accreted to the craton in four phases. The first phase, during the Tremadocian to Floian (485–470 Ma), consisted of emergent oceanic volcanic islands fringed by shallow-water tropical carbonates. That was followed by a 9 Myr hiatus until the Darriwilian, when the second phase of volcanism occurred, much of it submarine but some locally emergent, also with fringing carbonates. The third phase consisted of intrusives associated with regional uplift as well as the deposition of a carbonate platform. The final phase, of Katian to Early Silurian age (450–440 Ma), included the extrusion of lavas and the intrusion of porphyries with evolved shoshonite geochemistry (Percival & Glen, 2007). In contrast to the

Cambrian, no Ordovician or Silurian kimberlites or LIPs are known in Gondwana, probably because most of that very large continent was situated between the Tuzo and Jason LLSVPs.

Eastern Asia. Arc activity continued apace in many areas in eastern Asia. Within the wide Triassic and later Song Ma Suture Zone, which unites South China and Annamia (Indochina) today, there are remnants of a Cambrian to Ordovician island arc which accreted to South China before the Devonian (Cocks & Torsvik, 2013). The combined Annamia–South China continent continued its strike-slip progress northeastwards along the Gondwanan margin throughout the period (Fig. 6.4). The northern (Sulinheer) sector of North China is an arc–trench complex including Ordovician ophiolites and other volcanics intruded by latest Ordovician and earliest Silurian (448–438 Ma) plutons (de Jong et al., 2006). The Qaidam–Qilian volcanic arc was still active off the south-western margin of North China (Yan et al., 2010). In the Kitakami Terrane of Japan there was a Cambrian to Middle Ordovician island arc which was subsequently metamorphosed and then intruded by Late Ordovician (457–440 Ma) granites, but that arc was situated off the eastern margin of South China in the Ordovician (Isozaki et al., 2010). The western (Thai) sector of Annamia still did not exist throughout Ordovician time, although the Precambrian Kontum craton in the north-east of Annamia had adjacent Ordovician graptolite-bearing sediments deposited in a deeper-water ocean basin there (Ridd et al., 2011).

There are other significant Eastern Asian areas with Ordovician rocks, the chief of which is the Tarim Microcontinent, whose detailed situation is poorly constrained but whose fossil faunas place it near to the south of North China and east of Siberia (Fig. 6.4): a provisional location which differs from our previous papers (e.g. Cocks and Torsvik, 2013). Today's northern margin of Tarim includes Archaean rocks with unconformable Cambrian and Ordovician sediments, but the southern margin had an Early to Middle Ordovician Andean-type active margin, with progressive accretion of island arcs (de Jong et al., 2006).

Central Asia. The large area east of the Caspian Sea which includes most of Kazakhstan, as well as Uzbekistan, Turkmenistan, Kyrgyzstan, and Tajikistan, forms the western part of the Altaids or Central Asian Orogenic Belt (CAOB), which extends into south-west China and contains many tectonic units (Kröner, 2015). The north-eastern units (mostly grouped within 'Altai–Sayan') were parts of peri-Siberia (Fig. 7.7), the western units, including Tourgai (Unit 465) were peri-Baltica, and the southern units, including South Tien Shan, Tarim, and Ala Shan, were peri-Gondwanan. However, the remaining dozen or so central units (Fig. 3.8), most of which were themselves composite, and which are collectively known as the Kazakh Orogen, have controversial Palaeozoic tectonic histories. Many of them, notably Kokchetav–Ishim, North Tien Shan, Chu-Ili, Atashu–Zhamshi, Karatau–Naryn, and possibly Chingiz–Tarbagatai, have Precambrian rocks in them and were therefore probably independent microcontinents prior to the Cambrian and Ordovician, although their locations are poorly constrained in those times, which is why they are omitted from our earlier maps here. Other units, including North Balkhash, Stepnyak, Selety, and Boshchekul, were island arcs which grew up from ocean floors during the Ordovician, and Junggar–Balkhash is an Upper Palaeozoic accretionary sequence. However, following individual analysis of both the tectonics and the Late Ordovician benthic marine faunas of each unit, it now seems more likely that, although there were some unit mergers during the Ordovician, most remained separate within an equatorial archipelago, like the East Indies of today, until after the end of the Ordovician. Nevertheless, Chu-Ili and North Tien Shan, and Selety and Boshchekul had merged in the Late Cambrian, and Karatau–Naryn merged with the combined North Tien Shan and Chu-Ili microcontinental unit in the Sandbian at about 455 Ma (Popov & Cocks, 2017). Many authors, e.g. Willem et al. (2012), have asserted that most of the units had combined to form a 'Kazakhstania' continent before the end of the Ordovician; however, it was not until after more mergers had occurred in the Silurian and Devonian that a unified substantial Kazakhstania can be eventually identified (see Chapters 7 and 8).

Laurentia. Although the central part of the Laurentian Craton remained stable after the Floian until the end of the Ordovician, there was substantial tectonic activity on many of its margins (Cocks & Torsvik, 2011) (Figs. 6.5 and 6.6). On today's eastern seaboard of North America and also in its Palaeozoic extension into Scotland and Ireland, several island arcs were progressively accreted (Mac Niocaill et al., 1997). Prior to those accretions, some of the island arcs had been near the Laurentian continental margin, some on the Gondwanan or Baltica continental margins, and some were developed on mid-ocean sites within Iapetus. Also at the eastern Laurentian margin, the closure of the seaway in the Newfoundland area, which had started in the Late Cambrian, progressed steadily through subduction beneath the Dashwoods Terrane, and that narrow ocean finally closed in Floian times at about 470 Ma (Waldron & van Staal, 2001). The Taconic Orogeny has been defined in various ways as affecting the Appalachian area during the Middle and Late Ordovician, but it appears to have started by the final closure of the Humber Arm Seaway between the Dashwoods

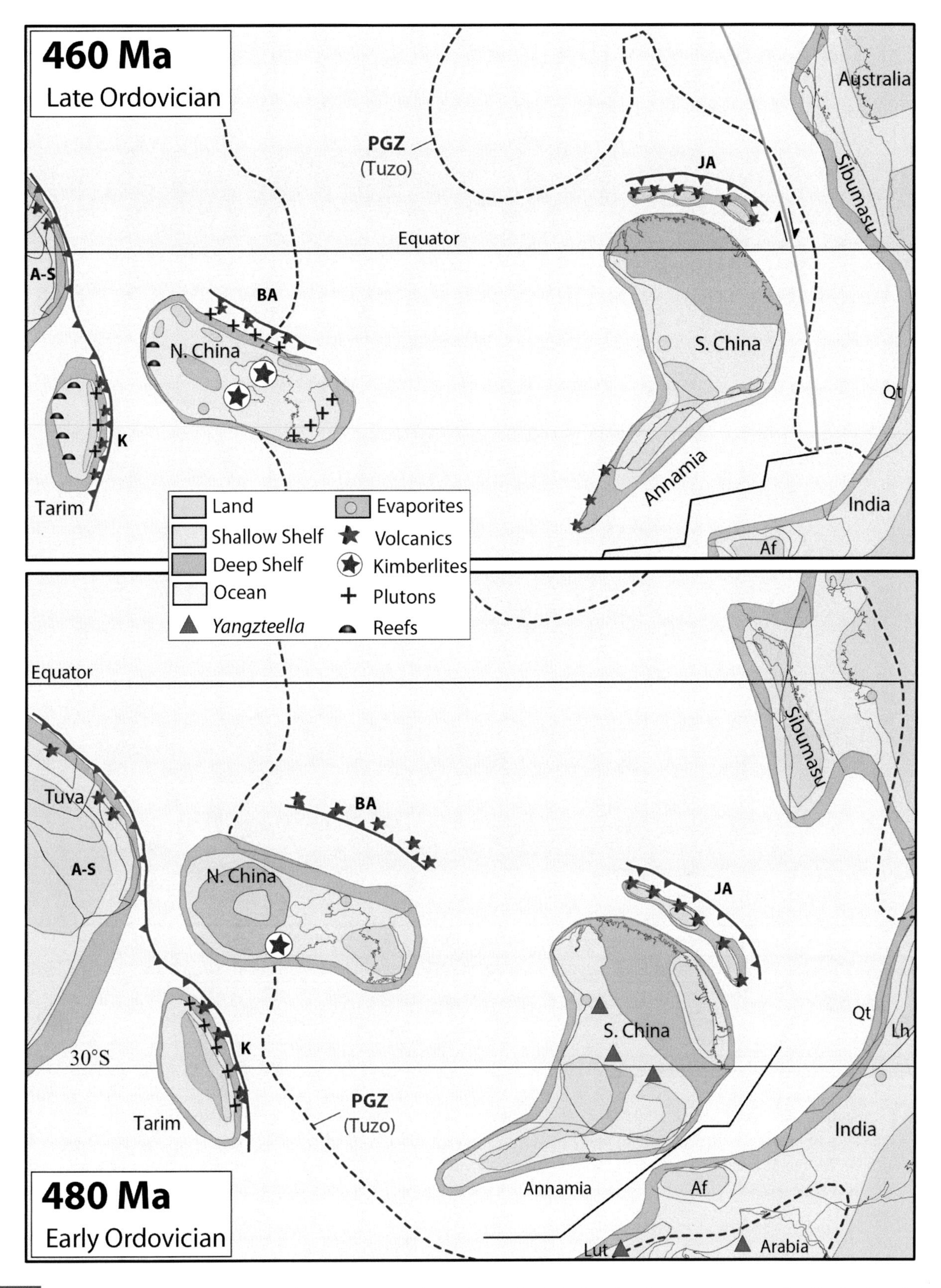

Fig. 6.4 (Bottom) Eastern Asia terranes and continents at 480 Ma (Tremadocian), showing the lands and seas, the distribution of the pentameride brachiopod *Yangzteella*, and the kimberlites known in North China. (Top) The same area at 460 Ma (Late Ordovician; Sandbian). Dotted red lines are the African plume generation zone (PGZ), solid blue lines are subduction zones with teeth on the upper plate, black lines are spreading centres and transform faults, and green line is transform plate margin. A-S, Altai–Sayan; Af, Afghan Terrane; BA, Bainaimiao Arc; JA, Japanese arcs; K, arcs now in the Kunlun Terrane (Unit 457); Lh, Lhasa Terrane; Qt, Qiantang Terrane.

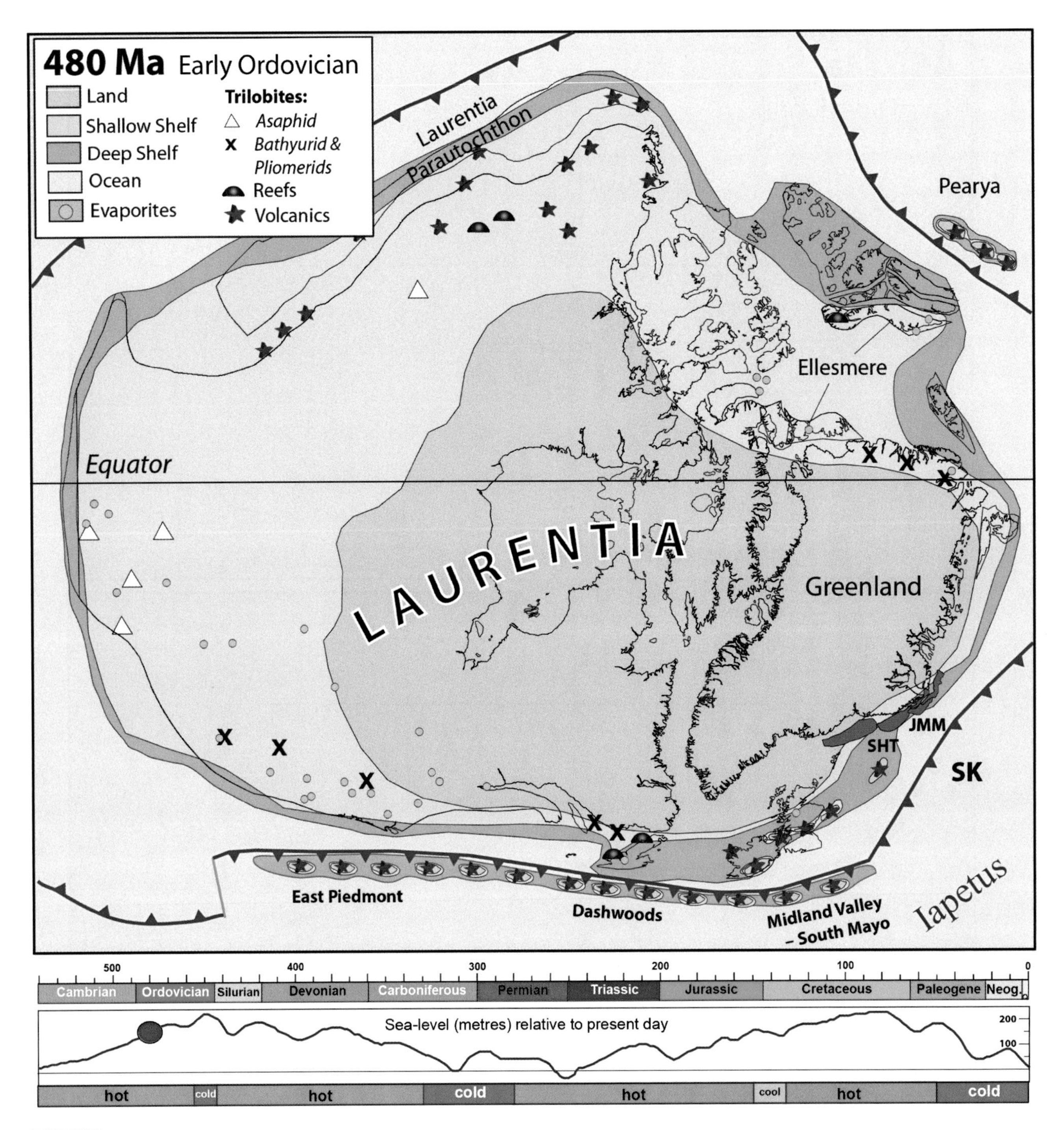

Fig. 6.5 Laurentian palaeogeography at 480 Ma (Early Ordovician: Tremadocian), showing the distribution of terranes, the lands and seas, and the sites of provincial trilobites. Solid blue lines are subduction zones with teeth on the upper plate, and green line is transform plate margin. JMM, Jan Mayen Microcontinent; SHT, Smøla–Hølanda Terrane; SK, Sunnhordland and Köli nappe complexes. Bottom: Phanerozoic time scale and sea-level variations (Haq & Al-Qahtani, 2005; Haq & Shutter, 2008) and icehouse (cold) and greenhouse (hot) conditions.

Microcontinent and the main Laurentian Craton, which triggered a second phase of Notre Dame Arc magmatism and metamorphism in the later Ordovician of Newfoundland. The two Ordovician phases of the Taconic Orogeny in the Notre Dame Arc are Taconic 2, from 470 to 460 Ma (Dapingian and Darriwilian), and Taconic 3, from 454 to 442 Ma (Katian and Hirnantian), both phases reflecting the accretion of separate island arcs to Laurentia (van Staal et al., 2007,

2009). At the same time as Taconic 2, a tholeiitic island arc preserved within Alabama was active: an arc which originated at 468 Ma in the extensional environment of a back-arc basin, to be followed at 460 Ma by the intrusion of plutons (Tull et al., 2007). In Pennsylvania there are Lower to Middle Ordovician trench-fill deposits which were allochthonously emplaced onto the craton at the end of the Ordovician or perhaps a little later. The Taconic Orogeny controlled the distribution of the turbidite deposition, which commenced in the Early Ordovician (Floian) *Isograptus victoriae* graptolite biozone in Newfoundland, and migrated south-westwards until the Middle Ordovician *pygmaeus* Zone of southern Quebec. That orogeny also resulted in crustal shortening of about 270 km.

The Late Ordovician (Fig. 6.6) saw the beginning of the crustal shortening in Greenland, with granites in south-eastern Greenland dated at 466 Ma (Darriwilian). Today's northern margin of Laurentia had been passive since the Late Neoproterozoic, but, at about the Ordovician–Silurian boundary time at 443 Ma, there was the creation and the initial filling of a turbidite basin in northernmost Greenland and Ellesmere Island, which heralded the arrival and subsequent accretion of the Pearya Terrane in the Silurian (see below and Fig. 7.3). However, also today within northern Ellesmere Island and to the south of the Pearya Terrane, there are the remnants of an Ordovician island arc, which must previously have been situated at an active subduction zone south of Pearya, rather than at the passive margin of Laurentia. The Arctic Alaska–Chukotka Microcontinent was still far away from the Laurentian margin, but the distance between them was closing, with a substantial strike-slip element (Cocks & Torsvik, 2011). The western Cordilleran margin of Laurentia remained passive, but there was submarine alkaline igneous activity in the extensional basins near the edge of the craton in the USA, particularly in the Early to Middle Ordovician (Goodfellow et al., 1995).

Siberia and Peri-Siberia. During the Ordovician, Siberia drifted over the Equator from south to north, and its latitudinal drift rate was rather high, particularly at the end of the Ordovician, and, from the Sandbian until the present day, Siberia has remained north of the Equator (Fig. 9.11). In contrast to the preceding Late Cambrian, there appears to have been a substantial land area over the Anabar Massif in the then south-west of the continent (Keller & Predtechensky, 1968), but most of the Precambrian craton was flooded by warm shallow epeiric seas (Fig. 6.7). Orogeny was active in the Altai–Sayan area throughout the Ordovician, during which many of the previously independent terrane units and island arcs accreted to the main Siberian Craton (Fig. 7.7); and there were high mountains on much of the then north of the main continent inboard from the active margin. The Mongol–Okhotsk Ocean still divided the enlarged Siberian continent from the various components of the Central Mongolian Terrane Assemblage, to the north of which there were active island arcs (Fig. 6.7). The Early Ordovician (Tremadocian and Floian) was a time of substantial tectonic activity on or near today's southern margin of Siberia. There was a series of tectonic events which may or may not have been directly linked to each other along that margin of the craton, although much of it was apparently passive. The combined Salair–Kuznetsk Basin, today bordering the south-west of the craton (Fig. 7.7), was deposited in a marginal sea, with maximum deposition rates near the Tremadocian–Floian boundary, followed by the transition from an active to a passive margin in that part of Siberia at about 475 Ma (Floian). Collision between Tuva–Mongolia and the West Sayan volcanic arc occurred in the Darriwilian at about 460 Ma, accompanied by the deposition of substantial olisthostromes (Sennikov, 2003) and the intrusion of granites (Dobretsov et al., 2003).

Baltica. Baltica was isolated on its own plate and surrounded by four separate oceans: the Iapetus, which was at its maximum width near the start of the Ordovician at about 490 Ma and which extended far to the west, the Ægir between Baltica and Siberia, the Ran, and the Tornquist (Fig. 6.11). Palaeomagnetic data from Baltica are very reliable for the Ordovician (in contrast to the Cambrian). Baltica was still undergoing the very fast rotation (Fig. 2.12) that had begun in the Cambrian and continued until the Middle Ordovician, a counter-clockwise rotation which totalled about 120° (Torsvik & Rehnström, 2001).

However, because Baltica was some way apart from its neighbouring continents, that rotation must have been achieved largely through strike-slip faulting, and little tectonism affected the central continental area during the whole period, as shown by the Ordovician rocks there, which are mostly relatively flat-lying and unmetamorphosed. Although it seems probable that much of the craton was originally covered by those sediments, in much of Norway and Sweden they have been lost through subsequent erosion to reveal the underlying Precambrian shield, apart from in the Oslo Region, where the Lower Palaeozoic sediments are only preserved through graben development in the Late Palaeozoic, and also in the Dalarna area of Sweden, where there is a ring of shattered Late Ordovician Boda limestones preserved as part of a Devonian meteorite impact crater.

The centre of Baltica was little affected by the Avalonia–Baltica collision; for example, in the uppermost Ordovician and Lower Silurian in central Sweden (Jämtland), the only changes in the sedimentary regime there were the regression

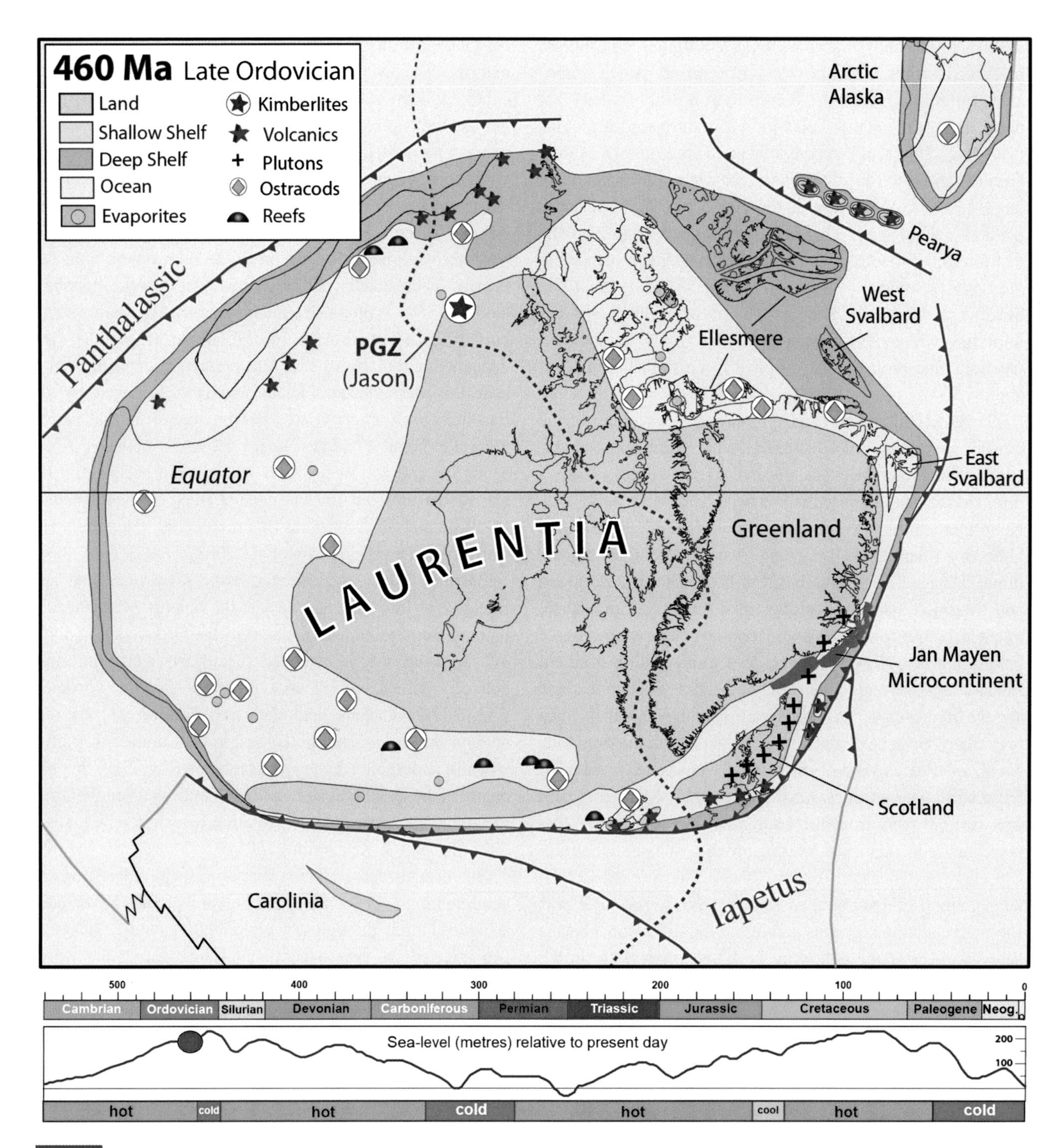

Fig. 6.6 Laurentian palaeogeography at 460 Ma (Late Ordovician; Sandbian), showing the distribution of lands and seas, kimberlites, and the distribution of key ostracod localities (see text). Solid blue lines are subduction zones with teeth on the upper plate, black lines are spreading centres, and green lines are transform plate margins. Dotted red line is the plume generation zone (PGZ). Bottom: Phanerozoic time scale and sea-level variations (Haq & Al-Qahtani, 2005; Haq & Shutter, 2008) and icehouse (cold) and greenhouse (hot) conditions.

and subsequent transgression representing the global Hirnantian Ordovician–Silurian boundary glacioeustatic event (Cocks & Torsvik, 2005). In the Early Katian at about 451 Ma, when Baltica was nearing Avalonia, there was an immense Plinian Andean-type eruption in or near north-east England, and the consequent ash, known as the Kinnekulle Bentonite, was deposited over much of western Baltica (Figs. 6.12 and 2.19), The thickness of that bentonite varies

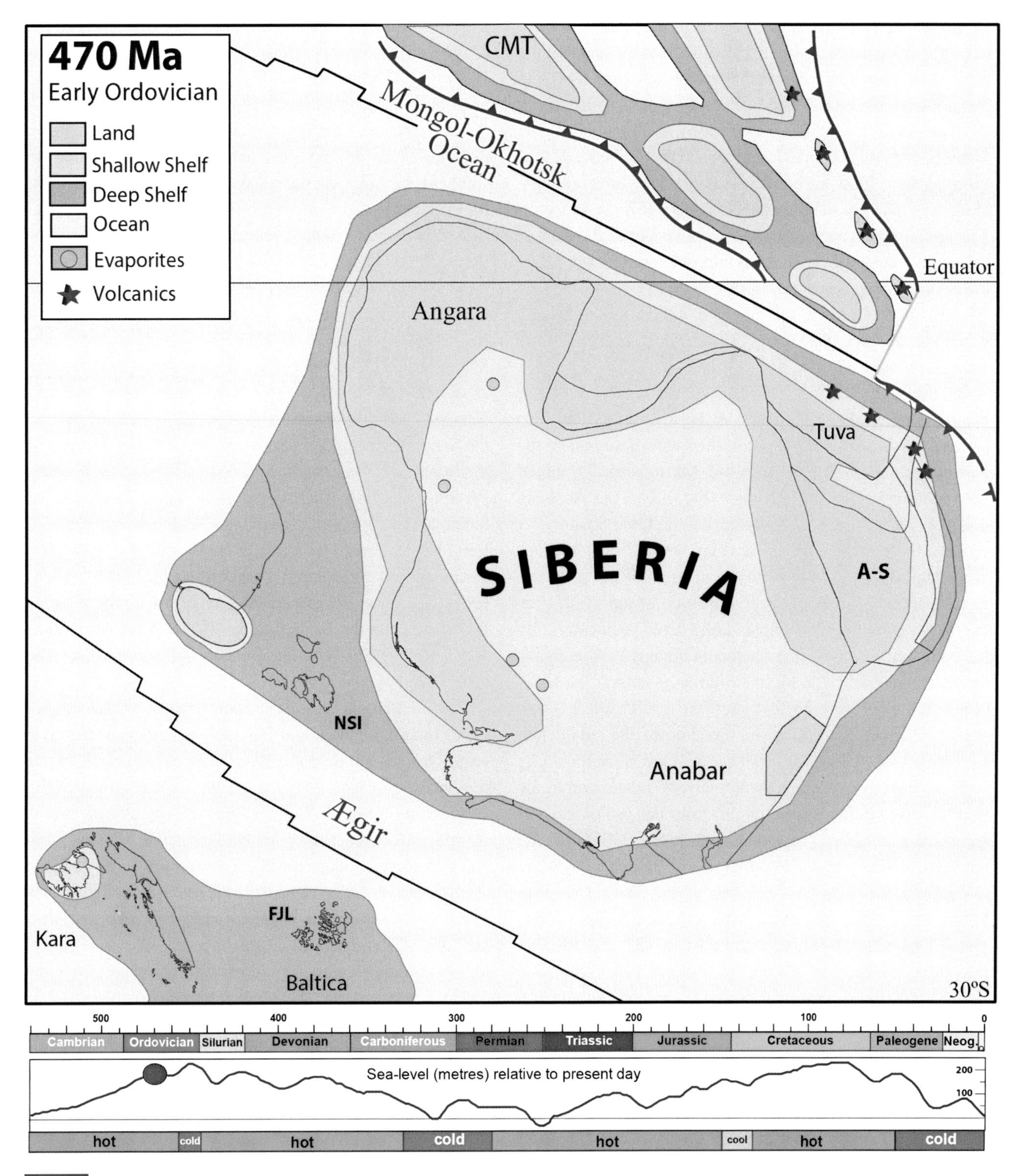

Fig. 6.7 Siberia, with the Central Mongolian Terrane area (CMT) to its north and part of Baltica to its south at 470 Ma (Middle Ordovician: Dapingian), showing the lands and seas. The string of islands shown is diagrammatic, since the extent of each unit within the arc and which units were above sea level are uncertain. Solid blue lines are subduction zones with teeth on the upper plate, black lines are spreading centres and transform faults, and green line is transform plate margin. A-S, Altai–Sayan; FJL, Franz Josef Land; NSI, New Siberian Islands. Bottom: Phanerozoic time scale and sea-level variations (Haq & Al-Qahtani, 2005; Haq & Shutter, 2008) and icehouse (cold) and greenhouse (hot) conditions.

from a maximum of over two metres in the south-west, for example at Kinnekulle Mountain in southern Sweden (which must have been some distance from the eruption itself), to just a few centimetres in the St Petersburg area of Russia.

Avalonia. The elongate continent of Avalonia is divided today by the Atlantic Ocean, and the microcontinent has thus been termed West and East Avalonia by some authors. Domeier (2015) has shown the Avalonian strip as united on a progressively growing single ocean plate from the start of the Ordovician until very near its end, at about 453 Ma, when a new spreading centre developed to divide West and East Avalonia just before the latter merged with Baltica. Avalonia was originally an integral part of Gondwana situated offshore from Amazonia, Florida, and North-West Africa. However, following Late Cambrian rifting, it drifted away from Gondwana at about 490 Ma (Cocks & Fortey, 2009), and progressed north-eastwards until colliding with Baltica near the end of the Ordovician at about 445 Ma to form the united new continent of Avalonia–Baltica (Fig. 6.2). Thus Avalonia was only an independent entity during the Ordovician.

Since the Avalonia–Baltica docking was oblique, it apparently caused little tectonism at the margins of the two continents, although some relatively localised unconformities in the Welsh Borderland of England may have been caused by the collision; however, that Shelvian Orogeny was insignificant compared with the Middle Silurian events of the Caledonide Orogeny some 15 Myr later (see next chapter). There is disagreement concerning the direction of the subduction following the docking with Baltica, with the latter shown by some authors as overriding Avalonia with subduction to the north. However, there is no doubt, from the geological and geophysical evidence (Torsvik & Rehnström, 2003), that the converse is true, with a substantial slab of Baltica today lying underneath Avalonia to the south of the Trans-European Suture Zone, following southwards subduction. Late Ordovician calc-alkaline magmatism between 457 and 449 Ma now buried below the Mesozoic in eastern England has been linked with the Tornquist Ocean closure and subduction beneath Avalonia.

Facies and Faunas

Climate and Sea Levels. Temperatures and sea levels varied greatly as the Ordovician progressed (Fig. 16.2), as testified by carbon and oxygen isotope curves and also by the radiations seen in the biota (Harper & Servais, 2013). There was an overall rise in both temperatures and sea levels during the first half of the period, and the highest global sea level (the second highest in the Phanerozoic: Fig. 16.2), and probably also the warmest temperatures, were at about 455 Ma in the Middle Ordovician (Sandbian) and slightly later. However, heavier $\delta^{18}O$ values and falling sea levels near the Dapingian–Darriwilian boundary in Baltica at 467 Ma indicate much rapid fluctuation prior to the Sandbian, leading Rasmussen et al. (2016) to conclude that the global temperatures had even dropped low enough for the formation of a southern ice cap, although there is no sedimentary evidence for the latter in the polar sector of Gondwana (Fig. 6.1b). The subsequent rises resulted in substantial transgressions and may well have been one of the triggers for the Great Ordovician Biodiversification Event (GOBE; Webby et al., 2004). Certainly, many of the cratons and other marginal lands were flooded, leading to more habitats for the marine benthic faunas and causing underlying unconformities in many areas, as well as distributing the distinctive graptolite faunas of the *Nemagraptus gracilis* Biozone, which mark the bases of the Sandbian Stage and also the Caradoc Series. But, during the subsequent Katian, the global temperatures fluctuated several times, and, after an initial dip immediately after the Sandbian, there was a relatively short global warming in the Early Katian termed the Boda Event (Fortey & Cocks, 2005), during which substantial bioherms, with newly radiated endemic faunas, developed over much of northern Europe from Ireland to the Urals (Fig. 6.12). The final large change was during the end-Ordovician Hirnantian glaciation (see below).

Faunal Provinces. Because the Iapetus Ocean between Gondwana, Laurentia, and Baltica was near its widest in the Early Ordovician, probably over 5000 km, distinctive faunal provinces are apparent in the benthic faunas on the shelves surrounding the ocean (Figs. 2.20a and 6.8), as well as at the margins of other continents further afield. In the North Atlantic region today, those provinces are most strikingly seen among the trilobites, with the tropical Bathyurid Province in Laurentia, the Megistaspinid Province in Baltica, and the Calymenacean–Dalmanitacean Province in the higher-latitude sectors of Gondwana (Cocks & Fortey, 1982). Further afield, the Megistaspinid Province extended as far eastwards as Siberia and North China, and there was also the separate Dikelokephalinid Province in the tropical parts of Gondwana. The latter reflects the substantial widths of the oceans between the various continents straddling the Equator. Although Early Ordovician articulated brachiopods were less diversified, many, such as the locally abundant *Syntrophina* and *Nanorthis*, appear to have been pan-tropical in distribution.

Provincialisation in the benthic faunas lessened as the Ordovician progressed; however, the brachiopods can be divided into four provinces even up to the Late Ordovician (Harper & Servais, 2013); the tropical East Gondwana

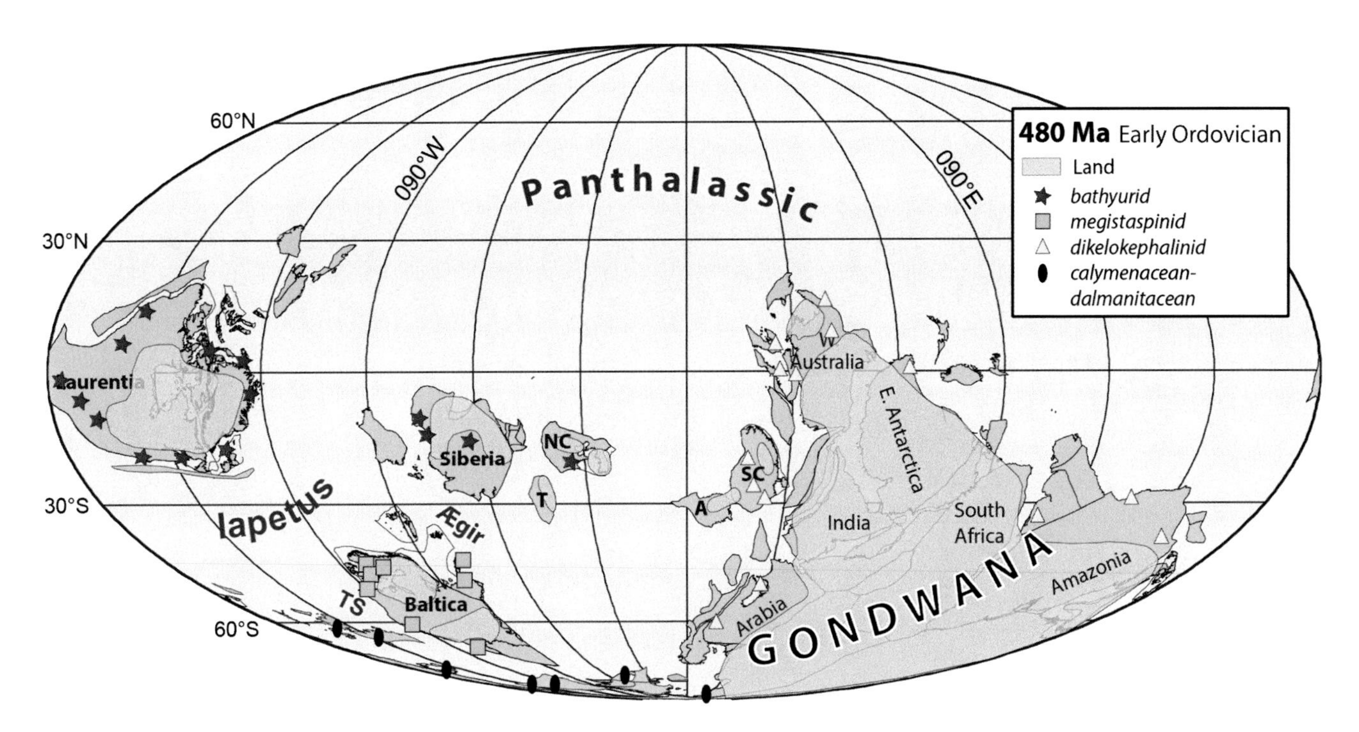

Fig. 6.8 Global distribution of Early Ordovician (Tremadocian–Floian) trilobite provinces (data points from Fortey & Cocks, 2003) on a new 480 Ma reconstruction.

Province (including Australasia, North and South China, and most of the Kazakh terranes and the rest of central Asia), the Baltica Province, the Laurentia–Siberia Province, and the Mediterranean Province in the higher latitudes of Gondwana. Four global provinces have also been defined within the bryozoa in the Ordovician (P.D. Taylor, in Webby et al., 2004), one of which is termed North American, and their distribution is very similar to the brachiopods.

In contrast to the provinces seen in the benthic faunas, the planktonic faunas had no distinctive provinces (although their latitudinal distributions were constrained by the ambient temperatures), and were able to be transported by currents across the oceans without hindrance (Fig. 2.20). The graptolites migrated quickly as they evolved and are thus the most useful correlation tools, with some graptolite zones lasting less than a quarter of a million years; much more precise correlation than that available from nearly all radiometric dates. Other zones using planktonic microfossils such as acritarchs and chitinozoa have also been identified and used for correlation, although they were not as short as the graptolite zones; however, those smaller fossils can often be found, for example in boreholes, in rocks lacking graptolites.

However, as noted below under the individual continents, as most of the oceans mostly became narrower, so the most substantial differences between the various benthic faunal provinces decreased, and by the end of the Ordovician they had become much less distinctive.

Gondwana. Because it stretched from the South Pole to well north of the Equator (Fig. 6.3), there was plenty of space in which to develop temperature gradients in the shelf seas bordering the vast landmass, and thus the higher latitudes hosted the Calymenacean–Dalmanitacean Fauna, particularly the trilobite *Neseuretus*, in the Early Ordovician (Cocks & Fortey, 1982), as well as the Mediterranean brachiopod fauna (Havlíček et al., 1994) during most of the Ordovician. In contrast, the lower-latitude tropical benthic faunas (including the dikelokephalinid Province) were very different and naturally much more diverse (Fig. 6.8). However, between the two latitudinal extremes, there was provincial mixing, as can be seen in parts of Central Asia and South America, which were at opposing margins of the main Gondwana continental landmass. Thus there were ecological clines which changed across the various latitudes; for example, the large and unmistakable Early Ordovician pentameride brachiopod *Yangzteella* was originally thought to be endemic to lower-latitude South China (Fig. 6.4), but was subsequently found in the temperate-latitude Taurus Terrane of Turkey (Cocks & Fortey, 1988), and is now also known from various medium to low latitudes in Gondwana and peri-Gondwana (see below), although not from the highest-latitude regions, which hosted the Mediterranean Province (Cocks & Torsvik, 2013).

In the Himalayan sector, there are diverse shallow-marine faunas with many endemic genera and species, such as the

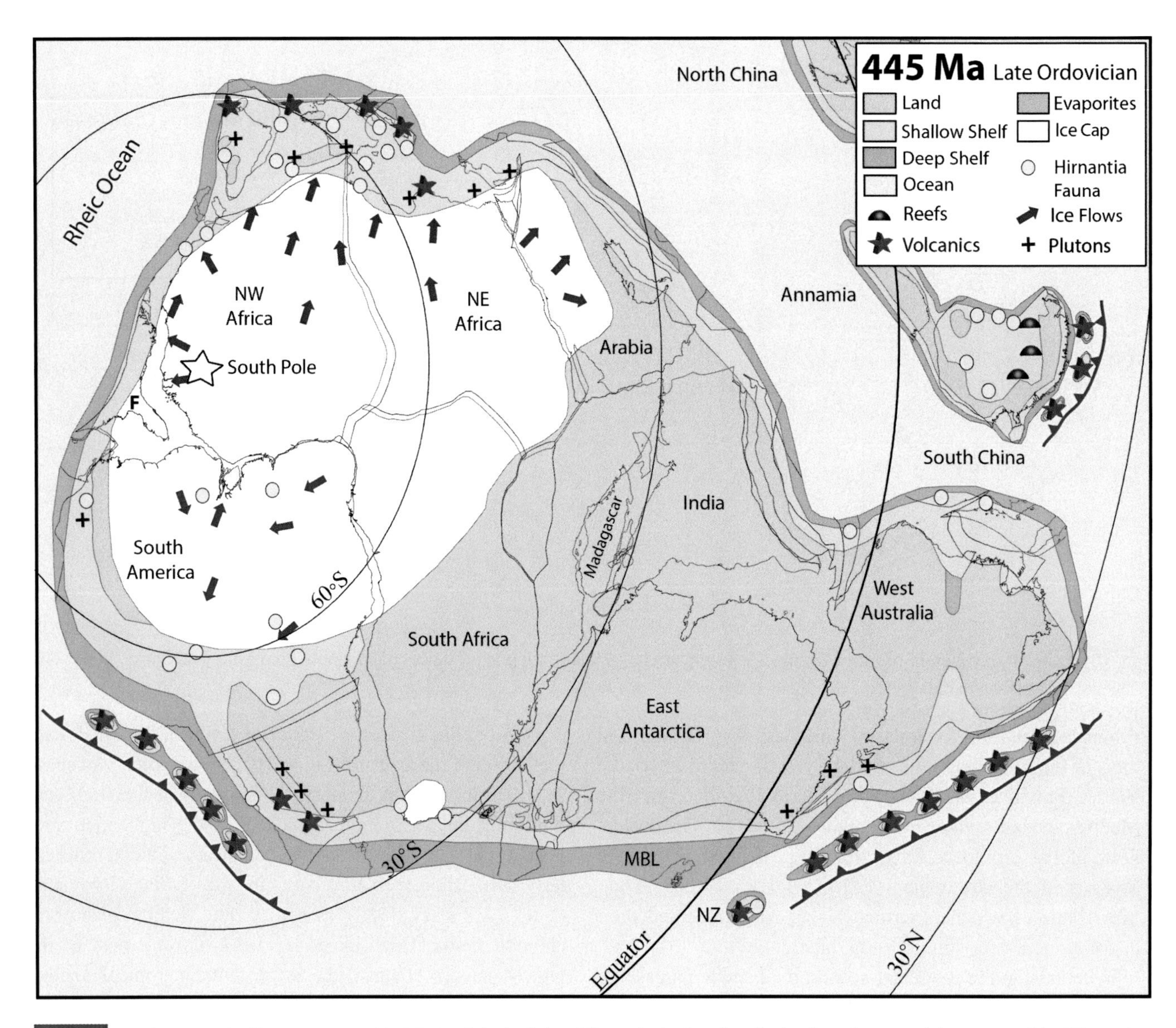

Fig. 6.9 Gondwana and adjacent areas at 445 Ma (end-Ordovician: Hirnantian), showing the lands and seas and the extent of the Hirnantian polar ice cap and its ice floes, and the distribution of the *Hirnantia* brachiopod Fauna. Solid blue lines are subduction zones with teeth on the upper plate. The strings of island arcs shown are diagrammatic, since the individual extents of the units within the arcs and which units were above sea level are uncertain. MBL, Marie Byrd Land; NZ, New Zealand.

Darriwilian brachiopods described in the nineteenth century from Spiti, north-eastern India, now preserved within the remnants of arcs that were separated by a deep basin from the main craton in the Ordovician (Zhu et al., 2013). Land in the eastern Australian sector of the Gondwana Craton was separated from the main continental Gondwanan land area to its west by the Larapintine Sea, a shallow-water and entirely intra-cratonic sea which stretched across Australia from today's north to south, some of which is preserved in the marine sediments in the Amadeus and Georgina basins of central Australia today (Fig. 6.3). In the Later Ordovician, the Larapintine Sea was more intermittent, and eventually dried up, with the result that extensive evaporites were deposited in the remnants of those basins (Fig. 6.9).

Outboard of the eastern Australian sector of Gondwana, there were island arcs that were hosts to a variety of benthic faunas, such as the brachiopods (Percival, 1991), many of which were endemic, and their diversity reflected their tropical location. The contrast between the warmer-water and cooler-water faunas is most dramatically seen in the Early Ordovician brachiopods from the now adjacent Cuyania (Precordillera) and Famatina terranes of Argentina (Benedetto, 1998), the former being much more diverse (and similar to the tropical Laurentia), and the latter with many fewer species, reflecting their higher palaeolatitude. The decreasing percentages of Laurentian brachiopod genera in the Cuyania Terrane during the Ordovician (from 52% in the Floian, down to 32% in the Darriwilian, and their virtual absence by the Katian) reflects the relatively speedy movement of that terrane from peri-Laurentia to Gondwana between 475 and 450 Ma (Fig. 6.2a), which agrees with the plate tectonic scenario of Domeier (2015).

Eustatic sea level rose and fell during the Ordovician (Fig. 6.5), as demonstrated in the flooding surfaces of various ages in the subsurface rocks through the entire Ordovician in Oman and the United Arab Emirates, which were within the passive margins of the Arabian sector of the Gondwanan Craton and therefore little prone to tectonic upheaval then. In South America, there are Middle Ordovician (Dapingian to Darriwilian) sediments in the Sierras Subandinas of north-western Argentina, representing the outermost exposures of the Central Andean Basin, and which are alternations of shallow-marine deltaic systems and estuarine environments.

Because of the range of its latitudes, Gondwana was much affected by the global temperature changes during the Ordovician, but in varying ways. For example, after an initial cooling after the Sandbian, there was the Boda warming event in the Early Katian (Fortey & Cocks, 2005), which is reflected in the patch carbonates, some of which were bryozoan bioherms, seen in the higher latitudes of Gondwana in Morocco, the Iberian Peninsula, Sardinia, and France. Although those bioherms were formed under cooler water than the larger reefs of lower latitudes, and are relatively thin, they stand out since the thick sequences in which they occur otherwise consist entirely of clastic rocks. The Boda Event was followed less than 10 million years later by the Hirnantian glaciation (Fig. 16.2) at the end of the Ordovician, which is best seen in Gondwana (see below).

Eastern and Central Asia. Nearly all of North China was uplifted to form land in the latest Ordovician, extensive bauxite deposits were formed there, and the land persisted until the Carboniferous (compare Fig. 6.4 with Fig. 7.4). Within the relatively widespread benthic provinces noted above, there were also distinctive endemic genera; for example, the locally abundant plectambonitoidean brachiopod *Spanodonta*, first described from north-western Australia, also occurs in comparable shallow-water equatorial limestones in Sibumasu (Cocks et al., 2005), but not in South China or Annamia (Fig. 6.4). Also shown on Fig. 6.4 is the distribution of the pentameride brachiopod *Yangzteella*, which was originally seen as endemic to South China, but which is now also known from Turkey, Karakorum, and Iran. Furthermore, analysis of Late Ordovician (Sandbian and Early Katian) brachiopod genera indicates that the diverse South China faunas did not have a high proportion of genera in common with North China, the Chu-Ili–North Tien Shan Terrane of Kazakhstan, or the Australian faunas from the island arcs in New South Wales.

The brachiopod endemism in South China (Zhan et al., 2011) grew from nothing in the Early Tremadocian (no endemics in the eight articulated genera recorded) to over a quarter (23–28% endemics within 52 articulated genera) in the Middle Ordovician (Dapingian to Darriwilian). Analysis of all the trilobite genera found in the various parts of China at five successive Ordovician time intervals (Tremadocian, Floian to Early Darriwilian, Middle to Late Darriwilian, Sandbian to Early Katian, and Late Katian to Hirnantian) showed that the faunas of Northern Xinjiang and Hinggan areas were peri-Siberian. However, although quite different from those in Siberia, the North China, South China, Tarim, and Qaidam–Qilian trilobites were closely related to each other throughout the Ordovician, and formed a single faunal province, which was essentially peri-Gondwanan. However, although that province was relatively uniform at the beginning and end of the Ordovician (Tremadocian and Late Katian to Hirnantian), in the intervening Middle Ordovician (Floian to Early Katian) two subprovinces can be recognised, one in South China, Tarim, and Annamia, and the other in North China, Sibumasu, southern Tibet (the Lhasa Terrane), and Tianshan–Beishan (Fig. 6.4). That confirms the scenario in which the Chingiz, Chu-Ili, Tien Shan, and Tarim terranes were grouped in the Sandbian on the basis of their contained shallow marine brachiopods and distinctive trilobites such as *Taklamakania*, which was originally considered as endemic to Tarim but was later also collected from the adjacent areas (Fortey & Cocks, 2003).

During the Early Floian the very large platform area in South China was largely drowned, accompanied by a shoreward expansion of the deeper-water outer shelf trilobite assemblages, a deepening that continued until the Early Darriwilian. The Dapingian–Darriwilian trilobites from the Chinese (Yunnan Province) sector of Ammania indicate that the latter was probably near Gondwana, and the bivalves from the same area also suggest affinity with South China and central Australia. Annamia was united only with South China in the Lower Palaeozoic (Cai & Zhang, 2009; Cocks & Torsvik, 2013), rather than being offshore of Sibumasu and part of the core Gondwanan Craton as suggested by, for example, Metcalfe (2011).

Analysis of Katian brachiopods from many localities in Kazakhstan and elsewhere indicates that the equatorial situation of the Kazakh Archipelago facilitated the dispersal of originally endemic genera to adjacent clusters through the prevalent westward South Equatorial and North Equatorial currents and the eastward-flowing Equatorial Countercurrent (Popov & Cocks, 2017).

In the Tarim Microcontinent (Fig. 6.4), a northward-dipping carbonate platform continued on from the Cambrian, which was fringed to the north by deeper-water facies that developed along the South Tienshan Mountains (Chen et al., 2010). The affinities of the diverse Tarim benthic faunas (Zhou & Dean, 1996) with other areas are noted above.

Laurentia. The continent remained equatorial for the whole Ordovician, with high temperatures and consequent high faunal diversity, and its craton was repeatedly flooded by epeiric seas (Fig. 6.5). The second half of the Ordovician saw the maximum marine transgressions, with consequent large sediment thicknesses building up in the marginal basins; for example, the 9 km thick Ordovician pile on today's Arctic margin (Trettin, 1998). The Bathyurid Province includes the abundant trilobites on the Laurentian Craton during the Early Middle Ordovican (Fig. 6.8); however, bathyurids had not yet diversified in the earliest part of the Ordovician (Tremadocian, locally termed Ibexian or Whiterockian), and endemic hystricurid trilobites took the bathyurids' place as provincially diagnostic taxa. There are some differences between the western and eastern Laurentian Craton trilobite faunas, with the western faunas including endemic asaphids such as *Aulacoparia* and *Lachnostoma* not found further east, where there are related but endemic genera such as *Bathyurellus*; however, there is no obvious land barrier between the east and west, and the reason for the faunal differentiation is unknown (Fortey & Cocks, 2003). There was a global decrease in the proportion of endemic trilobites towards the end of the Ordovician, when the Baltic and Avalonian faunas were gradually combined with those from Laurentia as the Iapetus Ocean narrowed.

However, there was much generic endemism in the Middle Ordovician (Dapingian–Darriwilian: also known locally as Whiterockian) brachiopods, reflecting global provinciality, with common species such as *Orthidiella longwelli* and *Ingria cloudi* (Fig. 6.10) in both North America and Scotland, but not outside Laurentia (Cocks & Torsvik, 2011). The east and west margins of the Laurentian Craton carried the same communities in the Middle and Late Ordovician (Fortey & Cocks, 2003), although only the shallower-water benthic assemblages (BA) 2 and 3 were present on the craton, in contrast to the wider BA2 to BA5 range developed on the margins. By the Late Ordovician (Sandbian and Katian) the distinctive and largely endemic Richmondian brachiopod faunas, notably *Megamyonia*, *Hypsiptycha*, *Hiscobeccus*, and *Lepidocyclus*, had become established in Laurentia. Those faunas are best known from the eastern parts of the craton, such as the Cincinnati area of Ohio, New York State, and also northwards in the Hudson Bay and Manitoba areas. Many of the brachiopods are large, abundant, and well preserved, such as *Hebertella*, *Strophomena*, and *Rafinesquina* (Fig. 6.10), and have been well known since the mid-nineteenth-century monographs of James Hall.

The bryozoa have five associations within a wider North American–Siberian Province in the Late Ordovician (Katian), including the Cincinnati–Maquoketa faunas, which inhabited the same area as the Richmondian brachiopod fauna; the bivalves did not reach Laurentia from Gondwana before the Late Ordovician, and their subsequent appearances are sporadic. The distribution of corals is complex, and, although some genera were cosmopolitan, many were local; for example, there are four biogeographically significant coral divisions in the Late Ordovician (Katian) of eastern Laurentia (Webby et al., 2004). Endemic Laurentian fish genera radiated during the Late Ordovician. Ostracods are relatively small mainly benthic arthropods and are abundant at many sites; Cocks & Fortey (1982) had originally considered them as provincially restricted between Laurentia, Avalonia, and Baltica until the Early Silurian, largely because the infant ostracods had no planktonic larval stage. However, although there were almost no Early Ordovician genera in common between the three continents, from Sandbian times onwards significant elements of the ostracod faunas crossed the Iapetus Ocean (Schallreuter & Siveter, 1985), and thus the ostracods appear to have been less endemic than the brachiopods.

Many of the fossils from marginal sites around Laurentia were destroyed by subsequent orogenic activities and those that remain are hard to find. However, a fauna from the Early Ordovician (Floian) of California (Fortey & Cocks, 2003) is characteristic of the deeper-water Nileid trilobite biofacies originally described from the marginal peri-Laurentian deposits at the other side of the continent in Spitsbergen; and a shelf to deeper-water sequence of Early Ordovician trilobite communities is preserved in re-deposited limestone boulders in the Cow Head Group of western Newfoundland. At the southern margin of the craton, there are also deeper-water basin sequences preserved, particularly in Texas, where there is a nearly complete succession of Ordovician graptolite zones.

Originally living offshore of the eastern Laurentian area, distinctive and often endemic shallow-water marine faunas, particularly the trilobites and brachiopods, have been found in rocks from many of the Iapetus Ocean island arc sequences, and some include Laurentian genera well known from the craton itself. However, at other localities, such as those in the Grangegeeth Terrane of central Ireland

Fig. 6.10 Distinctive provincially endemic Early and Late Ordovician brachiopods from Laurentia. (a), (b), (d) and (e) Early Ordovician (Floian) Pogonip Formation, Nevada: (a) and (b) *Orthidiella longwelli*, ventral and brachial interiors; (d) and (e) *Ingria cloudi*, ventral and brachial interiors. Late Ordovician: (c) and (f) *Apatomorpha pulchella*, Athens Formation (Sandbian), Tennessee, ventral and brachial interiors; (g)–(n), Trenton Group (Katian), Cincinnati, Ohio; (g)–(j) *Strophomena planumbona*, ventral interior and dorsal and posterior views of conjoined valves; (k) *Hebertella occidentalis*, ventral interior; (l)–(n) *Rafinesquina alternata*, exterior and dorsal interior.

(preserved within the Iapetus Suture Zone), there is a mixture of benthos with both Laurentian and Baltic–Avalonian affinities, and other distinctive and endemic genera are also known only from single sites. Those Ordovician–Middle Iapetus faunas have been termed the Celtic Province (Harper & Servais, 2013), although the distributions of the endemic genera are patchy and none of them have been found in all of the 'Celtic' faunal sites. The Arctic Alaska–Chukotka

Microcontinent between Siberia and Laurentia was close enough to both continents to allow some faunal mixing of the brachiopods and other megafossils as well as the conodonts. In contrast, among other faunas, the Chukotka and Seward sectors of that microcontinent have yielded *Monorakos*, an unmistakeable trilobite whose whole family is otherwise largely endemic to Siberia (Cocks & Torsvik, 2011).

Siberia and Peri-Siberia. In the Early Ordovician, Siberia also had many endemic faunas (Fortey & Cocks, 2003); for example, the trilobite family Monorakidae is not only confined to the continent (apart from the Chukotka occurrence mentioned above), but its relationships with the other families within the Order Phacopida are enigmatic. Considering Siberia's relatively low palaeolatitude (Fig. 6.7), its articulated brachiopods were of surprisingly low diversity, with only five cosmopolitan genera, *Apheoorthis*, *Archaeorthis*, *Finkelnbergia*, *Nanorthis*, and *Tetralobula*, as well as the endemic *Eosyntrophopsis*, known from the Tremadocian. Of those, *Finkelnbergia* and *Nanorthis* alone continued on up into the Floian and Dapingian, to be joined there only by the cosmopolitan *Syntrophopsis* and the endemic *Rhyselasma*. Thus, far fewer brachiopods lived in Siberia than would be predicted from its low palaeolatitudes (comparable to those of Laurentia), which further underlines its relative longitudinal isolation (Fig. 6.2).

The unusual number of endemic trilobite and brachiopod genera present in the Ordovician and Early Silurian of the Tuva area (Fig. 6.7) also indicates either geographical or climatic differences between Tuva and the surrounding areas (Cocks & Torsvik, 2007). The diversity on the Siberian Craton did increase from its Early Ordovician low, presumably reflecting both a decrease in Siberian isolation and also the global generic increase in the Ordovician radiation (Webby et al., 2004). Thus, in the Darriwilian, for example, there are 15 brachiopod genera, of which only one, *Evenkina*, is endemic. In the Late Ordovician (the Sandbian and Early Katian), there are slightly more diverse brachiopods, but with the strophomenoid *Maakina* and the orthide *Evenkorthis* as the only endemics (Rozman, 1978). The most common Upper Ordovician rocks over most of the Siberian Craton have abundant but low-diversity rhynchonelloid brachiopods (such as *Lepidocycloides*, *Evenkorhynchia*, and *Rostricellula*), apparently indicating the presence of only shallower water on the flooded shelves there. Not all the latest Ordovician is represented by rocks in Siberia, but, although the faunas from Taimyr in northern Siberia represent the Late Katian Boda global warming event (Fortey & Cocks, 2005), rocks of subsequent latest Katian and Hirnantian ages are absent from all of Siberia. Indeed, apart from Taimyr, Late Ordovician outcrops are confined to today's south-west of the continent.

The huge Late Katian carbonate platform, seen over much of the old Siberian Craton, extended south-westwards into the Salair and Gorny Altai areas, with the development of many bioherms (Sennikov, 2003). Whether the paraconformity between the latest Ordovician there and the earliest Silurian was caused by the eustatic sea-level fall which accompanied the Hirnantian global glaciation event, or by additional tectonic factors (or both), is unknown. Otherwise, there is no direct evidence in Siberia for the Hirnantian glaciation.

In the Sandbian and Early Katian, some of the shallower-water benthic trilobites, such as *Prionocheilus*, *Neseuritinus*, *Vietnamia*, and *Calymenia*, suggest that some faunal communication existed between Altai–Sayan and other continents such as South China (Fortey & Cocks, 2003), although it seems unlikely that those areas were close to each other. However, as the Ordovician progressed, Siberia apparently became closer to Baltica, and there was a steadily developing faunal interchange, so that by the Late Katian the brachiopods of the Taimyr Peninsula of Siberia and the Boda Limestone of Sweden (Baltica) were similar. In contrast, analysis of the various Katian brachiopod faunas from Gorny Altai and Taimyr, which were on opposite sides of Siberia, found only 19% similarity between the two, in contrast to a 30% similarity between Taimyr and South China, 29% between Gorny Altai and the Chu-Ili Terrane (now in Kazakhstan but then independent), and 27% between Gorny Altai and South China. That indicates a fair degree of separation and consequent endemism between those areas, which is also supported by analyses of trilobites and corals. However, all those regions were more similar to each other than peri-Gondwanan marine shelly faunas of comparable age from New South Wales, Australia, despite the fact that they too lived in similar tropical latitudes, although within ecological niches in island arcs.

Baltica. A large number of extensive Ordovician sequences are preserved on the main Baltica Craton, particularly within the Oslo Region, southern Sweden and the East Baltic, extending north-eastwards to the St Petersburg area of Russia, where, because of the lack of contemporary tectonism and the high ground from which clastic sediments could be eroded, most of the formations are extensive but thin. In the north of Baltica (Arctic Russia) there are substantial shelf deposits in Timan-Pechora, Pai-Khoi, and Novaya Zemlya, and also sporadic outcrops in the Urals (Fig. 6.11).

Baltica had travelled steadily northwards towards the Equator from the Middle Cambrian, and that movement

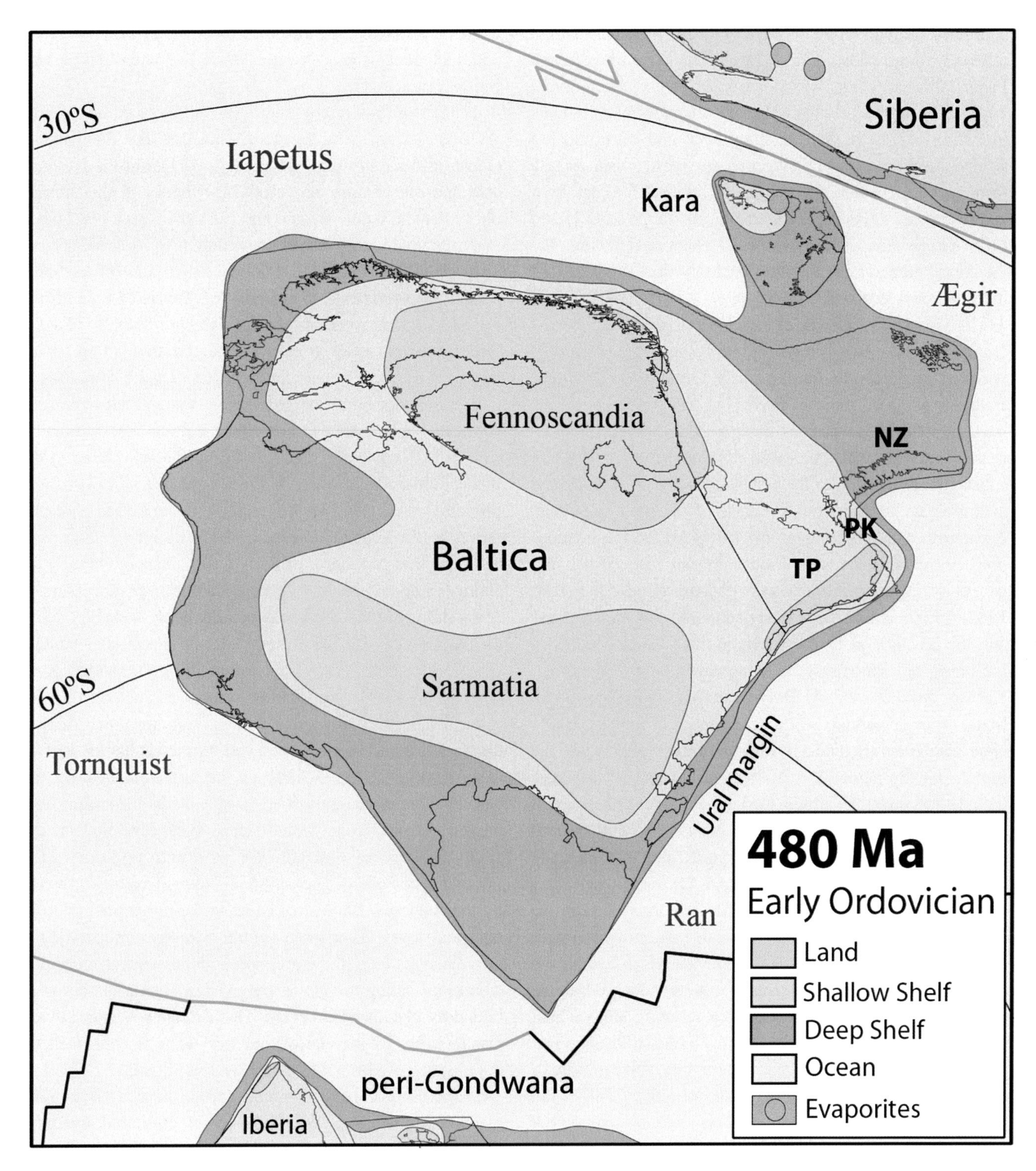

Fig. 6.11 Baltica and surrounding areas at 480 Ma (Tremadocian), showing the lands and seas, and the evaporites, which were confined to the more temperate latitudes in Siberia and Kara. Black lines are spreading centres, and green lines are transform plate margins. NZ, Novaya Zemlya; PK, Pai-Khoi; TP, Timan–Pechora.

continued throughout the Ordovician, with the result that the relatively thin-bedded Early Ordovician limestones in the East Baltic area are of cooler-water origin (Jaanusson, 1973). The only Early Ordovician carbonate mud mounds in Baltica are in the St Petersburg area, and were built up around accumulations of siliceous sponges, which mostly form in cooler-water environments (Fig. 6.12). That is in contrast to the much more diverse faunas in the Middle and Later Ordovician of the whole of Baltica, reflecting the increased temperatures through its steady drift to lower latitudes (Torsvik & Cocks, 2005).

The Iapetus Ocean was at its widest at about Cambrian–Ordovician boundary times, and the Ran Ocean had also widened significantly by the Early Ordovician, and thus Baltica was at its most isolated (Fig. 6.1) and the oceans must therefore have been wide enough to prevent the successful passage and ecological integration of larvae for a substantial proportion of the benthos, leading to independent evolution within the benthic faunas of the shelf seas then. The most abundant benthos, the trilobites and the brachiopods, were represented in Baltica by not just species and genera but even families that were endemic (Cocks & Fortey, 1982). They include the trilobite subfamily Megistaspidinae and the brachiopod family Lycophoriidae, which occurs in rock-forming abundance in Estonia, north-west Russia, Norway, Sweden, and the Holy Cross Mountains of Poland. Many other rhynchonelliform brachiopods in the East Baltic were also endemic, and, as well as the Lycophoriidae, the Floian (locally named the Billingenian Stage) assemblages include the endemic family Gonambonitidae with the distinctive and abundant genera *Antigonambonites* and *Porambonites*, and the earliest clitambonitids and endopunctate orthoides such as *Angusticardinia* and *Paurorthis*.

As the Ordovician progressed, the surrounding oceans to the south and west of Baltica became steadily narrower (Fig. 6.12), so that many faunal elements, which had ancestors in neighbouring continents but were new to Baltica, arrived. There are no known offshore oceanic islands which might have subsequently amalgamated with Baltica along its southern margins (the Tornquist Suture Zone and eastwards). However, on its western margin, and originally within the Iapetus Ocean, there are various suspect terranes, some now within the Scandinavian Caledonides, which carry Ordovician faunas, chiefly of brachiopods, and most of those brachiopods show little affinity with contemporary Baltic faunas elsewhere. One such Dapingian to Darriwilian fauna is from the Hølonda area, western Norway, in which 8 out of 13 (62%) brachiopod genera and 12 of 13 (92%) trilobite genera also occur in Laurentia, and the remainder are endemic to Hølonda (Harper et al., 1996). Thus that fauna must have lived on an island arc in the Iapetus Ocean near Laurentia and had better faunal contact with the latter rather than Baltica.

Today's north-western margin of Baltica faced northwards towards the northern Iapetus Ocean and the Panthalassic Ocean in the Early Ordovician (Fig. 6.11), but the Hølonda area and others were emplaced onto Baltica in the Silurian only after the whole terrane had rotated by about 90° when that margin by then faced Laurentia (Fig. 7.3) (Cocks & Torsvik, 2005). Thus the Hølonda fauna, and others transported eastwards later by nappes of Silurian age onto the Baltica Craton, represent faunas which originally lived a great distance away from the autochthonous Ordovician faunas of Baltica, despite the fact that they are found today in outcrops not far away from faunas transported within the nappes. Our Late Ordovician Baltica reconstruction (Fig. 6.12) shows an island arc in the Iapetus Ocean to the then north (today's west) of Baltica, whose volcanism was probably responsible for the increase in rare earth elements found in the organophosphatic shells of inarticulated brachiopods and conodonts of Early Ordovician age in the East Baltic, when compared with their Cambrian predecessors.

As Baltica's palaeolatitudes steadily decreased (Fig. 2.12), so the abundance and diversity of the successive benthic faunas increased as average ambient temperatures rose. However, there were marked climatic fluctuations; for example, in the Late Katian facies overlying much of Baltoscandia, limestones are followed on the shallower shelves by graptolite shales, which are in turn followed by the carbonate mud mounds such as the Boda Limestone discussed below. Deeper-water facies were deposited on the south and east of the Baltica Craton and are seen in the boreholes of Poland. The centre of Baltica was little affected by the Avalonia–Baltica collision; in the uppermost Ordovician and lower Silurian in central Sweden (Jämtland), the only changes were the regression and subsequent transgression representing the global Ordovician Hirnantian–Silurian boundary glacioeustatic event. That Baltica was at relatively low latitudes by the end of the Ordovician is confirmed by the absence of Hirnantian glaciogenic sediments.

In the early Late Katian, and before the end-Ordovician glaciation, there was the Boda Event of global warming (Fortey & Cocks, 2005), which led to the formation of substantial carbonate mud mounds (bioherms) with very diverse brachiopods, trilobites, molluscs, echinoderms, and bryozoans, and there were unusually large numbers of endemic genera and species, for example in the strophomenide brachiopods. Those mud mounds are best seen within the Boda Limestone in central Sweden, but also in Norway, Estonia, Novaya Zemlya, and the Urals, as well as in

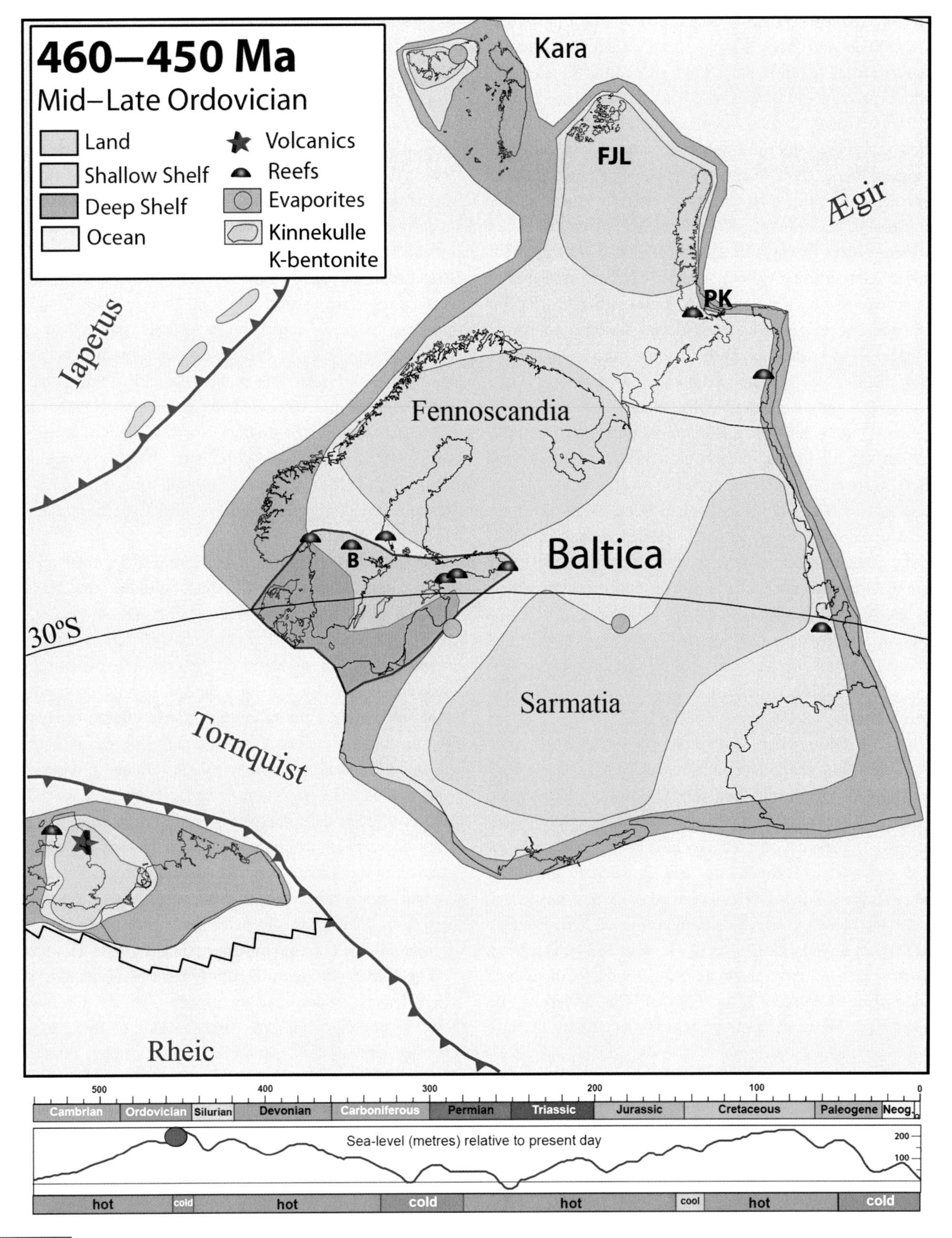

Fig. 6.12 Baltica, part of Avalonia and the approaching Iapetus Ocean arcs (shown diagrammatically) at 460 Ma (Sandbian) to 450 Ma (Katian), showing the lands and seas, and also the distribution of the extensive Kinnekulle Bentonite in the southern Baltic Sea area (boundary with thick red line), which came from the volcano shown in Avalonia. Solid blue lines are subduction zones with teeth on the upper plate, black lines are spreading centres. B, Boda, Sweden; FJL, Franz Josef Land; PK, Pai-Khoi. Bottom: Phanerozoic time scale and sea-level variations (Haq & Al-Qahtani, 2005; Haq & Shutter, 2008) and icehouse (cold) and greenhouse (hot) conditions.

adjacent Avalonia (Fig. 6.12). The Late Katian contrasts with the Sandbian and Early Katian, from which only smaller carbonate mud mounds are known in central Sweden and northern Estonia. No latest Ordovician bioherms are known, due to the Hirnantian global cooling and glaciation, and thus there was a substantial reduction in the number and variety of ecological niches available, which partly explains the faunal turnover and extinctions across the Ordovician–Silurian boundary.

Temperature Variations and the Late Ordovician Hirnantian Glaciation. Although various oxygen and carbon isotope excursions are known in some earlier Ordovician rocks, there were no Ordovician glacial intervals confirmed by glaciogenic sediments until the end-Ordovician Hirnantian glaciation (the Middle Ordovician event reported from North Africa can be discounted; Torsvik & Cocks, 2011). In the Late Ordovician (Early Katian) sediments on much of south-eastern Laurentia, there was a marked change from underlying warm-water limestones to overlying more temperate limestones. That change had been ascribed to the advent of ice caps near the poles. However, it is probably better explained as due to tectonism to the south-east in the Taconic Orogeny, which forced deep cool oceanic waters into the Sebree Trough and onto the adjacent Lexington Platform in the eastern USA where the carbonates were laid down, in contrast to the earlier warmer-water carbonates there which had been deposited without influence of that cooler influx.

The Late Ordovician South Pole lay under North-West Africa, and thus the surrounding areas were covered by a substantial ice cap during the Latest Ordovician (Hirnantian) glaciation. The glacial and peri-glacial deposits were extensive, and impressive striated pavements and other glaciogenic rocks and features are still preserved, originally illustrated from Algeria (Beuf et al., 1971), and have been plotted in wider areas in Africa (Ghienne et al., 2007) and the Middle East at 445 Ma (Fig. 6.9) (Torsvik & Cocks, 2011). This was one of only three glacial periods in the whole Phanerozoic (the others being Permian–Carboniferous and Pleistocene). Partly for tectonic reasons and partly because water was locked up in the ice caps, global sea levels dropped in the Hirnantian, with the result that there are Ordovician–Silurian unconformities in most places round the world. However, the Hirnantian glacial episode probably lasted for substantially less than a million years (Villas et al., 2006).

Where Hirnantian rocks are preserved, it can be seen that not only did the sea level drop, but also the oxygenated zone extended to deeper sea depths than usual. That deeper oxygenation enabled some opportunistic benthos to colonise deeper parts of the continental shelves than usual, in particular a group of brachiopods named the *Hirnantia* Fauna, whose principal members are *Hirnantia*, *Eostropheodonta*, and *Hindella*, as well as a few trilobites including *Mucronaspis*. The diversity of the *Hirnantia* Fauna varied from as few as three genera to as many as twenty, but its distribution was global (Rong & Harper, 1988), and we show some sites in and near Gondwana on Fig. 6.9. Although its development resulted from the glacial cooling, the *Hirnantia* Fauna stretched from high to low latitudes, and it is therefore less correct to interpret that fauna as 'cooler-water', as many authors (including ourselves) have done, rather than as one whose generally lower diversity reflects the limited proportion of benthic taxa that reacted quickly to the changing environments and colonised the newly oxygenated deeper sea floors. In the Oslo Region, Norway, there is a regressive sequence of brachiopod-dominated benthic communities from deep shelf to subtidal, conformably followed by rocks representing a transgressive deepening in the earliest Silurian (Brenchley & Cocks, 1982).

Although no Hirnantian glaciogenic rocks occur in Laurentia, since it was at low palaeolatitudes, the glaciation affected even that equatorial area; firstly, in the eustatic lowering of sea level (Fig. 16.2) which led to widespread unconformities, particularly on the craton, between rocks of Ordovician and Silurian ages, and, secondly, in the breakdown of many of the relatively fragile marine ecosystems. The most complete latest Ordovician sequences in Laurentia are to be found at the margins of the terrane, in particular at Anticosti Island, in eastern Canada (Copper & Jin, 2015). However, even there there is a minor paraconformity at the Ordovician–Silurian boundary, and it is noticeable that the basal Silurian (Llandovery: Rhuddanian) rocks of Anticosti contain almost no warmer-water limestones and no bioherms, in contrast to the patch reefs and other carbonates dominating the Hirnantian strata beneath the paraconformity.

The End-Ordovician Extinctions. The Hirnantian glaciation led to widespread extinctions at the Ordovician–Silurian boundary interval, such as that of the previously characteristic eastern Laurentian Richmondian brachiopod and coral faunas, in which many genera and higher taxa both in individual environments and in more widespread communities failed to adapt to the climatic changes. Unlike some other extinction events, all the phyla were involved, including vertebrates, invertebrates, and microfossils (Harper & Servais, 2013). This was the third-largest extinction event in the whole Phanerozoic, but is not linked to any known LIP eruption.

The Hirnantian glacial interval lasted for less than a million years, but two separate progressive extinction phases have been recognised within it. However, there were also a significant number of benthic families and higher taxa (so-called 'Lazarus taxa') which, although unknown from both Hirnantian and Early Silurian rocks (as reported for the brachiopods by Cocks & Rong, 2008), are recorded from both older Ordovician and younger Silurian or later rocks and thus must have survived that extinction in as yet undiscovered refugia.

7 Silurian

During the Early Silurian, the climate warmed gradually after the end-Ordovician glaciation so that by Late Silurian times tropical limestones and reefs abounded, such as these at Holm Hallar rocks on the Baltic island of Gotland, Sweden. Photo by Robin Cocks, © The Natural History Museum, London.

The Silurian has the shortest length of any of the Palaeozoic systems, only some 24 million years; however, within it occurred one of the most important events in the Phanerozoic, the collision of the substantial continents of Laurentia and the combined Avalonia–Baltica in the Caledonide Orogeny.

Roderick Murchison (in Sedgwick & Murchison, 1835) named the Silurian System after the Silures, an ancient Welsh Celtic tribe, but that early definition was imprecise on the limits of the system, and thus it was left to Lapworth (1879) to define its lower boundary immediately above his new Ordovician system and at the base of the Llandovery Series, also in Wales. That base is now recognised formally as at the base of the *Akidograptus ascensus* graptolite Biozone at Dob's Linn, Scotland, and is dated at 443 Ma. The global distributions of the continents and oceans at the beginning of the Silurian at 440 Ma (Early Llandovery: Rhuddanian Stage), 430 Ma (Late Llandovery: Telychian Stage), and at 420 Ma (Pridoli) are shown in Fig. 7.1. The Silurian System is divided into four series: Llandovery, named after a town in central Wales; Wenlock, and Ludlow, which are two towns in Shropshire, England; but the uppermost series, the Pridoli (which was for many years before the 1970s historically placed within the Devonian), has its type area in Bohemia,

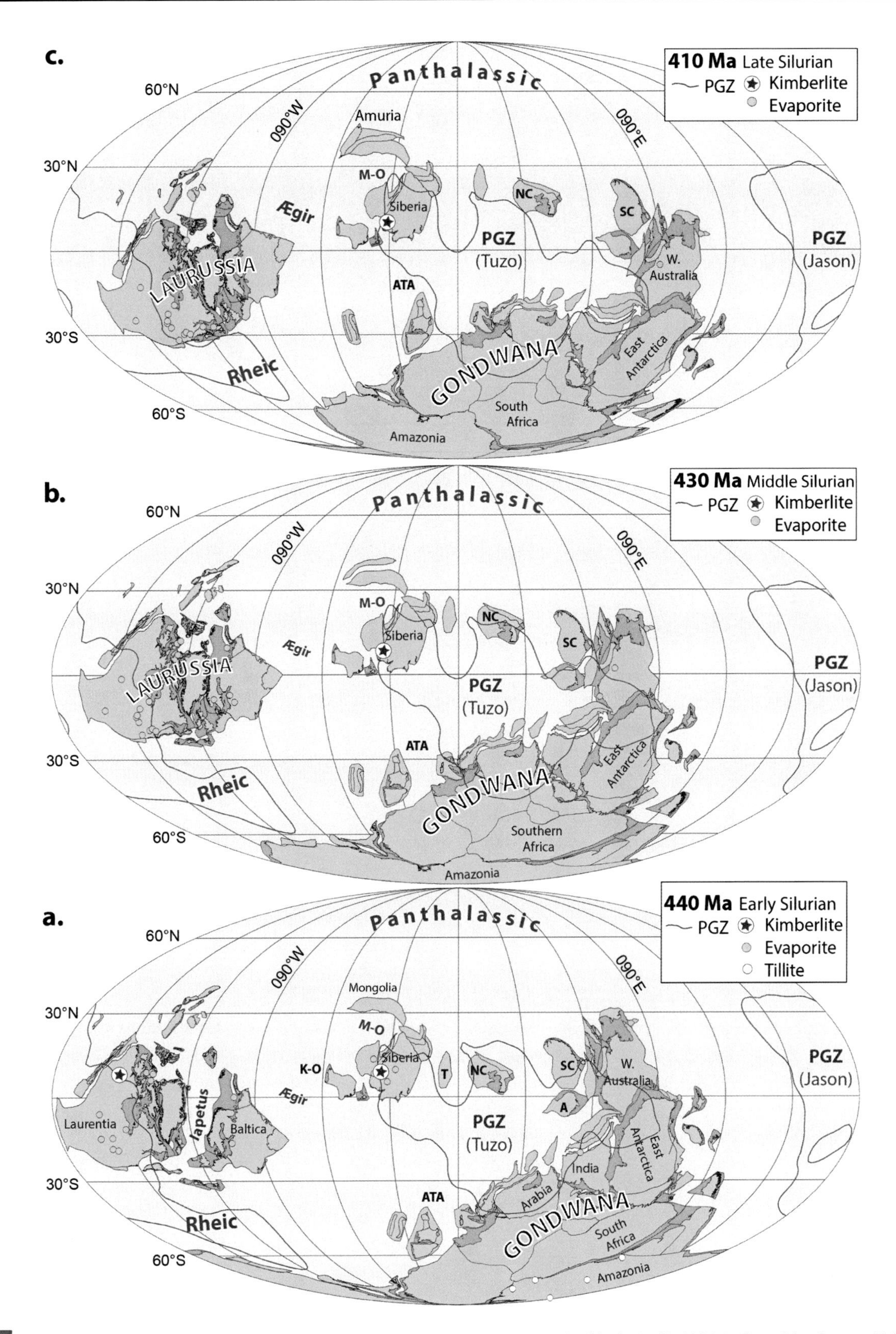

Fig. 7.1 Outline Earth geography in the Silurian at (a) 440 Ma (Early Llandovery: Rhuddanian), (b) 430 Ma (Late Llandovery: Telychian), and (c) 420 Ma (Pridoli), including the outlines of the major crustal units, and the more substantial oceans. Kimberlites are known in Siberia and Laurentia. A, Annamia; ATA, Armorican Terrane Assemblage; K-O, Kolyma–Omolon; M-O, Mongol–Okhotsk Ocean; NC, North China; PGZ, plume generation zone; SC, South China; T, Tarim.

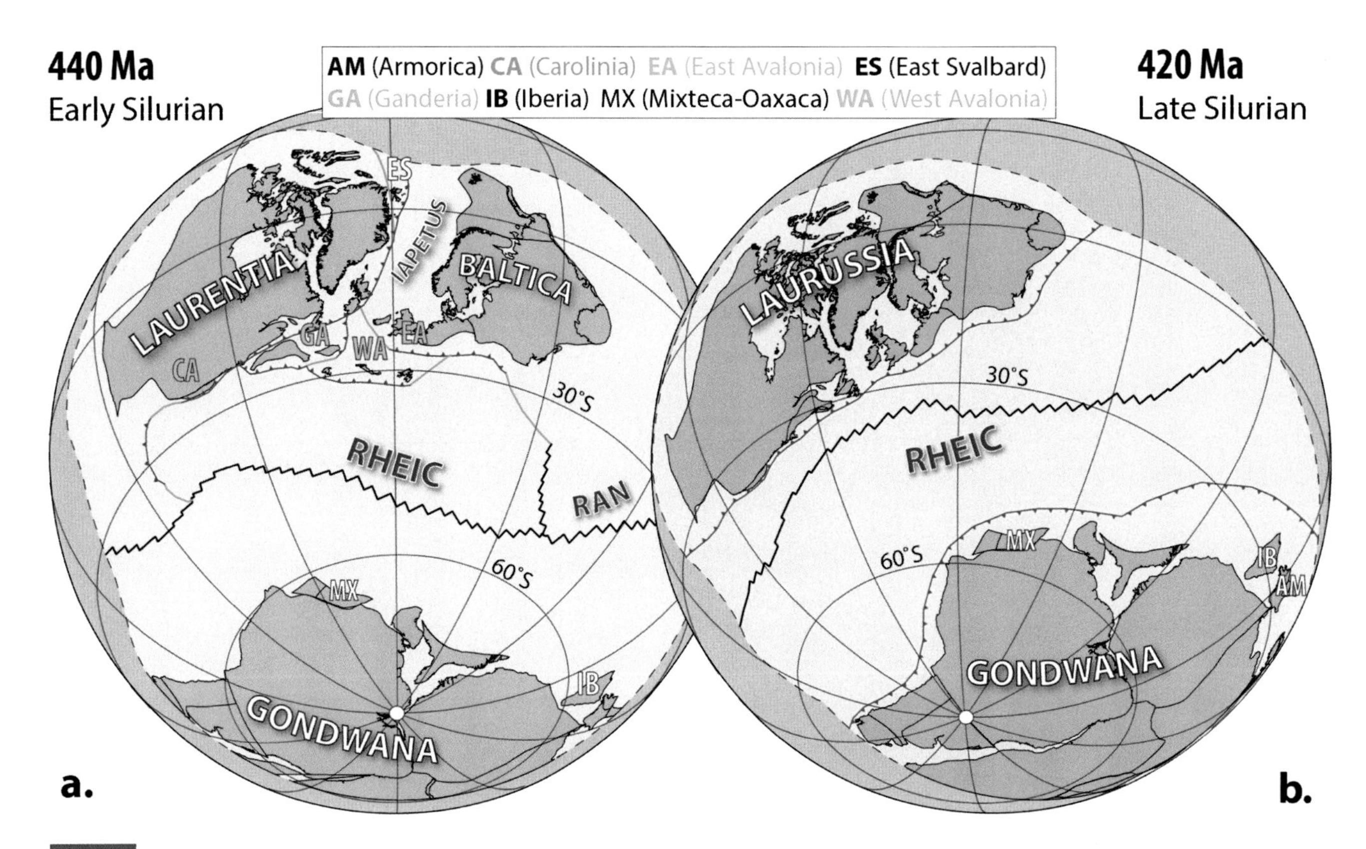

Fig. 7.2 (a) Early Silurian (440 Ma: Llandovery) reconstruction, with the dominant Rheic Ocean and a much-reduced Iapetus. (b) Late Silurian (Pridoli: 420 Ma) reconstruction after the Iapetus Ocean had closed in the Caledonide Orogeny. Solid blue lines are subduction zones with teeth on the upper plate, black lines are spreading centres, and green lines are transform plate margins. The dashed boundary to the grey areas in the reconstruction margins is an arbitrary perimeter (Domeier, 2015) marking the outer limit of the Iapetus and Rheic–Ran domains.

Czech Republic. The Llandovery Series represented more than half of Silurian time.

Tectonics and Igneous Activity

Oceans. The Panthalassic continued to dominate nearly half of the globe (Fig. 7.1). In today's Atlantic region, the processes which had become well under way during the Ordovician continued on into the Early Silurian. However, the previously mighty Iapetus Ocean had already dwindled to a fraction of its former size by the start of the period, and disappeared entirely during the Middle Silurian Caledonide Orogeny (see below). In contrast, the Rheic Ocean continued its expansion (Domeier, 2015), and remained the dominant ocean in the region until the Late Palaeozoic (Fig. 7.2). Further afield, there was the relatively small Mongol–Okhotsk Ocean between Siberia and the Central Mongolian Terrane area, which was narrow in the Late Silurian, as can be seen from the otherwise endemic temperate *Tuvaella* brachiopod Fauna on both its opposing sides (Fig. 7.7). Few oceanic tracts have been consistently named in that general area, although the Paleoasian Ocean is a term often used, but in various different ways, in the area today in Central and Eastern Asia.

Caledonide Orogeny. The northward movement of Baltica had continued steadily during the Ordovician, and thus the Iapetus Ocean was not wide at the start of the Silurian (Fig. 7.2a), and, as subduction progressed along its margins, it gradually shrank even more before completely vanishing (Fig. 7.2b). Since by the Middle Silurian, both Laurentia and Avalonia–Baltica were at those margins, they inevitably collided, and since that collision was direct, with only a little strike-slip motion involved, the resultant orogeny was very substantial (Fig. 7.3). Rocks representing that Caledonide Orogeny (named after the Caledonian Mountains of Scotland) outcrop extensively on both sides of the North Atlantic today from the Arctic and extending as far southwards as north-western Africa. The orogeny (often termed the Scandian Orogeny in the Baltic area) was at its peak during the second half of the Silurian. Although both the continents of Laurentia and the combined Avalonia–Baltica had themselves been substantial before they merged, the newly

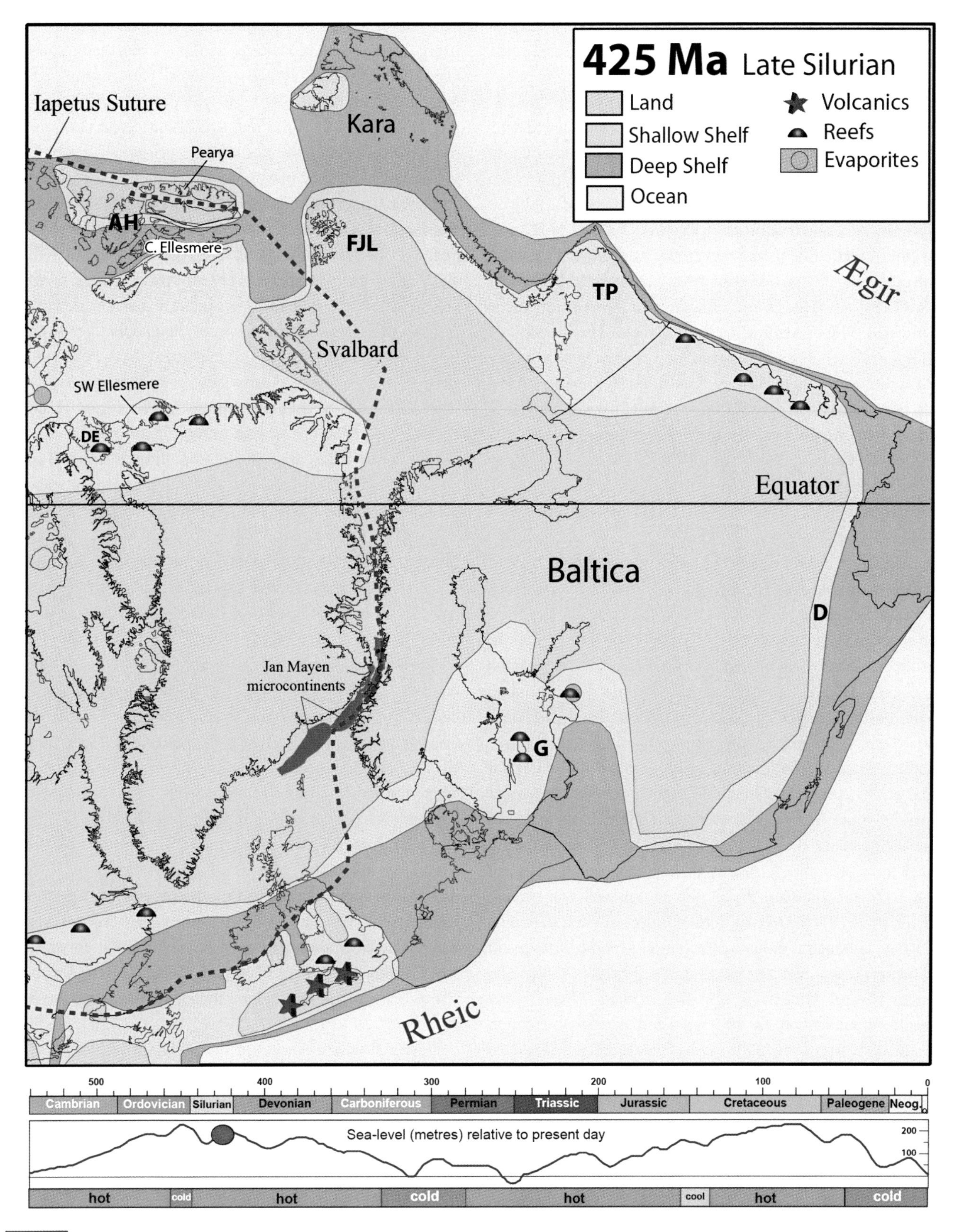

Fig. 7.3 The eastern and central Laurussian area at 425 Ma (Late Silurian: Ludlow), showing the lands and seas, and the distributions of reefs, evaporates, and volcanoes. Blue dashed line is the Iapetus Ocean Suture. The Jan Mayen microcontinental fragments, now mostly submerged between Norway and Greenland, are shaded dark grey. AH, Axel Heiberg Island; D, Dniester River, Ukraine; DE, Devon Island; FJL, Franz Josef Land; G, Gotland, Sweden; TP, Timan–Pechora, Russia. Bottom: Phanerozoic time scale and sea-level variations (Haq & Al-Qahtani, 2005; Haq & Shutter, 2008) and icehouse (cold) and greenhouse (hot) conditions.

combined and very large continent of Laurussia therefore stretched from the west coast of North America eastwards to the Ural Mountains of Russia.

There had been progressive collisions during the Ordovician and Silurian between all the various island arc chains which had been situated independently from each other within the Iapetus Ocean before the final collision between the two chief (and by then amalgamated) continents of Baltica–Avalonia and Laurentia (Torsvik & Cocks, 2005). In Baltica, the chief Caledonide nappe movement was towards the east, with elements of what had previously been parts of Laurentia, as well as some of the exotic terranes from the Middle Iapetus island arcs, both overriding the Baltic Craton. The Caledonide uplift and mountain-building resulted in the extensive terrestrial sediments of the Old Red Sandstone Continent: deposition of which started during the Late Llandovery in Scotland, and the Late Wenlock in Norway, and continued over a wide area into Devonian times (Friend & Williams, 2000).

The whole Caledonide Orogeny was prolonged and complex in detail; for example, on the Laurentian side (Fig. 7.5) the bimodal volcanism seen in New Brunswick shows that the Middle Silurian basalts and rhyolites present there erupted in an extensional rather than a compressional environment (Van Wagoner et al., 2002). Within and around the complicated suture zone representing the Iapetus Ocean closure, a large number of relatively small terrane units were eventually accreted to the newly enlarged continent. In the northern Appalachians and Newfoundland, the Ganderia Terrane collided obliquely with Avalonia in a phase locally termed the Salinic Orogeny at around 430 Ma (van Staal et al., 2009). The Meguma Terrane (shown as part of West Avalonia in Fig. 7.2) of south-eastern Canada might have accreted to the southern margin of Avalonia during either the Caledonide Orogeny or the Neoacadian Orogeny in the Early Devonian: the timing is not well constrained (Domeier, 2015). A probable further aspect of the Caledonide Orogeny, although it predated the central event in Britain and Scandinavia, was the compression by 200–400 km in the eastern part of Greenland, and the welding of the Franz Joseph Parautochthon to the Greenland part of the craton (Smith & Rasmussen, 2008), with granites there dated as Early Silurian.

Gondwana. It seems likely that incipient rifting took place on the North African margin of Gondwana before the end of the Silurian in advance of the opening of the Galicia–Moldanubian extension of the Paleotethys Ocean, as shown in Fig. 8.9, but that activity could not have been linked to the Caledonide Orogeny, since the Rheic Ocean between Gondwana and Laurussia had become several thousand kilometres wide by then (Figs. 7.1 and 7.2). As in the Ordovician, today's northern margin of Gondwana eastwards from Turkey was quiescent as far as northern Australia (Torsvik & Cocks, 2009). However, Gondwana's eastern margin continued to be very active, with the Macquarie Arc of New South Wales and neighbouring parts of Australia, which was originally of oceanic origin and Early Ordovician to Llandovery in age, being accreted to the main West Australia Craton during the later Silurian (Glen, 2005). To what extent that active subduction zone extended continuously south-westwards along the Antarctic, South African, and South American margins of Gondwana is uncertain: only oceanic rocks are known from Antarctica as Ordovician and Silurian fragments in the Antarctic Peninsula. Nevertheless, even though the subduction zone may not have been continuous, subduction was clearly active and continuing along many parts of that long Gondwanan margin (Vaughan et al., 2005). There was also tectonic activity and propagation of new terranes in an island arc setting offshore of south-western South America, a process which had commenced in the Cambrian and continued throughout the Silurian.

Central and Eastern Asia. In contrast to the areas affected by the Caledonide Orogeny, there does not seem to have been so much extensive tectonic activity in the Silurian of Asia as there had been in the Cambrian and Ordovician. In the Kazakh Orogen, there was still not yet a unified Kazakhstania continent in the Early Llandovery; however, the units within the archipelago there became progressively more unified by terrane amalgamation during the Silurian. Near the beginning of the Silurian, the Selety Terrane and the Boshchekul Microcontinent merged. The North Tien Shan Microcontinent, which had previously accreted with the Karatau–Naryn Terrane, was joined by the Chu-Ili Terrane in the Middle Silurian. Thus there still appear to have been five fairly substantial separate microcontinents in the Silurian: North Tien Shan (including the previously independent Karatau–Naryn and Chu-Ili units), Atashu–Zhamshi, Kokchetav (including Ishim), Boshchekul (including Selety), and Chingiz–Tarbagatai (Popov & Cocks, 2016). Tarim was also nearby and still remained independent during the Silurian (Fig. 7.1).

On the southern margin of the Solonker Suture Zone, the Sulinheer Terrane had become substantial enough through the mutual accretion of several island arcs to depict separately on today's northern margin of North China from the Early Silurian onwards (Fig. 7.4). At today's south-western edge of North China, subduction continued, with the relatively small South Qinling Terrane being overridden by North Qinling (already accreted to North China), with the concomitant intrusion of numerous granulites within North Qinling, from near the start of the Silurian at about 440 Ma to

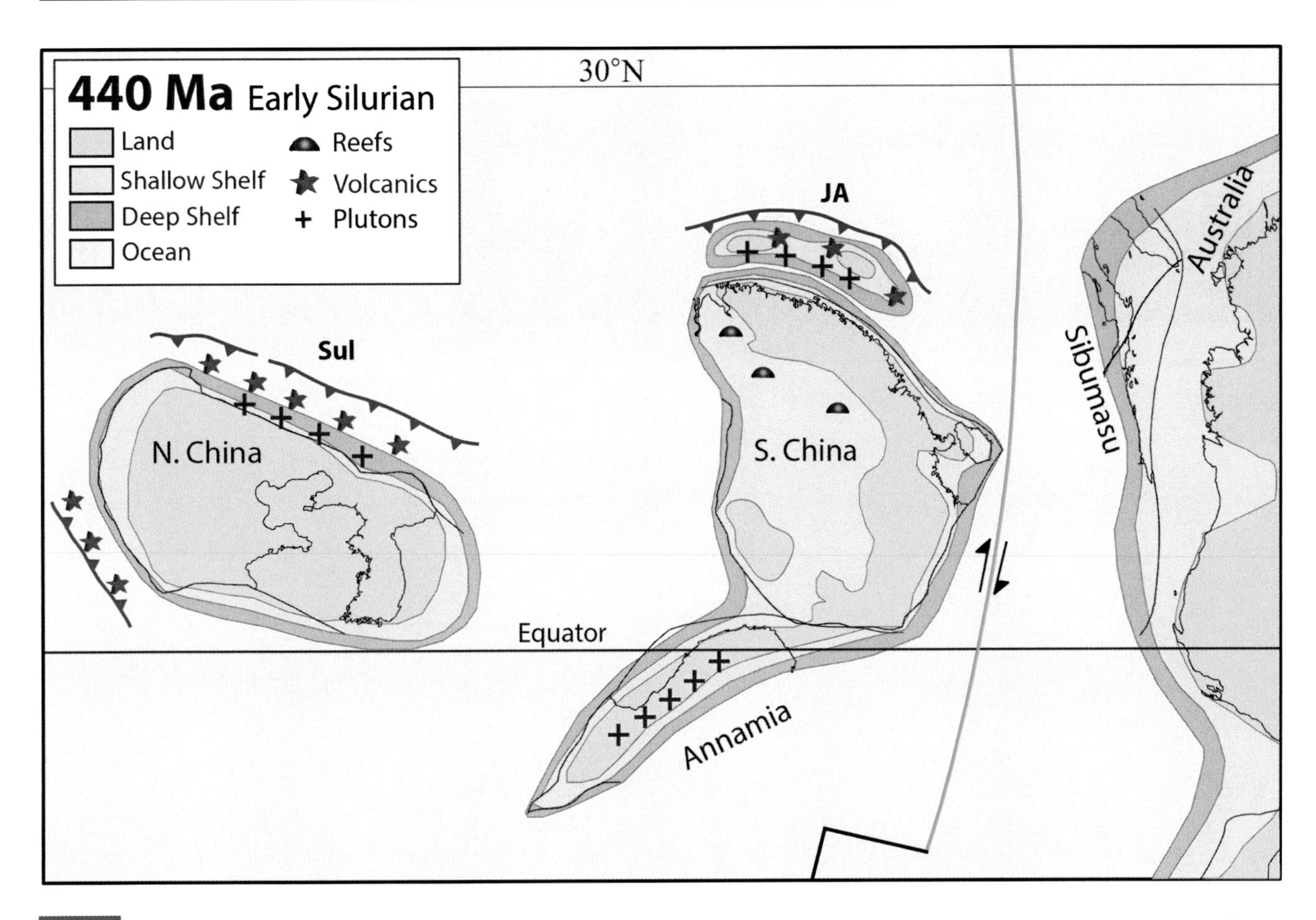

Fig. 7.4 Eastern Asia terranes and continents at 440 Ma (Early Llandovery: Rhuddanian), showing the lands and seas. Solid blue lines are subduction zones with teeth on the upper plate, black lines are spreading centres and transform faults, and green line is transform plate margin. JA, Japanese arcs; Sul, Sulineer Arc.

after its end at about 415 Ma (Lochkovian), after which metamorphism continued in the North–South Quinling junction area until the Early Devonian (Pragian) at about 400 Ma (Xiao et al., 2009a). Whether or not the strike-slip movement between Gondwana and the combined Annamia–South China continent continued throughout the Silurian is poorly constrained, but it seems likely, perhaps at a slower pace, since the rifting which heralded the opening of the Paleotethys Ocean between the two continents did not start until the Early Devonian. Late Silurian (424 Ma) granites of uncertain significance were intruded within the Vietnam (eastern) sector of Annamia, although there are no certain Silurian rocks known from the western (Thai) sector of that continent (Ridd et al., 2011), which therefore presumably remained oceanic. The Japanese Island Arc system still remained off today's eastern margin of South China.

In the Hutag Uul–Songliao Terrane (Unit 453 on Fig. 3.6), which spans the Chinese–Mongolia border and was later part of the Amuria continent (Fig. 8.7), evidence of Early Silurian (440 to 434 Ma) ridge subduction and subsequent microcontinent accretion is preserved in the Solonker Suture Zone (Jian et al., 2008).

Laurentia. In the Canadian Arctic islands (Fig. 7.6) the Laurentian Craton was deformed in the latest Silurian to form the (then) north–south trending Boothia Uplift, whose first phases may have caused local unconformities as old as Cambrian, but the cause of that uplift is uncertain: it reached its maximum at around Silurian–Devonian boundary time, and might have been due to post-Caledonide pressure causing readjustment of the Precambrian basement. The Pearya Terrane (Fig. 7.5) and equivalent rocks in North Greenland probably accreted to Laurussia in the Late Silurian (Trettin, 1998).

At today's western margin of Laurentia in the Cordilleran region, submarine alkaline igneous activity was much less evident in the Silurian than it had been in the Ordovician or Devonian (Goodfellow et al., 1995), but there are some diatremes near the craton margin in Yukon. Offshore of the craton in the Californian region, rocks preserved in the peri-Laurentian terranes in the Klamath Mountains and the Sierra

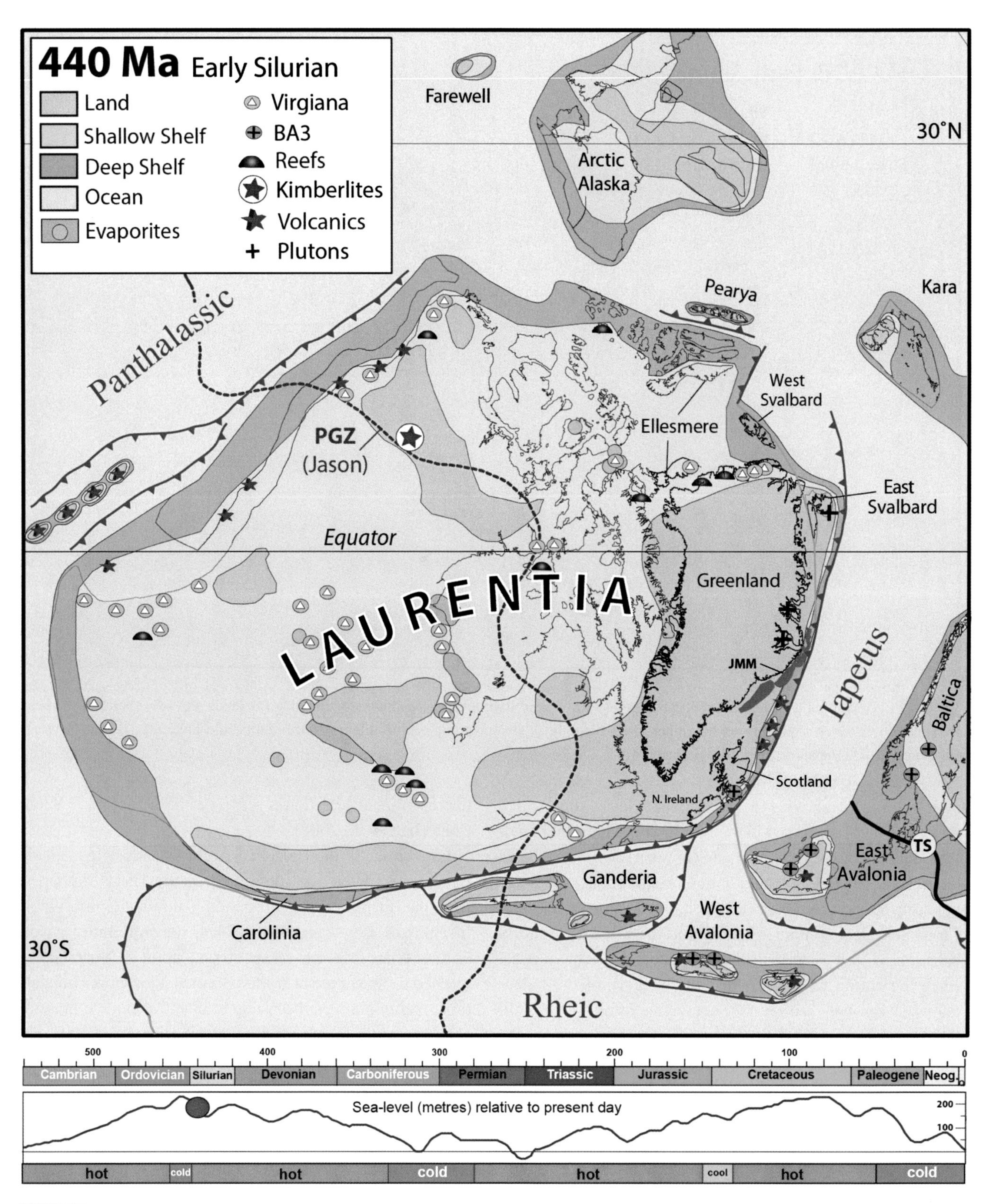

Fig. 7.5 Laurentia, Avalonia, and the Arctic Alaska–Chukotka area at 440 Ma (Early Silurian: Rhuddanian), showing the lands and seas and the distribution of the brachiopod *Virgiana* and other mid-shelf Benthic Assemblage (BA3) brachiopod communities without *Virgiana*. Solid blue lines are subduction zones with teeth on the upper plate, black lines are spreading ridges, and green lines are transform plate margins. Solid red line is the plume generation zone (PGZ). JMM, Jan Meyen Microcontinent; TS, Thor Ocean Suture. Bottom: Phanerozoic time scale and sea-level variations (Haq & Al-Qahtani, 2005; Haq & Shutter, 2008) and icehouse (cold) and greenhouse (hot) conditions.

Nevada represent remnants of one or more Ordovician and Silurian accretionary prisms which underlie the sub-Devonian unconformity there (Dickinson, 2000). At the end of the Silurian, the northern Alaskan area was affected by the Romanzof Orogeny, which was probably a preliminary phase of the Devonian Ellesmerian Orogeny, and which affected the Arctic Alaska sector of the Arctic Alaska–Chukotka Microcontinent (Fig. 7.6), as well as the Laurentian Craton margin further to the east in Arctic Canada. In the Alexander Terrane (Unit 165), now in the Cordillera, the Klakas Orogeny of Late Silurian to Early Devonian age represents a collision between two entities, but whether they were both island arcs or whether one was a microcontinent is uncertain; if the latter, it is poorly defined. However, Alexander, Wrangellia, and associated terranes were then situated together in the Panthalassic Ocean at some distance from Laurentia (Cocks & Torsvik, 2011), and their whereabouts is not constrained well enough to be shown on our maps until Mesozoic times.

Along the southern margin of Laurentia, former peri-Gondwanan terranes accreted during Late Ordovician and Silurian times (Figs. 7.5 and 7.6). Carolinia (Unit 172) probably collided with Laurentia at around 455 Ma (closing the Iapetus in that sector) whilst Ganderia (Unit 171) collided with the northern Appalachians during the Salinic orogeny, which is broadly coeval with the collision between Scotland and East Avalonia and Greenland–Norway. West Avalonia (including Meguma) probably accreted to Ganderia after peak Caledonian orogenesis and they were juxtaposed through a dextral oblique convergent boundary between 440 and 420 Ma (Domeier, 2015). In the Early Silurian (Figs. 7.2 and 7.5), the width of the Iapetus Ocean was still considerable between Greenland and Norway (as much as 1500 km), and the Iapetus Suture ran outboard of East Svalbard, through the Barents Sea region, and into the high Arctic (Pearya). East Svalbard was located along the north-east margin of Greenland during the Precambrian and Early Palaeozoic, but during the Silurian (probably after 440 Ma) East Svalbard was pushed northwards along a major sinistral strike-slip fault zone. It had almost fully accreted to West Svalbard at around 430 Ma (Early Wenlock), but there were more significant younger movements there, which culminated in the Late Devonian Svalbardian Orogeny at around 360 Ma. However, in the Silurian, the Caledonide orogenesis was very extensive and affected a zone more than 9,000 km long (red dashed line in Fig. 7.6).

Siberia and Peri-Siberia. The northward movement of both Siberia and peri-Siberia continued during the whole of the Late Cambrian, Ordovician, and Silurian, and the southern margin of the continent (today's northern margin) became clear of the Equator at some time near the Ordovician–Silurian boundary at 443 Ma (Fig. 7.7). Most of the centre and south-east of the Siberian Craton continued to be flooded by epeiric seas, but the continental land area was larger than in the Ordovician. For most of the Silurian onwards into the Late Palaeozoic, Siberia (including peri-Siberia) was the only continent to be situated entirely within the northern hemisphere, forming one of the margins of the immense Panthalassic Ocean.

As the Silurian progressed, more terranes became accreted to the already-enlarged Siberian Craton. In particular, the former West Sayan Terrane (Unit 431), which was largely composed of a series of accretionary wedges, gradually merged with today's south-western edge of the craton bordering the Angara landmass. However, the even larger Central Mongolian Terrane group (Unit 441: subsequently part of the continent of Amuria) remained on the other side of the Mongol–Okhotsk Ocean, but must have been close to Siberia, as can be seen from the presence of the distinctive *Tuvaella* brachiopod Fauna on both sides of the ocean in the Late Silurian (Fig. 7.7). The Mongol–Okhotsk Ocean grew steadily during the Late Palaeozoic (Figs. 8.1, 9.1, and 10.1), but Amuria did not collide with North China until the Early Mesozoic (Triassic) at around 240 Ma.

Facies, Faunas, and Floras

Global Climates and Sea Levels. After the end-Ordovician glaciation, the Earth grew gradually warmer (Fig. 16.2) during the first half of the Silurian, particularly during the long-lasting Llandovery Stage, although some glaciogenic deposits from as late as the Middle Silurian (Wenlock) are known from the high-latitude sector of Gondwana now in Brazil, and a substantial rise in eustatic sea level resulted from that melting. When the ice cap retreated from North Africa and Arabia, the transgressive seas incorporated many previously glaciogenic sediments and were poorly oxygenated (Ghienne et al., 2010), and that resulted in the extensive deposition of Early Silurian (Llandovery) 'hot' black shales there, which together form the world's largest area of hydrocarbon source rocks. Relatively abundant Middle and Late Silurian bioherms signalled that by then global temperatures had completely recovered after the ice age.

In the more temperate and warmer areas, the eustatic sea level rises resulting from melting ice caps and local isostatic readjustments also led to large transgressions in many areas. One of the most famous is the progressive transgression over the Midlands Microcraton of the former Avalonia from west to east over Wales, through the Welsh Borderland, and on

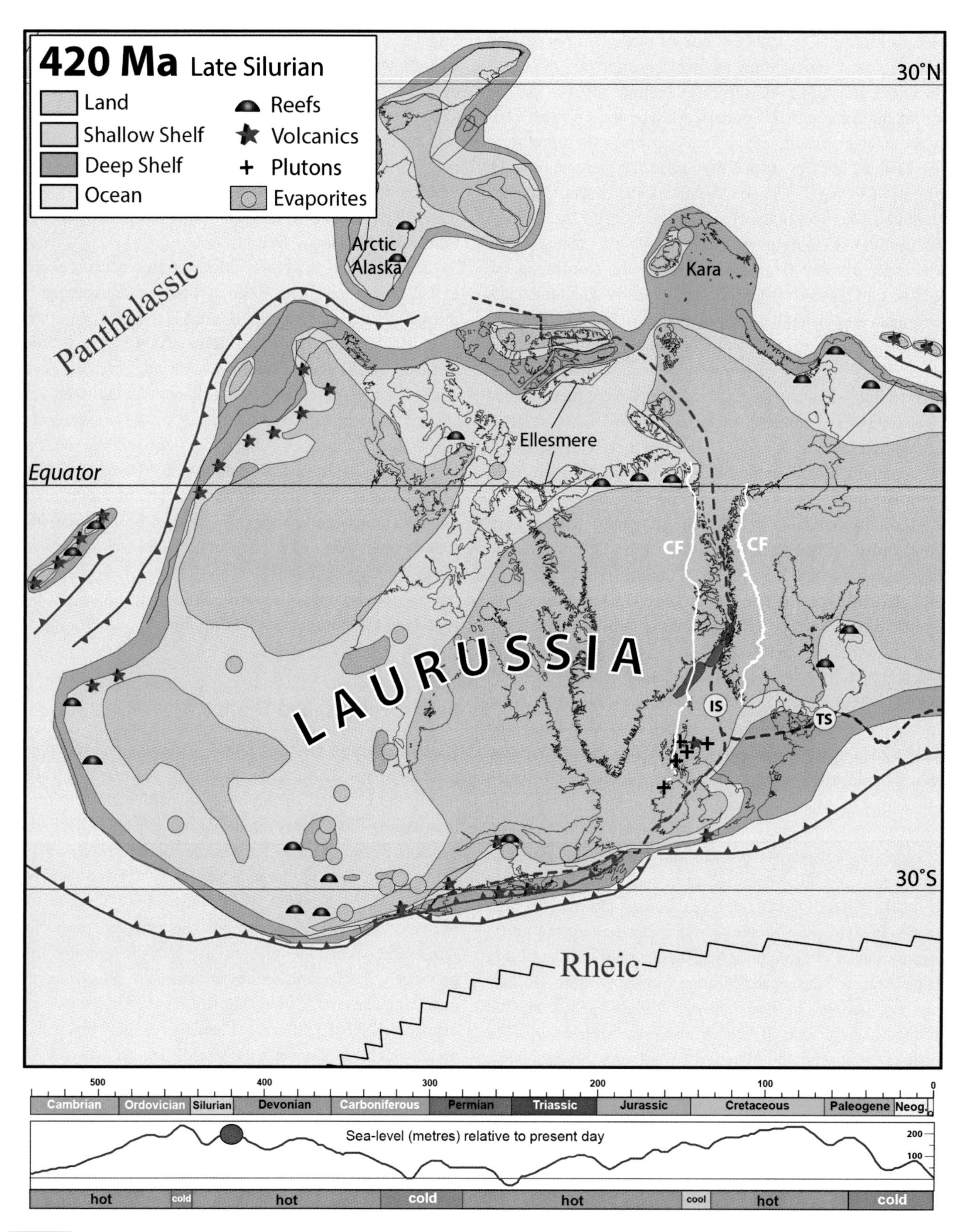

Fig. 7.6 The western and central Laurussian area at 420 Ma (Late Silurian: Pridoli), showing the lands and seas. Solid blue lines are subduction zones with teeth on the upper plate, and black lines are spreading ridges. Red dashed line is the Iapetus Ocean Suture (IS). TS, Thor Suture; CF, Caledonian Front (between the white lines). Bottom: Phanerozoic time scale and sea-level variations (Haq & Al-Qahtani, 2005; Haq & Shutter, 2008) and icehouse (cold) and greenhouse (hot) conditions.

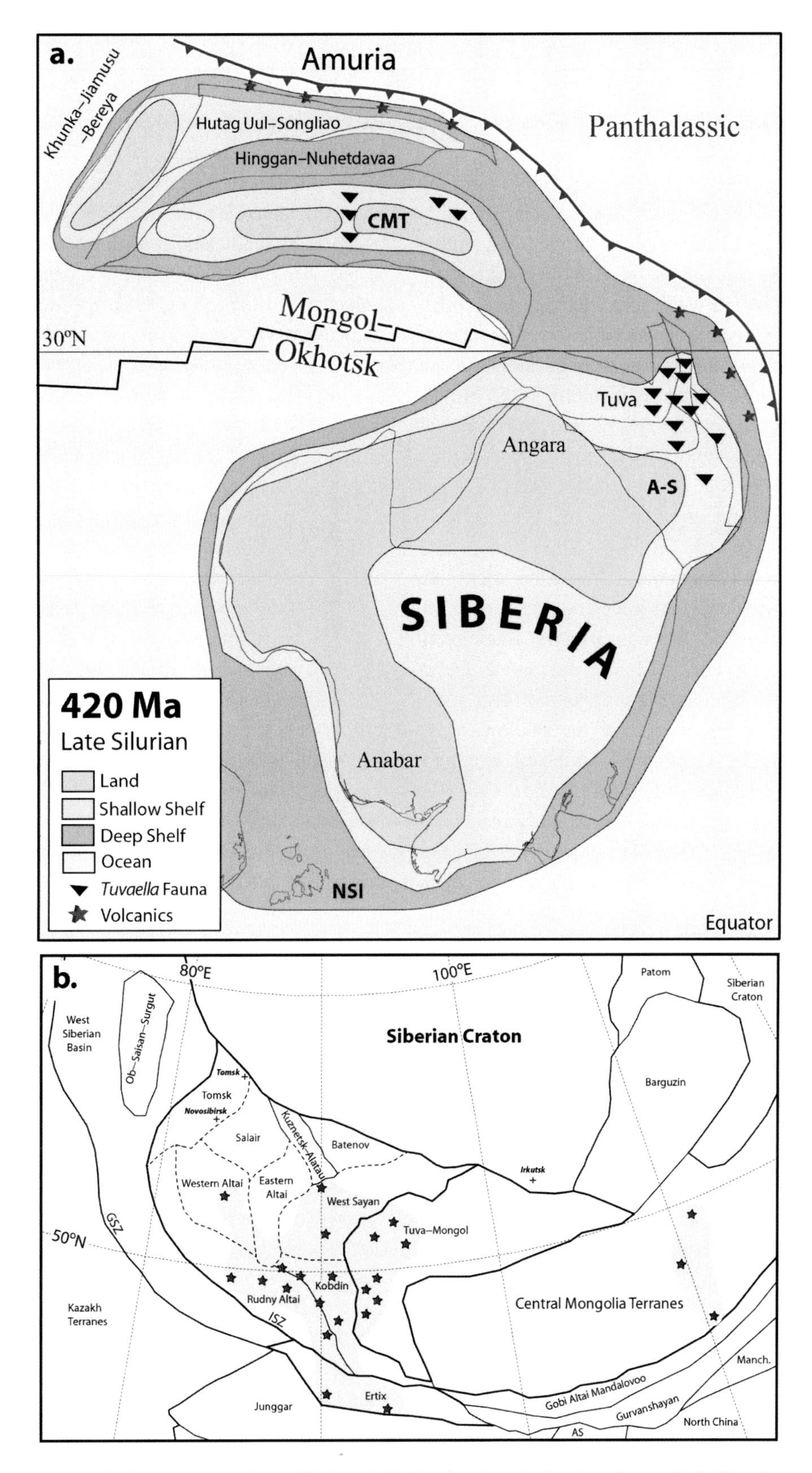

Fig. 7.7 (a) Siberia and peri-Siberia at 420 Ma (Late Silurian: Pridoli), showing the lands and seas. Solid blue lines are subduction zones with teeth on the upper plate, black lines are spreading centres and transform faults. A-S, Altai–Sayan; CMT, Central Mongolian Terranes; NSI, New Siberian Islands. (b) Today's south-western part of the Siberian Craton and adjacent peri-Siberian and adjacent Lower Palaeozoic terranes in Russia, Mongolia, eastern Kazakhstan, and north-west China. All of the Ob–Saisan–Surgat (whose margins are shown diagrammatically) and much of the Tomsk and Rudny Altai areas are obscured by Mesozoic to Recent cover. The units between the West Siberian Basin and the Tuva Mongol Terrane are together termed Altai–Sayan. AS, Ala Shan composite terrane; GSZ, Gornostaev Shear Zone; ISZ, Irtysh Shear Zone; Manch. Manchurides. Red stars show the distribution of the *Tuvaella* brachiopod fauna in the Late Silurian.

over central England, during the whole of Llandovery time mentioned in the following paragraph.

Faunal Provincialisation. The post-glacial slow warming was reflected in the global distributions and gradually expanding diversity of Llandovery brachiopods and other benthos. However, in contrast to most of the Ordovician; the major continental blocks were close enough to each other to enable invertebrate larvae to cross the intervening oceans (Fig. 7.1), and thus the benthic faunas were far more cosmopolitan than they had been earlier, particularly in the Middle Silurian. However, the dominant taxa were in many groups different from those in the Ordovician; for example, within the brachiopods, the benthic communities had much more abundant pentameroids and atrypides in the Llandovery than in previous times.

Those 'cosmopolitan' brachiopods in clastic rocks are distributed in many regions in six depth-related communities, originally defined in the transgressive Llandovery rocks in the Welsh Borderland of England noted above, which dominated the benthos on both clastic and more calcareous sediments in many regions: the shallowest the *Lingula* (now *Mergliella*) Community, outboard of which were the *Eocoelia*, *Pentamerus*, *Stricklandia*, *Clorinda* (now *Brevilamnulella*) communities, and a deeper-water, originally termed 'marginal *Clorinda*', community (Ziegler et al., 1968). *Stricklandia* is notable since it is a large pentameride and was named after Hugh Strickland, an Oxford don who had the dubious distinction of being the first geologist anywhere to be killed by a train, whilst working in a new railway cutting in 1853. Those six communities subsequently formed the original examples for the depth-related Benthic Assemblage (BA1 to BA6) terminology, which is now used for synecological assemblages of all ages (Boucot, 1975). In addition to those cosmopolitan communities, which lived on sea floors largely consisting of clastic rocks, there were more local pioneer communities that lived on rocky bottoms, as well as some varied communities that thrived on more calcareous substrates and bioherms, particularly as the latter became more common from Wenlock times onwards.

There are two exceptions to the general cosmopolitanism in the benthic faunas of the Early Silurian. One is the less diverse *Clarkeia* brachiopod fauna in the southern hemisphere, which occupied higher latitudes within Gondwana, which is found only in South America and parts of Africa (Cocks, 1972), a fauna which was the precursor of the Early Devonian Malvinokaffric Province in the same area. That the *Clarkeia* Fauna is found in the Cuyania (Precordillera) Terrane of South America confirms that the latter must have essentially finished its passage from peri-Laurentia to higher latitudes (Fig. 6.2), and had reached Gondwana near the start of the Silurian (Domeier, 2015). The other Silurian brachiopods which were notably different from the faunas in other continents lived in Siberia, which was then inverted and well away from Gondwana, and where the *Tuvaella* Fauna, which lasted from the Middle Llandovery until the Middle Ludlow, reflected its isolation in the temperate region in the northern hemisphere (Cocks & Torsvik, 2007) (Fig. 7.7).

However, as Silurian time progressed, many of the shallow-water marine faunas, including the brachiopods, showed increasing provinciality again in many parts of the world, a process which culminated in the very distinctive discrete faunal provinces seen in the Early Devonian and described in the following chapter.

In contrast to the brachiopods and many other phyla, the abundant small ostracods (Arthropoda), most of which did not have a pelagic larval stage to be carried far by ocean currents, were not dispersed so widely. Thus, although some ostracod genera had crossed the Iapetus Ocean during the Ordovician, there were still provincial differences in Silurian ostracods between the Laurentian sector of Laurussia and the combined Baltica–Avalonia (Schallreuter & Siveter, 1985). Eurypterid and phyllocarid arthropods had been rare in the Ordovician, but they subsequently diversified, and there are many endemic Laurentian eurypterids, some of giant size and well over a metre long, particularly in the shallow waters of lower salinity preserved in the Upper Silurian rocks of eastern North America, such as the Salina Group in New York State. However, following the extinction of many groups at the end of the Ordovician, Silurian trilobites were far less diverse than their ancestors had been in the earlier parts of the Palaeozoic.

The Silurian and Devonian fish faunas of North China, South China, Annamia, and Tarim were identified by Young and Janvier (1999) as representing a substantial unified continent termed the 'Asian Superterrane' offshore from Gondwana to its north-east and within the Panthalassic Ocean, chiefly because those continents are the only places from which an entire order of fish, the Yunnanolepidoidei, is known; and nearly all of another fish group, the antiarchs, with over 40 genera, are also recorded only from 'Asia', with 22 genera endemic to South China alone. However, the concept of that 'Asian Superterrane' is not supported by the rest of the geological evidence, and thus those fish must represent a faunal province, whose margins were perhaps confined by oceanic currents of variable temperatures.

The Advent of Plants onto the Land. The earliest complete land plants known as fossils date only from the Silurian, but evidence from fossil spores preserved in marine and deltaic deposits indicates that true plants may have existed since the Middle Ordovician. Land plants with trilete spores probably evolved from freshwater algae, which firstly became adapted to the occasionally desiccated non-marine environment near the margins of the oceans, but subsequently became able to survive for their whole lives without being permanently immersed in water.

Mats of algae and cyanobacteria are known from the Neoproterozoic onwards (see Chapter 4), but the first known record of fertile bifurcating axial true plants (included within the broad generic grouping of *Cooksonia*) which lived on the land are from the Late Wenlock of Ireland (reviewed by Edwards et al., 2015). Such plants were small and only a few centimetres high, and the oxygen which they produced would not have had much impact on the overall atmospheric composition in those times. Some authors think that precursor lichens may have facilitated the early evolution of vascular land plants. However, by the end of the Ludlow at about 424 Ma, the chance preservation of a palaeosol with many plant remains in Pennsylvania, USA, shows that by the Late Silurian the plants had become divided into three separate tiers above the ground, with the tallest over a metre. Below the ground level, bioturbation caused by fungal structures extended downwards to 2 m, although the vascular plant systems only went down to 80 cm, and there were also burrowing millipede arthropods. Wetland ground cover was more extensive than in nearby well-drained soils (Retallack, 2015). But it was in the subsequent Devonian that plant biomass became very much more substantial, including the advent of tree-size vegetation, so that those varied terrestrial floras then much influenced the atmosphere (see next chapter).

Eastern Asia. There are no Silurian rocks known from the main North China Craton, which then was still land (Fig. 7.4). The *Retziella* brachiopod Fauna was distributed over most of the rest of South and South-East Asia, including Tarim, South China, and Annamia, as well as on the eastern margin of the Australasian sector of Gondwana (Rong et al., 1995). Most of the earlier Llandovery rocks in South China are graptolitic shales, but the later sediments, particularly those in the Telychian, contain abundant and diverse shallow-water marine benthic assemblages. Telychian carbonates are confined mainly to the north-western and south-western margins of the Yangtze Platform and are absent from the central and eastern parts. There is a deltaic series in the Vietnamese sector of South China within which are occasional interbeds with plants similar to those from some of the Kazakh terranes, which range from Ludlow to Early Devonian (Pragian) in age, as well as occasional layers with brachiopods indicating marine incursions. The lands and seas of South China on our reconstructions are largely based on the Latest Ordovician and Silurian maps of Rong et al. (2003).

Siberia and Peri-Siberia. In contrast to the trans-equatorial position of Laurentia, Siberia was a continent separate from all others and in the northern hemisphere (apart from the adjacent Central Mongolian Terrane Assemblage: Unit 440) (Figs. 6.2 and 7.1). The facies shown follow Yolkin et al. (2003), and the faunas include mid- to deeper-shelf communities as well as shallower-water ones, indicating that the shelf seas over much of Siberia were of fair depth in many places (Cocks & Torsvik, 2007).

The temperate-latitude Silurian faunas now to be found in Altai–Sayan, Mongolia, and north-west China, as well as in political Siberia (all in the then northern parts of the Siberian Continent), were different from those elsewhere. Widespread Silurian red gypsiferous marls and gypsum beds confirm the palaeomagnetic data indicating movement of Siberia into more temperate palaeolatitudes (Fig. 9.11) and also a shift to more arid climates. Siberia (including peri-Siberia) was the only continent to host the *Tuvaella* Fauna (Fig. 7.7), whose fossils occur today in Russia, many parts of northern and central (but not southernmost) Mongolia, and north-western parts of China (Xinjiang, Heilongjiang, and Inner Mongolia provinces). Since the *Tuvaella* Fauna occurs not only in areas of Siberia which had accreted to the Siberian Craton in pre-Silurian times, such as Western Altai, but also in many areas which were not yet parts of the core Siberian Craton, it can be deduced that all of the Altai–Sayan area, the Mongolian Terrane collage, the Ertix Terrane, the Mandalovoo Terrane, and the Tuva–Mongol Fold Belt were parts of peri-Siberia and thus must have lain to the then north of core Siberia in the Silurian, which was bounded to the south by the major Gornostaev and Irtysh shear zones (Fig. 7.7).

The Mongol–Okhotsk Ocean divided Siberia from the Central Mongolian Terrane Assemblage. The then northerly margins of both Siberia and the Central Mongolian Terrane Assemblage were both active, and there were relatively high and mountainous lands within the Angara area. There is also evidence of land to the then west and south of the main craton area in the Anabar Massif area, and Fig. 7.7 shows continuous continental land area between the Anabar and Angara areas, although the shelf sea might have transgressed and extended further westwards at some times during the Silurian.

In contrast to the *Tuvaella* Fauna in today's southern Siberia, in the former more southerly part of Siberia (today's north), the Silurian rocks there have yielded more diverse brachiopods and other benthic faunas, a large proportion of which were cosmopolitan (Yolkin et al., 2003). It has been claimed that the *Tuvaella* Fauna was geographically separate from the more cosmopolitan Silurian faunas occurring in much of the rest of Siberia, but there are some Mongolian localities in which *Tuvaella* assemblages are interbedded with the more cosmopolitan brachiopods, suggesting a simple faunal gradient there. In addition, there are some Middle Llandovery faunas from the peri-Siberian part of southern Mongolia which include the endemic brachiopod genera *Templeella* and *Mongolostrophia* from deeper-water Benthic Assemblages 4 and 5. Such endemics are globally unusual for the Silurian and reinforce the conclusion that Siberia was relatively isolated. Lower Silurian fish demonstrate two faunal provinces in the Early and Middle Llandovery, termed 'Tuva' (corresponding to the *Tuvaella* brachiopod Fauna outcrop in Fig. 7.7) and 'Siberia', in today's northern part of the Siberian Terrane (Žigaitė and Blieck, 2006); however, by Late Llandovery and Wenlock times, those provincial distinctions had disappeared. In the shallower-water sedimentary facies and biofacies of the Lower Silurian of Siberia, the transgressions and regressions and local highstands there can be correlated with contemporary eustatic events in Laurentia, Baltica, and Avalonia, reinforcing the relative paucity of Silurian tectonic events affecting the Siberian continent.

On parts of the Siberian passive margin, there were large-scale carbonate buildups in the Altai–Sayan region, with barrier reefs on the outer shelf, which include a 600 km long, 10–50 km wide reef belt of Late Llandovery to Late Wenlock age in the Salair and Western Altai regions. Those reefs lie above Early Llandovery graptolitic shales which reflect both cooler global temperatures and higher sea levels. Extensive shallow-water siliciclastics were deposited in the Tuva–Mongol area (Yolkin et al., 2003), which also at least partly reflected the lower average temperatures in that then most northerly part of the peri-Siberian collage.

Laurentia/Laurussia. At the start of the Silurian, Laurentia and later the western sector of Laurussia remained at low latitudes, with the Equator passing through northern Greenland and northern Canada (Fig. 7.5), but, because most of the other major continents were at least partly at comparable latitudes, many of the faunas were similar over much of the world (Fortey & Cocks, 2003). Due to the preceding Hirnantian glaciation, sea-level stands were low in the earliest Llandovery (Rhuddanian), which resulted in the extensive Ordovician–Silurian boundary unconformities prevalent over most of the Laurentian Craton, although there are some earliest Silurian shallow-water rocks in Oklahoma and adjacent areas. Those unconformities seem to be due to the sediments never being laid down or perhaps eroded soon after deposition, since there is little evidence for substantial emergent land areas over much of the Laurentian Craton. Nevertheless, in some marginal parts of Laurentia, most notably Anticosti Island, Quebec, richly fossiliferous deposition of largely carbonate and fine clastic sediments extended from the Late Ordovician to near the end of Llandovery time with only minor paraconformities. By the end of the Rhuddanian, the majority of the brachiopods and other marine benthos, both in Laurentia and elsewhere, appear to have recovered well from the extinctions of the latest Ordovician Hirnantian glacial episode, even though abundances were less than average at most localities. At the Laurentian Craton margin and adjacent shelf in Nevada, an Early Silurian subtidal to peritidal carbonate ramp was drowned during the second half of the Llandovery. By the Late Llandovery, the faunas had become very cosmopolitan, with wide belts of the mid-shelf *Pentamerus* and *Pentameroides* Communities and the deeper-shelf *Stricklandia* and *Clorinda* Communities, in Iowa, Anticosti Island, and elsewhere (Cocks & Torsvik, 2011).

Numerous basins and highs have been identified on the Laurentian Craton, but it is difficult to be sure which highs were above sea level and thus forming land; for example, most of Wyoming lacks preserved Silurian rocks. However, since some of the rocks on the craton are of tidal or peri-tidal origin, many were probably deposited under shallow seas bordering low-lying landmasses. Successive faunal assemblages represent steady water deepening in the mid-Llandovery (Aeronian) of the substantial Michigan Basin, where there is a range from tidal flat deposits of BA1, dominated by ostracods and with just two brachiopods, *Hercotrema* and *Alispira*; a much more diverse and just subtidal BA2 dominated by corals and stromatoporoids and with 12 different brachiopod genera; three different BA3 mid-shelf assemblages dominated by stromatoporoids, corals, and the brachiopod *Pentamerus*; and a BA4–BA5 deeper-shelf assemblage dominated by crinoids and sponges.

Exceptionally substantial microbial reefs, some as much as 1.3 km thick, are found in the Canadian Arctic (de Freitas & Dixon, 1995), and, by the Wenlock, widespread bioherms had formed, particular in Kentucky, in Oklahoma, and in the Niagara area of New York State and Ontario, and that reef belt extended eastwards into the Avalonia–Baltica sector of Laurussia. As well as corals, the reefs had frameworks of various different organisms, such as stromatoporoids, algae, and bryozoans (Copper, 2002). In the succeeding Ludlow, there was extensive evaporitic deposition (Fig. 7.6), including the noted Lockport and Salina dolomites of the Michigan Basin and New York State. With the increasing ambient temperatures, large pentameride brachiopods, some endemic, forming densely packed beds on the sea floor comparable to oyster beds today, became abundant and widespread over the whole of Laurentia. A sequence of communities of large brachiopods from Wisconsin and surrounding areas commenced with the Middle Llandovery *Virgiana* Community, and continued with the Late Llandovery to Early Wenlock *Pentamerus* and *Pentameroides* Communities, and the Middle Wenlock and Ludlow *Kirkidium* and *Apopentamerus* Communities (Watkins, 1994). *Virgiana* and *Apopentamerus* were pentamerids which were largely endemic to Laurentia. There are comparable communities in the Williston Basin of Saskatchewan and Manitoba, and the *Virgiana* Beds are used locally in the recognition of the base of the Silurian in the many hydrocarbon exploration wells drilled there.

On Fig. 7.5 some of the records of *Virgiana* are plotted, together with contemporary Late Rhuddanian and Aeronian faunas of apparently the same environment (BA3) in Avalonia–Baltica and Scotland (Girvan), from which *Virgiana* is notably absent but where other pentamerids are common. Those include the BA1 to BA6 depth-related communities mentioned under 'Faunal Provincialisation' above, which were first described from the Welsh Borderland of Avalonia.

Baltica and Eastern Laurussia. In the Silurian, most of the Baltic Shield was relatively flat, and the rocks in Gotland (Sweden), the east Baltic, and Ukraine represent facies unaffected by tectonics which were deposited under shelf seas in relatively shallow basins (Baarli et al., 2003); however, in the west of Baltica, the Caledonide Orogeny was progressively generating uplift and mountains. From latest Llandovery times onwards, the continental deposits of the Old Red Sandstone were deposited on flood plains and wadis, although most of the Old Red is of Devonian ages (Friend & Williams, 2000). In central Scandinavia, the East Baltic, Ukraine (Dniester River), and Timan–Pechora basins, there was progressive deepening (Fig. 7.3), as reflected by the benthic assemblage zones, which are all truncated abruptly by the Tornquist Suture, indicating that much of today's southern margin of Baltica was lost in the Late Palaeozoic Variscan Orogeny.

However, still preserved to the south of the Tornquist Suture in the Holy Cross Mountains of Poland, there are thick (about 1,500 m) Silurian turbidite sequences, indicating deposition under deeper water close to the original border of Baltica. Those rocks, and also those in the Late Llandovery and Wenlock of the Oslo Region, together indicate a greater sediment supply and higher topographical relief inland than in the Ordovician. In contrast to the earlier two separate land areas of the Caledonides (Fennoscandia and Sarmatia), Fig. 7.3 shows a united land area which includes both the earlier areas and which extended westwards through Greenland and Scotland into the old Laurentia, and there is no evidence for any intervening seaways (Cocks & Torsvik, 2005).

The Silurian successions of the Baltica sector of Laurussia today lack any significant metamorphism (again, apart from in the Oslo region of Norway and also some dolomitisation in the East Baltic), and include some of the best and most well-known Silurian sections anywhere, for example on the island of Gotland, Sweden, whose abundant fossils have attracted attention since the eighteenth century work of the famous naturalist Linnaeus. There are superbly exposed carbonate mud mounds of Early Wenlock and later Silurian ages in Gotland (Sweden) and Estonia. Although there are a few endemic Baltic species and genera, they form only a small percentage of the Gotland fauna (Bassett & Cocks, 1974). That indicates that shallow seaways still lay between the two former continental areas of Laurentia and Avalonia–Baltica, even during the latest Silurian. Since most of Laurussia was then within tropical latitudes, extensive Bahamian-style limestones and massive reefal bioherms were developed in Gotland and elsewhere (Fig. 7.3).

On Vaigach Island (between Novaya Zemlya and the northern Urals of Russia), Silurian carbonates, including bioherms, are over 1,400 m thick (Baarli et al., 2003). To the south and east of the mainly shallow-water outcrops in Norway, Sweden, and Estonia, Silurian rocks are preserved extensively in the subsurface of Latvia, Lithuania, Poland, Belarus, and the Ukraine. The sediments there show progressive deepening southwards and westwards, for example in Lithuanian boreholes (Musteikis & Cocks, 2004). Much of Baltica was too deeply submerged to sustain benthos, and much of the Silurian consists of relatively thin graptolitic shales. However, in south-western Ukraine, above an unconformity with the Ordovician, there are substantial and impressive outcrops of relatively flat-lying rocks in the Podolia area, which are interbedded carbonates and shelf clastics with relatively shallower-water benthos, including diverse brachiopods, from the Wenlock to the Early Devonian (Cocks & Torsvik, 2005).

8 Devonian

Kvamshesten Middle Devonian basin, Western Norway. Devonian basins formed in response to post-orogenic extension of over-thickened Caledonian crust (Western Gneiss Region). The Kvamshesten Basin is situated in the hanging wall of the spectacular Nordfjord–Sogn Detachment Zone. Credit: Torgeir B. Andersen, CEED.

Although fossil vertebrates are now known to occur sporadically in earlier rocks, the Devonian is the oldest system in which they became common, particularly the fish that have been known for over two centuries from the Old Red Sandstone of northern Europe and North America. The period is also notable for the first substantial invasion of the land by more advanced organisms, both plants, including trees, and animals. Since the Devonian was more than twice as long as the preceding Silurian (60 Myr as opposed to 24 Myr), there was more time for tectonic activity, of which there was a great deal.

Adam Sedgwick and Roderick Murchison (1837) together named the Devonian System after the south-western English county of Devonshire, where, in contrast to the continental deposits of the Old Red Sandstone seen in most of Britain, marine rocks of the same age occur. However, the original concept of its definition and extent has been much modified since then, and the base of the system is now formally defined at the base of the *Monograptus uniformis* graptolite Biozone, which is at the base of the Lochkovian Stage at Klonk in the Czech Republic, and has been dated at 419 Ma. The global distribution of the continental units and oceans at three periods within the system at 410 Ma (Pragian), 390 Ma (Eifelian), and 370 Ma (Famennian), is shown in Fig. 8.1. The Devonian is divided into seven stages, in ascending order the Lochkovian, Pragian, Emsian, Eifelian, Givetian, Frasnian, and Famennian, named after areas in the Czech Republic, Belgium, France, and Germany

Tectonics and Igneous Activity

Oceans. The Panthalassic Ocean, like the present-day Pacific, was a vast ocean. But despite covering a hemisphere throughout the Late Palaeozoic, little is known about that composite basin due to the subsequent destruction of all its constituent plates. There was a near-continuous subduction zone which encircled the Panthalassic around most of its boundary with the continents, notably around most of Gondwana. To allow for that progressive convergence along the edge of the domain, Domeier and Torsvik (2014) suggested a simple and relatively stable triple junction of spreading ridges within the ocean (Fig. 8.1). However, the edge of the Panthalassic at the western margin of Laurentia exhibits no indication of an active margin there prior to the Middle to Late Devonian.

The Rheic Ocean between Gondwana and Laurussia reached its maximum width (~6,000 km) near the start of the Devonian, but that width gradually decreased as the Devonian progressed, at least in its western sector, where there were active subduction zones at the margins of both continents. The north-western arm of the Rheic, between Franconia–Thuringia and the previous Avalonia sector of Laurussia, closed in the Middle Devonian. Smaller oceans in today's southern European area, the Galicia–Moldanubian and Saxothuringian oceans, opened and closed during the various phases of the Variscan Orogeny (Fig. 8.9), reviewed further below.

Following initial rifting in the Late Silurian, the Early Devonian also saw the opening of the Paleotethys Ocean (whose western arm is locally termed the Galicia–Moldanubian Ocean) between Gondwana and the various terranes which now make up southern Europe (shown on Fig. 8.1 as the Armorican Terrane Assemblage: ATA). However, it is uncertain whether or not that European and North African rifting is the same tectonic event as also caused oceanic opening much further away to the east, in which the combined Annamia–South China continent moved away from near the margin of Gondwana. That doubt is partly because the detailed separate timings are poorly constrained, and partly because evidence of rifting within and between the intervening microcontinental blocks is not obvious in all of those two widely separated regions. However, rather than coin another term, we use the name Paleotethys Ocean in both areas. As the Paleotethys opened, its western end merged with the Rheic Ocean, although the latter became progressively smaller as the Armorican Terranes moved northwards.

Smaller oceans in Fig. 8.1 include the Turkestan Ocean between Tarim and Kazakhstania, the Mongol–Okhotsk Ocean between Siberia and Amuria (a newly combined continent chiefly made up of the previously amalgamated former Central Mongolia Terranes), and the Ægir Ocean between Laurussia and Siberia. The latter two oceans might also be considered as branches of the Panthalassic, but were separated by the Kolyma–Omolon Terrane.

The Start of the Variscan Orogeny. Unlike the Caledonide Orogeny, which was essentially the result of the collision of a pair of large continents, the Variscan Orogeny of Central Europe was a messy series of events involving many microcontinents and terranes and with a series of opening and closing oceans of varied sizes and varied lengths of time in existence. Thus it can be viewed as a linked series of orogenic events that lasted from the Silurian, through the Devonian and into the Carboniferous and even Permian, but authors have differed considerably on the terrane definition and boundaries and the timings and significance of those events. Since the original palinspastic realities of the many units, including the orocline in the Iberian Peninsula, are open to many interpretations (e.g. Shaw et al., 2012), we have simply drawn the terranes as variable lozenges reflecting their modern areas on most of our pre-Carboniferous maps, and largely kept what may in many cases have been composite terranes together for lack of hard data (e.g. Fig. 8.1). However, in Fig. 8.9 we present a new and more mobilistic model, including the opening and closure of some smaller oceans (Franke et al., 2016).

Palaeozoic palaeomagnetic poles from the Armorican Terrane Assemblage (Armorica, Iberia, Saxothuringia, and Bohemia) are few and of variable quality, and there are time gaps with no data (Fig. 8.10); however, collectively they appear to represent palaeolatitudes compatible with a location near the North-West African margin of Gondwana. But that margin of Gondwana was far away from Laurussia in Early Middle Carboniferous times (Fig. 9.1a,b), at the peak

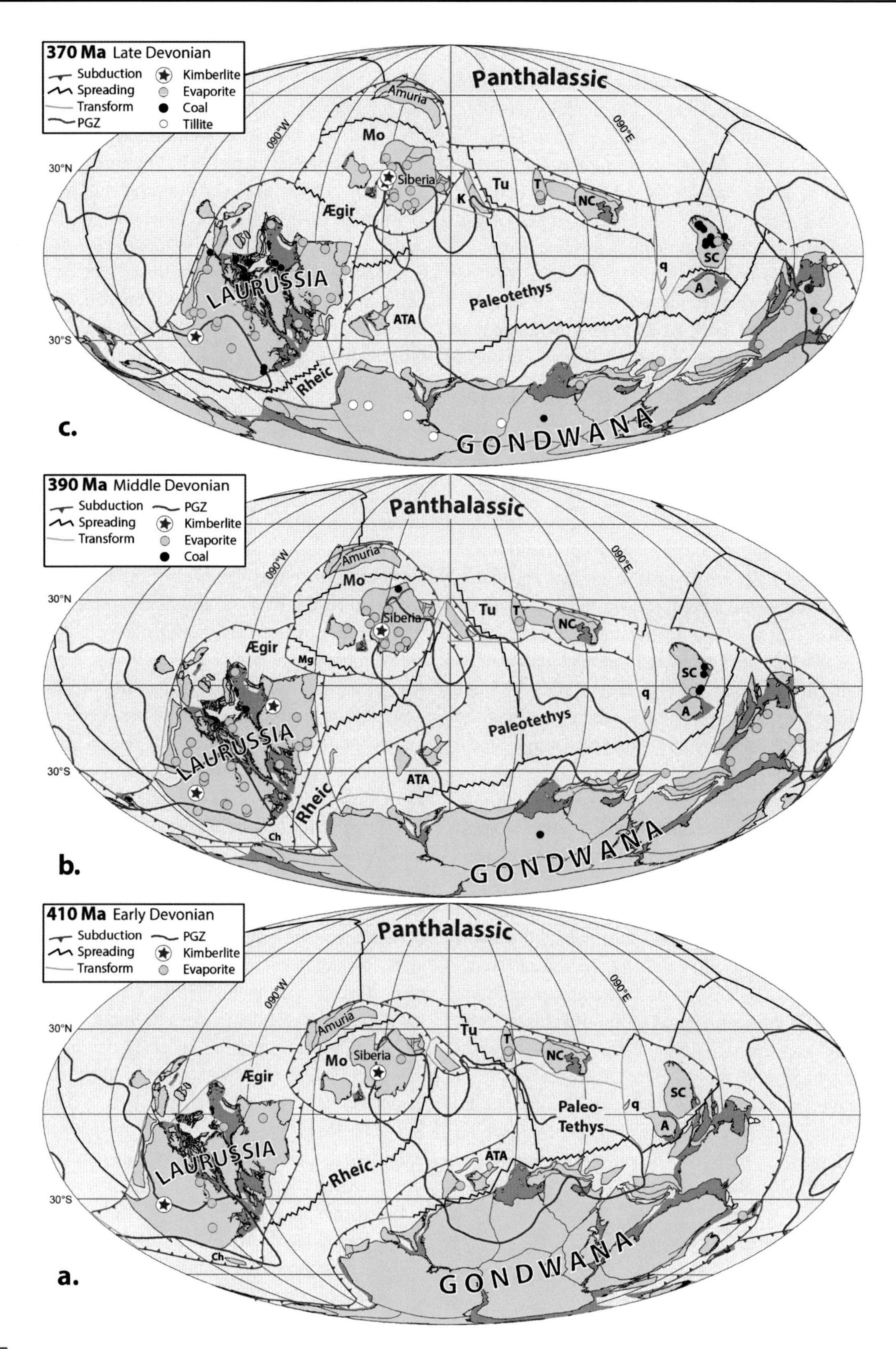

Fig. 8.1 Outline Earth geography at (a) 410 Ma (Pragian), (b) 390 Ma (Eifelian), and (c) 370 Ma (Famennian), including the postulated plate boundaries, outlines of the major crustal units, and the more substantial oceans. A, Annamia; ATA, Armorican Terrane Assemblage; Ch, Chilenia; K, Kazakhstania; Mg, Magnitogorsk Arc; Mo, Mongol–Okhotsk Ocean; NC, North China; PGZ, plume generation zone; q, South Qinling; SC, South China; T, Tarim; Tu, Turkestan Ocean.

of Variscan metamorphism. Thus the ATA cannot be considered as a promontory of Gondwana, and some or all of those terranes must therefore have drifted away from North-West Africa in the Galicia–Moldanubian Ocean by the Early Devonian so that they could later collide with Laurussia in the Variscan Orogeny.

At some unconstrained time before the Devonian (following e.g. Franke, 2006), and probably in the Late Ordovician soon after 460 Ma, the Saxothuringian Terrane had left a peri-Gondwanan position near the ATA, leaving behind it a widening Saxothuringian Ocean as a north-eastern arm of the Rheic Ocean. Soon afterwards, Bohemia also rifted away from peri-Gondwana, across the Galicia–Moldanubian Ocean (Fig 8.9d). The Early Devonian saw the progressive closure of the north-western arm of the Rheic Ocean, with the result that the Saxothuringia Terrane approached the south-eastern margin of the Avalonia sector of Laurussia and collided with it at the start of the main Variscan Orogeny at around 400 Ma (Emsian) (Fig. 8.9e).

The progressive situations in the Late Devonian at 380 Ma (Frasnian) and 360 Ma (Late Famennian) are also shown on Fig. 8.9f,g, in which the Saxothuringian and the western end of the Galicia–Moldanubian oceans had narrowed substantially, and the former disappeared entirely. As usual, there were consequent intrusions of a variety of granites and other igneous rocks following all those intricate collisions and ocean closures, as well as nappe thrusting and other tectonic and metamorphic activity; for example, the Mid-German Crystalline High, which lies between the main Rheno-Hercynian and Saxothuringian terranes, appears to be the much metamorphosed remains of a Silurian volcanic island arc (Franke et al., 2016). The further Variscan phases during the Carboniferous are described in the next chapter.

West Gondwana. In the south-western (South American) sector of Gondwana, the Achalian Orogeny was essentially the collision between the Chilenia Terrane and the Cuyania (Precordillera) Terrane in the Devonian (both now within Unit 290: Colorado), with much thrusting and sinistral strike-slip movements in the Sierras Pampeanas of Argentina, together with the subsequent intrusion of plutons in the Emsian to Frasnian from 403 to 382 Ma (Dahlquist et al., 2013). In central Chile, at around 39° S, there was an active margin until the Late Devonian (about 385 Ma), but that margin was subsequently passive, although with much strike-slip faulting, until the Early Carboniferous at about 340 Ma (Visean), after which there was the start of a subduction-related accretionary prism, which continued its activity until the end of the Permian (Hervé et al., 2013). In the Patagonian sector of the proto-Andes (Unit 291), substantial plutons were intruded from the Middle to the Late Devonian (Emsian to Famennian: 401–371 Ma). In Central America, there was a 35 Myr period of flat-slab subduction under the Acatlán Complex of southern Mexico (within Unit 216) which ended in the Late Devonian at about 365 Ma, during which the rocks were taken to about 40 km below the surface and subsequently rose back up again in a subduction erosion–intrusion cycle (Keppie et al., 2012). The relative positions of Gondwana and the Laurentian sector of Laurussia do not appear to have changed much between the Early and Late Devonian (compare Fig. 8.2 and Fig. 8.3), with the Rheic Ocean between them. However, between the Late Devonian and the Middle Carboniferous and before those two superterranes united to form Pangea, there must have been an enormous amount of strike-slip faulting between them so as to bring round today's southern Laurentia to face the north-western sector of Gondwana. That very substantial amount of lateral faulting also helps to explain the fragmentary and tectonised nature of the many smaller terranes now in Central America, often collectively grouped as the 'Mexican terranes'.

East Gondwana. The northern Gondwanan margin remained largely passive, although Late Devonian (Famennian) to Early Carboniferous (Tournaisian) granitoids were intruded into the central Himalaya area and adjacent Tibetan terranes (Fig. 8.3). In contrast, the eastern margin of Gondwana continued to be very active, with the Tabberabberan Orogeny, which had started in the Middle Silurian, continuing on until the Middle Devonian (430–380 Ma) in the New England and Thomson orogenic areas of eastern Australia. That included the accretion of yet more volcanic arc material to enlarge the area of the Gondwana Craton, associated with the intrusion of more granites. The various tectonic units within Tasmania amalgamated in the Early Devonian at about 400 Ma (Veevers, 2004); and the Melbourne Zone of Victoria was much shortened along a décollement during the Tabberabberan Orogeny. That was followed by the Middle Devonian to Carboniferous Kanimblan Orogeny, which began with rifting, followed by substantial terrestrial sedimentation inboard of a convergent margin, which can be grouped with the succeeding Hunter–Bowen Orogeny, a series of events that included the accretion of yet more island arcs and extended from the Late Devonian to the Triassic (Glen, 2005).

Kimberlites were intruded into Western Australia in the Late Devonian, between 382 and 367 Ma (Torsvik et al., 2014), and were sourced from the Pacific plume generation zone (Jason). In addition, numerous granitoids were intruded into the active southern margin of Gondwana in the Late Devonian at around 375 Ma (Frasnian) as parts of the Ross Orogeny, most notably in the Transantarctic Mountains of

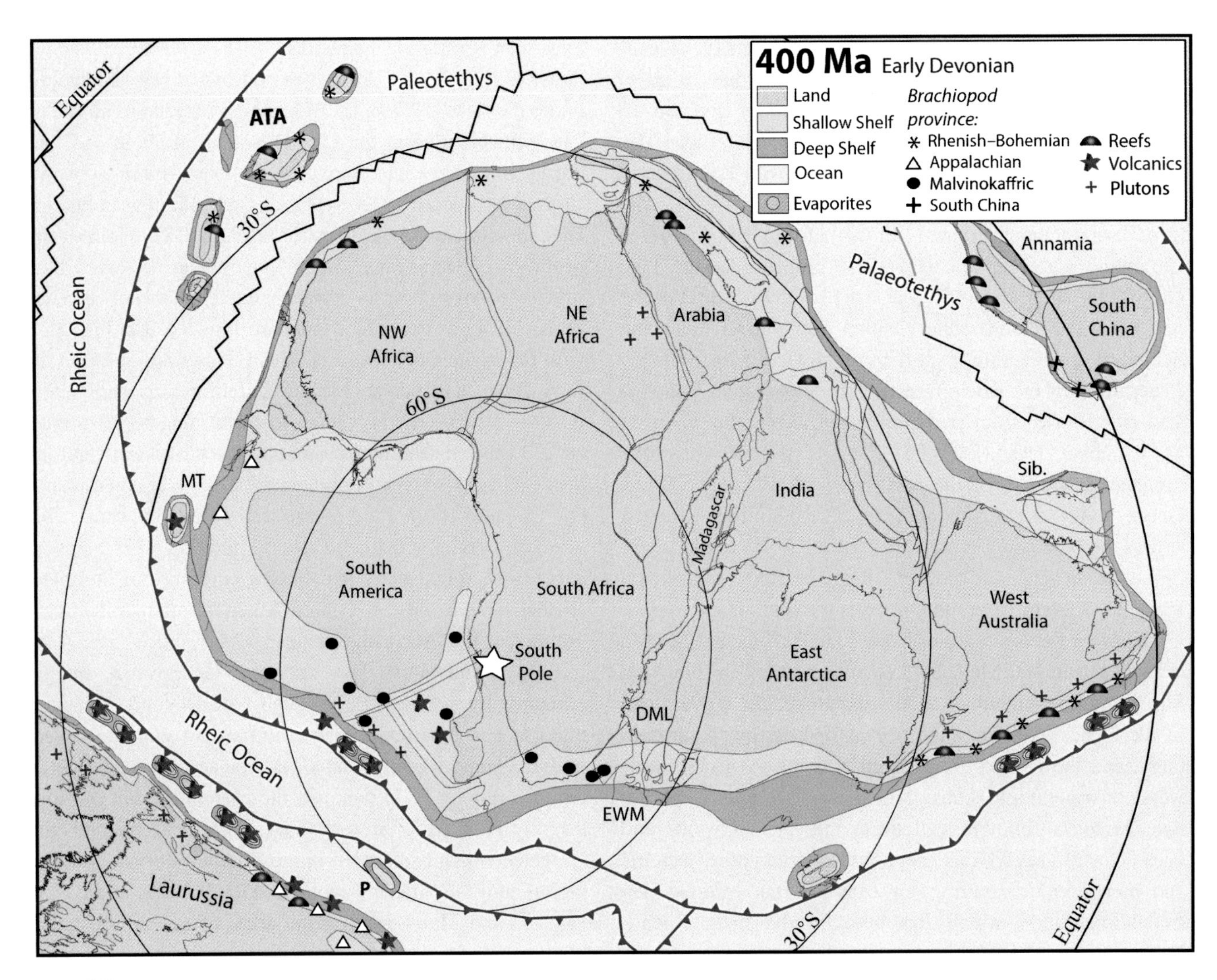

Fig. 8.2 Gondwana and adjacent areas at 400 Ma (Emsian), showing the contemporary pattern of lands and seas and the distribution of the brachiopod provinces (updated from Cocks & Torsvik, 2002). Solid blue lines are subduction zones with teeth on the upper plate, black lines are spreading centres, and green lines are transform plate margins. The strings of island arcs shown are diagrammatic, since the individual extent of each unit within the arcs and the extent to which each unit was above sea level are uncertain. ATA, Armorican Terrane Assemblage; DML, Dronning Maud Land; EWM, Ellsworth–Whitmore Mountains; MT, Mexican terranes; P, Precordillera (Cuyania) Terrane, Argentina; Sib, Sibumasu.

Antarctica, Marie Byrd Land, Tasmania, and south-western Australia (Elliot, 2013).

Eastern Asia. The most significant tectonic event in the area was the opening of the eastern sector of the Paleotethys Ocean in the latest Silurian and Early Devonian (Fig. 8.2), following initial Late Silurian rifting in the Australian sector of the north-eastern margin of Gondwana, which probably continued westwards as far as the north-western margin of Africa (Torsvik & Cocks, 2011), as noted under 'Oceans' above. However, the extensive area of Gondwana forming the southern shores of the Paleotethys, which included what was much later to become the microcontinent of Sibumasu, the Tibetan terranes, and those smaller terranes fringing India and further west in Afghanistan and elsewhere, continued to remain an integral part of Gondwana until the Permian.

Tarim, Annamia (Indochina), South China, and North China have been depicted (e.g. by Metcalfe, 2011) as forming a single elongate and unified continent to the north of the widening Paleotethys from the Early Devonian. However, whilst Annamia and South China had been together as a single unit since at least the Cambrian, there is no good reason to link those combined continents physically with North China or Tarim, particularly since the palaeomagnetic data, although somewhat poor for the Middle Palaeozoic, suggest that North and South China were at substantially different latitudes then. The unified Annamia–South China

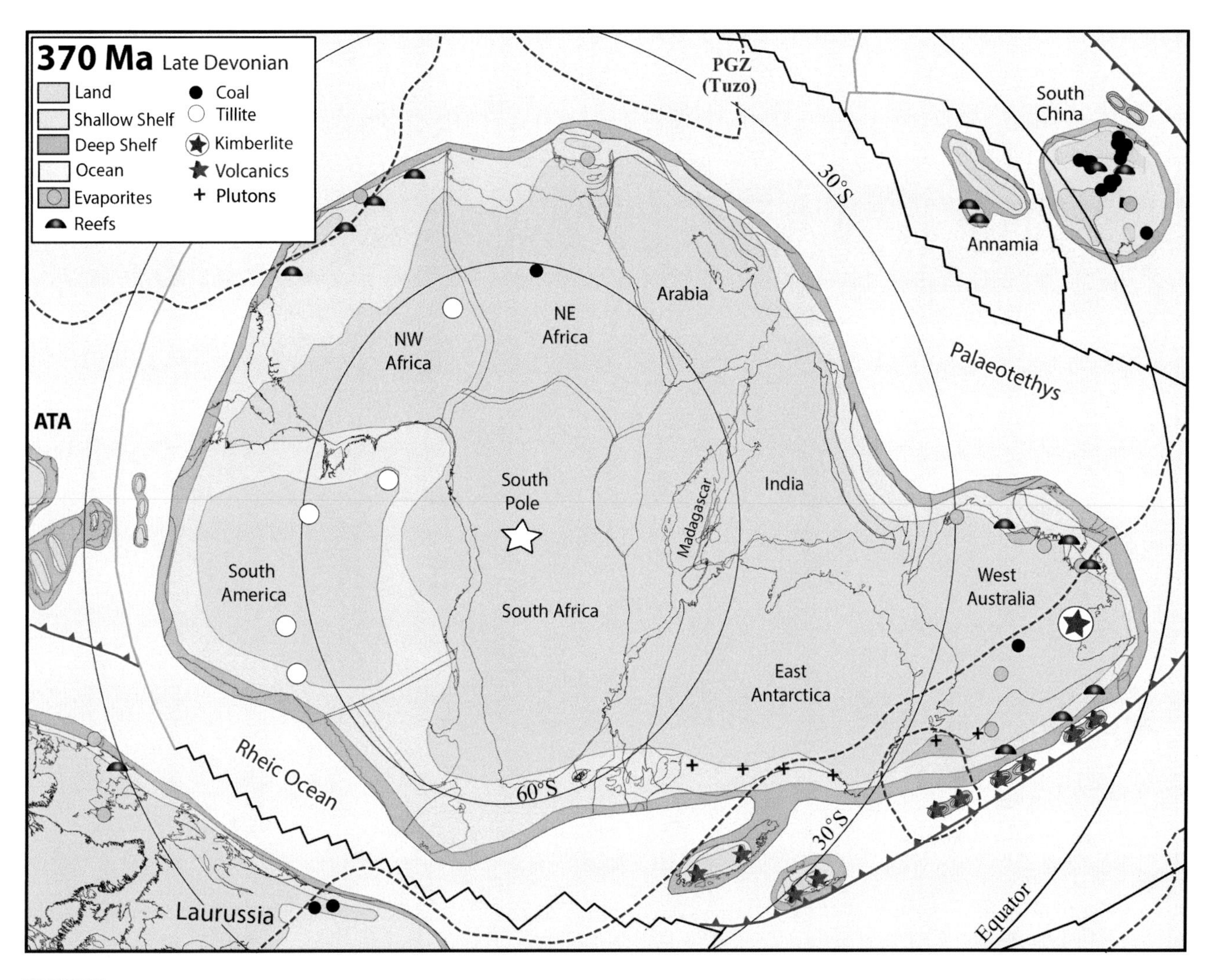

Fig. 8.3 Gondwana and adjacent areas at 370 Ma (Famennian), showing the contemporary pattern of lands and seas. Dotted red lines are the plume generation zones (PGZ). Solid blue lines are subduction zones with teeth on the upper plate, black lines are spreading centres and transform faults, and green lines are transform plate margins. ATA, Armorican Terrane Assemblage.

continent broke up later during the Devonian (Cai & Zhang, 2009), although the Song Ma Suture today between North and South China was not formed until the Late Triassic to Early Jurassic. The Paleotethys Ocean appears to have opened before Annamia and South China separated; however, the depiction of the relative situations of the two units on our maps (e.g. Fig. 8.4) is somewhat speculative, since, as noted above, it is not clear whether or not the geometry and tectonics of that Devonian separation in eastern Asia were linked to the Paleotethys opening in Europe, North Africa, and the Middle East.

In addition, the small South Qinling Block left the northwestern margin of the South China continent at some time near the end of the Silurian or the start of the Devonian and remained an independent unit (although without evidence of it ever rising far enough above the ocean floor to become land) before remerging with the other elements of the Qinling–Dabie Orogenic Belt in central China at about the end of the Permian.

The sparse Devonian palaeomagnetic data from North China and Tarim appear to indicate that those blocks occupied comparable mid to low latitudes in the northern hemisphere. The neighbouring Qilian Orogenic Belt (Unit 456) was assembled by the latest Silurian and the composite Qinling–Dabie Orogen is also correlative with the Qilian and Kunlun orogens. Therefore, to minimise the number of spreading centres and subduction zones in the region, we show Tarim and North China drifting as one united continental block from the Early Devonian onwards (Fig. 8.1). Throughout the Devonian, the passive northern margin of

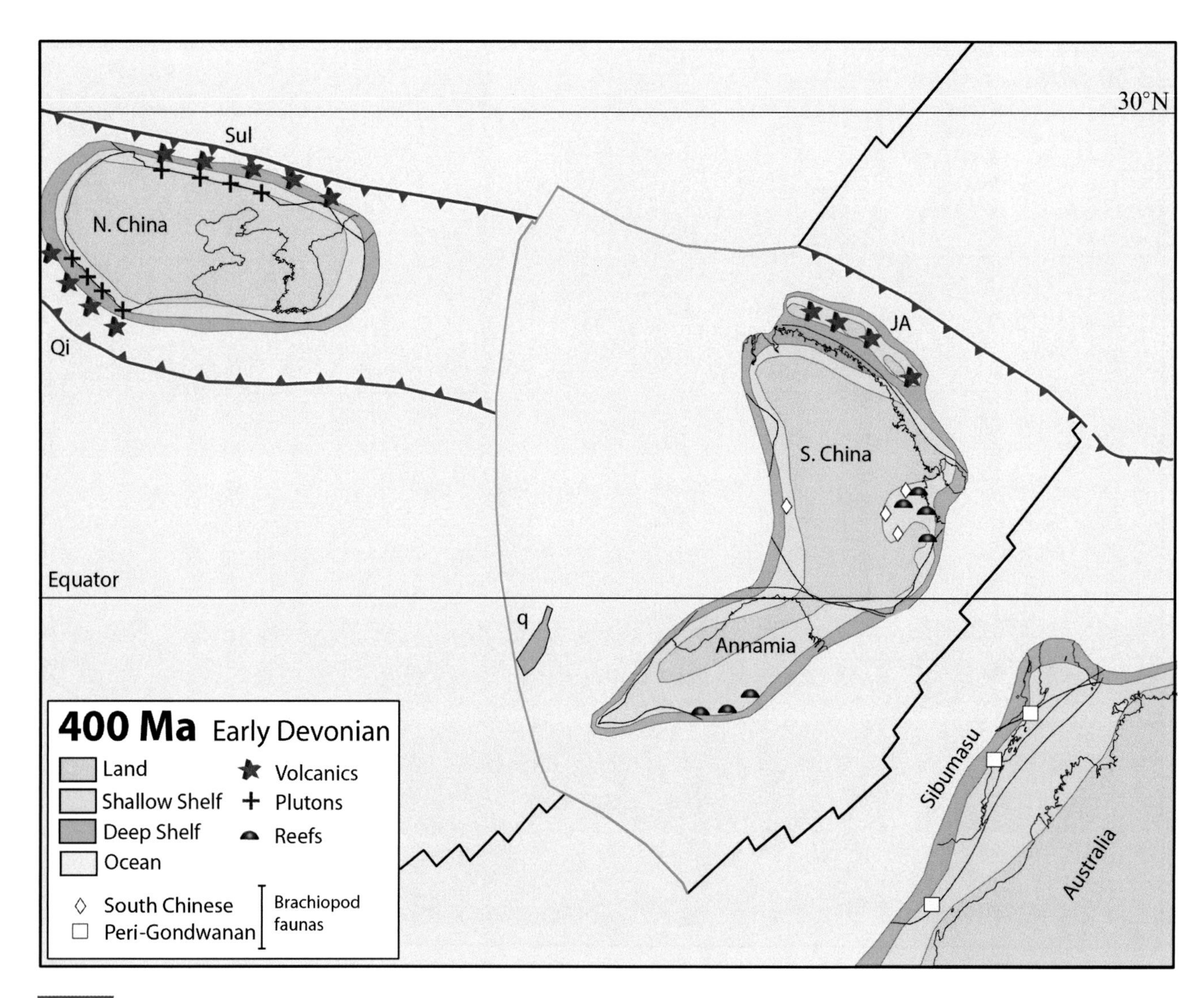

Fig. 8.4 The eastern Asia area at 400 Ma (Emsian), showing the contemporary pattern of lands and seas and the distribution of some brachiopod localities. All the latter were within the Old World Realm, but include sufficient endemic genera to define various regions (see text). Solid blue lines are subduction zones with teeth on the upper plate, black lines are spreading centres and transform faults, and green lines are transform plate margins. The strings of island arcs shown are diagrammatic, since the individual extent of each unit within the arcs and the extent to which each unit was above sea level are uncertain. JA, Japanese island arcs; q, South Qinling; Qi, Qaidam–Qilian: Sul, Sulinheer.

Tarim faced the Turkestan Ocean and was being subducted westward beneath Kazakhstania and Siberia. Along the shared Kunlun–Qinling–Dabie margin of North China–Tarim, north-dipping subduction of the Paleotethys Ocean was continuous throughout the Devonian, following a brief interval of minor transform motion in the earliest Devonian. Simultaneous southward-directed subduction of the Panthalassic Ocean occurred to the north beneath northern North China and Beishan until the Late Devonian at 370 Ma, when the neighbouring intra-oceanic subduction zone of the South China Plate passed by northern North China, converting its active margin into a transform boundary (Domeier & Torsvik, 2014).

South-dipping subduction of the Qilian oceanic plate in the north-west of China, which had started in the Cambrian, continued until the Early Devonian (410–375 Ma), when collision of the Qilian Arc and the North China continent occurred, associated with obduction of the Qilian Melange Complex over the passive margin of North China, and there was consequent Early Devonian metamorphism (Song et al., 2009b); however, Xiao et al. (2009a) placed the timing of that accretion as Late Devonian. Whichever is correct, we show Qaidam–Qilian as an integral extension of North China in our Late Devonian and subsequent maps. In the Yili area, today in the south-western Tien Shan Mountains, there are substantial volcanics of Late Devonian and Carboniferous ages in the Atashu Zhamshi Terrane (Unit 461). There was also active island arc volcanism in the Gurvansayhan Terrane of Mongolia (Unit 452), but that unit did not accrete to the neighbouring Ala Shan Terrane (Unit 451) until the Carboniferous.

Palaeomagnetic, palaeontological, and sedimentary data all suggest that South China was not far from north-east Gondwana in the Middle Palaeozoic, although it did not appear to have formed an integral part of that large continent. However, Early Devonian regional strike-slip faults and extensional basins in South China may be linked to the initiation of the Paleotethys spreading centre nearby. In addition, along the south-western margin of South China and the adjacent north-eastern margin of Annamia, an Early Devonian unconformity and Middle Devonian to Permian ophiolitic and passive margin rocks are interpreted to signify a near-contemporaneous separation of Annamia and South China (Jian et al., 2009a,b). Thus there may have been a rift system that separated the combined South China and Annamia from north-east Gondwana in the Early Devonian (Fig 8.1a), and by the Late Devonian that system had progressed to include the separation of South China and Annamia (Fig. 8.1c).

Given the broad spatio-temporal concurrence of those events with the development of the Paleotethys Ocean much further to the west, it is tempting to directly link those rift systems. However, the palaeomagnetic (though sparse) and geological data from South China are enough to demonstrate that it cannot have moved passively within the eastern part of the Paleotethys, and thus Domeier and Torsvik (2014) concluded that the two were decoupled then, with an intra-oceanic north–south-running transform boundary (Fig 8.1a).

Central Asia. The core of the continent of Kazakhstania had been established during the Silurian, but during the Devonian several island arcs from various previously independent terranes became accreted to it. There was very substantial volcanism during the Early Devonian, which today forms an arcuate series of outcrops often interpreted as an orocline. Those volcanic rocks are the earliest in the Kazakh Orogen to have yielded many reliable palaeomagnetic data (Abrajevitch et al., 2007, 2008).

Although the Stepnyak volcanic island arc (Unit 470) had accreted to Kokchetav–Ishim (Unit 458) in the Ordovician, the arc remained active until the Early Devonian, and there were also significant Devonian granitoids in both sectors (Kheraskova et al., 2003), probably also related to its union with Chingiz–Tarbagatai (Unit 462) to enlarge the Kazakhstania continent (Popov & Cocks, 2017). In the Chingiz area itself, Late Pragian to Frasnian terrestrial rocks with volcanics lie unconformably upon the Silurian, above which there is a Famennian shallow-water marine sequence. In Junggar–Balkash (Unit 464), there are deeper-water sediments interbedded with abundant lavas at many levels from the Eifelian to the Famennian (Daukeev et al., 2002).

The Tarim Microcontinent became welded to the substantial Junggar Terrane (Unit 450), today to its north, during the Devonian and earlier Carboniferous (Charvet et al., 2007): Junggar had previously been a composite unit made up of Cambrian and later volcanic island arcs and accretionary prisms.

Western Laurussia. During the Devonian, the Rheic Ocean between the Laurentian sector of Laurussia and Gondwana, which had been at its widest in Middle Silurian time (Fig. 7.1), steadily closed (Torsvik & Cocks, 2004). Near the beginning of the Devonian, the Acadian Orogeny of the Appalachians, including the Meguma Terrane (Unit 170: part of West Avalonia in our reconstructions), was characterised by polyphase deformation and regional metamorphism. That was accompanied by voluminous magmatism, which ceased between 395 Ma (Emsian) and 380 Ma (Frasnian) and was followed by a period of quiescence (Murphy et al., 1999). There was diachronous migration of Acadian deformation over the whole Appalachian area from about 415 Ma (Lochkovian) in the south-east to about 370 Ma (Famennian) in the north-west, and that extended for more than 600 km into the continental interior: it may have been caused by the migration of Laurussia over a mantle plume. Tectonic activity in the Northern Appalachians can be divided between an Acadian Orogeny, which had begun in the Latest Silurian at 421 Ma and continued until the Emsian at about 400 Ma, and a Neoacadian Orogeny, which lasted from the Emsian to about 360 Ma, near the end of the Devonian (van Staal et al., 2009). There was also substantial activity further south in the Appalachians; for example, there was Devonian metamorphism in the Ordovician Hillabee Greenstone island arc of Alabama as it was thrust westwards onto the Laurentian margin, and that was closely followed by the intrusion of 369 Ma (Famennian) granite plutons (Tull et al., 2007). The Acadian Orogeny is also recognised in the British Isles, where there was widespread Middle Devonian deformation which formed the slate belts of much of north-western England and Wales lying to the north of the Variscan Front. The latter is largely the result of Carboniferous and later tectonism that affects only the region south of a line stretching from southernmost Ireland, through southern Wales and the Bristol Channel south of the former Midlands Microcontinent to north of the Brabant Massif in Belgium (Woodcock & Soper in Brenchley & Rawson, 2006).

The western (Cordilleran) margin of the Laurentian sector of Laurussia had been passive since the breakup of Rodinia in the Neoproterozoic, but that ceased in Early Devonian (Pragian to Early Eifelian) times at about 395 Ma (Fig. 8.5). At least one new island arc formed in the Yreka Terrane, today preserved in the eastern Klamath Mountains of California. Late Devonian (Frasnian–Famennian) to Earliest Carboniferous arcs also formed in the northern Sierra Nevada Mountains of California and eastern Nevada, including the extrusion of volcanic lavas over 5 km thick there. That extensional tectonism was subsequently reversed, which resulted in the Late Devonian Antler Orogeny, in which the Roberts Mountains Allochthon was thrust onto the Laurentian Craton (Dickinson, 2000). A great clastic wedge, the Antler Flysch, was shed eastwards from the resulting highlands into a broad foredeep that included much of eastern Nevada and extended into Utah. Comparable relationships are seen in the Pioneer Mountains of central Idaho and to the south-west in roof pendants in the Sierra Nevada Batholith. Opinions have differed on the cause of the Antler Orogeny, but it seems most likely to have been due to the collapse of a back-arc basin following the change from a passive to an active continental margin, and the subsequent accretion of the arc to Laurussia. Further north in the Cordillera, there was a further pulse of alkaline magmatism in the Early Devonian due to extension in the craton (Goodfellow et al., 1995), and ophiolites were intruded along much of the Cordillera during the Middle and Late Devonian. Arc magmatism began in the Kootenay Terrane of the Western Cordillera in the Late Devonian, and that magmatism extended onto the neighbouring Laurussian Craton margin in the Earliest Carboniferous.

A spreading centre developed in the Late Devonian between the Laurussian Craton and the Kootenay and associated terranes which developed as time progressed into the substantial Carboniferous and later Slide Mountain Ocean (Colpron & Nelson, 2009). In the northern end of the Cordillera, at the north-western Arctic margin of Laurussia, the Ellesmerian Orogeny was caused by the oblique collision of the eastern end of the Arctic Alaska–Chukotka Microcontinent with the Laurussian Craton, and the deformation there continued on past the end of the Devonian. The relationship between the Romanzov Orogeny, which was largely Early to Middle Devonian (and mostly consisted of shortening within the Arctic Alaska–Chukotka Microcontinent) and the generally later Ellesmerian Orogeny further to the east is not entirely clear: we provisionally conclude that they were probably parts of essentially the same series of events. Since the microcontinent was positioned at right angles from its orientation today, from the Middle to the Late Devonian (Fig. 8.6) a thick clastic wedge filled the resultant foreland basin which developed into the present-day Arctic islands of Canada between the microcontinent and the Laurussian Craton (Embry, 1991). There was also deformation in the Arctic islands around Ellesmere Island itself, where the Ellesmerian Orogeny was originally defined for Late Devonian (Famennian) to Early Carboniferous disturbances there (Trettin, 1998). The causes of the Ellesmerian Orogeny in northern Canada are unresolved, but it may have been due to convergence between northern Laurussia and its conjugate plates (Siberia and the Panthalassic).

At the western margin of Laurentia, strike-slip motion occurred along the Laurentia–Panthalassic boundary margin until the Middle Devonian, when relative convergence coincided with the first appearance of arc-related magmatism there (Fig. 8.6). In the Late Devonian, the relative motion became more obliquely convergent, in anticipation of the latest Devonian–Carboniferous opening of the Slide Mountain and Angayucham Oceans. The Innuitian margin of northern Laurentia is different. Bearing in mind the Baltic faunal affinities of the Pearya Terrane (Unit 134) and its Late Silurian to Early Devonian arrival by sinistral transpression, its accretion with the main Laurentian sector of the Laurussian continent seems most likely to have been a northern continuation of Caledonide orogenesis. Thus we consider the Pearya Terrane to have been a small, unified plate that became detached from Baltica when the latter collided with Laurentia in the Middle Silurian, only to continue slowly drifting for 20 million years until it collided with Laurentia in the Devonian (Fig. 8.6). In the same region, the two halves of Spitsbergen, East Svalbard (Unit 311) and West Svalbard (Unit 309), became united in the relatively local Late Devonian Svalbardian Orogeny.

In addition to the units described above, within the Cordillera of north-west Canada and Alaska, there are the Alexander and Wrangellia (Unit 165), and the Angayucham and Goodnews terranes, all of which are composite. The Alexander Terrane has a Precambrian core with Cambrian to Devonian rocks and fossils and accreted arc volcanics, within which have been recognised a Cambrian Wales Orogeny and a Middle Silurian to earliest Devonian Klakas Orogeny. The others also have a variety of Palaeozoic rocks (Cocks & Torsvik, 2011). Alexander and Wrangellia are stitched by a Late Carboniferous granite. The varied faunas are neither explicitly Siberian nor Laurentian in affinity, and palaeomagnetic data from both Alexander and Wrangellia suggest palaeolatitudes further to the south than their neighbouring terranes today. We therefore agree with Nokleberg et al.'s (2000) analysis that the terranes probably made up a separate microcontinent originally in a mid-oceanic position, but its location is not constrained well enough to place it securely on our Palaeozoic maps, and thus it is omitted from them.

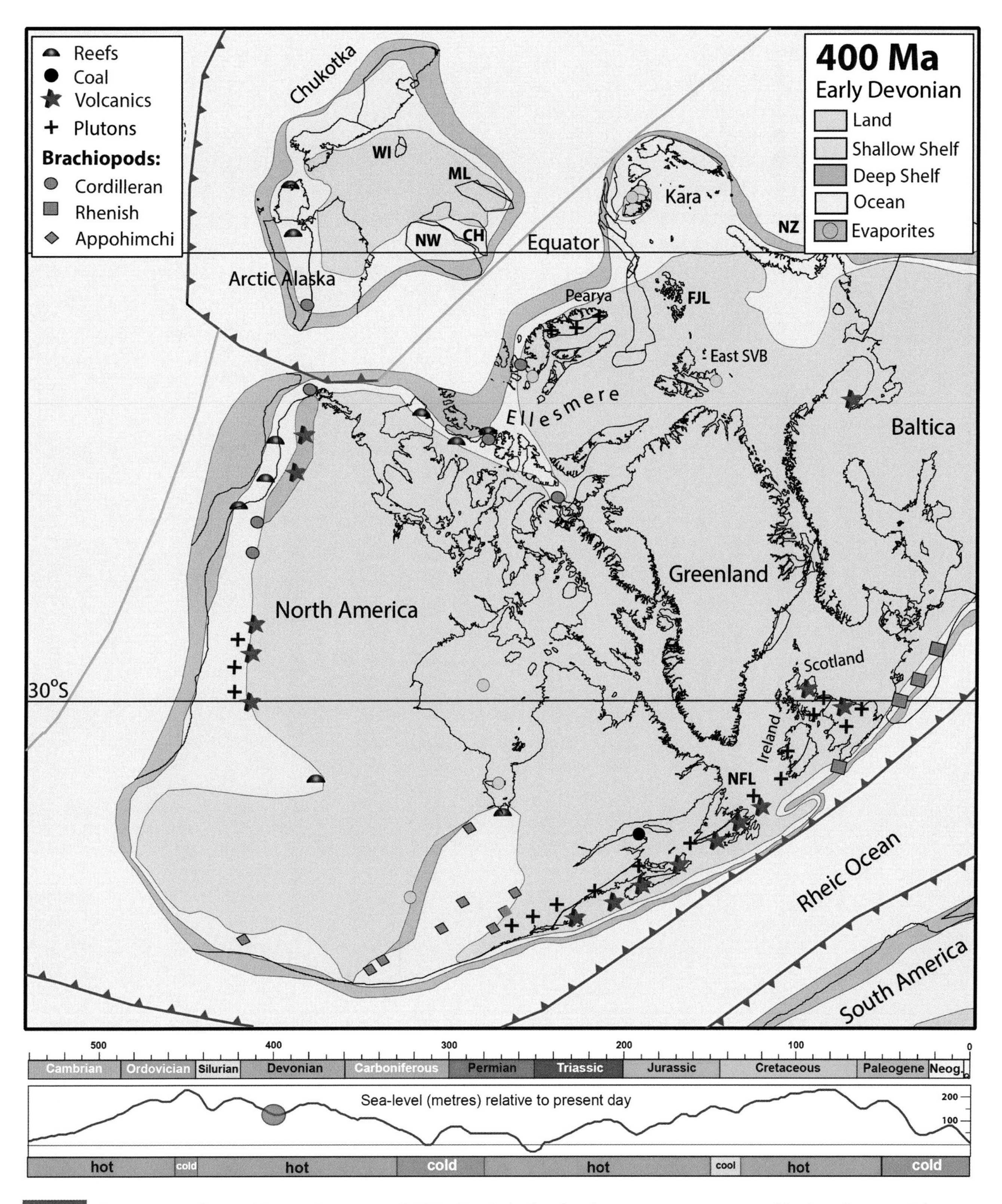

Fig. 8.5 The western and central Laurussian area at 400 Ma (Emsian), showing the contemporary pattern of lands and seas and the distribution of brachiopod provinces. Solid blue lines are subduction zones with teeth on the upper plate, and green lines are transform plate margins. CH, Chukchi; FJL, Franz Josef Land; ML, Mendeleev; NFL, Newfoundland; NW, Northwind; NZ, Novaya Zemlya; SVB, Svalbard; WI, Wrangel Island. Bottom: Phanerozoic time scale and sea-level variations (Haq & Al-Qahtani, 2005; Haq & Shutter, 2008) and icehouse (cold) and greenhouse (warm) conditions.

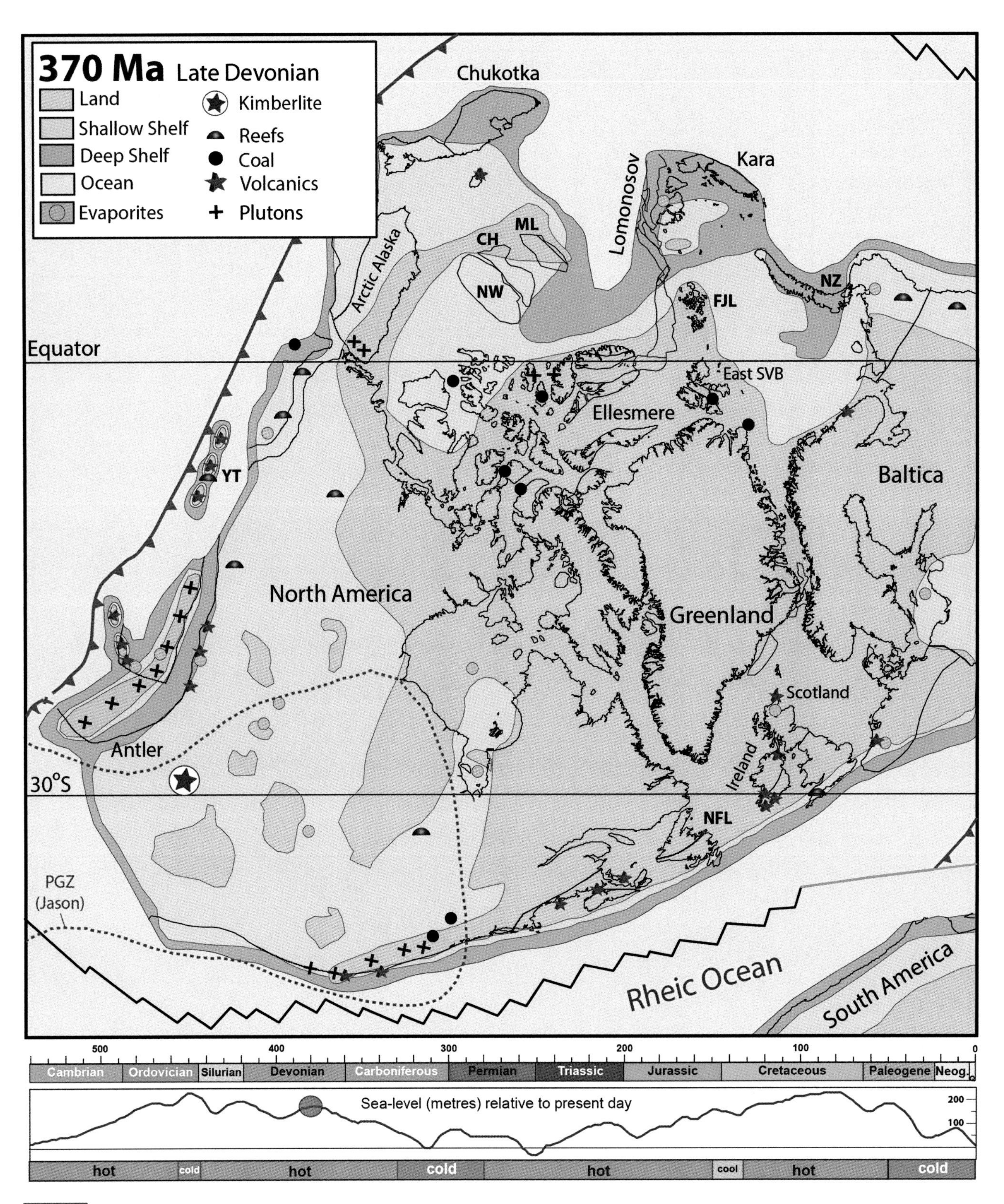

Fig. 8.6 The western and central Laurussian area at 370 Ma (Famennian), showing the contemporary pattern of lands and seas. Solid blue lines are subduction zones with teeth on the upper plate, black lines are spreading centres and transform faults, and green lines are transform plate margins. Dotted red lines are the plume generation zones (PGZ). CH, Chukchi; FJL, Franz Josef Land; ML, Mendeleev; NFL, Newfoundland; NW, Northwind; NZ, Novaya Zemlya; SVB, Svalbard; YT, Yukon–Tanana. Bottom: Phanerozoic time scale and sea-level variations (Haq & Al-Qahtani, 2005; Haq & Shutter, 2008) and icehouse (cold) and greenhouse (warm) conditions.

Eastern Laurussia. At the global scale of most of our maps, the former continental plates of Baltica–Avalonia and Laurentia were essentially united before the start of the Devonian to form Laurussia, which included a substantial land area characterised by the Old Red Sandstone facies, which was chiefly deposited in non-marine basins of substantial size. Those basins resulted from a variety of mechanisms, chiefly lithospheric, flexing, and wrenching, which were associated with the Caledonide, Variscan, and Ellesmerian orogenies at the margins of the continents (Friend & Williams, 2000). In the centre of Laurussia, the Caledonide Orogeny was reflected in its final stages in the Late Silurian and Early Devonian (after 425 Ma) by the widespread intrusion of substantial granites in the Southern Uplands region of Scotland, which lies just to the north of the suture marking the Silurian closure of the Iapetus Ocean (Fig. 7.6), often collectively termed the 'Late Caledonian granites' (although in Scotland they are confusingly named the 'Newer Granites'). Those are largely I-type granites, which are characteristic of post-collisional settings in which the melting of the country rocks has resulted from decompression following crustal thickening, and are particularly seen near Aberdeen, in the Southern Uplands of Scotland, and the Lake District of England, including the 392 Ma (Eifelian) Shap Granite, which is noted for its distinctive large rhomb porphyroblasts and is much favoured for use as ornamental building stone.

The eastern (Uralian) margin of the former Baltica progressively approached the northern margin of Siberia (which was south-facing in the Early Devonian) through the Devonian. The accretion of the substantial southern Uralian Magnitogorsk Island Arc to Baltica occurred in the Middle to Late Devonian, when the adjacent Tagil Island Arc in the central Urals (Fig. 9.1a) was still offshore (Brown et al., 2011). To reconcile those varied observations, we place the Devonian Magnitogorsk Arc above a north-dipping intra-oceanic subduction zone. The northern Caspian Sea Basin has more than 25 km of sediments, below which geophysical exploration indicates that its basement consists of Middle Devonian oceanic floor (Zonenshain et al., 1990).

Siberia and Peri-Siberia. In the Early Devonian, Siberia was at low latitudes in the northern hemisphere (Fig. 8.1a), and most of the continent seems to have continued to be tectonically relatively quiet, as it had been in the Silurian. Although the palaeolatitude of Siberia is almost entirely interpolated for the Devonian (Fig. 9.11), since there are no reliable palaeomagnetic data between the Middle Silurian at 430 Ma and the latest Devonian at 360 Ma (Cocks & Torsvik, 2007; Torsvik et al., 2012), nonetheless those data continue to indicate that Siberia was 'upside down' (azimuthally inverted) at the beginning of the Devonian and centred at about 15° N. As the Devonian progressed, Siberia slowly rotated clockwise and drifted north to become centred at about 30° N by 360 Ma (Fig. 8.7), and continued to be the largest landmass in the northern hemisphere. Early Devonian kimberlites in east Siberia and the 400 Ma (Pragian) Altai–Sayan LIP in its south-west support the longitudinal placement of that continent above the north-west arm of the African (Tuzo) LLSVP. Abundant Late Devonian kimberlites and the 360 Ma (Late Famennian) Yakutsk LIP in east Siberia (Fig. 9.9) indicate that the continent remained over the Tuzo Plume Generation Zone throughout the Devonian, perhaps drifting slightly eastwards. By applying the longitudinal calibration of Torsvik et al. (2014) to Siberia, our reconstructions have been revised, and are now very different from earlier authors, including Cocks and Torsvik (2007).

Today's south-eastern margin of Siberia (north-west-facing in the Early Devonian) was passive through nearly the entire Devonian (Fig. 8.1), facing the Mongol–Okhotsk Ocean, which had opened in the Silurian or earlier (Bussien et al., 2011), and that ocean slowly widened throughout the Devonian as Amuria drifted away from Siberia. In the latest Devonian (at about 360 Ma) that passive margin collapsed and south-dipping subduction commenced beneath Siberia (Fig. 8.7). The margin of eastern Siberia was likewise passive throughout the Devonian, apart from a Middle to Late Devonian episode of extension in the Viljuy Basin (see below).

The small amount of palaeomagnetic data from the Devonian of Amuria is untrustworthy, and thus the location of Amuria (which included the former Central Mongolian Terrane Assemblage) is poorly constrained in those times; however, we assume that westernmost Amuria was loosely contiguous with the Altai–Sayan region of Siberia and have allowed the widening Mongol–Okhotsk Ocean to slowly separate their eastern margins (Fig. 8.1). That ocean must have been expanding during the Devonian, since its north and south margins were both passive until the latest Devonian. Today's northern margin of Siberia was also passive during the Devonian, according to observations in central and southern Taimyr (Torsvik & Andersen, 2002).

The accretion of the various Altai–Sayan and Tuva–Mongol (Unit 432) terranes and island arcs to the main Siberian Craton was complete by the start of the Devonian (Fig. 8.8). However, the margin was still very active along the then north-east part of Siberia (the Altai–Sayan area), and there was also much activity on the northern margins of the Central Mongolian Terrane Assemblage (Unit 440), by then an integral part of the Amuria Microcontinent, including the Mandalovoo Island Arc (within Unit 441).

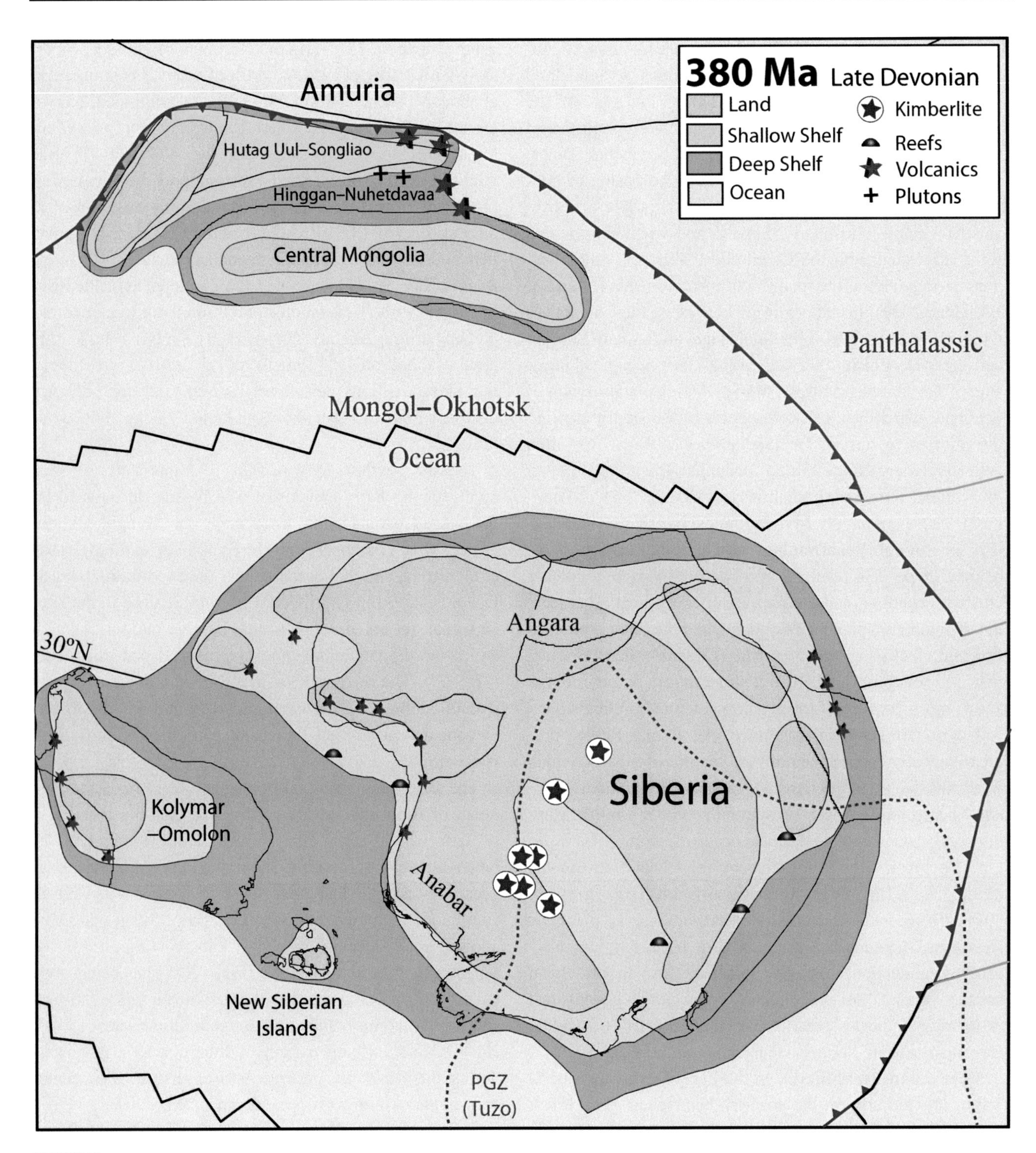

Fig. 8.7 The Siberia and Amuria continents at 380 Ma (Frasnian), showing the contemporary pattern of lands and seas. Solid blue lines are subduction zones with teeth on the upper plate, black lines are spreading centres and transform faults, and green lines are transform plate margins. Dotted red line is the plume generation zone (PGZ).

From the Middle Devonian onwards there was significant rifting and magmatism, particularly in today's east of the craton area, with the main rift system being in the Viljuy Basin in eastern Siberia (Cocks & Torsvik, 2007). That rifting started with flood basalt magmatism and basalt igneous dyke swarms. The second and largest phase of tectonic activity ran from the Late Devonian and continued on into the earliest Carboniferous (Tournaisian), and included the

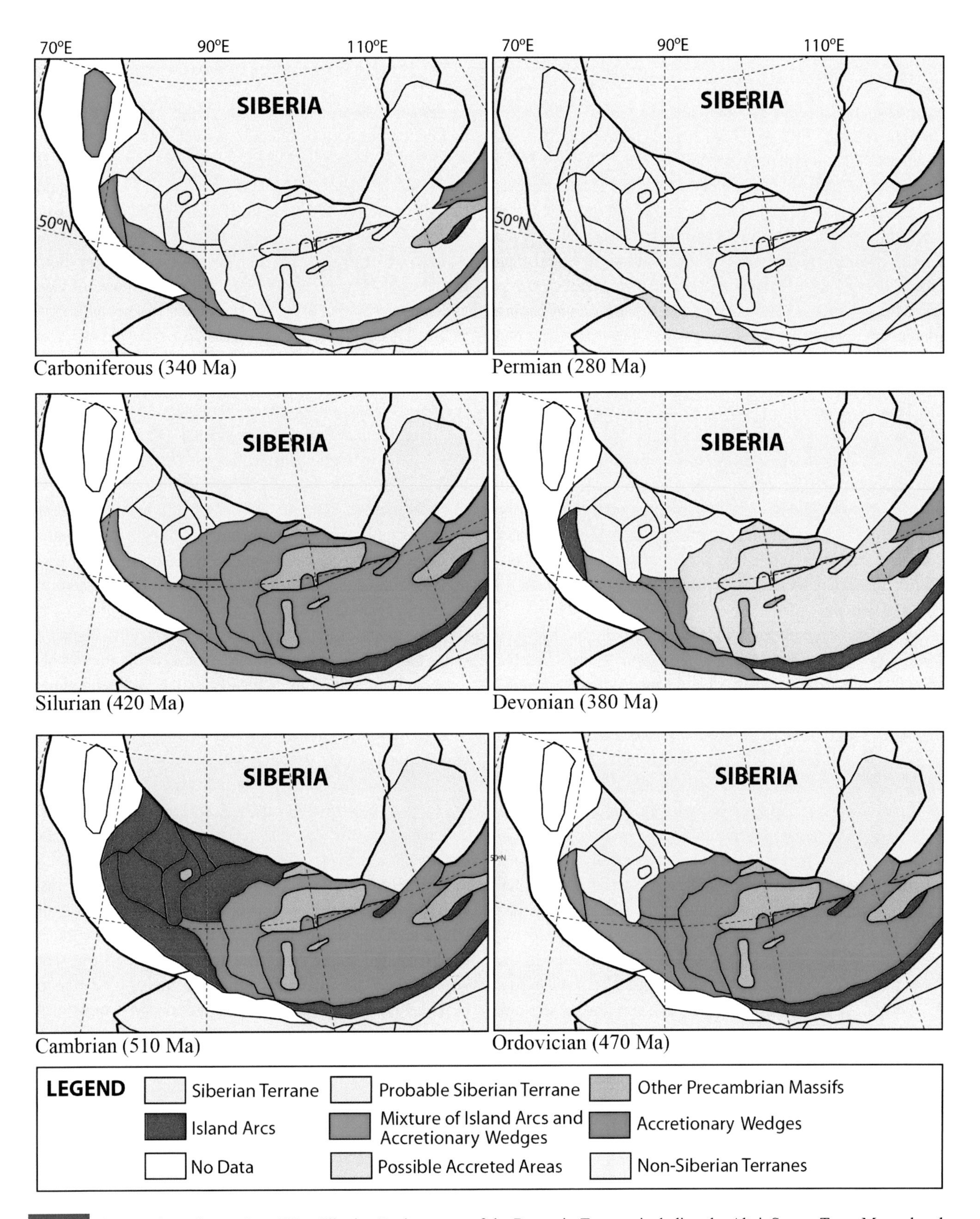

Fig. 8.8 The area from the southern West Siberian Basin to east of the Barguzin Terrane, including the Altai–Sayan, Tuva–Mongol and much of the Central Mongolia Terrane Assemblage area (names of the individual units are shown in Fig. 7.7). The six maps show the terrane units in different colours reflecting their dominant aspect during each period, and also show how the Siberian Continent increased by accretion. The Barguzin Terrane did not completely accrete to the Siberian Craton until the Late Devonian, but may have formed part of it in the Proterozoic (see text). The generalised outlines of the Precambrian massifs (chiefly in Tuva–Mongolia and Central Mongolia), which did not form part of Siberia in the Precambrian, follow Cocks and Torsvik (2007).

Yakutsk Large Igneous Province eruptions at 360 Ma (latest Devonian) (Fig 9.9), but had ceased before the Visean. Numerous diamondiferous kimberlites were emplaced into Archaean nuclei within the Siberian Craton (Fig 8.7), reflecting the effect of Middle to Late Devonian plume events. It is interesting that the same Viljuy aulacogen had opened in the Late Precambrian (Riphean in local terminology) and then closed again in the subsequent Vendian at about 550 Ma (Zonenshain et al., 1990), indicating a zone of continuing fundamental tectonic weakness within the Archaean and Early Proterozoic rocks in that area of the Siberian Craton.

In addition, Middle to Late Devonian rifts also formed in the Tunguska Basin, and along today's northern margin of the craton, and there was very substantial strike-slip faulting in the Altai–Sayan area, which reached its maximum in Late Devonian times. The cause of that tectonism is uncertain, but, apart from the accretion of the largest parts of the Central Mongolian Terrane Assemblage, there were no major continent–continent collisions which could explain those rifts or the extensive Devonian magmatism present over so much of Siberia.

Intra-plate magmatism was widespread in the Early to Middle Devonian in southern Siberia, where large basaltic and alkaline volcanic fields formed in at least three separated depressions in the Altai–Sayan region (Zonenshain et al., 1990). In western Altai–Sayan, the volcanic activity started in the Early Devonian (Emsian), peaked in the Middle Devonian (Early Givetian), and carried on into the Late Givetian in the south-west part of the area (Yolkin et al., 2003). Subaerial volcanics were also extensive in Altai–Sayan in the Late Devonian. There was also considerable Devonian igneous activity in the Okhotsk Massif, much of Mongolia, and elsewhere in peri-Siberia: that in the Mandalovoo Terrane, Mongolia, at today's southern margin of peri-Siberia, included Late Devonian pillow basalts and andesites. In the former Tuva–Mongol and Central Mongolia Terrane sectors of Amuria, a high proportion of the previously separate terrane areas have Devonian volcanics, in contrast to those other Mongolian terranes in today's south of peri-Siberia, in which the volcanic activity is of a much greater variety of geological ages. Many granitoid batholiths were also intruded into peri-Siberia, the most spectacular of which is the immense Late Devonian Barzugin granodiorite–tonalite, whose outcrop today occupies a high proportion of the area of the Barzugin Terrane (Unit 436).

Facies, Faunas, and Floras

Climates. There was a greenhouse global climate during most of the Devonian (Fig. 16.2), and thus the average temperatures were exceptionally high, although near the end of the period a progressive switch to a cooler environment led to significant changes. That cooling was probably the result of an eventually dramatic decrease in atmospheric carbon dioxide near the end of the Devonian which continued on into the Early Carboniferous. That decrease was perhaps chiefly caused by the absorption of carbon dioxide by the massively increasing numbers and volumes of plants, and particularly the advent of large trees, on the land, as well as by the development of extensive peat deposits, which locked up substantial amounts of carbon. Following a low-stand near the Silurian–Devonian boundary, global eustatic sea levels rose substantially during Early Devonian and Middle Devonian times (Lochkovian to Givetian, from about 420 Ma) to a high-stand within the Late Devonian (Late Frasnian) at about 380 Ma. That rise caused an expansion in availability of new ecological niches, which in turn led to faster evolution and an expansion in biodiversity. In particular, the largest development of reefal systems in many regions known in Earth history occurred in the Devonian, and those reefs are estimated to have covered areas of perhaps as much as 5 million square kilometres, almost ten times the areas of comparable reef ecosystems found today.

In addition, there was clearly much differentiation in equatorial to polar temperatures during the Early Devonian. That is reflected in the development of provinces in the benthic faunas such as the brachiopods, which were largely developed at different palaeolatitudes. The provinciality rose to a peak in the Emsian (Fig. 8.11) and subsequently dwindled so that the differences between the provinces became progressively less obvious. After the Emsian, the biodiversity firstly levelled out and then started to decrease, at first slowly and then at a much faster rate after the Eifelian, with a marked extinction event in the marine invertebrates at the Frasnian–Fammenian boundary at about 383 Ma. There was another, although less important, series of extinctions near the Devonian–Carboniferous boundary, which started earlier in the marine realm than the terrestrial.

Colonisation of the Land. Land plants with trilete spores probably evolved in the Ordovician from freshwater algae (see Chapter 6) which adapted firstly to occasionally desiccated non-marine margins of the oceans, but which subsequently became able to spend their whole lives without being permanently immersed in water. However, it was not until the Devonian that the land plants evolved from small weeds into substantial trees which eventually formed forests, which in turn profoundly influenced and changed both the climate and the relative amounts of both carbon dioxide (Fig. 16.2) and oxygen in the Earth's atmosphere. A further great change came as the increasing area of vegetation cover progressively

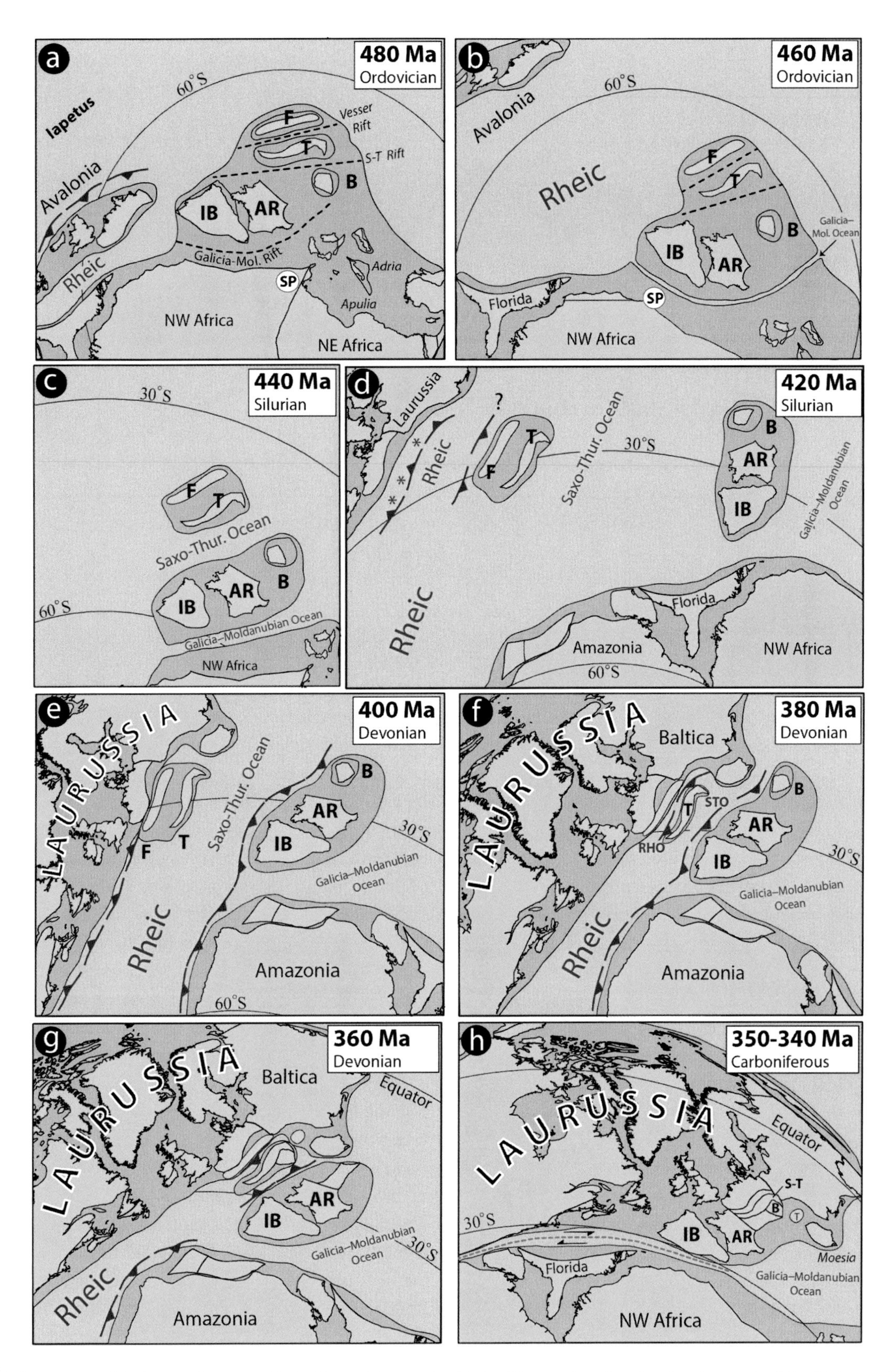

Fig. 8.9 Palaeogeographical maps from the Early Ordovician to the Carboniferous showing the development of the Variscan Orogeny in the European area and beyond (see text). AR, Armorica; B, Bohemia; F, Franconia; IB, Iberia; RO, north-western arm of the Rheic Ocean; SP, South Pole; S-T, Saxothuringia; STO, Saxothuringian Ocean; large T, Saxothuringian Terrane; small T, Tisia Terrane (Unit 337).

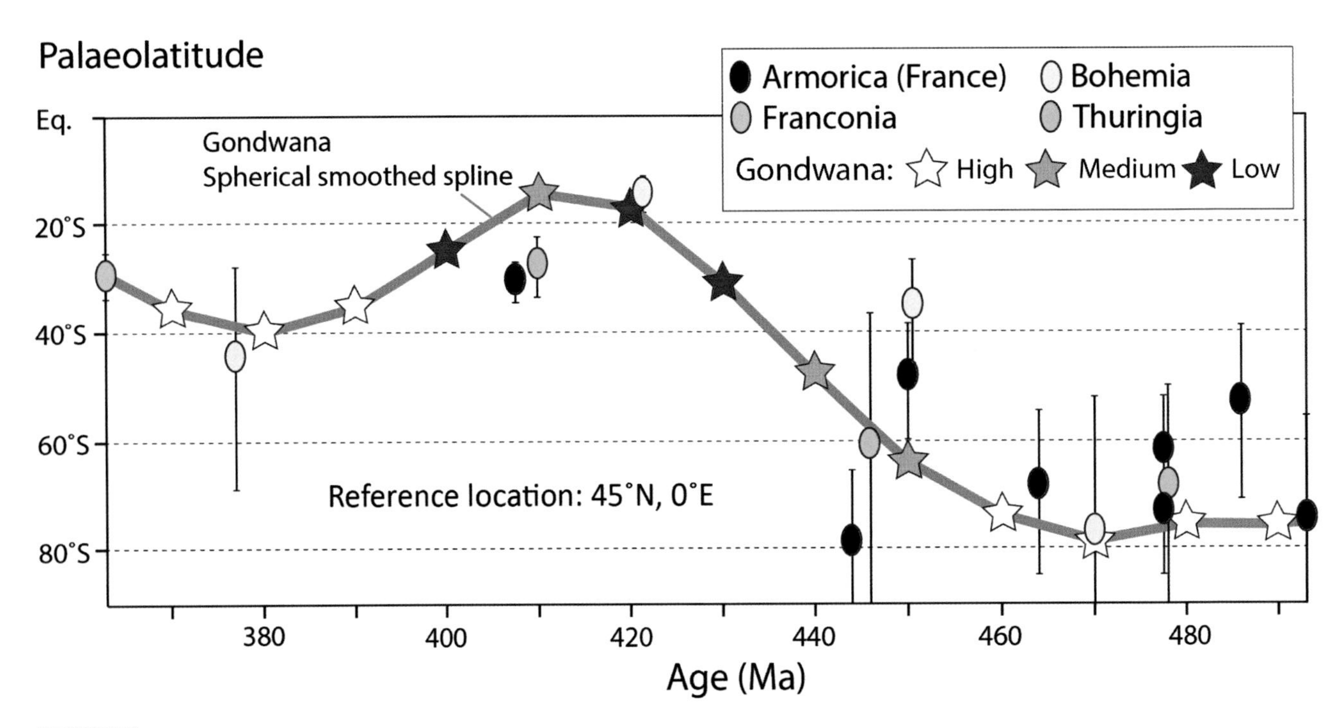

Fig. 8.10 Palaeolatitudes with 95% uncertainties for Armorica, Thuringia, Franconia, and Perunica (Bohemia) calculated to a common reference site (45° N and 0° E) and compared with palaeolatitudes calculated for Gondwana using the same reference location. Note Silurian and Devonian gaps in the data between 444 and 421 Ma, and 407 and 377 Ma.

slowed down the eroded particle runoff due to rainfall across all the lands, and that slower runoff enhanced the development of new styles of meandering streams and larger rivers, which in turn affected the nature of the sedimentary deposits. The rise in the abundance of muddy floodplain and coastal plain deposits indicates the increased storage of fine sediments in the rivers of lowland areas, and the incoming of channel-braided architecture also reflects the increasing amount of mud, which made the sediments more cohesive (Gibling et al., 2014). In addition, the incoming of wood and substantial logs provided a whole new suite of sedimentary particles after individual disintegration, some of which eventually merged to become coal deposits.

The Early Devonian (Pragian) Rhynie Chert of north-east Scotland, which consists of siliceous deposits from neighbouring hot springs, has entombed within it an exceptional variety of lichens, fungi, plants, and animals remarkably preserved in three dimensions. The development of early fungi was also particularly important, since they were the principal decomposers of organic matter on which other terrestrial organisms could feed. In the Rhynie Chert, there has been found the first direct evidence of rhizoid root systems, mycorrhizal fungi, and fungal–algal symbiosis (Kenrick et al., 2012). Germinating plant spores, and even sperm in the process of release from the male fertile organ of a gametophyte plant, are known. All but one of the plants were leafless, although some had minute hairs. The rise in the quantity and variability of the terrestrial plants progressively provided many new ecological niches which were colonised by a wide variety of animals, the latter mostly arthropods of many orders and classes. Curiously, the animals in the earlier niches were mostly carnivores or detritus-feeders, and eaters of living plants (herbivores) were rare (Kenrick & Davis, 2004). In the Rhynie Chert, as well as terrestrial biota (arthropods, including spiders, insects, and centipedes), aquatic forms such as freshwater crustaceans and many others have also been found. Those earliest insects (apterygotes) had no wings, which did not develop until the Early Carboniferous, together with the ability to fly.

Much later in the Devonian, in the Frasnian at about 375 Ma, the first tetrapod vertebrates, the amphibia, evolved from lobe-finned fish and proceeded landwards. From rocks of Famennian age in East Greenland, many complete skeletons of the well-known amphibians *Acanthostega* and *Ichthyostega* have been found and described; however, those animals probably lived largely in freshwater. In contrast, the less famous *Tulerpeton* from contemporary rocks in Russia, which is known only from a single specimen, has much more in common with later fully terrestrial amphibians (Clack, 2002).

Gondwana. The sea-level rises during the Early Devonian led to transgressions over many Gondwanan cratonic areas,

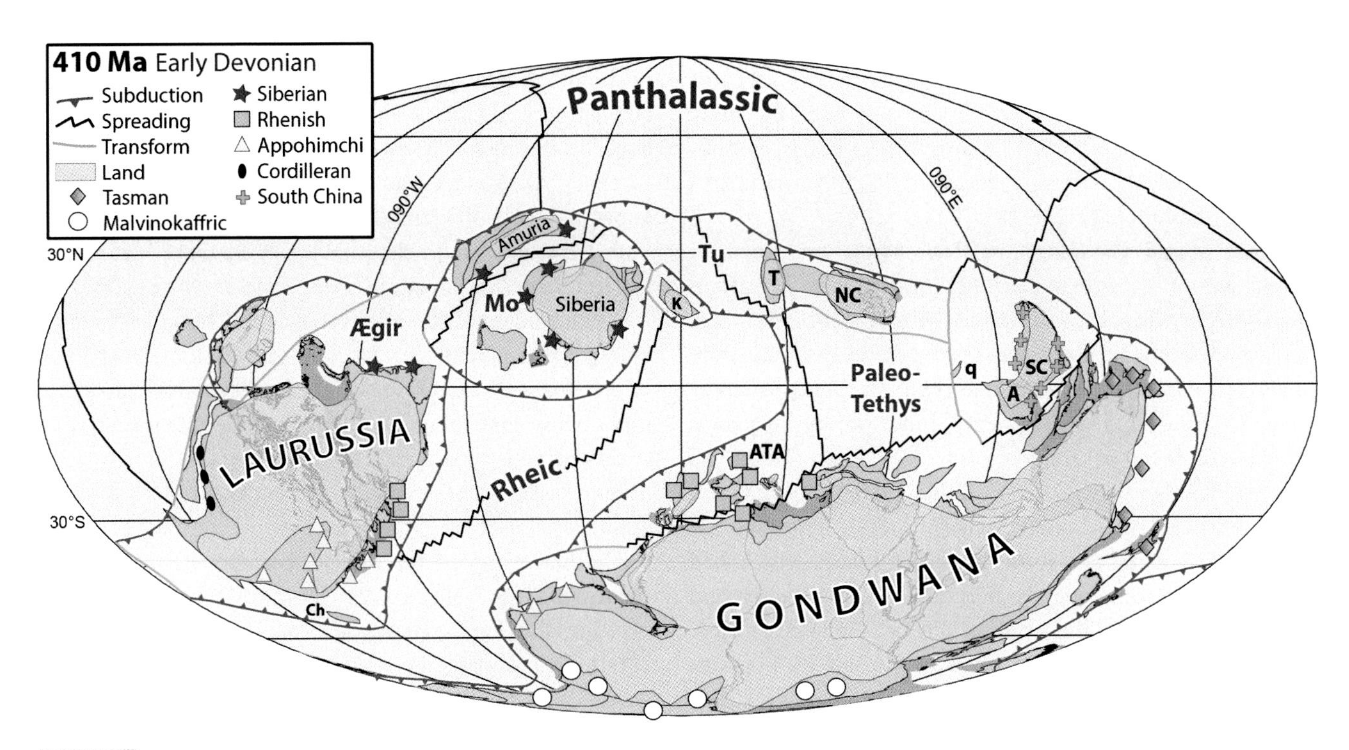

Fig. 8.11 Global distribution of brachiopod provinces in the Emsian at about 400 Ma. A, Annamia; ATA, Armorican Terrane Assemblage; Ch, Chilenia Terrane; Mo, Mongol–Okhotsk Ocean; NC, North China; q, Quinling Terrane; SC, South China; T, Tarim; Tu, Turkestan Ocean. New diagram, including sites plotted from Boucot et al. (1969) and Boucot & Blodgett (2001).

for example the substantial southward extension of the shallow sea over much of North Africa. At the opposite Gondwanan margin, in Antarctica, the shoreline passed the Ellsworth Mountains and Ohio Range and unconformably transgressed the Lower Palaeozoic granitoids and deformed Lower Palaeozoic sediments there. In the Transantarctic Mountains there are thick largely non-marine sediments of the Beacon Group, which is still relatively undeformed today, although there are sporadic marine interbeds, some yielding Mavinokaffric Province brachiopods (Elliot, 2013). As well as oceanic and climatic differentials, those transgressions were probably one of the factors causing a notable increase in the provinciality in the shallow-marine benthic faunas over the same period, which was one of the greatest in the whole Palaeozoic, particularly in the brachiopods (Boucot & Blodgett, 2001), and those provinces are shown on Fig. 8.11 for the Emsian, when they were at their most diverse. As in the preceding Silurian, the higher latitudes of Gondwana (over 60° S) were colonised by the low-diversity Malvinokaffric Province, principally in the centre and south of South America and the south-western parts of Africa, as well as in Antarctica. That period of maximum provinciality in the Emsian may have been partly enhanced by a global regression between the Late Pragian and the Early Emsian; for example, in the Paraná Basin of Brazil there were substantial faunal extinctions before the Late Emsian. The faunal diversity in the Pragian to Eifelian successions in the Paraná Basin was generally low or medium across successions ranging from shoreface to distal tempestites; and inarticulated lingulide brachiopods can be used as indicators of regular transgressive and regressive phases throughout the period since they were (and still are) the only brachiopods which can tolerate living in waters with any reduction in salinity. However, the Paraná Basin remained at the inboard margin of an epeiric sea throughout that period.

A cladistical analysis of the global distribution patterns shown by Early and Late Devonian fish indicates that those found around Gondwana differed substantially from those around Laurussia in the Early Devonian, but, as the Rheic Ocean narrowed, there was less faunal differentiation (Young, 1990). Fish faunas in north-west Gondwana and the whole of Laurussia show that the Rheic Ocean was still wide enough in the Early Devonian (Lochkovian) at 415 Ma to be divisible into separate fish provinces, but that by Frasnian times (380 Ma) there was no substantial difference between the fishes found in the two regions (McKerrow et al., 2000). The temperate and equatorial latitudes hosted many diverse warm-water faunas and sedimentary features, including bioherms (Copper, 2002). Although the climates were relatively warm during nearly all of the period, the latest Devonian saw the first, although relatively minor, phase of the long-lasting but intermittent Upper Palaeozoic glaciations which continued right through until the Early Permian (Fig. 16.2); however, those earliest glaciogenic tillites are found only in the Famennian of South America, then near the South Pole (Fig. 8.3).

Eastern and Central Asia. There were two separate faunal provinces within the Old World Realm in eastern Asia: North China, Qaidam, and Tarim on the one hand, and South China on the other. The brachiopod faunas from the Hutag Uul–Songliao Terrane (Unit 453) and Khanka–Jiamusu–Bureya Microcontinent (Unit 454), by then parts of the continent of Amuria (Fig. 8.7), were within the North China provincial area. Although the North China continent itself was largely land (Fig. 8.4), it formed the centre of the marine North China Province, which had many endemic brachiopods and corals in the Pragian and Emsian; however, the proportion of endemic genera dwindled from the Givetian onwards, so that by Frasnian times the whole area was within a single faunal province. In the Early Devonian, in addition to the higher-latitude Malvinokaffric Realm brachiopods seen in the Gondwanan area (Fig. 8.11), for example in New Zealand, the remainder of eastern Asia lay within an overarching Old World Realm, although there is some overlap at the margins between the two realms in eastern Australia (Boucot & Blodgett, 2001).

However, within the Old World Realm, various different regions can be recognised, including a separate South China Region, whose area also extended south-westwards from the South China continent to include part of the Kunlun Terrane assemblage (Unit 457). That fauna is also known from the northern Vietnamese sector of South China. We follow the facies maps for South China in the Early and Middle Devonian by Cai and Zhang (2009). A key Lower Emsian brachiopod fauna from Guangxi Province, South China, includes many endemic genera, notably *Dicoelostrophia*, *Eosophragmophora*, and *Parathyrisina*, and 30% of the 81 brachiopod genera in the various Emsian localities in the South China Region were endemic. The South China Region Emsian faunas plotted on Fig. 8.4 differ substantially from those in the 'Balkhash–Mongol–Okhotsk' Region of Boucot and Blodgett (2001) to the north, which was part of peri-Siberia (Fig. 8.7). The latter include the brachiopods described from the Hinggan Massif in the Hutag Uul–Songliao Terrane and the Khanka–Jiamusu–Bureya Microcontinent, by then parts of Amuria. Those brachiopods differ again from the Gondwanan faunas in the Australian and Sibumasu sectors, for example the Early Devonian fauna in southern Thailand (Boucot et al., 1969).

As can be seen from our maps (Fig. 8.1), the relative positions of the major palaeocontinents did not change substantially during the Devonian, and thus the faunal differences in the earlier Devonian must have been due primarily to changing ocean currents or differing gradients between equatorial and polar temperatures, which varied substantially with time and would thus have caused more varied local differentiation in climates. In the Late Devonian (Frasnian and Famennian) all the eastern Asia and Australian regions lay within one faunal province; for example, the brachiopods of the Bonaparte and Canning Basins of north-western Australia show that the reasonably diverse assemblages there consisted entirely of cosmopolitan genera, although a high proportion of their species were endemic (Wright et al., 2000).

The narrowing of the Rheic Ocean to the south-east of Laurussia was reflected in the progressive similarities between the terrane-diagnostic faunas (McKerrow et al., 2000). For example, the Rhenish Province of Europe, North Africa, and the eastern side of North America, which was chiefly originally defined on brachiopods and other macrofauna, and which had not previously been plotted on a plate-tectonically based map, resulted from differing palaeolatitudes, rather than being simply restricted to particular continents, as seen in Fig. 8.11.

In Central Asia, there is a huge variety of Devonian rocks and fossils in the many Kazakh terranes. For example, within the composite Atashu–Zhamshi (Unit 461) alone, in one sector there are volcanic rocks from the Lochkovian to the Frasnian (415–385 Ma), followed by Famennian terrestrial rocks and marine rocks, largely shallow-water carbonates with abundant benthic marine faunas not far away; whilst in another sector deposition of shallow-water marine rocks persisted throughout the Devonian from the Lochkovian to the Famennian. In Chu-Ili (Unit 460) and North Tien Shan (Unit 459), by then integral parts of the Kazakhstania continent, there are Lochkovian to Fammenian terrestrial sediments interbedded with many volcanic layers (Daukeev et al., 2002).

Western Laurussia. The continuing equatorial position of the Laurentian and Western European parts of Laurussia during this period (Figs. 8.5 and 8.6) ensured that the shallow seas were warm enough to support a very varied shelf benthos, including the largest trilobites known, which were over a metre in length. To judge by the distribution of reefs through the period, the maximum warmth was during the Frasnian; and that has been confirmed by oxygen isotope studies (Joachimski et al., 2009). Detailed work on the Middle and Late Devonian rocks covering the western half of the USA shows that the shelf–basin boundary at the craton margin remained remarkably stable from the Ordovician to the Early Devonian. Most of the north-western part of the Laurussian Craton consisted of the Old Red Sandstone continent, a land area that extended eastwards to include much of the former Avalonian sector and other parts of northern Europe as well as dominating the central sector of Laurussia (Ziegler, 1989), and which included many wadi deposits from tropical flash floods (Friend & Williams, 2000). Similar red sandstone facies, which are usually stained with the iron and manganese compounds that give the rocks their name, are also seen in the Baltica sector of Laurussia, although during the Middle Devonian there appears to have been an extensive marine embayment, running approximately north–south along the approximate site of the North Sea today, between much of the two large Old Red Sandstone land areas on both its sides.

The earliest Devonian was a time of relatively low sea-level stands, which caused much emergence and a significant number of land barriers within the craton areas of Laurussia. This resulted in niche partitioning which ensured that there was considerable provincialism within the shallow-marine benthic faunas, both globally and inside Laurentia, as shown by the brachiopods (Boucot et al., 1969). In the Pragian at 410 Ma, the macrofauna of Laurentia, particularly the brachiopods, were divided between a Nevada Subprovince, in the western half of the craton, and an Appohimchi Subprovince, in the east, the latter stretching as far round to the south of the continent as Sonora in Mexico. Those subprovinces were separated in the centre of the craton by a substantial landmass termed the Transcontinental Arch, and their subsequent Emsian distribution is shown in Fig. 8.5, in which the Appohimchi Province is apparently confined to the large embayment west of the Appalachians. However, those subprovinces broke down and were united into a single biogeographical unit by the transgression termed the Taghanic Onlap near the end of the Middle Devonian (Givetian). An analysis of Middle and Late Devonian brachiopod, bivalve, and phyllocarid faunas showed how the land areas on the craton (often termed 'rises' in the literature) much affected the distribution of the biota.

The Old Red Sandstone continent was host to a variety of fish faunas at its margins (Blieck & Cloutier, 2000), but it is uncertain to what extent those fish were able, like modern salmon, to thrive in both marine and non-marine environments. As noted above, amphibian tetrapods evolved from fish in the Late Devonian, and the oldest are from terrestrial deposits in Greenland, but their known distributions are too patchy to be palaeogeographically significant. Among Upper Devonian (Frasnian and Famennian) and Lower Carboniferous strata there are substantial black shales which reflect the deep-water anoxia present in many of the cratonic basins, particularly in Illinois and the Appalachians within western Laurussia, but there are some black shales as far north as Alberta. Analyses of molybdenum, and other trace elements, have led to the conclusion that those shales were the result of high plankton productivity, occurring at the same time as organic buildups, which in some places were extensive. The palaeosols present in the Famennian of Pennsylvania indicate deposition under alternately semi-humid and sub-arid conditions, but only the semi-humid climates could have supported the tetrapods *Hynerpeton* and *Densignathus*, whose remains are entombed there.

Sedgwick and Murchison's original type area of Devonshire, England, saw continuous sedimentation during Devonian times, with alluvial Old Red Sandstone clastic sediments in the north prograding southwards into marine deposits in the Cornubian basins of Devon and the adjacent Cornwall by about 385 Ma (Givetian). However, those marine sediments contain much evidence of contemporary tectonic and volcanic activity due to the Variscan Orogeny, as well as coral and stromatoporoid reefs associated with brachiopods and other benthos interspersed with basins filled with deeper-water clastics. That mixture of volcanic and sedimentary rocks persisted to the end of the Devonian (Leveridge & Shail, 2011), and extended over much of central Europe (Figs. 8.5 and 8.6).

Eastern Laurussia and Siberia. All of today's eastern Laurussia straddled the Equator, and, like western Laurussia, much of it was occupied by the Old Red Sandstone continent for the entire Devonian. However, the seas at its margins hosted rich faunas, and there were many reefs. At the eastern end of the continent, the flooded craton in Podolia, formerly part of Poland and now in the Ukraine, there is a famous thick and relatively undeformed sequence, continuing on upwards from the Silurian without unconformity into the Early Devonian (Lochkovian and Pragian), which has yielded over 50 different brachiopod genera, and many other benthic and planktonic invertebrate and vertebrate fossils.

In the Late Silurian and Early Devonian, there was maximum marine regression on the Siberian Craton (Fig. 7.6); however, in contrast, there was a major transgression eastwards in the Altai–Sayan sector of the larger Siberian continent during Early Devonian times (Yolkin et al., 2003). The tectonically quieter areas in the western part of Altai–Sayan allowed substantial reefs to develop on the flooded craton in places, including the massive accumulations of the thick-shelled strophomenoid brachiopod *Megastrophia uralensis* found in the Emsian of Salair (Unit 430).

There are many Lower and Middle Devonian brachiopods known from central Mongolia, then part of Amuria, and the 'Balkhash–Mongolia–Okhotsk' is a distinctive local assemblage within the wider Old World Realm (Boucot & Blodgett, 2001). The endemic brachiopods from that region include *Khangaestrophia*, *Xingjianospirifer*, and what has been termed '*Paraspirifer*', although the Mongolian form differs from that essentially European genus.

9 Carboniferous

Part of a branch from a substantial Carboniferous lycopod *Lepidodendron* tree, whose abundant remains formed much of the Coal Measures of England.

It was during the Carboniferous that the majority of previously independent continents merged to form the only Phanerozoic supercontinent, Pangea. 'Carboniferous' means 'coal-bearing' and the name was coined by William Conybeare and William Phillips in the 1820s for such rocks in northern England: coal formed the major source of power for the Industrial Revolution of the eighteenth century. It was one of the first two geological systems to receive a name which is still in current use (the other is the Cretaceous, which also reflects a common rock type, rather than a geographical area like most of the others). However, the concept of the system has subsequently been much changed and enlarged, and its base is now defined at the base of the Tournaisian by the start of the *Siphonodella waubaunsensis* conodont Biozone in the Montagne Noire, France, which is dated at 359 Ma. The global distribution of the continental blocks and oceans at three intervals within the period is shown in Fig. 9.1 at 350 Ma (Tournaisian), 330 Ma (Serpukhovian), and 310 Ma (Moscovian).

Because of historical disagreements between North American and European geologists, now resolved, the Carboniferous is the only system in the Phanerozoic to be divided into two subsystems: the older one the Mississippian, in which is included the Tournaisian, Visean, and Serpukhovian stages; and the younger one the Pennsylvanian, which includes the Bashkirian, Moscovian, Kazimovian, and Gzhelian stages. The Serpukhovian to Gzhelian stages have their type sections in the Moscow Basin, Russia. However, many authors, particularly in western Europe, still use the terms (in ascending order) Namurian, Westphalian, and Stephanian for the periods after the Visean, which, like the Tournaisian and Visean, are all defined in France, Belgium, and Germany, although the age of the base of the Namurian is not the same as that of the base of the Serpukhovian, which is at a younger level within the Namurian.

Tectonics and Igneous Activity

Pangea Assembly. It was during the Carboniferous that Gondwana and Laurussia ceased to be independent superterranes, since they merged to form the supercontinent Pangea at around 320 Ma (Bashkirian). The collision was heralded by the Early Carboniferous down-warping of the Ouichita Basin in the southern USA, and the subsequent Ouachita Orogeny in that area is reflected directly in the compressional deformation first seen in the Middle Carboniferous of Oklahoma. Orogenic activity peaked in the Late Carboniferous and the final phase of union between the two continents was complete by the earliest Permian. Those two very substantial continents collided obliquely, as shown by the considerable evidence of strike-slip faulting in the Ouachita Mountains and nearby. The reason for that strike-slip can also be understood by comparing the situation in the Late Devonian (Fig. 8.1c) with that in the Carboniferous (Figs. 9.2 and 9.3), during which time Laurussia had moved a very long distance by lateral strike-slip faulting (~4,000 km) and had also rotated considerably. Probably because of the sheer bulk of Laurussia as well as the strike-slip faulting, the Gondwana–Laurussia union seems to have caused few large tectonic ripples for any great distance to the north of the Ouachita orogenic zone in Laurentia and similarly for much distance inboard within the old Gondwana Craton in South America and North Africa. In contrast, as well as the continuation and merger of the Variscan–Alleghanian Orogeny in Europe and North America described below, in Mexico there was substantial plutonic and other igneous activity in the Early Carboniferous, as well as faunal exchange between the margins of the area between Gondwana and Laurussia (Keppie et al., 2008), which was followed by much disturbance subsequently as the two superterranes became united and the intervening much smaller and previously independent Central American terranes suffered extreme tectonism.

There are many published differences in the details of the earlier geographical relationship between Gondwana and Laurussia and in the way that those two large continents came together, and we use what is termed the 'Pangea A' fit at 320 Ma (Torsvik & Cocks, 2004; Domeier et al., 2012).

Oceans. The Panthalassic Ocean was still dominant throughout the period (Domeier & Torsvik, 2014). The north-western arm of the Rheic Ocean between Gondwana and Laurussia had closed as part of the Variscan Orogeny during the Late Devonian, but the southern sector remained during the Early Carboniferous (Fig. 9.1a), although by the middle of the Carboniferous it had evolved to become one of the many sectors within the Panthalassic (Fig. 9.1b). The Mongol–Okhotsk Ocean was progressively closing from west to east between Siberia and Amuria along the Mongol–Okhotsk Suture (Badarch et al., 2002), and the substantial Adaatsag Ophiolite (which is in Russia near the Mongolia–China border) was intruded within that suture zone as it closed there in the earliest Carboniferous. The Turkestan Ocean, which lay between Siberia, Kazakhstania, Tarim, and a corner of Amuria, became steadily smaller (Fig. 9.1). There was also a Paleoasian Ocean between North China and the Annamia and South China continents, although the term has been used in different ways in the literature, with some authors also using the name for the oceanic area between Laurussia and Siberia. The

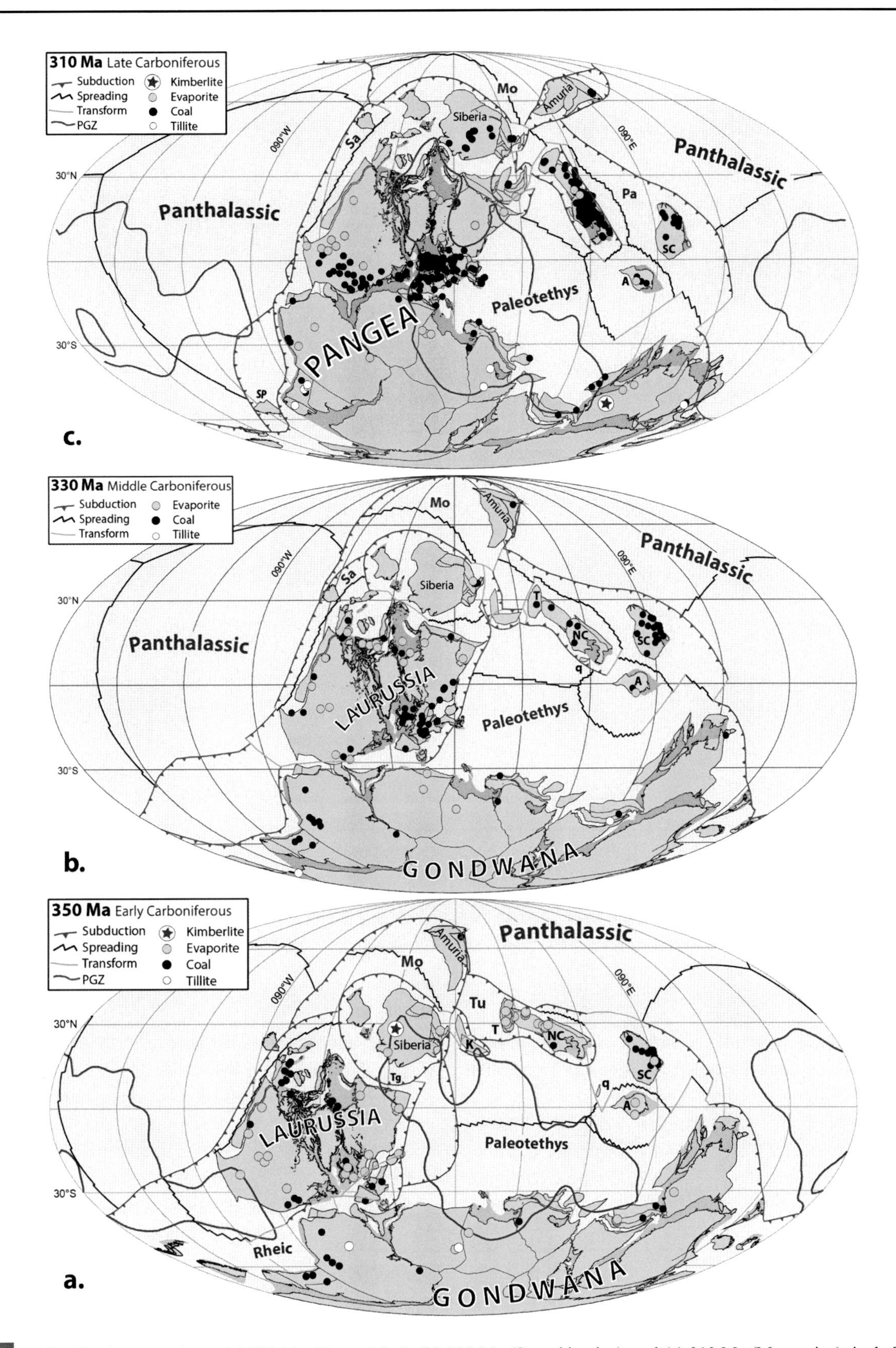

Fig. 9.1 Outline Earth geography at (a) 350 Ma (Tournaisian), (b) 330 Ma (Serpukhovian), and (c) 310 Ma (Moscovian), including the postulated plate boundaries, outlines of the major crustal units, and the most substantial oceans. A, Annamia; K, Kazakhstania; Mo, Mongol–Okhotsk Ocean; NC, North China; PGZ, plume generation zone; q, South Qinling; Pa, Paleoasian Ocean; Sa, Slide Mountain–Angayucham Ocean: SC, South China; SP, South Patagonia; T, Tarim; Tg, Tagil Island Arc; Tu, Turkestan Ocean.

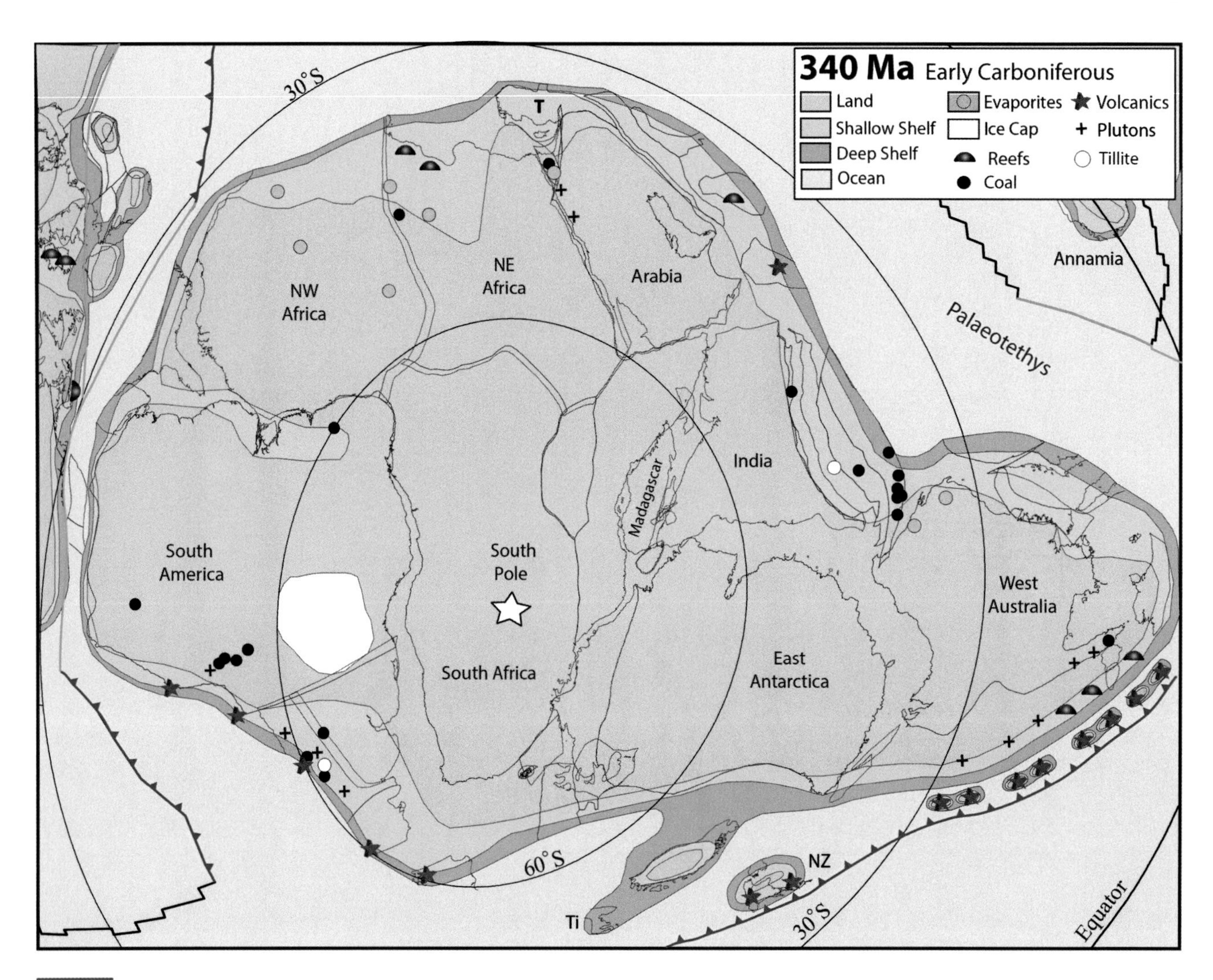

Fig. 9.2 Gondwana and adjacent areas at 340 Ma (Visean), showing the contemporary pattern of lands and seas. Solid blue lines are subduction zones with teeth on the upper plate, black lines are spreading centres and transform faults, and green lines are transform plate margins. The strings of island arcs shown are diagrammatic, since the individual extent of each unit within the arcs and the extent to which each unit was above sea level are uncertain. NZ, New Zealand; T, Taurides Terrane of Turkey; TI, Thurston Island, Antarctica.

Paleotethys Ocean was still the largest ocean apart from the Panthalassic, and remained substantial throughout Carboniferous times.

Variscan Orogeny. The complex Variscan Orogeny, which affected much of Europe south-west of the old Baltica Craton, had started near the end of the Silurian and continued on to reach its maximum activity in the Late Devonian and Early Carboniferous. Continuing the narrative from the previous chapter, the Galicia–Moldanubian Ocean narrowed and closed in the Early Carboniferous. By the end of the Visean at 340 Ma (Fig. 8.9h), most of the original Variscan movements were complete, although there was still much igneous activity in the region (Franke et al., 2016). However, the Variscan subsequently continued on into a further phase, with the integration of the area to form part of Pangea, as well as the intrusion of batholiths.

Although the major collision between Laurussia and Gondwana to form Pangea was not in the Central European area, prior to that collision there had been very substantial strike-slip movement between the two major continents, which further complicated the situation, as may be seen by comparing the differing sectors of Gondwana in (Fig. 8.9g,h). For example, the double orocline, usually considered as within the Variscan, seen in the Iberian Peninsula today, was formed at about the Carboniferous–Permian boundary time and seems to have represented a 1,300 km-long sector of the Gondwanan margin which was completely distorted as a slow consequence of the Gondwana–Laurussia collision

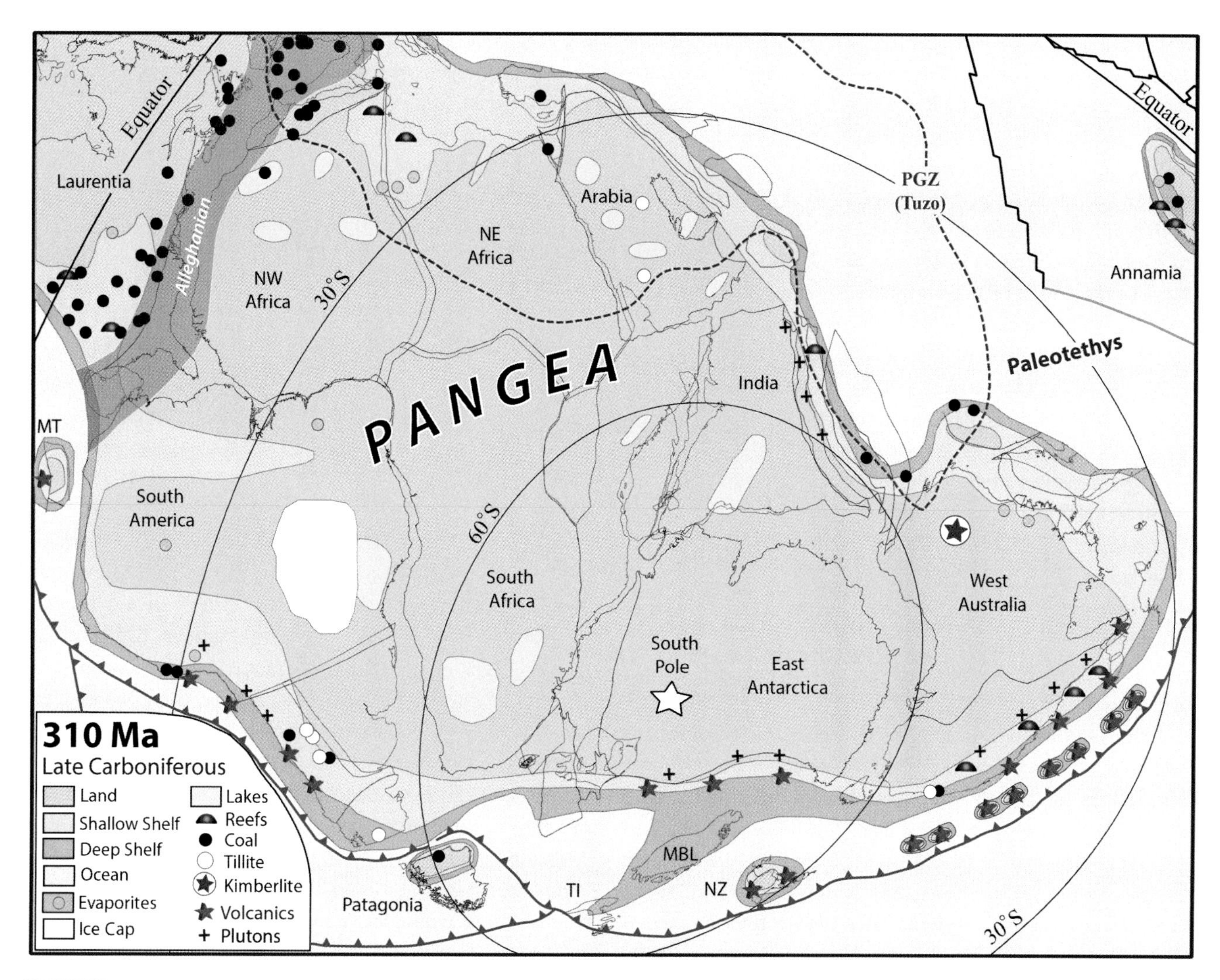

Fig. 9.3 Gondwana and adjacent areas at 310 Ma (Moscovian), showing the contemporary pattern of lands and seas. Solid blue lines are subduction zones with teeth on the upper plate, black lines are spreading centres and transform faults, and green lines are transform plate margins. The strings of island arcs shown are diagrammatic, since the individual extent of each unit within the arcs and the extent to which each unit was above sea level are uncertain. Dotted red line, plume generation zone (PGZ). MBL, Marie Byrd Land; MT, the Mexican terranes of Mixteca–Oaxaquia and Sierra Madre; NZ, New Zealand; TI, Thurston Island, Antarctica.

(Shaw et al., 2012). That collision was preceded in Iberia by deepening basins on both sides which were filled with turbidites with different zircons on their opposing sides, indicating that the Ossa–Morena Zone of central Iberia had originated on the Gondwana margin of the Rheic Ocean, but that the now adjacent South Portuguese Zone had previously been on the Laurussian side. At today's Atlantic margins, and extending eastwards as far as Egypt, the Variscan Orogeny was at its maximum (compare our latest Devonian and Carboniferous maps: Figs. 8.9g and 8.9h, respectively), including the closure of the Rheic Ocean and turbulent events in Europe (Franke et al., 2016). Early Carboniferous extensional tectonics caused grabens such as the Midland Valley of Scotland and the Rhine Graben in Germany. The Variscan is represented in eastern North America in the Appalachian area, termed there the Alleghanian Orogeny, which involved much crustal shortening and other complex tectonic activity (Hatcher et al., 1989), extending northwards into New England.

West Gondwana. In southern Europe and northern Africa, the Laurussia–Gondwana collision was one of several factors which led to the substantial and prolonged Variscan Orogeny. But nearly all of southern Europe had left the main Gondwanan continent well before the Carboniferous, and thus it is only in north-western Africa that its effects can be seen in Gondwana itself: in Morocco, following the development of Late Devonian to Early Carboniferous transtensional sedimentary basins there. There was also a smaller tectonic phase in the Visean in the eastern and western Meseta boundary area of Morocco, and a much more substantial event in the Late Carboniferous, which saw a regional shortening which affected the entire Meseta as well as the Anti-Atlas Mountains (Hoepffner et al., 2005), all to the south of the Paleotethys Ocean.

In south-western South America, there was emplacement of Early Carboniferous granites to the west of the Precordillera Terrane in Argentina (Dahlquist et al., 2013), from 357 Ma (Tournaisian) to 322 Ma (Serpukhovian) times, which peaked at about 341 Ma (Visean), as well as the intrusion of the largest batholith in the region, the Achala granite in the Sierras Pampeanas at 349 Ma (Tournaisian), which is interpreted as a post-collisional/slab-breakoff A-type granite inboard from the passive Gondwanan margin (Domeier & Torsvik, 2014). In the Late Carboniferous there was retroarc volcanism in the same western Gondwana marginal zone near Mendoza, Argentina, which was extruded over both subaerial and submarine facies. In central Chile, the substantial Coastal Batholith there was intruded during the Bashkirian and Moscovian between 320 and 300 Ma (Hervé et al., 2013).

East Gondwana. In contrast to West Gondwana, the south-central and eastern sectors of Gondwana were little affected by its union with Laurussia, since the northern margin in north-eastern Africa and the Middle and Far East of Asia remained largely passive. However, the eastern margin continued to be active in the Australian sector (Fig. 9.2). In eastern Australia, there was further arc accretion and crust enlargement in the Yarrel and New England orogens, including Late Devonian to Carboniferous deformation (Glen, 2005). During the Late Carboniferous (Fig. 9.3), kimberlites were intruded in Australia at 305 Ma (Torsvik et al., 2014) which had originated at the Tuzo Plume Generation Zone. Although the Lhasa Terrane of Tibet has been depicted as leaving the Himalaya before the Visean (Zhu et al., 2013), we have found no evidence that rifting and subsequent sea-floor spreading occurred there at that time, and thus conclude that the Neotethys Ocean did not open in the region until the Early Permian.

Although most of Gondwana's northern margin was passive, in contrast, all round today's southern and south-western margins of the continent, orogenesis caused by the very extensive subduction zone there continued on from earlier times, and that activity, sometimes termed the Gondwanides Orogeny, was quite independent of the Variscan Orogeny. For example, in East Antarctica, the substantial Devonian–Carboniferous Admiralty Granite was intruded and the subaerial Gallipoli Volcanics extruded, both of which straddle the boundaries between the Gondwana Craton and the Robertson Bay, Bowers, and Wilson terranes, all within the Ross Orogen of North Victoria Land (Tessensohn & Henjes-Kunst, 2005).

Eastern Asia. The Paleotethys Ocean between South China and Annamia to the north and Gondwana (including Sibumasu, which was still then an integral part of the Gondwana Craton) to the south continued to widen throughout the period (Fig. 9.4). Within Annamia there is a regional unconformity beneath the Middle Carboniferous which has been identified as representing the first phase of the relatively local Indosinian Orogeny (Ridd et al., 2011). During the Middle to Late Carboniferous, the Central Tien Shan–Beishan region of northern China accreted to the Northern Xinjiang region, indicating the initial accretion of the Kazakhstania continent to peri-Siberia (Fig. 9.10). Substantial subduction and thrust faulting occurred to the south of the South Tien Shan area, particularly in the Early Carboniferous between 360 and 320 Ma, which caused much metamorphism there and culminated in the collision between South Tien Shan and Tarim in the latest Carboniferous or Early Permian (Zhou et al., 2001a; Zhang et al., 2008). Island arc volcanoes were active in the Gurvansayhan Terrane, to the north of which there were highly effusive plateau volcanics in the southern margin of peri-Siberia (the Gobi Altai area) at 323 Ma (Serpukhovian), giving a date for the accretion and continental assembly within Pangea for the south-eastern Mongolian area.

In the Japanese terranes, which still lay outboard of the South China continent, new arc activity started after the relative quiescence of the later Devonian. Granite plutons ranging in age from 302 to 304 Ma (Zhang et al., 2007) indicate the presence of an Andean-style continental arc along the northern margin of the North China continent (Fig. 9.5). The south-western margin of North China, along which Qaidam–Qilian had previously been accreted, does not appear to have any active volcanic arcs offshore, although there was an active subduction zone further seawards (Fig. 9.9).

Western Laurussia. As noted above, the initial collision of Laurussia with Gondwana was an oblique and relatively soft docking, and thus the orientation of the Laurentian sector in the north appears to have been little changed, as can be seen from Fig. 9.6, even though the prime collision zone is located in the south of the previously independent Laurussia. However, those collisional tectonics contrast greatly with the earlier scenario, since the southern margin of Laurentia had previously remained passive since the Late Neoproterozoic (Bradley, 2008). The latest Carboniferous final phase is reflected in the Ouachita Mountains of Texas, where there are olisthostromes and volcanoclastic units. On

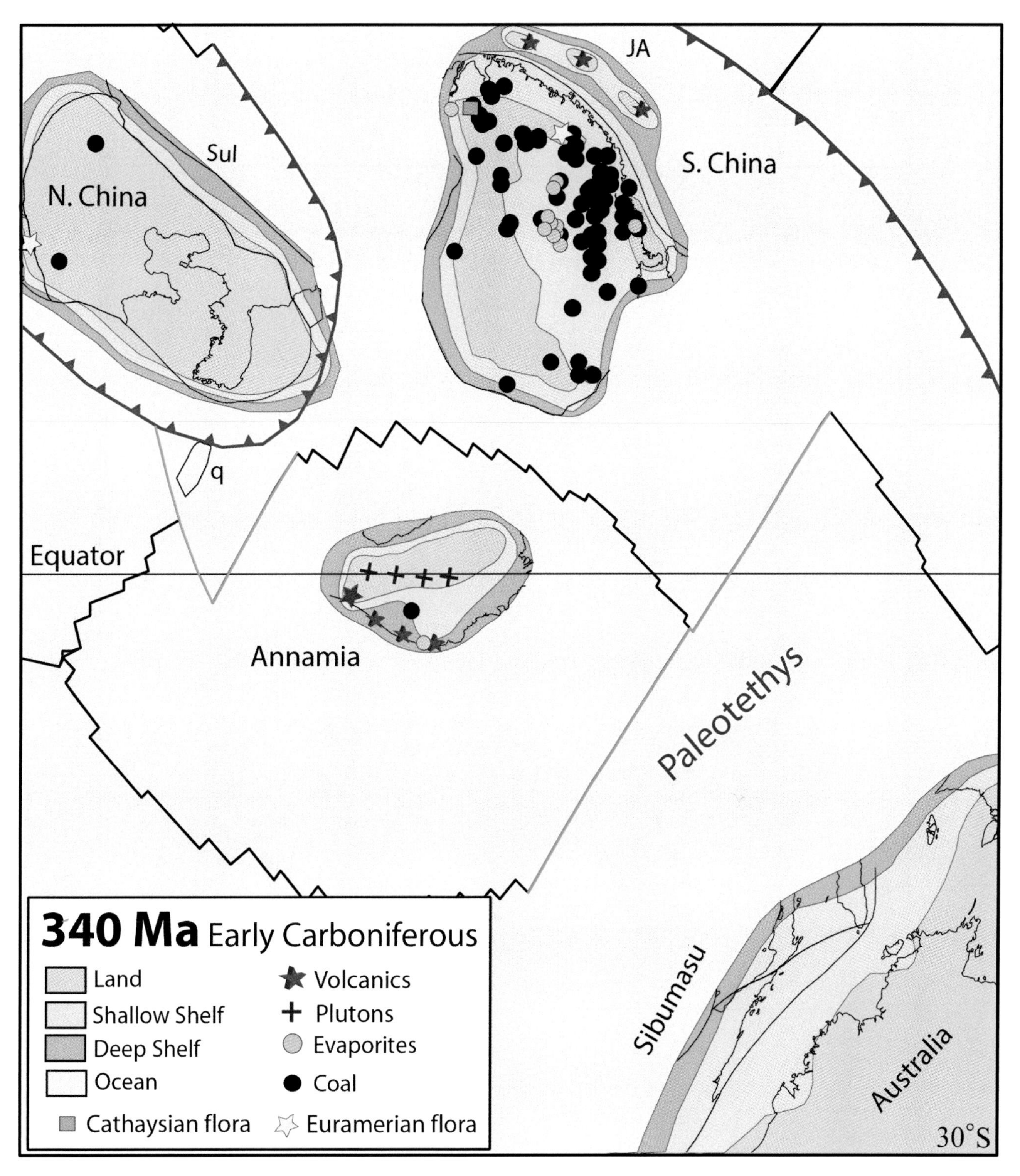

Fig. 9.4 The eastern Asia area at 340 Ma (Visean), showing the contemporary pattern of lands and seas and the distribution of floral provinces. Solid blue lines are subduction zones with teeth on the upper plate, black lines are spreading centres and transform faults, green lines are transform plate margins. The strings of island arcs shown are diagrammatic, since the individual extent of each unit within the arcs and the extent to which each unit was above sea level are uncertain. JA, Japanese island arcs; q, South Qinling; Sul, Sulinheer.

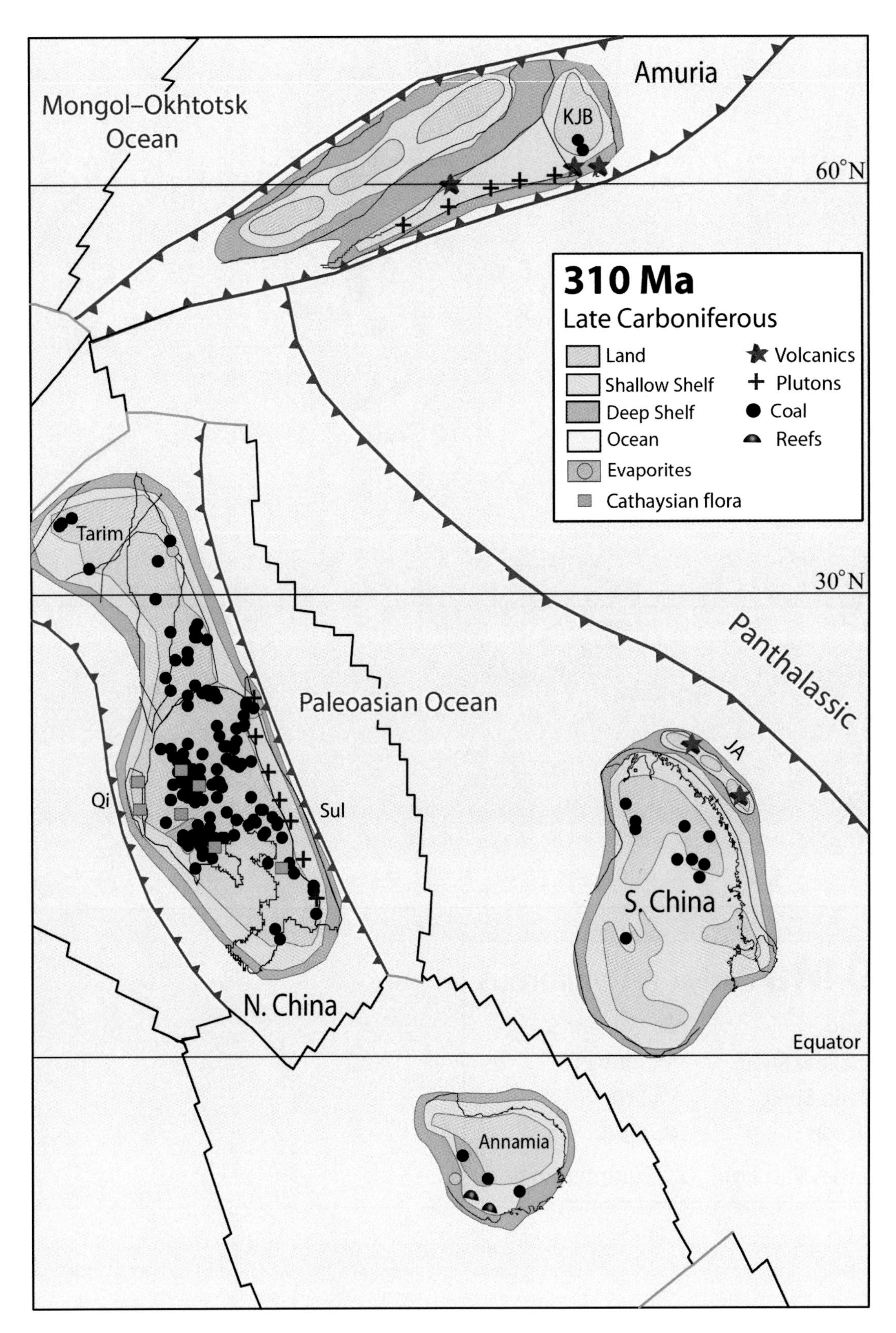

Fig. 9.5 The Amuria, North and South China, and Annamia continents at 310 Ma (Moscovian), showing the contemporary pattern of lands and seas and the sites of the Cathaysian floras. North China includes the now combined (but earlier separate) cratons of North China and Tarim, as well as the area between them previously termed the Ala Shan Terrane area. Solid blue lines are subduction zones with teeth on the upper plate, black lines are spreading centres and transform faults, and green lines are transform plate margins. JA, Japanese arcs; KJB, Khanka–Jiamusu–Bureya block; Qi, Qilian; Sul, Sulineer.

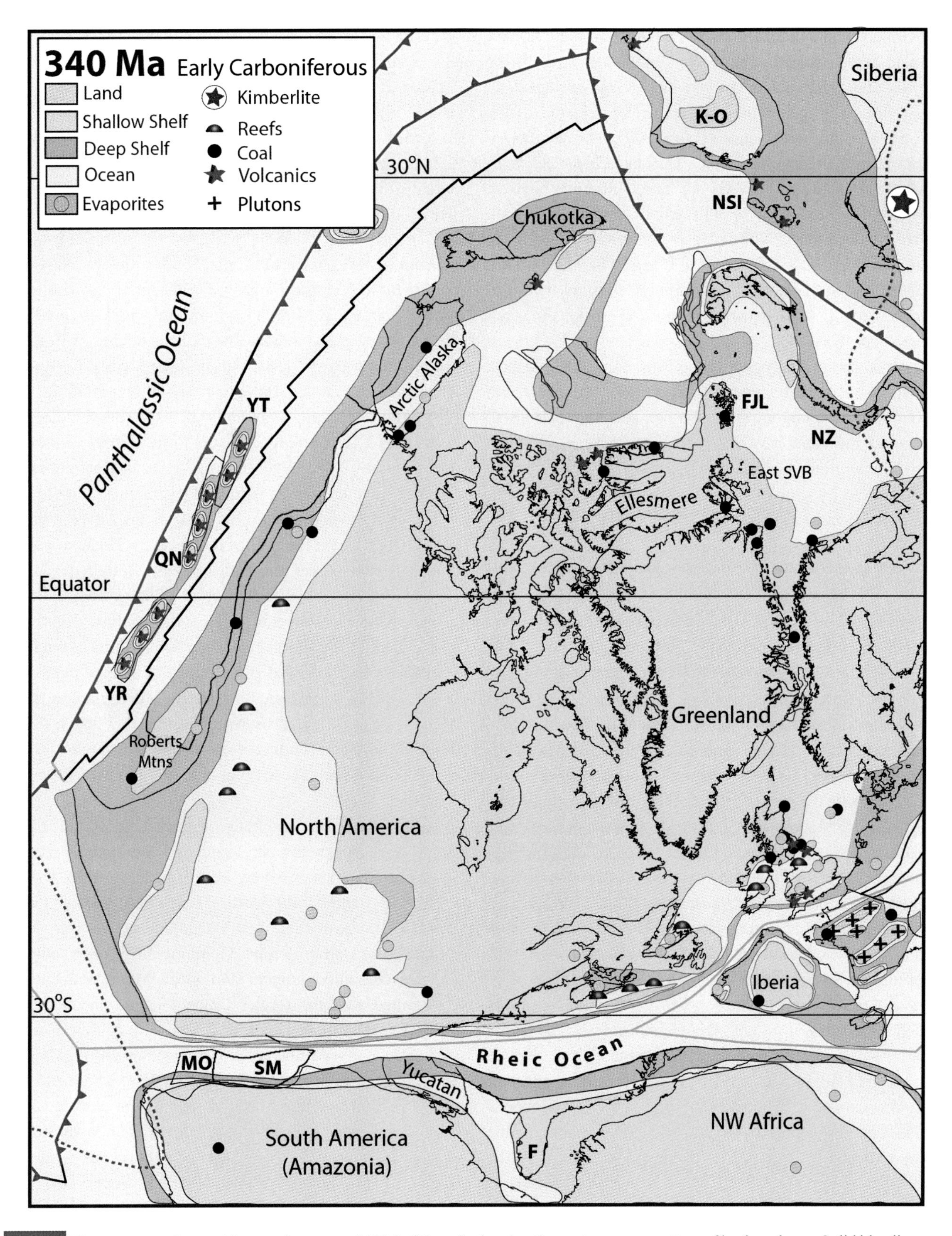

Fig. 9.6 The western and central Laurussian area at 340 Ma (Visean), showing the contemporary pattern of lands and seas. Solid blue lines are subduction zones with teeth on the upper plate, black lines are spreading centres and transform faults, and green lines are transform plate margins. F, Florida; Fa, Farewell Terrane; FJL, Franz Josef Land; K-O, Kolyma–Omolon; MO, Mixteca–Oaxaquia, Mexico; NSI, New Siberian Islands; NZ, Novaya Zemlya; QN, Quesnellia; SM, Sierra Madre, Mexico; SVB, Svalbard; YR, Yreka Terrane; YT, Yukon–Tanana Terrane.

the Gondwanan side of the suture, in Mexico, there are the Acatlán and Granjeno continental rises to the north of the Chortis and Oaxaquia terranes, which were tectonised by much strike-slip movement just before the collision of the two main continents (Nance et al., 2009). The amount of lateral translation involved in the Ouachitas had been minimally estimated as between 50 and 100 km (Nielsen, 2005), but we assess it as substantially greater than that, probably more than 2,000 km over the whole suture zone (compare Fig. 9.6 with Figs. 9.7 and 9.8). The south-western margin of Laurussia, including California, was also much affected, with substantial sinistral strike-slip fault movements there.

In the north of Laurussia, in the Canadian Arctic islands, the Sverdrup Basin continued its steady development, with the Ellesmerian Orogeny continuing on into the Tournaisian there, but the latter was over before the Visean, although there was further subsequent deepening of the Sverdrup Basin caused by rifting (Trettin, 1998), as well as the intrusion of associated volcanics of Serpukhovian age at about 325 Ma. In the Cordilleran area to the west of the former Laurentian Craton, the sea-floor spreading, which had started in the Late Devonian, continued. A variety of island arcs (Fig. 9.7), some preserved today within the North Sierra, Klamath, Quesnellia, and Stikinia terranes, were positioned on the western margin of the widening Slide Mountain–Golconda Ocean (Nokleberg et al., 2000), an ocean which subsequently reached its maximum extent in the Early Permian. Late Carboniferous and Early Permian voluminous basalts indicative of sea-floor spreading are preserved in the Slide Mountain Terrane of Yukon and British Columbia, and there was metamorphism as high as eclogite grade in the Cordilleran Yukon–Tanana Terrane (Colpron & Nelson, 2006; Cocks & Torsvik, 2011).

Eastern Laurussia. The Variscan Orogeny was its height over much of central Europe (see above). The north-western arm of the Rheic Ocean had closed during the Devonian, but its more substantial southern half remained distinguishable until the Early Carboniferous, when the Iberian Peninsula joined south-eastern Canada (Newfoundland and Meguma) and the Rheic merged with the Panthalassic (Fig. 8.9h). All those varied terranes underwent much distortion, together with the extrusion of volcanics, the intrusion of igneous rocks, and the sedimentation of a variety of rocks in the many basins within them, as well as on the shelves of the shallower and deeper seas surrounding them. However, much of Europe was relatively stable (McCann, 2008), which enabled sedimentary deposition within extensive basins there, as described under 'Facies, Floras, and Faunas' below.

Further eastwards, rifting in western and central Europe is also seen in the 1,500 km long Pripyat–Dnieper–Donets–Dunbar–Karpinsky Rift in Russia and Ukraine (Nikishin et al., 1996). On the eastern margin of Laurussia (which formed the eastern margin of Pangea from Middle Carboniferous times onwards), the Uralian Orogeny was at the height of its activity during the Late Devonian and Carboniferous (see under Siberia below).

Central Asia. Some Carboniferous accretions in the Kazakh terrane area took place after the Kazakhstania continent had already become substantial, as confirmed by the continental Middle Carboniferous rocks unconformably overlying the Lower Palaeozoic in the Kokchetav–Ishim sector (Unit 458). To the north-east of Atashu–Zhamshi (Unit 461), there are ophiolites dated as Late Carboniferous at approximately 325 Ma (Serphukhovian), as well as Late Carboniferous to Early Permian volcanics and molasse (Biske & Seltmann, 2010). In Karatau–Naryn (Unit 466), there was a Middle Devonian to Middle Carboniferous passive margin, as well as an active volcanic arc situated on an Early Carboniferous to Permian continental margin, where the volcanics are interbedded with marine carbonates (Windley et al., 2007). In the south-eastern sector of that unit, there was southward thrusting of island arc rocks over the carbonate platform in the Early Carboniferous, followed by laterite and bauxite deposition there in the Bashkirian and Early Moscovian at about 320–315 Ma, followed by overthrusting and flysch deposition in the south, ending with olisthostromes and further nappes in the Late Moscovian at about 310 Ma (Belousov, 2007).

Near today's border between Kazakhstan and China, ocean floor between near Tarim and the Junggar Terrane (Unit 450) had progressively been subducted under Junggar (including Central and North Tien Shan), mostly during Moscovian time at about 310 Ma, and those two were united by the latest Carboniferous. Opinions differ on whether or not that combined Junggar and Tarim Microcontinent was also united with the former Qaidam–Qilian and Ala Shan terranes (Dumitru & Hendrix, 2001). The Middle Carboniferous to Early Permian saw the development of a pile of nappes between Tarim and South Tien Shan which ended in their eventual amalgamation, following the closure of the Turkestan Ocean between the two areas. The lower nappes were originally parts of the passive northern margin of Tarim, which was subducted. North and South Tien Shan merged in the same general area progressively during the Middle and Late Carboniferous, followed by latest Carboniferous molasse deposits and stitching Early Permian Type-A granites.

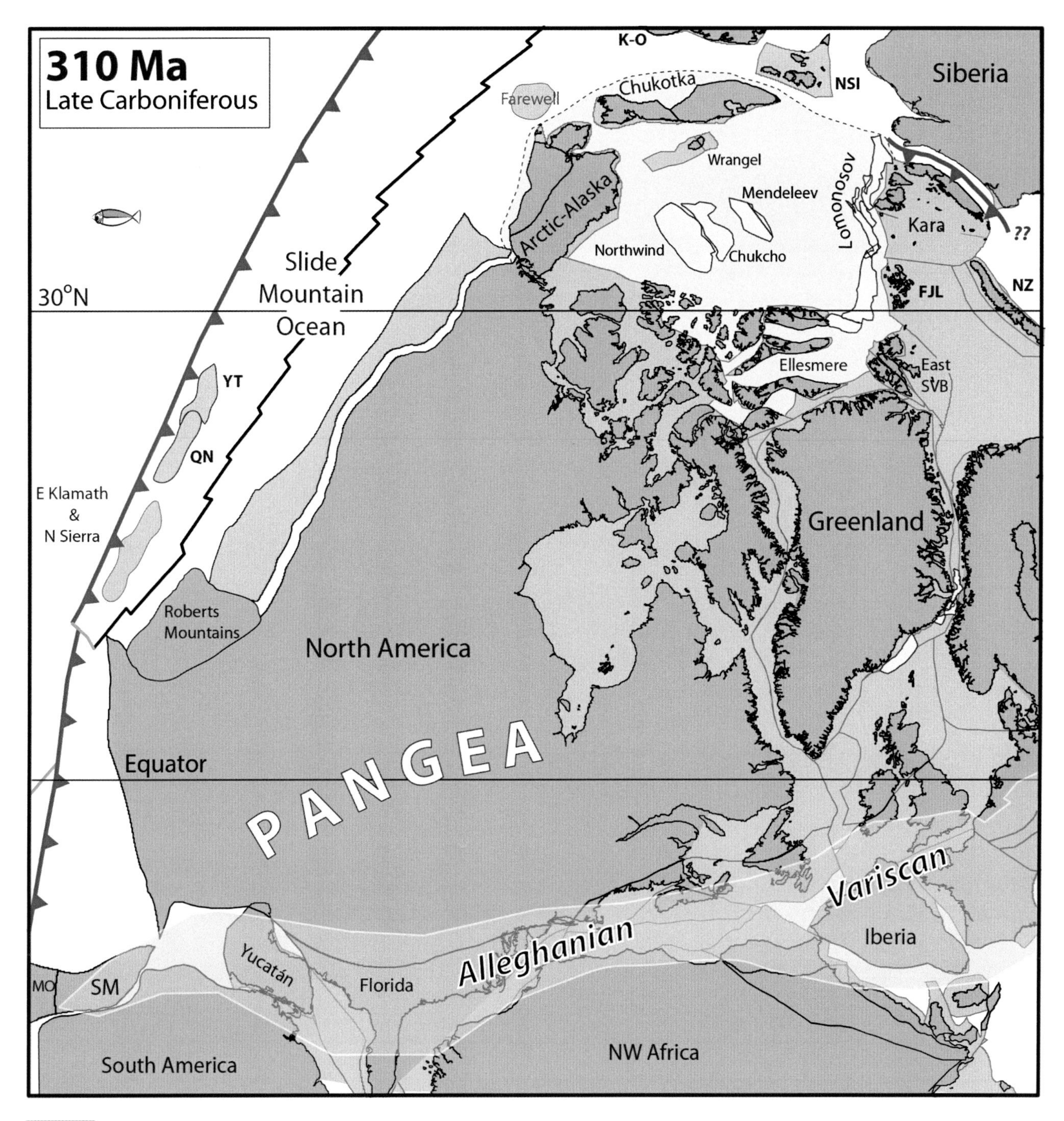

Fig. 9.7 The western and central Laurussian sector of northern Pangea and adjacent part of Siberia at 310 Ma (Moscovian), showing the outlines of the contemporary continental and terrane blocks, and the areas affected by the Variscan and Alleghanian orogenies (light brown). Solid blue lines are subduction zones with teeth on the upper plate, black lines are spreading centres and transform faults. Possible subduction beneath Kara is indicated. The yellow shading represents probable continental areas and terrane extensions in the Arctic area, and the lightly shaded areas within it are terranes now submerged. FJL, Franz Josef Land; K-O, Kolyma–Omolon; MO, Mixteca–Oaxaquia, Mexico; NSI, New Siberian Islands; NZ, Novaya Zemlya; QN, Quesnellia; SM, Sierra Madre, Mexico; SVB, Svalbard; YT, Yukon–Tanana Terrane.

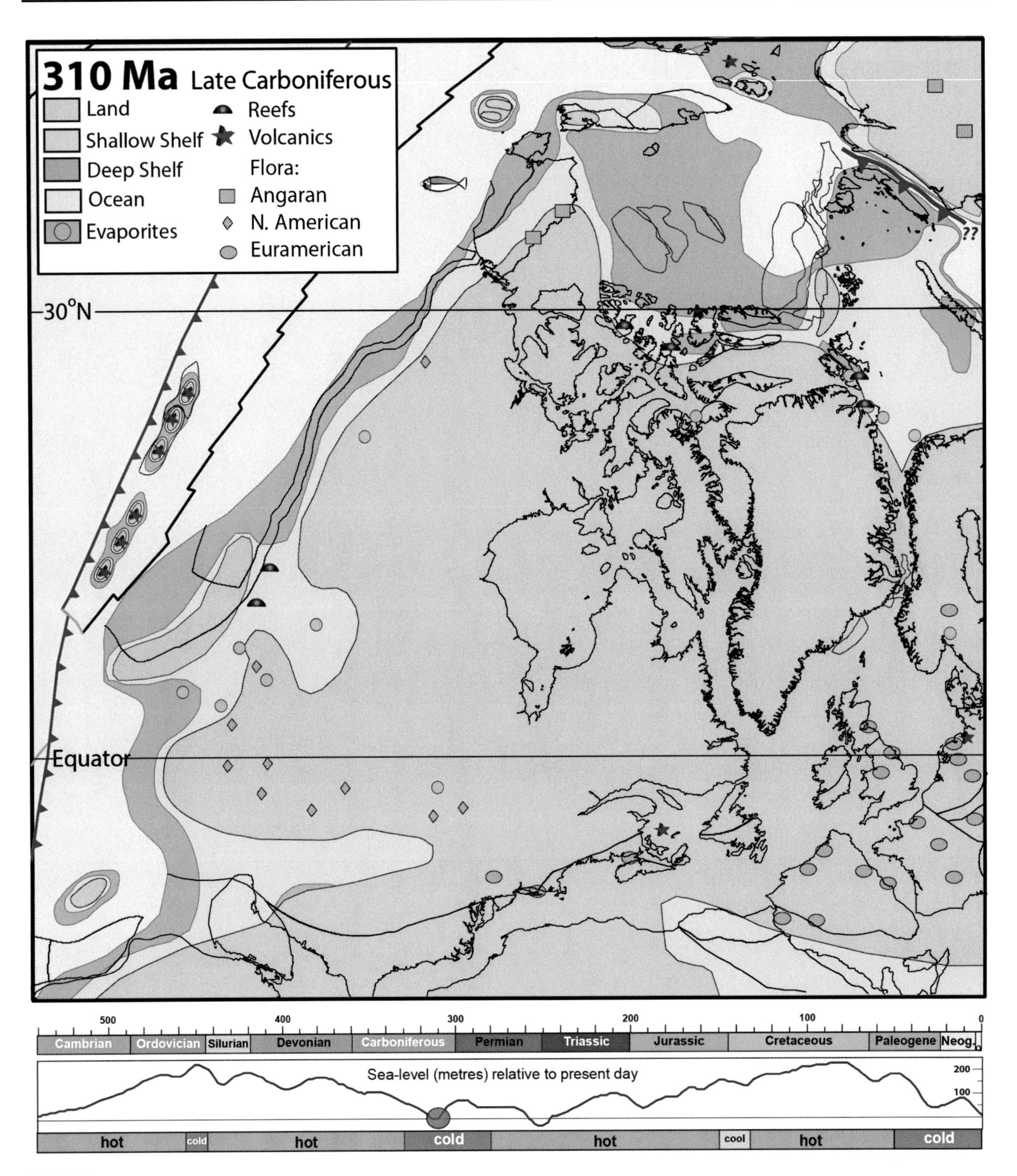

Fig. 9.8 The western and central Laurussian sector of northern Pangea and adjacent part of Siberia at 310 Ma (Moscovian), showing the contemporary pattern of lands and seas, and the varied floral provinces. Location names are on Figs. 9.6 and 9.7. Solid blue lines are subduction zones with teeth on the upper plate, black lines are spreading centres and transform faults, and green line is a transform plate margin. Bottom: Phanerozoic time scale and sea-level variations (Haq & Al-Qahtani, 2005; Haq & Shutter, 2008) and icehouse (cold) and greenhouse (warm) conditions.

Siberia. An abundance of Early Carboniferous kimberlites in east Siberia (continuing on from the numerous Devonian occurrences) suggests that the continent lingered above the north-west arm of the Tuzo Plume Generation Zone then (Fig. 9.9). There are also Middle Devonian to Early Carboniferous calc-alkali volcanics in Altai–Sayan, but the island arc in the Rudy Altai sector (Unit 433) ceased activity by the end of the Visean, and that unit was subsequently stitched to the adjacent Kobdin sector (Unit 434) by Late Carboniferous granites (Cocks & Torsvik, 2007).

The relative motion between Siberia and Laurussia was obliquely convergent in the Early Carboniferous, following an interval of transcurrent relative motion in the Late Devonian. Destruction of the basin between Siberia and Laurussia was achieved initially by north-easterly-dipping subduction beneath the Tagil Island Arc in the Urals and the then-inverted ridge that once lay behind the Magnitogorsk Island Arc, which was south of Tagil. By the Early Visean at 345 Ma the Tagil Arc had accreted to the Uralian margin of Baltica (Puchkov, 2009; Brown et al., 2011) and the polarity of subduction had flipped to allow closure of the remnant (back-arc) basin by subduction beneath Baltica. However, that phase of convergence was short-lived, and by the Middle Visean at 340 Ma motion along the boundary had become transcurrent to weakly divergent, producing a minor but long-lived basin between Baltica and west Siberia which later filled to form the extensive triangular West Siberian Basin. To the north-west along that boundary, transcurrent to transpressive motion from 340 to 320 Ma (Middle Visean to Early Bashkirian) between the 'Kara Terrane' (kept relatively coherent with Baltica in our reconstruction, following Lorenz et al., 2008) and north Siberia would have been responsible for the Late Palaeozoic deformation in Severnaya Zemlya and Taimyr, which would therefore have been distinct from the main Uralian Orogeny to its south. In the Later Carboniferous, from 320 Ma (Bashkirian), relative motion along the Siberia–Laurussia plate boundary slowed. Although transform motion between Siberia and Laurussia continued into the Early Mesozoic, particularly along the massive Yenesei Fault along the southern and south-western margins of the enlarged Siberia, where basalt was extruded sporadically from the Carboniferous to the Early Triassic, we consider that tectonism as intra-plate deformation. By the latest Carboniferous to Early Permian, the Uralian Ocean had closed completely as the continent–continent collision between Laurussia and Kazakhstania was complete, further enlarging Pangea (Brown et al., 2011).

During the Carboniferous, Siberia still remained as a discrete major entity in temperate to intermediate northern palaeolatitudes. In the Lower Carboniferous, the Verkoyansk branch of the Devonian rift system was transformed into the passive margin of the oceanic basin in today's east of the continent. The Mongol–Okhotsk Ocean continued its progressive closure from west to east along the Mongol–Okhotsk Suture (Badarch et al., 2002), and the substantial Adaatsag Ophiolite (in Russia near the Mongolia–China border) was intruded within the suture zone in the earliest Carboniferous (Tomurtogoo et al., 2005). Figure 9.9 shows the Siberian area in the Lower Carboniferous (the Early Visean) at about 340 Ma, and Fig. 9.10 shows it at about 300 Ma, the very latest Carboniferous (the Carboniferous-Permian boundary was at 299 Ma).

Siberia has no reliable palaeomagnetic data between 360 Ma and 275 Ma, and thus its position is based on interpolation for most of the Carboniferous and Early Permian. Our reconstructions for Siberia follow the Alternative B APW path discussed in Cocks & Torsvik (2007), and which is shown in Fig. 9.11.

Facies, Floras, and Faunas

The Late Palaeozoic Glaciation. Following the global change from a warmer to a cooler environment near the end of the Devonian (Fig. 16.2), there was a prolonged but intermittent series of ice ages through much of the Carboniferous and into the Early Permian from 330 to 290 Ma (Fielding et al., 2008). As well as some Late Devonian glaciogenic rocks in the Parnaíba, Solimões, and Parecis basins of South America, the first phase of Carboniferous glacial diamictites also range in age from the Tournaisian to the Late Visean (about 355–325 Ma) there. However, it was not until well into the Carboniferous, at the base of the Serpukhovian (and concomitant Namurian) at about 330 Ma, that glaciogenic rocks were deposited in more widespread areas over the Gondwana continent and beyond, associated with more temperate climates and a massive increase in coal deposition. That marked the start of the main glaciation, which was very substantial throughout the Late Carboniferous, but peaked in the Early Permian, and which extended over many palaeolatitudes. The effects were felt far beyond the glacial area; for example, in the Mid-Continent of North America, there are numerous cyclothems preserved in the Moscovian to Gzhelian rocks which directly reflect the far-field effects of the Gondwanan glacial episodes, when phosphatic black shales reflect the maximum flooding events caused by eustatic sea-level rise from interglacial melting periods, and interbedded regressive shallowing-upwards marine limestones and occasional palaeosols reflect the glacial peaks. This was by far the most long-lived series of

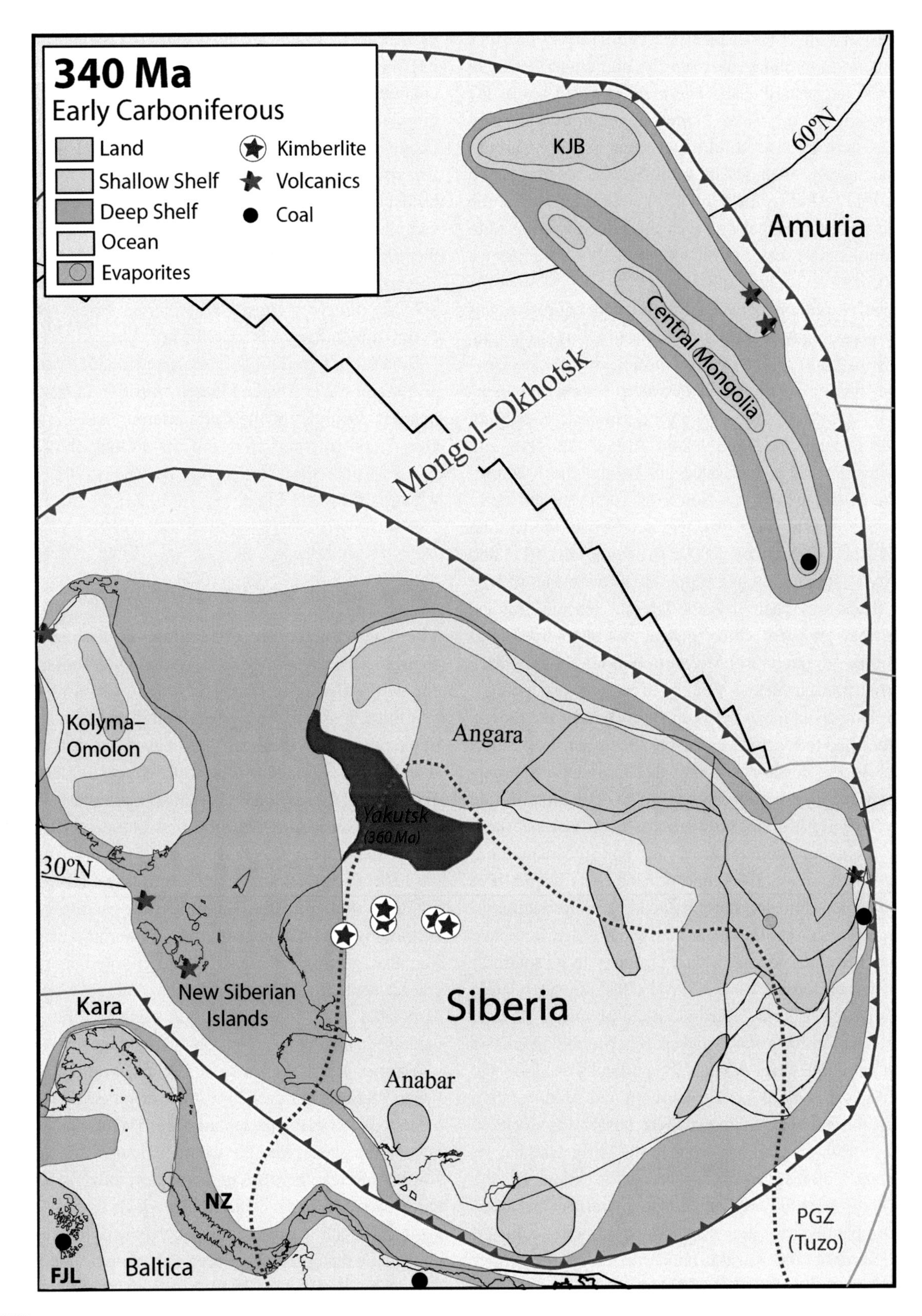

Fig. 9.9 The Siberia and Amuria continents and part of the Baltica sector of Laurussia at 340 Ma (Visean), showing the contemporary pattern of lands and seas, and also the area of the earlier (360 Ma: Late Triassic) Yakutsk Large Igneous Province in red shading. The Angara and Anabar massifs of Siberia are named. Solid blue lines are subduction zones with teeth on the upper plate, black lines are spreading centres and transform faults, and green lines are transform plate margins. Dotted red line is the plume generation zone (PGZ). FJL, Franz Josef Land; KJB, Khanka–Jiamusu–Bureya block; NZ, Novaya Zemlya.

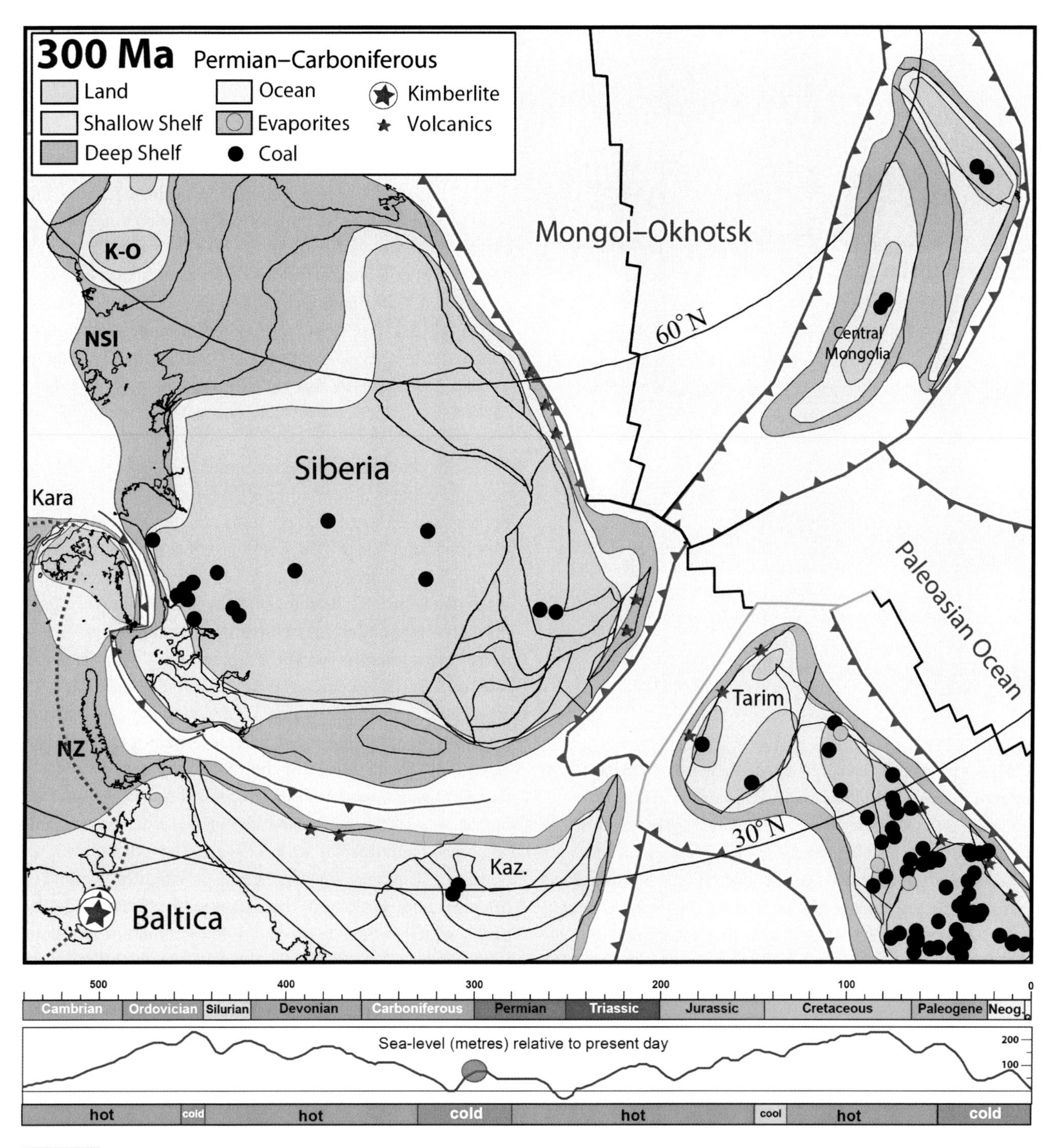

Fig. 9.10 Siberia at the time near its unification with the eastern Laurussian sector to become an additional part of the Pangea supercontinent, as well as the barely independent Tarim continent just before its amalgamation with Amuria (including North China and central Mongolia at the bottom right-hand corner) at 300 Ma (about the Carboniferous–Permian boundary time), showing the contemporary pattern of lands and seas. Solid blue lines are subduction zones with teeth on the upper plate, black lines are spreading centres and transform faults, and green lines are transform plate margins. Dotted red line is the plume generation zone. Kaz., the former Kazakhstania sector of Pangea; K-O, Kolyma–Omolon; NSI, New Siberian Islands; NZ, Novaya Zemlya. Bottom: Phanerozoic time scale and sea-level variations (Haq & Al-Qahtani, 2005; Haq & Shutter, 2008) and icehouse (cold) and greenhouse (warm) conditions.

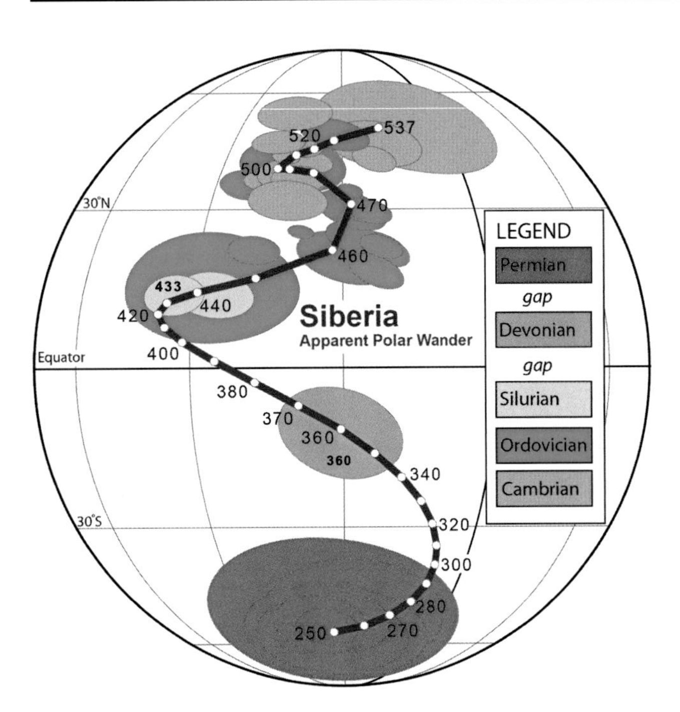

Fig. 9.11 Apparent Polar Wander path for Siberia. Although there are significant gaps in reliable palaeomagnetic data between the Late Silurian and near the end of the Devonian and from the end of the Devonian and the Permian, the Siberian poles are well established for the rest of the Palaeozoic. After Cocks & Torsvik (2007).

glacial events and development of their associated ice caps known in the whole Phanerozoic.

There was extensive and prolonged glaciation over all the higher latitudes of Gondwana. In South America, there are extensive glacial deposits in the Paraná and Sanfransicana basins of Brazil and north-east Argentina (Rocha-Campos et al., 2008) (Fig. 9.2). The glacial rocks there range in age from the Serpukhovian to the Gzhelian (about 325–300 Ma), a span of more than 25 Myr, and that ice cap may have extended over as far as Africa (Namibia) in the Late Carboniferous, where it is termed the Windhoek Ice Sheet.

Since Gondwana was so large, the intensity of the glacial events varied across its regions. For example, in the Alborz Mountains of Iran, the glaciogenic rocks were deposited in only two episodes in the Carboniferous, the first in the Bashkirian at about 320 Ma, and the second across the Carboniferous–Permian boundary from the Gzhelian to the Sakmarian (305–290 Ma) (Gaetani et al., 2009). It was the presence of glacial sediments in Gondwana and adjacent areas (including Sibumasu and the Tibetan terranes), and their absence from the Chinese continents, Tarim, and Annamia, that confirmed the site of the Neotethys Ocean Suture there (Metcalfe, 2006). There is variable evidence as to the extent to which the equatorial regions were affected by those much colder climates at higher latitudes, indicating that the global temperature gradients must have been much more diverse than in the preceding greenhouse periods, and perhaps not dissimilar to the present day.

Partly due to the evolution of larger trees and the formation of the world's largest coal deposits in the many low-lying areas near seas and lakes, which created a new organic carbon reservoir; and partly due to the increasing glaciation, the atmospheric concentrations of carbon dioxide decreased from about 1,500 parts per million in the Early Carboniferous to about 350 ppm in the Middle Carboniferous, comparable to modern levels (Berner, 1997). Contemporary changes are also recorded in the isotopic compositions of marine carbonates, which reflect the chemistry, circulation, and temperatures.

Floral Provinces. Although some floras of later Silurian ages have been described, and they became quite diverse during the Devonian (see previous chapter), it was not until Carboniferous times that floral provinces can usefully be distinguished. Ferns appeared for the first time. Unlike many modern plants, the Palaeozoic flora was pollinated mostly through wind action, rather than by animals such as arthropods. By at least the Middle Carboniferous, four provinces, reviewed by DiMichele et al. (2005), can be identified, the most well known of which is the Gondwanan Province (formerly termed the *Glossopteris* Province), which was essentially characteristic of the higher southern latitudes in the Gondwanan sector of Pangea, and whose recognition was one of Wegener's chief supporting planks when he originally recognised continental drift (1915). The others were the Angaran Province (in the northern latitudes centred in Siberia), and, both near the Equator, the Cathaysian Province, which is best developed in the Permian of North and South China (Fig. 10.4), but elements of which are known from both China and neighbouring areas of Gondwana in the Late Carboniferous; and the Euramerian Province, which was widespread over the equatorial regions of Europe and North America, and also extended to the more temperate fringes of Gondwana. The distribution of those floral provinces is shown here for eastern Asia (Fig. 9.4).

Thus those floral provinces were apparently more latitudinally climate-controlled rather than terrane-specific. That may be seen from the distributions of the *Monilospora* miospore flora, which was widespread in the more equatorial parts of Laurussia and stretched from north-west Canada to Scandinavia, in contrast to the more southerly temperate *Grandispora* Flora, which was prevalent in the eastern USA and southern Europe (McKerrow et al., 2000). During the Late Carboniferous, paralic coals were formed abundantly, originally as peat, in a wide belt reflecting climate and palaeolatitude across Laurussia, particularly in eastern North America, Britain, and central Europe (Fig. 9.1), and, as the Carboniferous progressed, the number of coal forests also increased in both North and South China. The coals reached their maximum in the Moscovian and, although present in the succeeding Kazimovian and Gzhelian stages, were less substantial then. That dwindling was probably due to an increase in aridity caused by global warming. Most of the trees were large club mosses, which could grow to over 30 m in height.

The Siberian plants include *Koretrophyllites*, *Paracalamites*, *Angaropteridium*, and *Angaridium*, which are largely endemic to the region and typify the Angaran Province. There are no plant species at all in common between the Angaran and Euramerian floras, and that endemic separation was well established by as early as Tournaisian times: however, as a result the floral biozones in Siberia cannot be correlated accurately with the ammonoid zones recognised in the marine rocks and used as biozones in the rest of Russia. The boundary between the Angaran and Cathaysian floras within Mongolia and China in the Upper Carboniferous (Yue et al., 2001) is clearly marked by the Hongshishan Suture to the west (from 95° to 105° E) and the Hengenshan Suture to the east (from 105° to 130° E): there is no doubt that the two areas and the floral provinces were originally widely separated from each other during the Carboniferous, and that both the North China and South China continents were at some distance from Siberia.

On the Early Carboniferous reconstruction of East Asia at 340 Ma (Fig. 9.4), Euramerian floras, including the characteristic *Lepidodendron* and *Paripteris*, are plotted as occurring in Annamia, South China, and North China, and the first Cathaysian flora is recorded from South China. However, in the Late Carboniferous at 310 Ma, only Cathaysian Province floras are known from the region, although there were Angaran Province floras in Amuria, which lay then to the north at much higher latitudes (Fig. 9.5).

Marine Benthic Faunas. The benthic invertebrate faunas changed from being overall of a cosmopolitan nature to becoming largely provincial, particularly at the species level, as the Carboniferous progressed. However, as with the floras, those faunal differences were due only partly to physical separation of the terranes but more importantly to latitudinal differences, which were in turn linked to local temperatures, which fluctuated a great deal, particularly in the periods with larger ice caps. Thus the western Pangean margin of Gondwana had the Midcontinent–Andean Province and the eastern margin the Tethyan–Uralian–Franklinian Province, largely defined by fusuline foraminifera. From Early Permian times onwards the further evolutionary separation of those faunas caused an upgrade in the biogeographical classification in the two sectors from provinces to realms (Ross & Ross, 1983). The global distribution of key end-Carboniferous to Permian ostracods show how dissimilar those of eastern Asia were from the North America part of Pangea, and indicate that the prevalent ocean currents at that time probably flowed from west to east (Lethiers & Crasquin-Soleau, 1995).

The brachiopods, whilst diverse in generic numbers, were more cosmopolitan than might have been expected. However, the outcrops yielding brachiopods are relatively restricted, with few found in Gondwana; for example, no brachiopod faunas at all are known in Australia from most of the Middle Carboniferous (Serpukhovian and Bashkirian). In the Tournaisian and Visean, only the high-latitude Gondwana Realm, with endemic genera such as *Chilenochonetes* and *Septosyringothyris*, based on relatively few sites in Chile and Argentina, can be differentiated from a Palaeoequatorial Realm in the rest of the world. However, as the intervening seaways between the continents closed to form Pangea, by Serpukhovian times, at about 330 Ma, the Palaeoequatorial Realm became divided into a Boreal Realm for the Siberian faunas, a North American Realm to the west of Pangea, and a Palaeotethyan Realm to the east of Pangea, the latter including North and South China and Tarim, each with characteristic endemic genera, although about half the brachiopods there were more cosmopolitan (Qiao & Shen, 2014).

Terrestrial Animals. The Carboniferous invertebrate faunas were dominated by the arthropods, although other phyla, such as the various groups of worms and nematodes, which are seldom preserved as fossils, were probably also present in quantity. Eurypterids and scorpions were the largest arthropods, sometimes nearly a metre long, although some lived in lakes, and there were also numerous spiders, millipedes, and myriapods. The first insects had made their appearance in the Devonian Rhynie Chert, but they did not fly, and the development of insect flight took place in the Middle Carboniferous within the Palaeodictyopora, a group which itself became extinct in the Permian, but not without relatively explosive radiation in and after the Carboniferous to give rise to the many orders which have persisted to the present day (E.A. Jarzembowski in Selley et al., 2005).

The vertebrates had gained access to the land in the Devonian, but development of those amphibians was relatively slow in the Early Carboniferous, and their lengths were mostly less than a metre. However, by the end of the Carboniferous, there were over 40 different families, including the earliest amniotes, the group which includes reptiles, birds, and mammals (Benton, 2005).

Western Laurussia. Progressive palaeogeographical maps of the north-eastern area of Laurussia, including eastern Greenland, Spitsbergen, and Norway (Figs. 9.6, 9.8, and 10.5), during Carboniferous and Permian times show that the facies over that substantial area underwent a progression from huge humid flood plains in the Early Carboniferous through shallow warm seas in the Middle Carboniferous to Middle Permian to cooler environments in the Late Permian (Stemmerik, 2000). Those changes reflected substantial shifts in palaeoclimatic and subsidence patterns, which were related to the northward drift of the area as well as to ongoing rifting. Further to the north-west, in today's Canadian Arctic area, substantial sedimentation occurred in the rapidly deepening Sverdrup Basin, and there are extensive Late Carboniferous (Moscovian) bryozoan reefs (which, in contrast to coral reefs, always tend to develop most substantially in temperate latitudes) on the basin margin there (Trettin, 1998). Some biotic distributions of plants and reptiles reflect the latitudinal belts which crossed the whole supercontinent. Successive Carboniferous ostracod faunas demonstrate that some originally North American forms spread eastwards as far as Hungary, Egypt, and Oman (Lethiers & Crasquin-Soleau, 1995).

Although Fig. 9.8 shows much of the former Laurentian Craton as emergent land, sporadic melting of high-latitude ice caused many sea-level changes, with the result that parts of that craton became sporadically flooded in a comparable way to the northern England sector of Laurussia. The same non-marine bivalves are present in the Late Carboniferous of both North America and Europe. Striking sedimentary cycles are seen in the Late Carboniferous rocks of the Appalachians, where U/Pb dating indicates that the average maximum duration of each cycle was about 100,000 years. That may support the possibility of short eccentricity-driven influences on sedimentation there (Greb et al., 2008); however, the chief variations in sedimentation seem to have been caused more by tectonic activity in the commencement of the Alleghanian Orogeny, with the glacioeustatic changes also having a subsidiary role.

The final phase of the Laurussian–Gondwanan unification in the earliest Permian is recorded in the rocks of Texas, where 12–14 km thicknesses of Carboniferous sediments were deposited in the Ouachita Mountains, including marine olistostromes and several thin volcanoclastic units, which are succeeded upwards (after the Pangea unification) by deltaic deposits of latest Carboniferous age. There is an exceptionally preserved Lagerstätte deposit at Mazon Creek, Illinois, with numerous plants and terrestrial arthropods, some with their soft parts preserved as fossils, among other faunas and floras.

Eastern Laurussia and Siberia. Figure 9.9 shows Siberia and Amuria, as well as parts of Baltica, which then formed the north-eastern sector of Laurussia, and which remained separate from Siberia until the Permian, and includes most of the type areas for the stages. Although there are unconformities, there is a largely complete marine succession through the system in the Moscow Basin of Russia, and much deposition to its south in the Donets Basin of the Ukraine, which also has paralic marginal facies including many coals.

In the Altai–Sayan area of peri-Siberia there was only non-marine deposition from the Middle Carboniferous onwards (Yolkin et al., 2003). In Ob–Saisan–Surgat, in the West Siberian Basin at today's western margin of the craton (Unit 435), there is a Devonian to Lower Carboniferous accretionary complex followed unconformably by Middle Carboniferous marginal marine clastic sediments and coals, and Upper Carboniferous terrestrial deposits. Most of the Siberian Craton was flooded for most of the Early Carboniferous, largely with shallow-marine limestones and lagoonal sediments, but there was one large continental land area in its north termed Angara, as well as a variety of smaller, but still sizeable, islands elsewhere on the craton (Fig. 9.9). Those limestones were followed upwards by Middle Carboniferous siliciclastic deposits (comparable with the Namurian Millstone Grit of Britain), with occasional plant horizons, but rare coals, whilst the thick Upper Carboniferous beds have many coals, deposition of which continued into the Early Permian. On the south-west part of the Siberian Platform, in the Anabar area, the Upper Carboniferous consists of 350 m of mainly terrestrial siliciclastics with plant remains of the Angaran Province floras, as well as fresh- and brackish-water invertebrates (Cocks & Torsvik, 2007).

Central Asia. During the Carboniferous, many of the terranes which had previously been independent constituent parts of the Kazakh Terrane Assemblage accreted so as to substantially enlarge the Kazakhstania continent, and that combined collage became progressively closer to Siberia; but, because of the very variable published views on its accretion, and even indeed its identity, it is only shown schematically on our figures. In Karatau–Naryn (Unit 466) there was an extensive carbonate platform, which was situated on the passive Kazakhstania margin, and which evolved from shallow-marine reef and sand shoals in the Late Devonian to deeper-water ramps and skeletal mounds in the Tournaisian and Early Visean, and to skeletal mounds and sand-shoal-rimmed margins from the Middle Visean to the Bashkirian (Cook et al., 2002).

The Tarim Microcontinent was also near Siberia: the deposition of marine carbonates persisted along the north-west margin of Tarim (Fig. 9.10) until the latest

Carboniferous or Early Permian (Zhou et al., 2001). The Carboniferous of Tarim includes a variety of turbidites and volcanic rocks, with a stable carbonate platform of marine deposits in the Tarim Basin in the south-western part of the palaeocontinent (Zhang et al., 2003).

Eastern Asia. In North China, Lower Carboniferous rocks are absent, and there was bauxite deposition in the centre of that continent immediately above the regional Middle Carboniferous unconformity which overlies rocks of Cambrian and Ordovician ages. Most of the Upper Carboniferous of North China is non-marine, with many coals there, some of high commercial quality (Fig. 9.5) (Cope et al., 2005). In the eastern sector of North China, Late Carboniferous (Moscovian) brachiopod associations from the Taebaeksan Basin of Korea include the diverse *Choristites* Assemblage, which, although generically similar to comparable communities from South China, Annamia, and Tarim, has numerous species which are endemic to North China (Lee et al., 2010).

In today's Thai sector of Annamia (Figs. 9.4 and 9.5), there are Visean brachiopods and foraminiferans of no particular provincial affinity within Famennian to Tournaisian faunas from seamounts which were deformed before Upper Carboniferous clastic rocks were unconformably deposited upon them (Ridd et al., 2011). The latter include coals and gypsum beds which are interbedded with marine rocks containing brachiopods and foraminiferans, deposition of which continued into the Permian.

Gondwana. As well as the evidence of glaciation noted above, since the South Pole was under central Africa, there were also extensive marine transgressions, with the shoreline retreating southwards over north-central Africa as far south as Chad (Torsvik & Cocks, 2011). However, there were several subsequent periods of more local regressions and transgressions, as reflected in interbedded marine and continental rocks in much of southern Algeria and Libya. There were many lakes and freshwater lake deposits, particularly in the Moscovian of North Africa. In the Arabian Peninsula, the Palmyrides Trough was developed across central Syria, where sedimentation continued until the end of the Cretaceous (Brew et al., 2001).

In south-western South America, substantial molasse sediments were deposited both in the Precordillera and in the adjacent Sierras Pampeanas following tectonism in the area. Although Carboniferous sedimentary rocks are known from the region, which are largely turbidites in the west and shallow marine sediments in the east, they were much affected by subsequent Andean tectonics (Moreno & Gibbons, 2007).

In the Australasian sector of Gondwana and subsequently Pangea, separate as it was from most other substantial tropical landmasses in Carboniferous times (Fig. 9.1), there were many diverse faunas, with a high proportion of endemic genera in some groups, for example the corals and foraminifera.

10 Permian

Leaves of *Glossopteris* from India, the characteristic plant found in today's separate continents which were all together in Gondwana before its breakup, and which Alfred Wegener used to substantiate his original concept of continental drift.

The Permian, which lasted for a substantial 47 Myr (from 299 to 252 Ma), is most notable for being the time when the supercontinent of Pangea, more than three-quarters of the Earth's total land area, was at its maximum extent. Close to the end of the era, at 251 Ma, there was an immense outpouring of basalts in Russia, a large igneous province known as the Siberian Traps, which even today covers over 40% of the area of political Siberia. The combination of that LIP and the climatic crises which it undoubtedly caused, together with the unusually meagre total combined lengths of the world's coastlines, were the chief triggers for the largest ever biotic extinction event of the Phanerozoic. That Permian–Triassic (P/T) Extinction Event affected the whole globe.

The Briton Roderick Murchison, the German Alexander von Keyserling, and the Frenchman Edouard de Verneuil together named this system in the 1840s following extensive field work near Perm, a city near the northern Ural Mountains of Russia, where there is a distinctive and largely complete sequence of marine rocks. The base of the Permian System (Asselian Stage) is now formally defined at the base of the *Streptognathus sulcata* conodont Biozone in northern Kazakhstan, which is dated at 299 Ma. The global distribution of the continents and oceans in two periods within the Permian at 290 and 270 Ma, and near the Permian–Triassic boundary interval at 250 Ma, is shown in Fig. 10.1. There are three formal series, the Cisuralian (within which are the Asselian, Sakmarian, Artinskian, and Kungurian stages), defined in Russia, the Guadalupian Series (Roadian, Wordian, and Capitanian stages), defined in North America, and the Lopingian Series (Wuchiapingian and Changhsingian stages), defined in China.

Tectonics and igneous activity

Oceans. The Neotethys Ocean (including what has been termed the Mesotethys by some authors) opened progressively during the Early Permian from about 275 Ma within the north-east rim of the Gondwanan sector of the Pangea Craton, following earlier rifting there (Domeier & Torsvik, 2014). At some stage in the first half of the period, sea-floor spreading ceased in the Paleotethys Ocean, which lay to the north of the Neotethys and the Cimmerian Chain (Fig. 10.1), and changed to subduction, and thenceforward that ocean dwindled progressively in size (Fig. 10.2). However, the precise timing of that change is poorly constrained, although Late Permian calc-alkaline granites indicating subduction are known along the Dian–Qiong Suture Zone between Annamia and South China (Cai & Zhang, 2009).

Bivergent subduction of the Mongol–Okhotsk Ocean continued during the Permian, but, as in the Carboniferous, sea-floor spreading outpaced subduction and thus the ocean basin progressively widened. Along its north-eastern edge, that wedge-shaped ocean rode over the Panthalassic Plate upon an intra-oceanic subduction zone which steadily lengthened throughout the Late Palaeozoic. The southern margin of Amuria also remained active throughout the Permian, first consuming the Panthalassic there and later the Paleoasian Ocean, until terminal closure of the latter, which completed the accretion of Amuria to North China at the end of the Permian (Figs. 10.3 and 10.4).

Pangea. Pangea is the only supercontinent to have existed during the Phanerozoic, but nevertheless it was never 'complete' at any single time during its existence, since some of its sectors had broken away before others had arrived (see also the previous chapter). The continental land area within Pangea stretching from Siberia to Eastern Europe was originally termed Angaraland by Wegener (1915), but that term is now seldom used.

Although the bulk of Pangea (chiefly the former Gondwana and Laurussia continents) had already assembled before the end of the Carboniferous, the continent of Kazakhstania and its adjacent island arcs did not complete its union with Laurussia until the earliest Permian. Within Kazakhstania, in Chingiz–Tarbagatai (Unit 482), North Tien Shan (Unit 459), and North Balkhash (Unit 463), there are Permian terrestrial sediments interbedded with tuffs, ignimbrites, and various other volcanic rocks. One of the island arcs within the composite Junggar Terrane (Unit 450) continued to be separate and active in the Early Permian, although there are also some terrestrial rocks interbedded with volcanics in other parts of that terrane (Daukeev et al., 2002).

The Uralian Orogeny was caused by the Laurussia–Kazakhstania collision, which had started in the Middle Carboniferous, and continued on through the Early Permian until around the beginning of the Artinskian at 290 Ma. In contrast, although it had neared north-east Baltica and the Kazakh sectors of Pangea before the Permian (Fig. 9.10), Siberia probably did not reach its eventual position as a fixed part of Pangea until the Early Mesozoic, during the Triassic. Thus, although most of the major sectors of Pangea had unified prior to the start of the Permian, all through the Permian there was a great deal of continuing rotational movement along important strike-slip faults, including those which bordered Siberia and the former peri-Siberia.

At the end of the Permian, a high proportion of the continents were still clustered within the united Pangea Supercontinent (Fig. 10.1); however, although the bulk of Pangea remained united throughout the period, there was breakup at some of its margins, most notably as the Neotethys Ocean opened in the Early Permian.

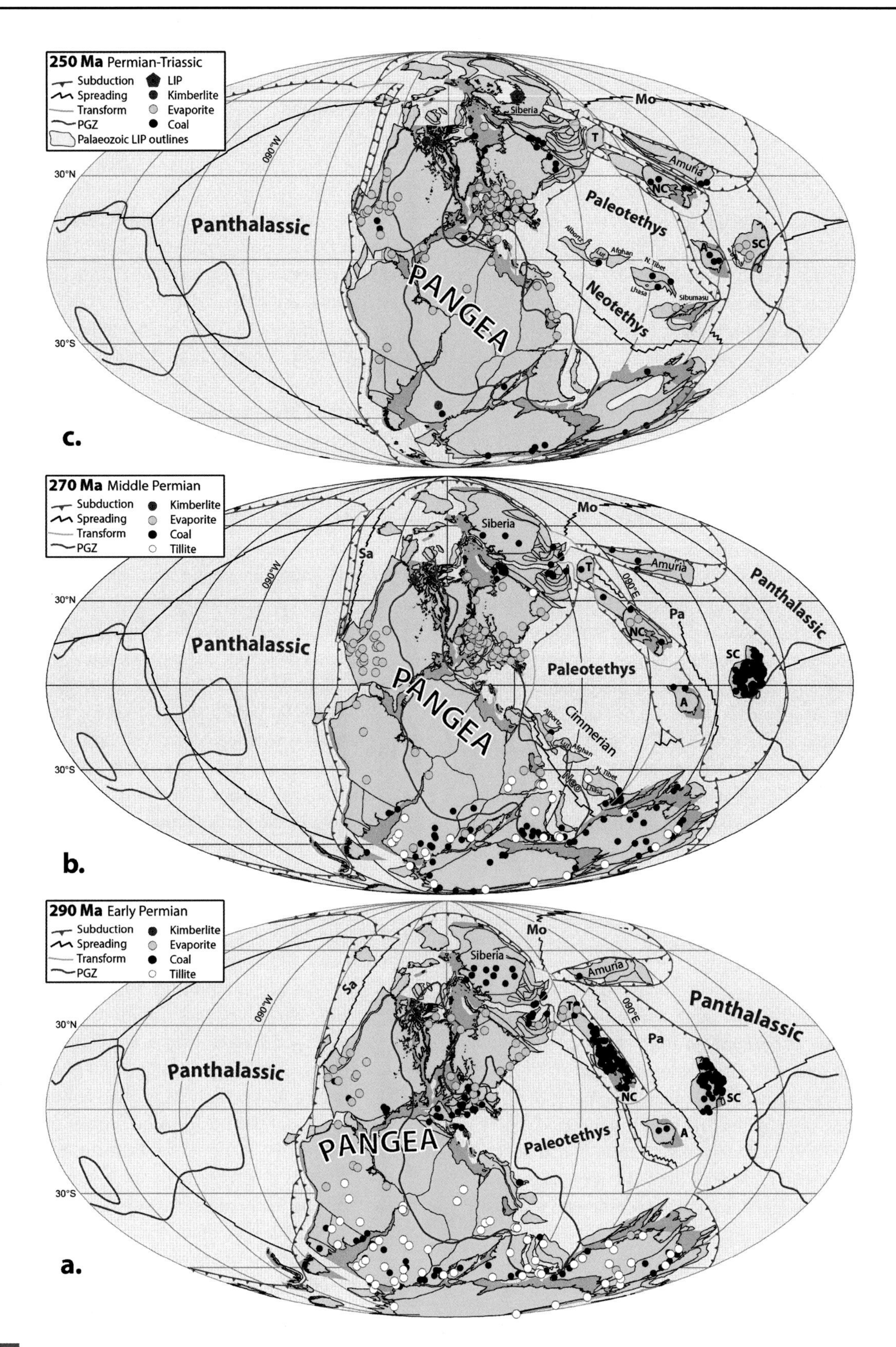

Fig. 10.1 Outline Earth geography at (a) 290 Ma (Artinskian), (b) 270 Ma (Roadian), and (c) near the Permian–Triassic boundary time at 250 Ma, including the postulated plate boundaries, outlines of the major crustal units, and the most substantial oceans. Solid blue lines are subduction zones with teeth on the upper plate, black lines are spreading centres, and green lines are transform plate margins. A, Annamia; LIP, large igneous province; Mo, Mongol–Okhotsk Ocean; NC, North China; Pa, Paleoasian Ocean; PGZ, plume generation zone; Sa, Slide Mountain–Angayucham Ocean: SC, South China; T, Tarim.

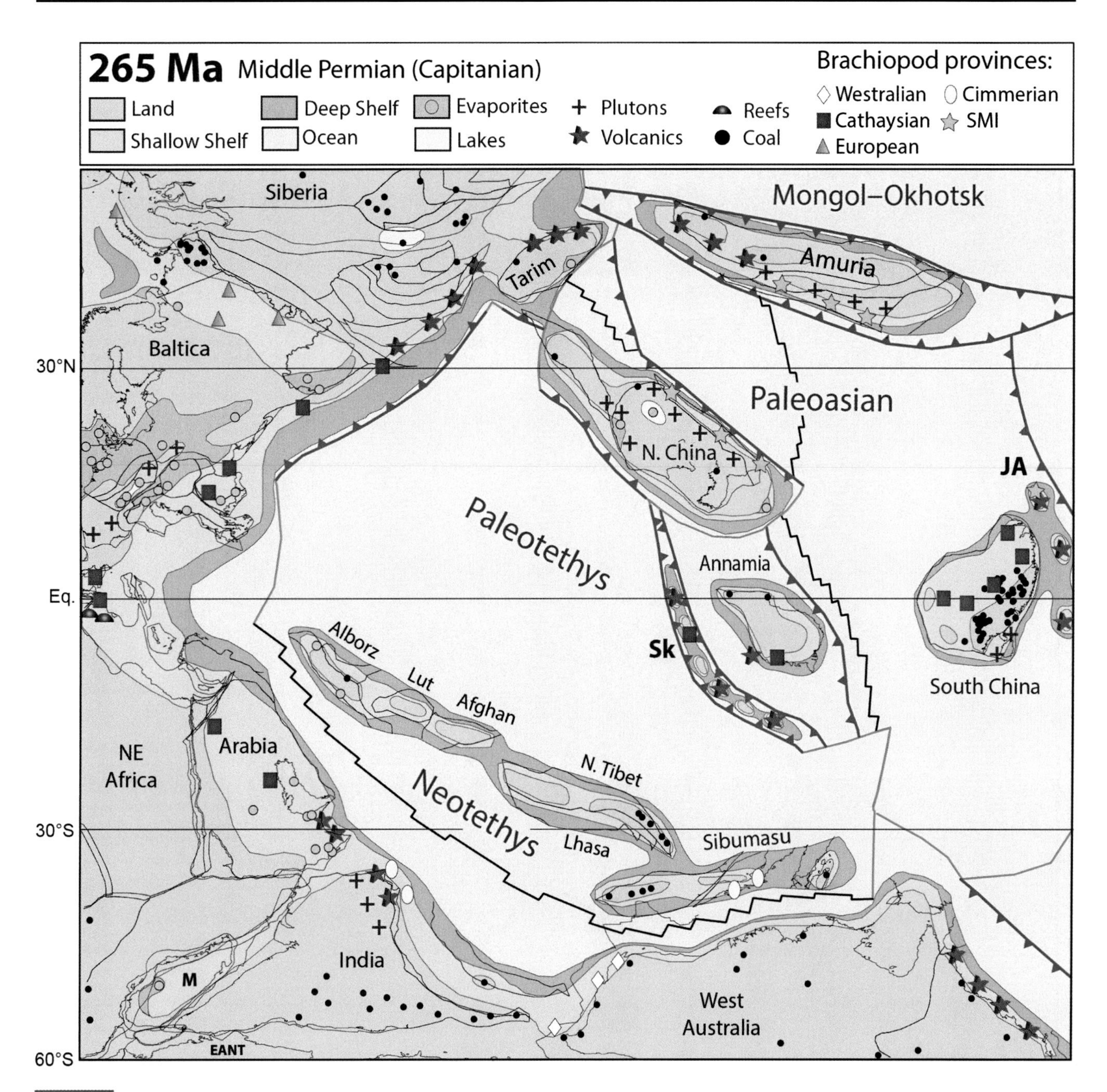

Fig. 10.2 The geography of Europe, southern and eastern Asia, and parts of Gondwana in the Middle Permian (265 Ma, Capitanian), including the opening Neotethys Ocean, the still large but dwindling Paleotethys Ocean, and parts of the Paleoasian, Mongol–Okhotsk, and Panthalassic oceans. The Cimmerian Chain is the string of terranes from Alborz to Sibumasu, all of which had been parts of the northern Gondwanan Craton margin. As well as the igneous and sedimentary facies, the sites of some of the brachiopod provinces documented by Shen et al. (2009) are plotted. EANT, Eastern Antarctica; JA, Japanese arcs; M, Madagascar; Sk, Sukothal Arc.

North-central Pangea. In most of Europe, the latest Permian and the earliest Triassic saw regional regression of the Arctic Sea and the start of a multi-directional rift system which transected the Variscan Fold Belt, as well as the continuously subsiding Northern and Southern North Sea basins such as the Viking and Central grabens, which today host substantial oil reservoirs (Ziegler, 1990). The Variscan Orogeny (see Chapters 8 and 9) was largely over before the Permian, but substantial post-orogenic granites were intruded, particularly in the south-west of England, where an enormous batholith over 250 km long was intruded during the latest Carboniferous and Early Permian between about

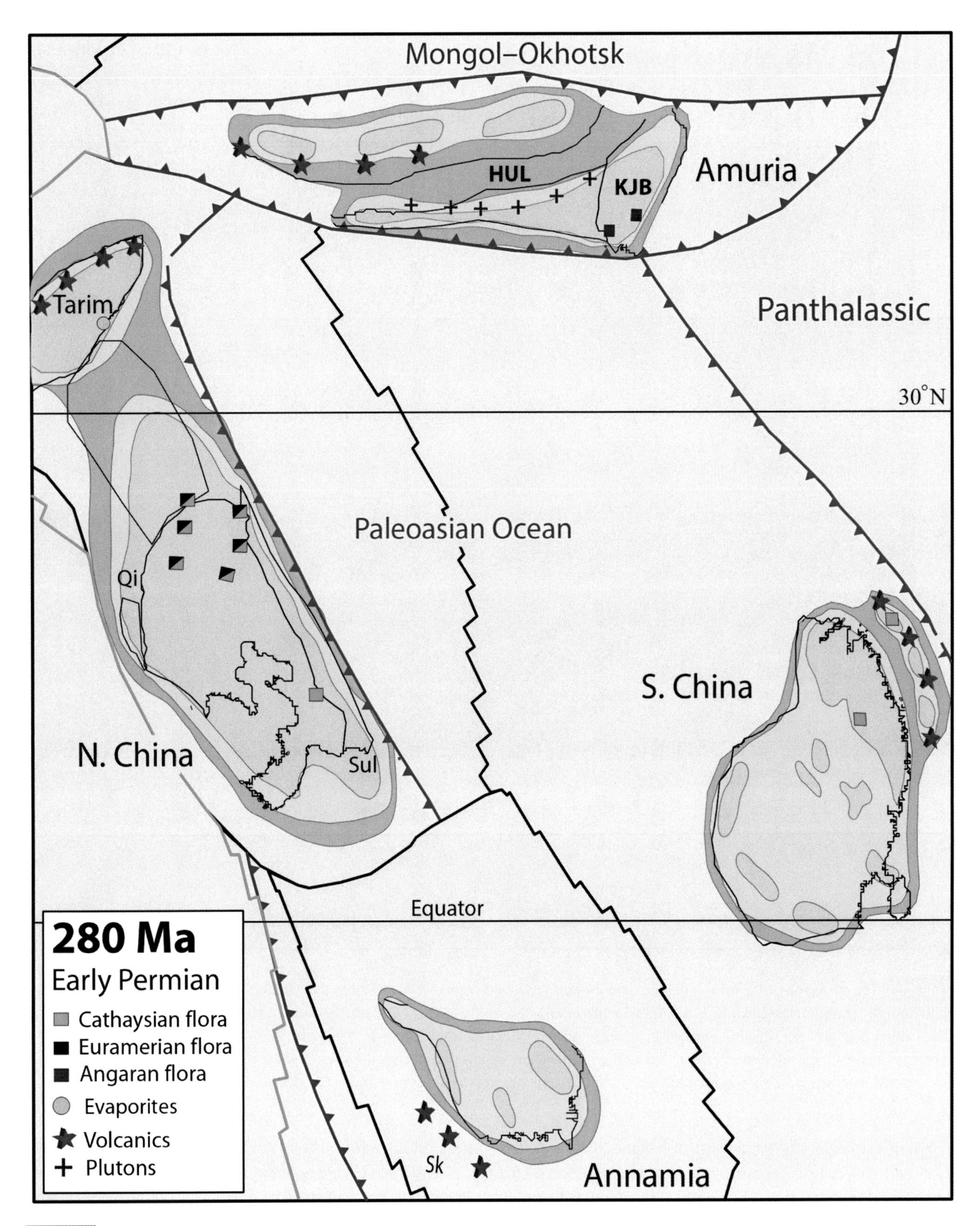

Fig. 10.3 The Amuria, North and South China, and Annamia continents at 280 Ma (Artinskian), showing the contemporary pattern of lands and seas and the sites of provincial floras. North China includes the old cratons of North China and Tarim, as well as the area between them previously termed the Ala Shan Terrane area (Unit 451). Solid blue lines are subduction zones with teeth on the upper plate, black lines are spreading centres, and green lines are transform plate margins. HUL, Hutag Uul–Songliao; KJB, Khanka–Jiamusu–Bureya block; Qi, Qilian; Sk, Sukothai and related island arcs; Sul, Sulinheer.

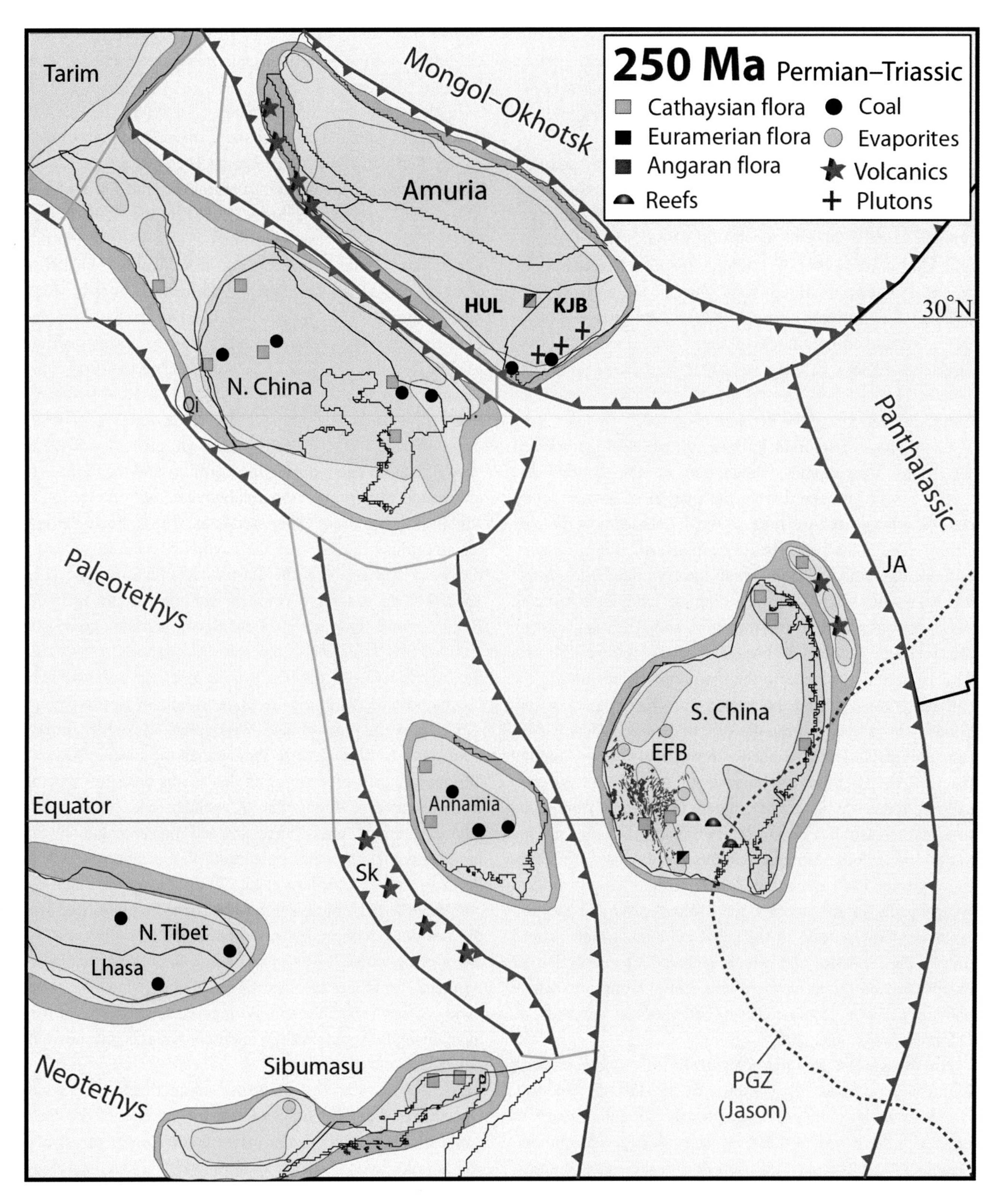

Fig. 10.4 The Amuria, North and South China, Annamia, Sibumasu, and other East Asian continents and terranes near the Permian–Triassic boundary time at 250 Ma, showing the contemporary pattern of lands and seas and the sites of provincial floras. North China includes the old cratons of North China and Tarim, as well as the area between them previously termed the Ala Shan Terrane. The future Solonker Suture Zone lies between North China and Amuria. Solid blue lines are subduction zones with teeth on the upper plate, and green lines are transform plate margins. Dotted red line is the plume generation zone (PGZ). EFB, Emeishan Flood Basalt Large Igneous Province; HUL, Hutag Uul–Songliao; KJB, Khanka–Jiamusu–Bureya block; JA, Japanese island arcs; Qi, Qilian; Sk, Sukothal and related island arcs.

290 and 275 Ma. However, that batholith is only united at depth, and on the surface there are five separate granitic outcrop areas with their tops variably eroded, in a line which stretches from the Isles of Scilly, through Cornwall, and into Dartmoor, Devonshire.

Baltica and Laurentia had been integral parts of Laurussia prior to the latter's amalgamation within Pangea, and included Franz Josef Land, today in the Barents Sea (Dibner, 1998), as well as the mainland parts of Baltica (Nikishin et al., 1996). The region between Greenland and Norway underwent rifting. Kara, now in the north-west of political Siberia, had been an independent terrane not far from Baltica during much of the Lower Palaeozoic and Devonian (Cocks & Torsvik, 2005), and had approached Siberia progressively during the Carboniferous before its accretion to it during the Permian (Fig. 9.9).

Gondwanan Sector of Pangea. In the latest Carboniferous, south-west-dipping subduction of the Panthalassic Ocean jumped outboard from the margin of former south-east Gondwana in Australia to the intra-oceanic Gympie–Brook Street island arc. However, in about 270 Ma (Guadalupian), the remnant basin behind the Gympie–Brook Street arc began to collapse by west-dipping subduction beneath the continental margin of Antarctica and Australia, and by the end of the Permian the basin had been entirely consumed. The arc terrane accreted to the margin of Gondwana in the earliest Triassic during the Hunter–Bowen Orogeny (Glen, 2005). Along the west margin of Gondwana, subduction of the Panthalassic continued throughout the Permian (Fig. 10.1). An important episode of dextral transpression affected the entire margin of southern Gondwana (from Chile to eastern Australia) during the Permian, producing a wide variety of tectonomagmatic features there which are sometimes grouped as a rather nebulously defined 'Gondwanides' Orogeny. In South America, there was the start of the pre-Andean tectonic cycle in the latest Permian, which peaked during the Triassic, and which included the evolution of extensional basins in northern and central Chile into which were intruded volcanic rocks of intermediate composition (Moreno & Gibbons, 2007).

North-western Pangea. The fusion of Gondwana with Laurussia was essentially complete by the start of the period. However, in the south-east of the former Laurentian sector of Laurussia there was further consolidation in the Marathon–Ouachita–Appalachian Fold Belt during the Early Permian (Cocks & Torsvik, 2011).

In the Appalachian area in the east of the USA, deformation in the Alleghanian Orogeny continued until somewhat after its companion Variscan Orogeny in Europe had ended. Changes in the detrital-zircon-age population there correlate with a shift from earlier transpressionally inspired oblique deformation to foreland-vergent contraction in the Early Permian (Becker et al., 2006).

In the south-west of the former Laurentia, the sinistral movement continued on from the Carboniferous, accompanied by Permian thrusting of earlier rocks (the Last Chance Allochthon) over the western margin (the Bird Spring Shelf) of the craton (Stevens & Stone, 2007). In the southern parts of the North American Cordillera, Early to Late Permian island arcs with diachronous calc-alkaline volcanism developed, such as those preserved in the Klamath Mountains and Sierra Nevada of California, arcs which had developed offshore in a previous phase of extensional tectonism (Dickinson, 2000). The arcs reached their maximum distance from the Laurussian Craton (Fig. 10.5) before the intervening Slide Mountain Ocean (in the same region which eventually hosted the Cache Creek Ocean in the Triassic) started to close again during the Permian, and the terranes on either side were probably reamalgamated with the craton by Middle Triassic times (Shephard et al., 2013). Early Permian metamorphism occurred in the Central Metamorphic Belt at the western margin of the Eastern Klamath Terrane (Unit 161). The faunas there confirm that the Eastern Klamath, Stikinia, and Quesnellia Cordilleran terrane group lay 2,000–3,000 km away from the Laurussian Craton at the time (Belasky et al., 2002), which gives an approximation for the possible width of the Slide Mountain Ocean there.

The previously combined Wrangellia–Alexander Terrane (Unit 165) had less diverse faunas than the Eastern Klamath Terrane group, which suggests that it was probably separate from the others. Wrangellia, Alexander, and some smaller adjacent terrane areas have yielded palaeomagnetic data indicating that their palaeolatitude was at about 15° N in Permian time (Nokleberg et al., 2000). However, since that combined terrane, big enough to be termed a microcontinent, did not dock with North America until the Cretaceous, we are not sure where it was in relation to Laurussia even by the Permian, let alone at older times in the Palaeozoic when many of the Alexander and Wrangellia faunas were at their most distinctive, and thus the microcontinent is not shown on our Palaeozoic maps.

As a consequence of the same subduction regime which closed the Slide Mountain Ocean by the end of the Early Mesozoic, in Nevada the Havallah sequence was thrust eastwards onto autochthonous shallow-water Upper Palaeozoic strata along the Golconda Thrust in the Late Permian to earliest Triassic Sonoma Orogeny (Stevens & Stone, 2007).

The Middle Permian appearance of a magmatic arc and high-pressure metamorphic rocks on the east side of the Yukon–Tanana Terrane within the Cordillera (Unit 162)

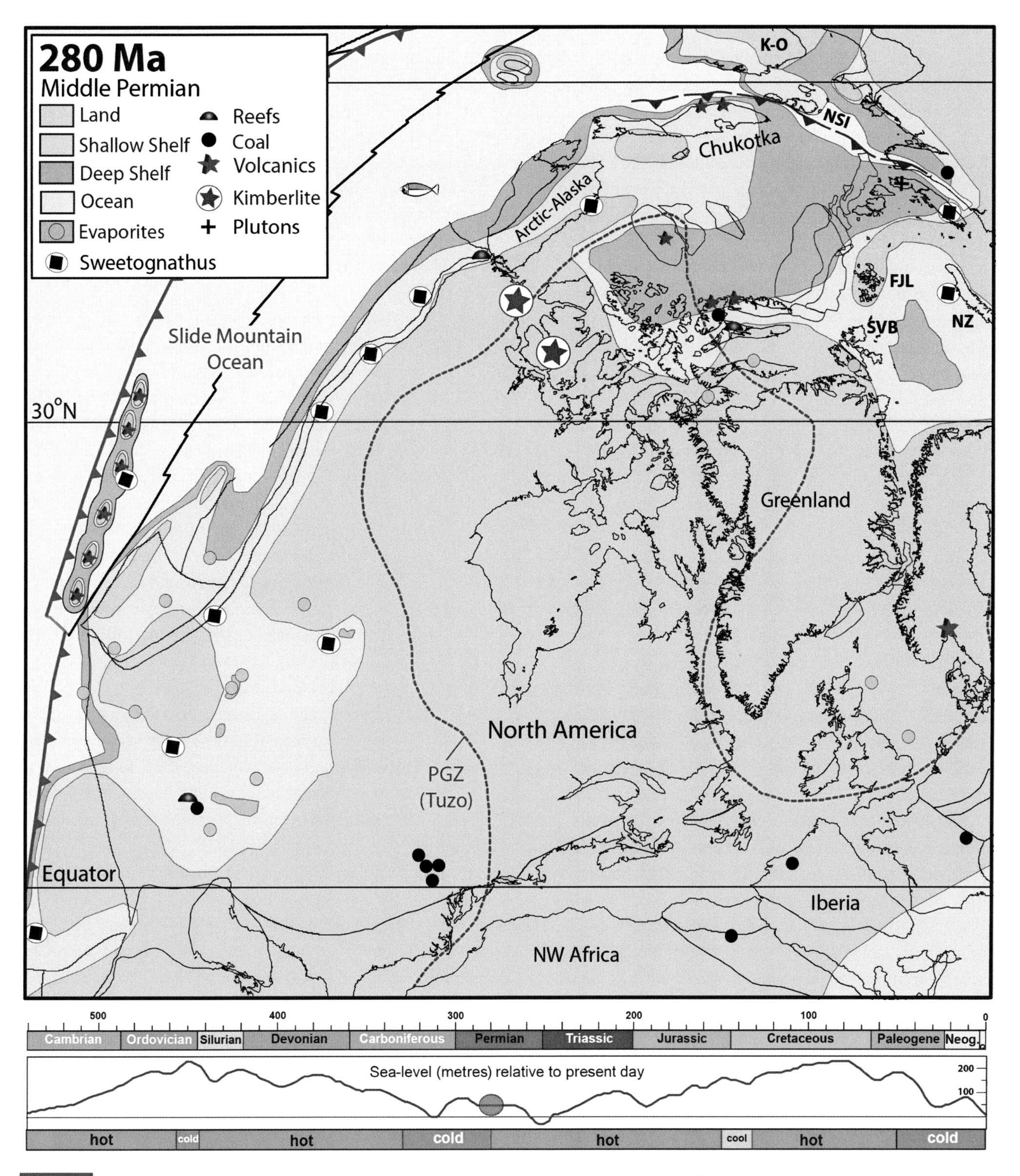

Fig. 10.5 The western and central Laurussian sector of northern Pangea and adjacent part of Siberia at 280 Ma (Artinskian), showing the contemporary pattern of lands and seas, and the distribution of the conodont *Sweetognathus* (see text). Solid blue lines are subduction zones with teeth on the upper plate, black lines are spreading centres and transform faults. Dotted red line is the plume generation zone (PGZ). FJL, Franz Josef Land; K-O, Kolyma–Omolon; NSI, New Siberian Islands; NZ, Novaya Zemlya; SVB, Svalbard. Bottom: Phanerozoic time scale and sea-level variations (Haq & Al-Qahtani, 2005; Haq & Shutter, 2008) and icehouse (cold) and greenhouse (warm) conditions.

indicates that the formerly passive margin had collapsed by then, and that the Slide Mountain Ocean had begun subducting to the west (Fig. 10.5). By the end of the Permian, the Slide Mountain Ocean had been largely consumed, and the arc terranes of the upper plate were thrust eastward onto western Laurentia during the Sonoma Orogeny (Fig. 10.6) (Dickinson, 2009). In contrast, passive margin sediments deposited in the Angayucham Ocean to the north reveal that it remained open until the Jurassic, suggesting either that the two systems had become decoupled in the Middle Permian (i.e. west-dipping subduction of the Angayucham Ocean did not commence in the Middle Permian) or that subduction was interrupted in the north before the basin was entirely destroyed. We have adopted the latter scenario and speculate that west-dipping subduction of the Angayucham Ocean ceased together with subduction to the south in the latest Permian to Triassic, leaving a remnant basin to the east that survived until the Middle Mesozoic (Fig. 12.1). We have adjusted the Middle-to-Late-Permian relative motion between the Yukon–Tanana Arc and the Panthalassic Plate margin so as to be nearly orthogonal in the south and highly oblique in the north; thus the accretion of the southern segment of the arc terrane might have disrupted the continuation of subduction to the north.

Today's Farewell Terrane in the Cordillera (Unit 155) was affected by the Browns Fork Orogeny, of uncertain cause, in the Early Permian at 285 Ma, but, although Farewell is shown on our maps, its position is arbitrary within the Panthalassic Ocean since it is very poorly constrained until the Jurassic.

South-western Pangea. On the western margin of the former Gondwana there was substantial volcanic and plutonic activity, as can be seen in the Mendoza area of Argentina, where the marine basin there closed in a tectonic phase in the Middle Permian between 284 and 276 Ma (Artinskian to Kungurian). In central Chile, below 39° S, the subduction-related accretionary prism continued its activity, inboard of which there were plutons dated at 294–281 Ma (Sakmarian to Artinskian) in the North Patagonian Massif (Hervé et al., 2013). In Central America, by this time also part of the unified western margin of Pangea, a continental magmatic arc stretching from Guatemala to southern California continued its activity from before the start of the Permian until the Middle Permian at about 263 Ma.

In the former eastern Gondwana sector, in the Early Permian (Sakmarian: 289 Ma), a LIP, the Panjal Traps (sometimes called the Tethyan Plume), was extruded in the north-western Himalayan area and has been correlated with fragments of flood basalts in other parts of the Himalaya, as well as in the Lhasa Terrane of Tibet and Oman in the Middle East (Shellnut et al., 2011). The many fragments of the Panjal Traps are allochthonous within the Himalayan Orogen today, but they are shown as united on our Early Permian map (Fig. 10.7). If the Panjal Traps are truly a LIP sourced by a plume from the core–mantle boundary, we link it to the African Plume Generation Zone (Tuzo), and it is possible that the Panjal Trap eruptions may have triggered the subsequent opening of the Neotethys Ocean (Fig. 10.2). It was in that north-eastern sector of Gondwana where the Neotethys Ocean opened after rifting in the Early Permian (Figs. 10.7 and 10.2), see below.

Eastern Asia. As a consequence of earlier rifting, a string of microcontinents and terranes, notably Sibumasu and the Tibetan terranes, as well as many smaller units from Turkey eastwards through Alborz, Iran, Karakorum (including the South and Central Pamirs, which had been stitched to Karakorum by Middle Devonian lavas), Lut, Sanand, Afghanistan, and Pakistan, all moved away from the northern rim of the Gondwanan sector of the Pangea Craton as the Neotethys Ocean opened during the Early and Middle Permian at about 260 Ma. Fringing the northern flank of the Neotethys were the united Tibetan terranes of Qiantang and Lhasa, as well as Sibumasu (Metcalfe, 2006). In the Himalayan area there are many Lower Permian volcanics marking the rifting, as well as granitoids, which continued to be intruded into the Qiantang Terrane up to the close of the Permian (Zhu et al., 2013). It is uncertain whether the eastern and western groups of terranes were united to form a single very elongate continent, often termed Cimmeria, or whether they formed several separate microcontinental blocks, but it seems probable that at least some were joined to their neighbours as continuous land areas, as we show on Fig. 10.2.

North China, South China, and Annamia appeared to have stayed in the same general relationships to each other as they all slowly drifted northwards during the Permian (Figs. 10.3 and 10.4); however, Annamia and South China both rotated gradually during the period prior to their union in today's configuration along the Ailaoshan and Song Ma Suture Zone during the Triassic. There are Lower Permian granites in the Chanthaburi Terrane at the margin of Annamia (Sone & Metcalfe, 2008), and also Upper Permian arc rocks which represented tectonic activity during the first phases of the largely Triassic Indosinian Orogeny.

Near the end of the Permian, the Emeishan Large Igneous Province was intruded into South China between 262 and 257 Ma, at a time corresponding approximately to the Middle to Late Permian boundary (Fig. 10.4). Those massive Emeishan basalts, which vary from 3 to 5 km in thickness, changed the topography of South China so much that the sedimentary regime of that entire continent was affected

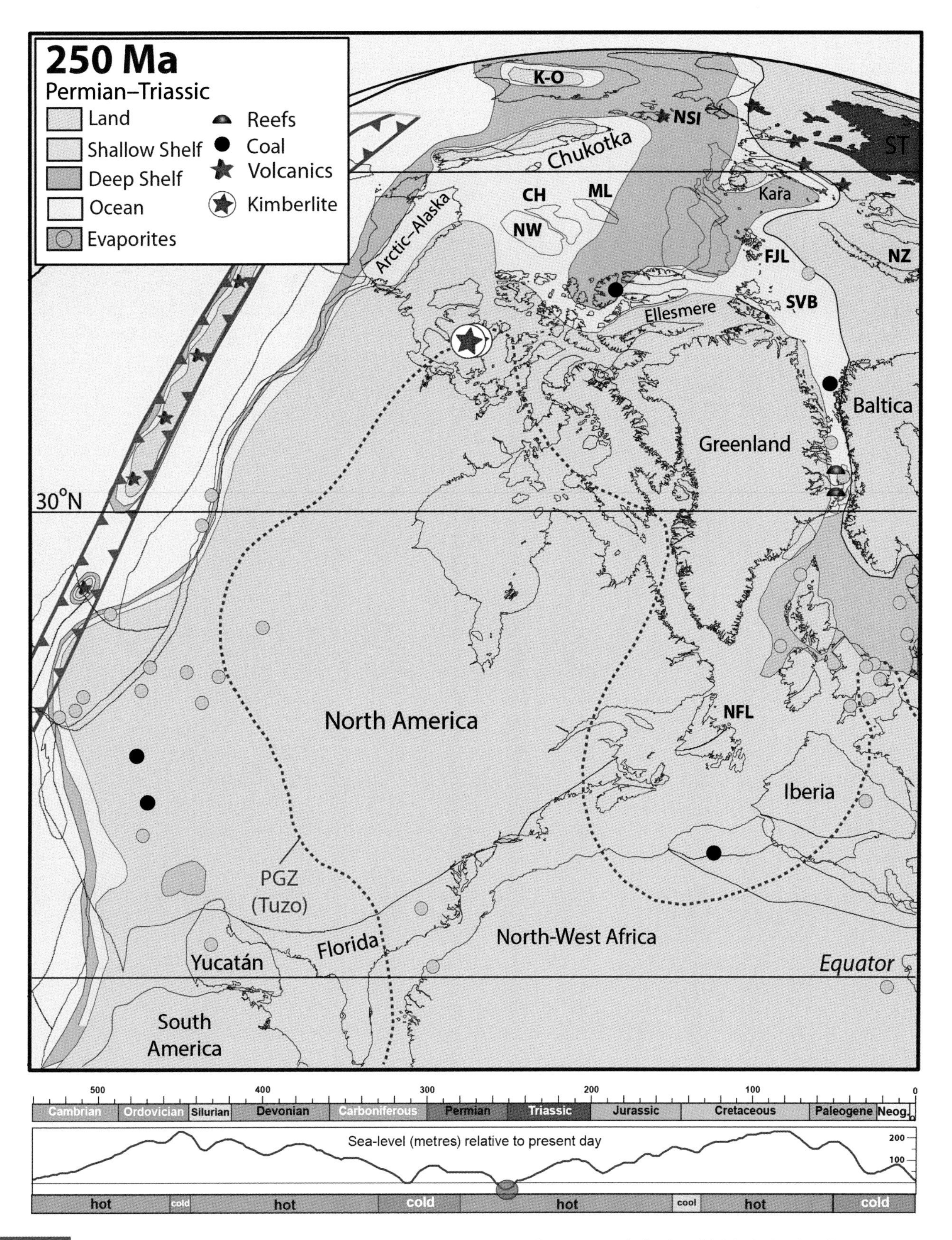

Fig. 10.6 The western and central Laurussian sector of northern Pangea and adjacent part of Siberia, which includes the Siberian Traps LIP, at 250 Ma (Permian–Triassic boundary time), showing the contemporary pattern of lands and seas. Solid blue lines are subduction zones with teeth on the upper plate, black lines are spreading centres, and green lines are transform plate margins. Dotted red line is the plume generation zone (PGZ). CH, Chukchi; FJL, Franz Josef Land; K-O, Kolyma–Omolon; ML, Mendeleev Ridge; NFL, Newfoundland; NSI, New Siberian Islands; NZ, Novaya Zemlya; NE, Northwind; ST, Siberian Traps; SVB, Svalbard. Bottom: Phanerozoic time scale and sea-level variations (Haq & Al-Qahtani, 2005; Haq & Shutter, 2008) and icehouse (cold) and greenhouse (warm) conditions.

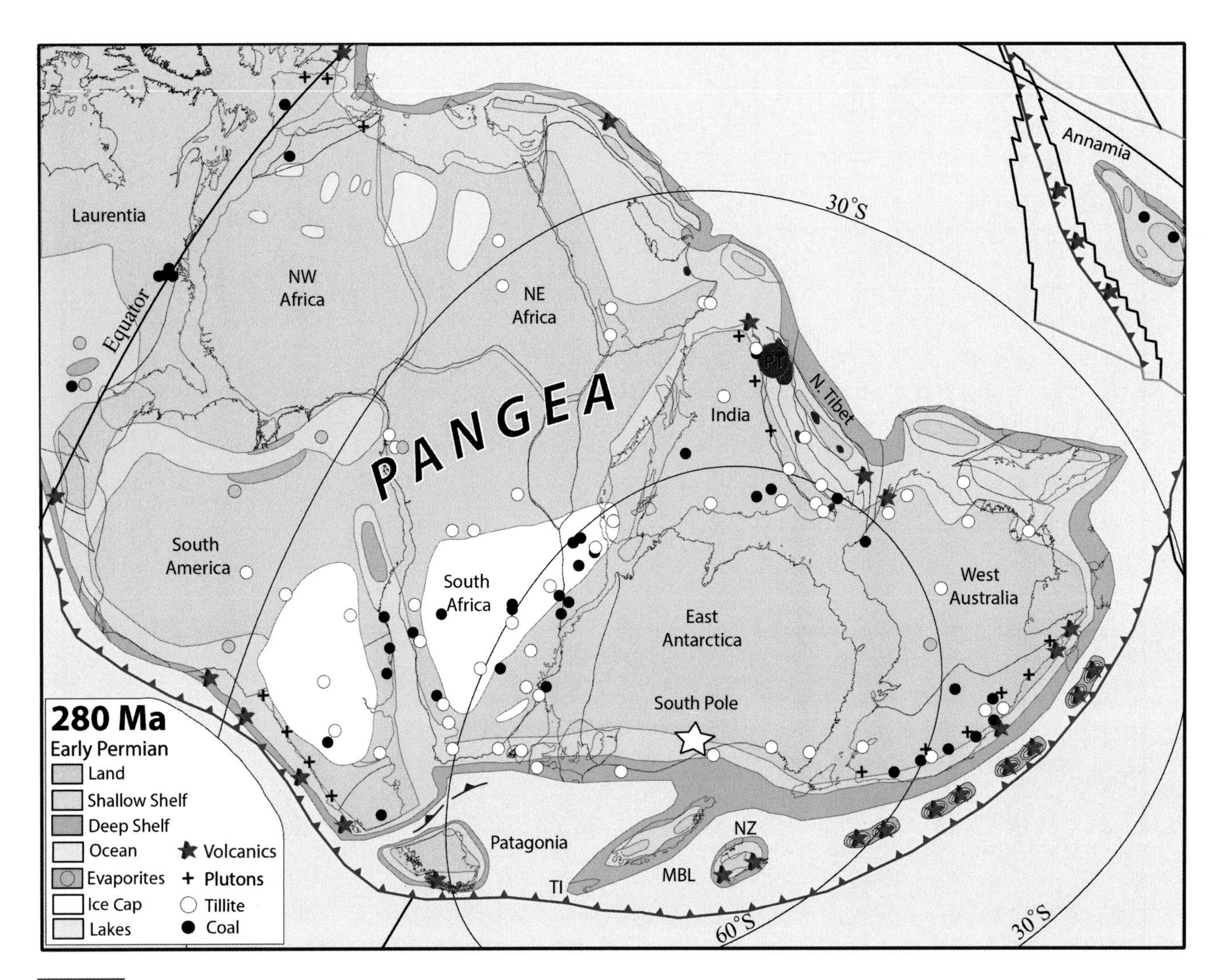

Fig. 10.7 The Gondwana sector of Pangea and adjacent areas at 280 Ma (Artinskian), showing the contemporary pattern of lands and seas. Solid blue lines are subduction zones with teeth on the upper plate, black lines are spreading centres, and green lines are transform plate margins. The strings of island arcs shown are diagrammatic, since the individual extent of each unit within the arcs and the extent to which each unit was above sea level are uncertain. MBL, Marie Byrd Land; NZ, New Zealand; PT, Panjal Traps Large Igneous Province; TI, Thurston Island.

during the Late Permian and Early Triassic, and which changed from a within-plate mafic-dominated source to the north-west to a mixed source involving magmatic arc and recycled orogenic detritus that lay to the west and east (Yang et al., 2014).

There was Permian narrowing of the Paleoasian Ocean between North China on the one side and Junggar, South Gobi, and the Amuria Microcontinent on the other side, with the Middle Permian Suolunshan Ophiolites between North China and the Khinggan–Bureya sector of Amuria defining both the position and the timing of the closure of the eastern Paleoasian Ocean there (Li, 2006). However, some authors (e.g. Xiao et al., 2008) show a united Siberia and Kazakhstania joining Tarim as the Paleoasian Ocean finally closed there, and identified the sea to the east of Tarim and Siberia and between them and North China as the Paleotethys Ocean. The faulting in the region between Siberia, Junggar, and Tarim, which centred round the Yili Block at the eastern end of the Atashu–Zhamshi Terrane, demonstrates that the major Erqiz Fault between Siberia and Junggar was sinistral, but the strike-slip faults to both north and south of the Yili Block were dextral, with lateral displacements of 600 and 1,000 km respectively (Wang et al., 2007), further confirming what a complex series of processes and events was involved during the final adjustments of Siberia's place within Pangea (Fig. 10.9). Despite some

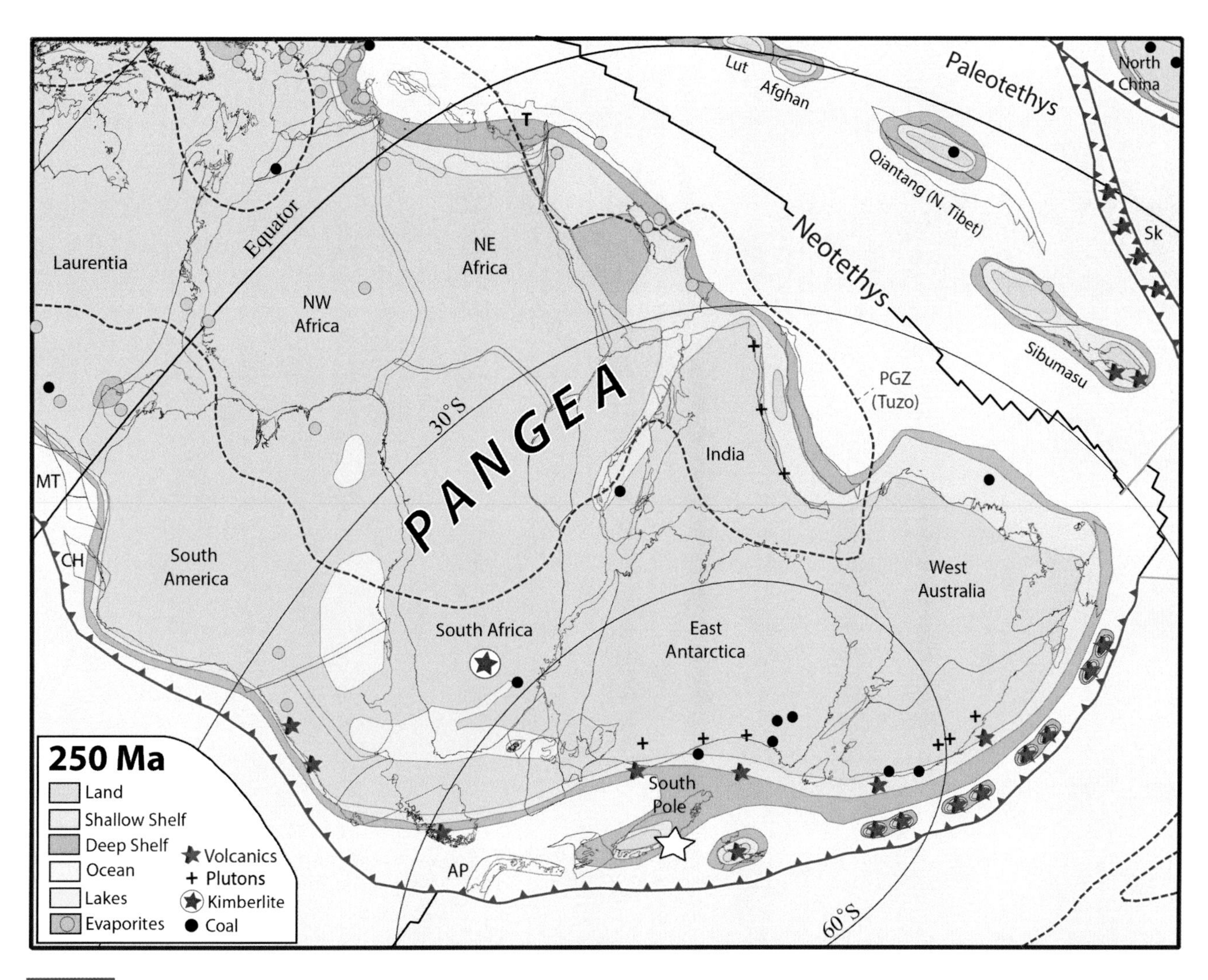

Fig. 10.8 The Gondwana sector of Pangea and adjacent areas near the Permian–Triassic boundary time at 250 Ma, showing the contemporary pattern of lands and seas. Solid blue lines are subduction zones with teeth on the upper plate, black lines are spreading centres, and green lines are transform plate margins. The strings of island arcs shown are diagrammatic, since the individual extent of each unit within the arcs and the extent to which each unit was above sea level are uncertain. AP, Antarctic Peninsula, CH, Chortis Terrane, Mexico; MT, the Mexican terranes of Mixteca–Oaxaquia and Sierra Madre; SK Sukothal Arc; T, Tauride Terrane of Turkey.

authors concluding that the ocean between Siberia and North China finally closed at some time in the Sakmarian soon after 290 Ma, the palaeomagnetic data indicate decisively that that closure did not occur until well into the Mesozoic, probably during the Late Jurassic at about 150 Ma.

South China and Annamia remained completely separate from Pangea for all of the Permian. North China, Tarim, and Amuria performed a complex tectonic dance to become united with each other, probably with their then western ends attached to eastern Pangea, together forming the north-eastern margins of the slowly dwindling Paleotethys Ocean (Figs. 10.2 and 10.3). The Khingan–Bureya Massif, whose modern locality is shown as Unit 454 in Fig. 3.7, Hutag Uul–Songliao (Unit 453), and the former Central Mongolia Terrane group (Unit 440) were all parts of the Amuria continent.

Tarim had finally become welded to Siberia by the Middle Permian, but the progressive Late Palaeozoic interaction between the two resulted in the orocline seen in Kazakhstan (Abrejevitch et al., 2007, 2008). The Kazakhstania continent and the other previously independent Kazakh terranes had mostly undergone consolidation earlier, and before their accretion to Siberia during the Early Permian. There was cessation of marine sedimentation and andesite volcanism at that time, as well as the initiation of uplift of the Permian Zaisan Mountains to the west of the Siberian Craton (Ziegler et al., 1997).

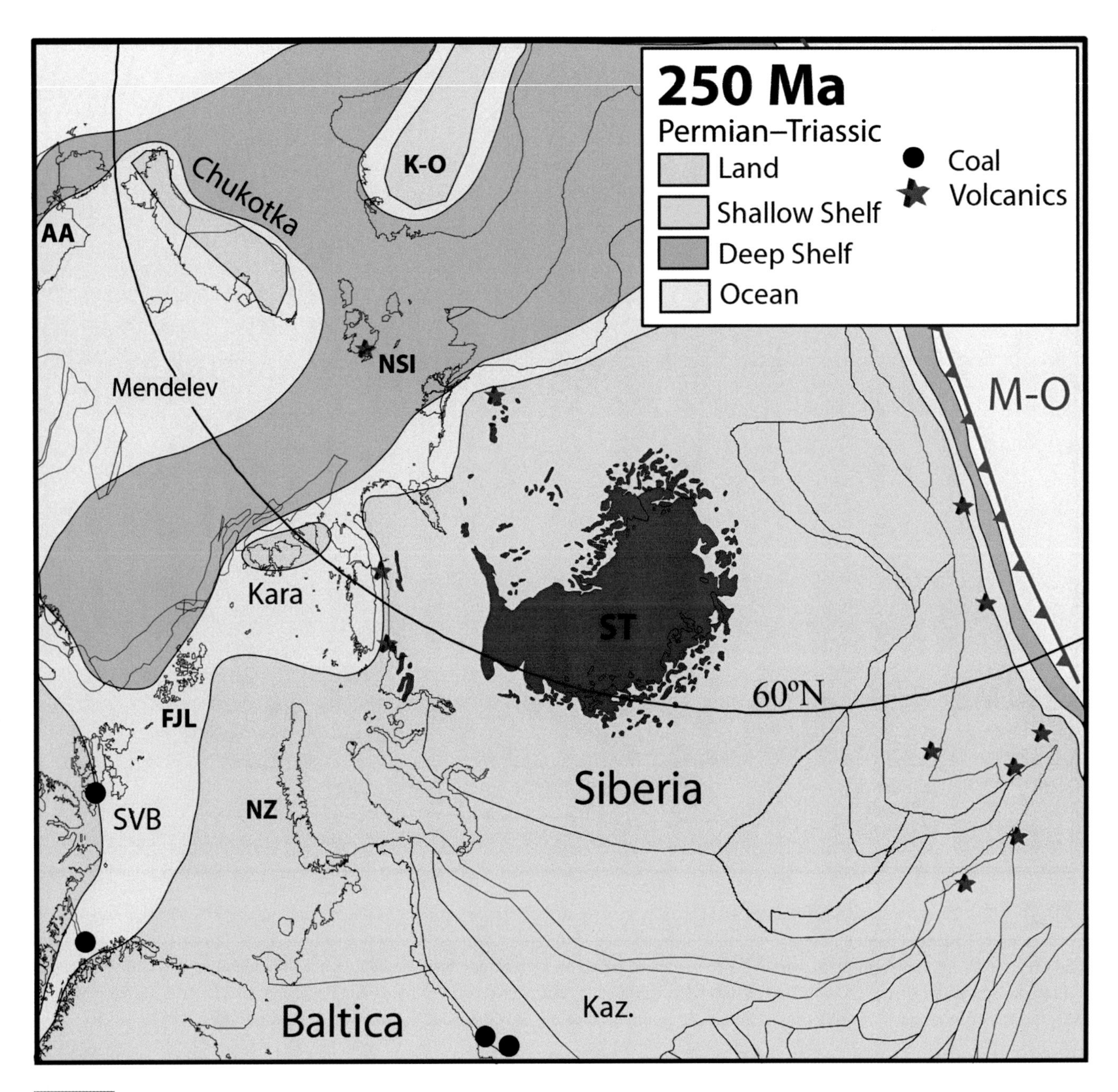

Fig. 10.9 Siberia and neighbouring sectors of northern Pangea at 250 Ma near Permian–Triassic boundary time at 250 Ma, showing the contemporary pattern of lands and seas. AA, Arctic Alaska; FJL, Franz Josef Land; Kaz., former Kazakhstania; K-O, Kolyma–Omolon; M-O, Mongol–Okhotsk Ocean; NSI, New Siberian Islands; NZ, Novaya Zemlya; ST, Siberian Traps; SVB, Svalbard.

Siberia and Adjacent Parts of Pangea. There are many high-quality palaeomagnetic data (Fig. 9.11) from rocks at around 250 Ma, such as those from the Siberian Trap basalt flows. Siberia came near to being an integral part of the supercontinent of Pangea as the Permian progressed, as demonstrated by maps of the region at about the Carboniferous–Permian boundary time at 300 Ma (Fig. 9.9) and at the end of the Permian at about 250 Ma (Fig. 10.9), but Siberia was not completely united with Pangea until the Early Triassic. The many peri-Siberian terranes had completed their accretion to the main Siberian Craton before the start of the Permian (Cocks & Torsvik, 2007). Although there were significant shear zones, including the Gornostaev Shear Zone at today's south-western margin of the Siberian continent and the Irtysh Shear Zone on the southern margin (Fig. 7.7), because of the thickness of the crust there it is notable how undeformed most of the old Siberian Craton has remained: in many places Neoproterozoic (Riphean and

Vendian) and Lower Palaeozoic rocks are still as flat-lying as when they were originally deposited on that craton.

The Tunguska Basin, which stretches across a large part of the west of the Siberian Craton, developed during the Permian, and it was there that the enormous flood basalts of the Siberian Traps were erupted and deposited immediately before the end of the period at 251 Ma (Bowring et al., 1998). As with other large igneous provinces, the eruptions were no doubt triggered by a plume from the deep mantle. The distinctive analytical fingerprint of the Siberian Traps basalts indicates that the traps appear to have extended northwards as far as the New Siberian Islands in the Arctic Ocean (Fig. 10.9).

Continents and terranes also shown on Figs. 9.9, 10.2, and 10.9 which were separate from Siberia at the start of the Permian include the united Kazakhstania and Laurussia, Kara, North China (which had enlarged to include Tarim and various formerly independent terranes now in the Altaid Fold Belt of central Asia), and Amuria. However, some other Early and Late Permian reconstructions of the region as it developed (e.g. Li, 2006) show the various blocks between peri-Siberia and North China as far less integrated within the Pangea Supercontinent than in our maps here for those times.

Facies, Floras, and Faunas

Our depictions of the Permian land, shallow- and deeper-water shelves, and oceans (although not the positions or shapes of the underlying continental and terrane margins) have drawn heavily on those of A.M. Ziegler et al. (1997) and P. Ziegler (1989). There was substantial global climate change at about 280 Ma (Sakmarian time) from 'icehouse' to 'greenhouse' conditions, which involved the general cessation of the extraordinarily long-lived major Permian–Carboniferous glacial period, although there were two subsequent short and local Middle Permian glacial intervals in Antarctica, the latest at 265 Ma (Guadalupian). Subsequently the planet enjoyed or endured above-average temperatures until the end of the Permian, when there were the massive biological extinctions described below.

Facies. What eventually became the New Red Sandstone Continent at the centre of Pangea had originally started its formation during the Devonian in Laurussia, and its successor continued as a very substantial landmass beyond the end of the Palaeozoic time. That continent was already large at the end of the Carboniferous, stretching from the Arctic Islands of Canada southwards into South America and Africa, and it expanded even further during the Permian, mirrored by a decrease in the extent of the epeiric seas which had hitherto covered so much of the former Laurentian Craton. But it was mostly during the Late Permian and Triassic that very extensive desert dune sandstones were deposited, and the ferric iron oxide included within them forms the brightly coloured and distinctive group of facies termed the New Red Sandstone over much of modern Europe and North America (see below under the various regions).

Permian sedimentary facies were perhaps more diverse than in any other period. In addition to the widespread desert New Red sandstones, there were also substantial Late Permian coals, often interbedded with carbonates which included reefs, in today's Arctic area of North America, as well as in South China (Shao et al., 2003) (Fig. 10.1), and also various extensive carbonate platforms, such as in the Thai sector of Annamia (Ridd et al., 2011). In eastern Brazil, a sequence of sediments, all deposited under arid climates in the intra-cratonic Recôncavo Basin, displays a regression during the Permian from shallow marine to isolated evaporitic, through continental sabkha, to sporadic lacustrine deposits (Silva et al., 2012).

Floral Provinces. As the global icehouse conditions persisted from the Late Carboniferous into the Early Permian, the floral provinces also continued with little change across the system boundary. As in the Late Carboniferous, the Permian plants around the world were differentiated into four groups, of which the best known is the Gondwanan Province (informally termed the *Glossopteris* Province), which was characteristic of the higher southern latitudes in the Gondwanan sector of Pangea (Torsvik & Cocks, 2004). The Angaran Province (including what some have termed the Verkolyman Province) occupied what we recognise as Siberia and peri-Siberia in the northern latitudes, with some elements extending eastwards into Annamia (Fig. 10.4). The more temperate Cathaysian Province floras colonised both North China (Stevens et al., 2011) and South China (Cai & Zhang, 2009), as well as Annamia. The Euramerian Province occupied warmer latitudes on either side of the Equator (DiMichele et al., 2009).

In contrast to modern forests, in which evergreen trees dominate in the Arctic and Subarctic regions, in the Late Permian the high-latitude forests were a mixture of opportunistic deciduous and evergreen trees, since deciduous trees seem to have adapted better in those times to high-disturbance areas caused by alternations of cooler and warmer climates and which were light-limited for much of the year (Gulbranson et al., 2014). Although the *Glossopteris* Flora had become absent from the Paraná Basin of Brazil before the earliest Permian (Holz et al., 2010), it persisted with much diversity throughout the Permian in India, South Africa, Australia, and Antarctica after the glaciers had largely gone at the end of the Early Permian, until the flora eventually became very much affected, and reduced due to the end-Permian extinction. Widespread coals were laid down over much of the Gondwanan sector of Pangea, particularly during the first two-thirds of the Permian (Fig. 10.7).

Siberia was the chief constituent area of the North Temperate climatic zone, in which substantial coals were also deposited, particularly in the Tunguska and Kuznetsk basins. In the Sakmarian (at about 295 Ma), the Angara Province (which was chiefly situated in Siberia) was not so well developed as it subsequently became; for example, nearly 30 Myr later (in the Wordian), the Angaran Province flora, dominated by Cordaite and Sphenophyte genera, not only had the highest number of plant genera on the planet but was also the most clearly differentiated in composition by comparison with the Cathaysian, Gondwanan, and other provincial floras (Rees et al., 2002).

Although there are no glaciogenic deposits known in North and South China since they were largely in the tropics, the area was directly affected by the Permian–Carboniferous glacial episodes; for example, in North China the earlier Permian continental deposits were coal-bearing and fluvial strata rich in plant fossils, in contrast to later fluvial red beds with many calcitic palaeosols. However, after the change from icehouse to greenhouse conditions, a xenophytic and more cosmopolitan floral assemblage started to dilute the identity of the Cathaysian Province (Cope et al., 2005). That province was originally seen chiefly in only North and South China, but it had spread into adjacent areas before the beginning of the Permian, and continued to expand even more widely as the period progressed (Stevens et al., 2011; Cocks & Torsvik, 2013). By the Early Permian at 280 Ma (Fig. 10.3), the Cathaysian Province flora, characterised by plants such as *Cathaysiopteris*, had spread from North and South China to the north-eastern margin of Gondwana, where it occurs in Sibumasu and the Tibetan terrane area; however, a Euramerian Province flora is also known in Sibumasu.

In reality, the floral provinces were not so rigorously delimited in the Permian as are shown on many maps; for example, in today's western sector of North China, many plant localities have yielded a mixture of both Cathaysian and Euramerian genera. In the Khanka–Jiamusu–Bureya (Unit 454) and the Hutag Uul–Songliao (Unit 453) sectors of the Amuria continent, only the higher-latitude Angaran Province floras are found in the Early Permian (Fig. 10.3). However, in contrast, by the end of the Permian at 250 Ma (Fig. 10.4), a mixture of Cathaysian and Angara floras, including the distinctive Cathaysian *Gigantopteris*, are known from Amuria. In addition, there are a few places within South China where both Cathaysian and Euramerian plants occur together.

The floras in Laurussia were distinct from those in Gondwana, China (Cathaysian Province), and Siberia (Angara Province), although there are a few Angaran elements in the Permian of the Cordillera. For example, although an Early Permian flora from the Mystic Subterrane of the Farewell Terrane in Alaska was identified by earlier workers as purely Angaran, further collecting and evaluation of that flora from the same Mount Dall Formation shows it to contain a mixture of both Angaran and Euramerian genera (Blodgett & Stanley, 2008). However, there are no Permian floras known from western Laurussia between the Farewell Terrane and 40° N in the USA today, and thus the boundary between the south-western American and Angaran floras probably reflected, like the preceding floras in the Carboniferous, a palaeotemperature-controlled cline, accompanied perhaps by fluctuations in aridity, rather than an abrupt line demarkating a clear separation between the two provinces.

Benthic Marine Faunas. Permian brachiopods were also provincial (Shi, 2006). The warm-water Boreal Realm faunas, including *Neochonetes*, *Costispinifera*, *Calliprotonia*, *Juresania*, and many others, flourished in the seas of Iran and adjacent Arabia, and that realm extended northwards to the Ural and Russian Platform sectors of Pangea. Those brachiopods contrast sharply with the lower-diversity and colder-water faunas to be found from Karakorum and Afghanistan all the way round to Australia, a contrast attributed to the existence of an oceanic gyre by Angiolini et al. (2007), but which may simply have been due to the different palaeolatitudes (Fig. 10.7).

After the earliest Permian glacial interval was over in the Early Sakmarian, a substantial and steep climatic gradient was developed during the Asselian to Sakmarian along Gondwana's northern margin, which is exemplified by the contrasting brachiopod and fusuline foraminiferal faunas. Five of the ten global brachiopod provinces recognised in the Middle Permian by Shen et al. (2009) in the Asian area are shown on Fig. 10.2, and it can be seen that some primarily resulted from their latitudinal position, some through geographical separation, and others from a combination of those two factors. The 141 known genera of Late Permian (Changhsingian) brachiopods can be divided between five provinces, of which three are found along the Neotethys shelf of Gondwana in a cline from the higher-latitude Austrazean Province in New Zealand (descendants of a province which previously occupied a considerable area of Gondwana in the Early Permian), through the temperate Himalayan Province, to the Western Tethyan Province in the Middle

East, which was at an equatorial latitude. However, by the Late Permian, the Neotethys had widened enough to have allowed the separate development of the Cathaysian Province in Sibumasu as well as South China (Fig.10.4), in contrast to the Euramerian floras (see above) found in Pangea on the other side of the ocean.

In most of Siberia, a Middle Permian Verkolyma Province has been identified, based on brachiopods (Shi, 2006), as well as a Panthalassic Province in far-eastern Siberia (the Sikhote–Alin Fold Belt and Ekonay Terrane) and parts of Japan and north-east China. Between those two provinces there was also a transitional Sino-Mongolian–Japanese Province, which occupied much of the eastern Asian area, including the margins of the North China continental area (Manankov et al., 2006). The latter includes brachiopod faunas intermediate in composition between the Verkolyma Province of Siberia and the more traditional definition of the Cathaysian Province which lay over and near the South China continental area. The Panthalassic Province includes a number of Permian faunas which inhabited offshore oceanic environments and which have mostly been found within allochthonous blocks in accretionary prisms of terranes of Jurassic ages, including the Mino Terrane of Japan, the Heilonjiang Terrane of north-east China, and the Sikhote–Alin Terrane of Far East Russia (Unit 437). All those faunas probably inhabited relatively isolated mid-oceanic seamounts or island arcs within the Panthalassic Ocean during the Permian.

In addition to those Asian faunas, there were many other rich and diverse Permian brachiopod faunas elsewhere, some mentioned in the geographical sections below, as well as others which were much less diverse, such as the hypersaline faunas found in the Magnesian Limestone of northern England.

Terrestrial Faunas. Most of the sites at which Carboniferous and Early Permian vertebrates have been found are in the northern hemisphere today, underlining how inhospitable the climate of most of the southern hemisphere (largely Gondwana) must have been there during the glacial period. However, after the Middle Permian amelioration, diverse Late Permian animal fossils are known from South Africa and elsewhere in Gondwana. The evolutionary division of primitive tetrapods into the two groups anapsids (various extinct groups as well as modern turtles) and diapsids (most of the reptiles, including dinosaurs, and birds, and, later, from the Triassic onwards, the mammals) had occurred earlier, but it was in the Permian that there was more extensive radiation (Benton, 2005). Most of the mammals were small, although some herbivores were much larger, such as the pareiasaur *Scutosaurus*, which was a formidable hippo-sized animal covered with bony excrescences (Benton, 2008). Some families ranged widely; for example, dicynodont amphibian fossils have been found in the Late Permian in Annamia, which are unlikely to have occurred unless there was some land connection with the mainland of Pangea at that time, but which sector of Pangea is uncertain (Metcalfe, 2006).

Radiation in other terrestrial animals, mostly arthropods, which progressively became much more diverse, also reflected the ambient temperatures, and thus increased greatly during the Middle and Early Late Permian.

Northern Pangea. Although some global provinciality existed in the Carboniferous and Permian, for example in the brachiopod faunas noted above, most of the former Laurussia then formed part of the extensive Tethyan Realm. For example, over most of northern and central Europe, the Lower Permian strata are non-marine red beds (the New Red Sandstone), which were partly due to the low sea levels caused by the glaciation elsewhere. In contrast, the Upper Permian (Zechstein) deposits are chiefly evaporites, with only a few marine incursions, in which the faunas, such as in the Magnesian Limestone of north-east England noted above, represent ecologically stressed animal communities which have relatively few species and genera. Eustatic variations led to periodic exposure of the Zechstein carbonate platforms and there was much contemporary diagenesis due to sporadic heavy monsoon rains.

The southern part of the western Laurussian sector of the Pangea Craton (Fig. 10.5) had shallow-water marine seas which have yielded an enormous number of well-preserved fossils, particularly the many and diverse silicified brachiopod faunas found in the Glass Mountains of Texas. During the Early to Middle Permian there were also many substantial reefs, such as the famous El Capitan reef in Texas.

There were extensive Early Permian (Artinskian) bryozoan mounds in the more temperate Canadian Arctic islands. A coral belt, typified by *Thysanophyllum*, stretched all the way from northern Greenland, through the Canadian Arctic Islands, and down the western edge of the craton into South America (Stevens & Stone, 2007). Early Permian benthic foraminifera termed the McCloud Belt, originally identified from a distinctive assemblage in the McCloud Formation of California in the Eastern Klamath Terrane (Unit 161), are useful in confirming the integrity of the terranes in that area (Nokleberg et al., 2000; Colpron & Nelson, 2009). Brachiopods, fusulinid foraminifera, and corals are abundant in the Early Permian of western Laurussia, and indicate that the faunas of the Eastern Klamath, Stikinia, and Quesnellia terranes in the Cordillera were by then similar to each other (Belasky et al., 2002). Conodonts were also widespread (Mei & Henderson, 2001), and we show the distribution of

Sweetognathus in the Artinskian (Fig. 10.5), indicating that, despite the contemporary glaciation at higher latitudes, the climate across the Laurentian sector of Pangea was probably not sharply differentiated enough to have formed many latitudinally distinctive belts within the oceans, and that the equatorial regions were at least as warm as those of the present day.

Siberian and East Asian Sectors of Pangea. Siberia then included today's American Plate areas of Kolyma and Omolon and parts of eastern peri-Siberia as well as the ancient Siberian Craton itself. Because it was still the only substantial continental area extending to the higher northern latitudes (Fig. 10.1), various faunal and floral provinces have been recognised there (see above), although at the margins there were clines between the provinces since the previously independent Kazakh terranes had become attached to Siberia before the Permian and the eastern parts of Laurussia were also nearby (Figs. 10.2 and 10.9). Most of the central parts of the region were continental during the Permian. Within Kazakhstania, in the western part of Kokchetav–Ishim (Unit 458), Middle Carboniferous to Middle Permian basins include substantial non-marine rocks and evaporites which unconformably overlie the Lower Palaeozoic (Windley et al., 2007).

Southern Pangea. In the Paraná Basin of Brazil there are various rocks which are largely fluvial in origin, with some glaciogenic deposits in the Early Permian (Early Sakmarian to Middle Artinskian), and with a single marine incursion in the Paraguaçu Member of the Rio Bonito Formation. Those are followed by sediments representing deposition within shallow marine shoreface embayments, which are followed in turn by Late Artinskian to Kungurian sediments of largely terrestrial origin which have yielded many identifiable plants (as well as coals), fossil reptiles, and fish in sediments which include evaporites, in addition to a few thin layers possibly representing marine incursions (Holz et al., 2010).

In Africa, the non-marine Karoo Supergroup, which was also deposited within large intra-continental lakes, was host to a wide variety of tetrapod vertebrates, together with interbedded volcanic ashes from which Late Permian absolute ages have been measured.

End Permian Extinctions. At the end of the Permian, and largely coincident with the Permian–Triassic boundary, there occurred the most substantial biological extinction event in the whole geological record, when approximately 75% of the animal and plant species perished (Wignall, 2007). In the shallow-water marine realm, the larger foraminifera, larger corals, trilobites, and most of the brachiopods all disappeared, many of which had hitherto been very significant as palaeogeographical indicators. The effects were worse on the land, where only one-quarter of the 48 tetrapod families found in the Late Permian (Tartarian) survived the extinction event to radiate again during the Triassic.

The prime cause of the extinctions was the very negative atmospheric and associated temperature effects (Fig. 16.2) caused by the colossal outpourings of the Siberian Traps Large Igneous Province at 251 Ma (Fig. 10.9), but the extinctions were also facilitated by the steadily reducing amount of coastline available for the establishment of the varied biological niches around Pangea and other continents which had been caused by the Pangea amalgamation. More niches would have been necessary to make homes for a larger number of more diverse faunas, particularly of benthic marine invertebrates. In addition, the fact that the Paleotethys (and to some extent the adjacent Neotethys) Ocean had become to some degree enclosed (Figs. 10.1 and 10.2), caused saprotrophic bacteria to increase in steadily reducing depths there, which, together with an increase in salinity, caused widespread anoxia and massive gas eruptions during the Lopingian, which are reflected in the widespread outcrops of black shale seen in the Upper Permian rocks in the areas surrounding both oceans (Şengör & Atayman, 2009).

As a further complicating factor, the overall global climate had steadily deteriorated for some time before the dramatic Permian–Triassic boundary events, as can be seen from the diminishing numbers of bioherms and other carbonate rocks in the Late Permian, in contrast to the Early and Middle Permian, when they were much more abundant and widespread.

The end-Permian extinction event is not seen directly in many places within Pangea, including Great Britain (Fig 10.6), where the area was mostly land and well away from the margin of the Pangea continent. The only substantial Permian deposits in Britain are the largely non-marine desert sand dunes and other rocks of the New Red Sandstone, within which the Permian–Triassic boundary is impossible to trace in detail. In contrast, the Laurentian sector of Pangea was as badly affected as elsewhere by the end-Permian events. However, the hitherto varied and abundant fusuline benthic foraminiferans became extinct in North America after the Capitanian transgression in the Permian at about 260 Ma, earlier than elsewhere in the world. Those Middle Capitanian extinctions have been linked to the eruptions of the Emeishan LIP flood basalts in China (Fig. 10.4) at 253 Ma.

11 Triassic

Distinctively coloured desert sand dunes preserved in the Triassic New Red Sandstone near Birmingham, England. Photo by Robin Cocks,

Recovery of the biota was initially slow after the great Permian–Triassic boundary extinctions, but then gathered pace quickly, with many different groups of animals and plants becoming newly important. At the end of the Triassic there was intruded one of the largest-known large igneous provinces, the Central Atlantic Magmatic Province (CAMP), which assisted the opening of the central part of the Atlantic Ocean, thus splitting the main area of the supercontinent into northern and southern Pangea, with the latter often termed Gondwana again (Fig. 12.1), as in the Palaeozoic. CAMP was also the prime cause of yet another biotic mass extinction.

Frederich von Alberti coined the name Triassic (or Trias), meaning 'threefold', in the 1830s to recognise the major group of rocks which has three divisions in Germany, then termed the Bunter Sandstone, Muschelkalk, and Keuper Marl: names which were used by Adam Sedgwick soon afterwards for rocks in Britain. The base of the Triassic is now defined at the base of the *Hindeodus parvus* conodont Biozone (Induan Stage) in Zhejiang, China, which is dated at

252 Ma. The global distributions of the continents and oceans at the start of the period near the Permian–Triassic boundary (250 Ma) are in the previous chapter (Fig. 10.1c), and those for 230 Ma (Carnian) and 210 Ma (Late Norian) are shown in Fig. 11.1. The Triassic lasted for 52 Myr, from 252 to 201 Ma, and within it are seven stages, the Induan, followed upwards by the Olenekian, Anisian, Ladinian, Carnian, Norian, and Rhaetian, named after various places in Europe and Siberia. The last three are all grouped within the Late Triassic, which occupied nearly two-thirds of Triassic time; indeed, the first two stages, which formally comprise the Early Triassic, are very short, and are together only 4 Myr long.

Tectonics and Igneous Activity

Large Igneous Provinces and Kimberlites. In Late Palaeozoic and Early Mesozoic times, Pangea lay over Tuzo, one of two major LLSVPs near the core–mantle boundary. There are only two known LIPs in the Triassic, one at the Permian–Triassic boundary (the Siberian Traps, Fig. 10.9) and the other the Central Atlantic Magmatic Province (CAMP), which was intruded at around 201 Ma near the end of the period. That LIP extended over a vast area (~10 million km^2; McHone, 2002), and is exposed today in many rocks in widely separated regions in South America, north-eastern Africa, eastern North America, and south-western Europe (Fig 12.1a). The CAMP plume head probably impinged on the lithosphere beneath the southern tip of Florida and had previously been sourced at the Tuzo Plume Generation Zone (Fig. 11.2a), and propagated horizontally along the base of the lithosphere for several thousand kilometres. Few kimberlites are known from the Triassic, but they are found in Canada, Greenland, Africa (Botswana), South America (Brazil), and Siberia. Except for the last, which were possibly sourced from the Perm Anomaly at the core–mantle boundary beneath Siberia, they were also sourced from the margins of Tuzo (Fig. 11.2).

Oceans. As in the Palaeozoic, the Panthalassic Ocean continued to dominate the world's palaeogeography (Fig. 11.1), but the three large tectonic plates within it (Izanagi, Phoenix, and Farallon), as well as a smaller Cache Creek Plate (Shephard et al., 2013), are all synthetic (Fig. 11.1; Appendix 2). Therefore, Triassic plate tectonic and geodynamic modelling is only as variably reliable as for parts of the Palaeozoic, with 'world uncertainty' about 60–70% (Fig. 12.3a).

The Farallon Plate was partly a continental plate, but a presumed ridge-jump in the Early Jurassic (~185 Ma) transferred the Alexander and Wrangellia terranes to the adjacent Cache Creek Plate, and the subsequent subduction of the Cache Creek Plate beneath the north-western margin of North America eventually brought those terranes into collision with that margin during the Early Cretaceous. The Mongol–Okhotsk Ocean, which had stretched in the Late Palaeozoic between Siberia and Amuria (Domeier & Torsvik, 2014), still occupied a discrete ocean plate of its own (Figs. 11.1 and 11.2b,c), but in the Triassic that plate apparently formed an integral sector of the Panthalassic Ocean, and thus a Triassic navigator would not have recognised it as separate.

At the start of the Triassic, the southern margin of the Paleotethys Ocean, to the north of the chain of Cimmerian terranes, was apparently passive, and that ocean was still substantial (Figs. 11.1 and 11.2b,c). In contrast, complex and rapid subduction, whose details are still poorly understood, along the northern boundary of the Paleotethys adjacent to North China and Annamia led to its consistent diminution in size. Although much of the Paleotethys continued to remain open in the sector between Sibumasu and Annamia, the onset of the Cimmerian Orogeny much affected the Paleotethys at its western end (see below), which closed there at the very end of Triassic time as well as along the Inthanon Suture Zone in South-East Asia (Sone & Metcalfe, 2008). However, other parts of the Paleotethys persisted until near the end of the Jurassic (Fig. 12.1), and perhaps also into the Cretaceous.

In contrast to the dwindling Paleotethys, the Neotethys Ocean between the passive margin of northern Gondwana and the south of the Cimmerian terrane chain continued to widen during the Triassic. It was also probably during the Late Triassic that subsidiary rifting enlarged the Neotethys at its western end between the Pontides and Taurides terranes of Turkey (Fig. 11.2c).

Central Pangea. Intense transtensional tectonics occurred in the later Triassic all along the southern margin of Europe, close to the future axis of the Alpine Tethys (Fig. 11.5). That tectonic activity extended westwards all the way across the future Atlantic area, where rift basins in America (the Newark Basins) were formed during the Norian at about 205 Ma. Between the small terranes making up the eastern Mediterranean area, relatively small but deep-water seaways, some probably floored by oceanic crust, have been recognised (Schandelmeier & Reynolds, 1997). However, the main Pangea supercontinent remained united throughout the Triassic (Fig. 11.1).

North Pangea and Rifting. As the Triassic progressed, a very large area of North Pangea was subjected to continuous extensional tectonics. That included most of Europe, from the Polish Trough in the north-east to the Atlas Mountains of Morocco and Algeria in the south-west (more than 2,000

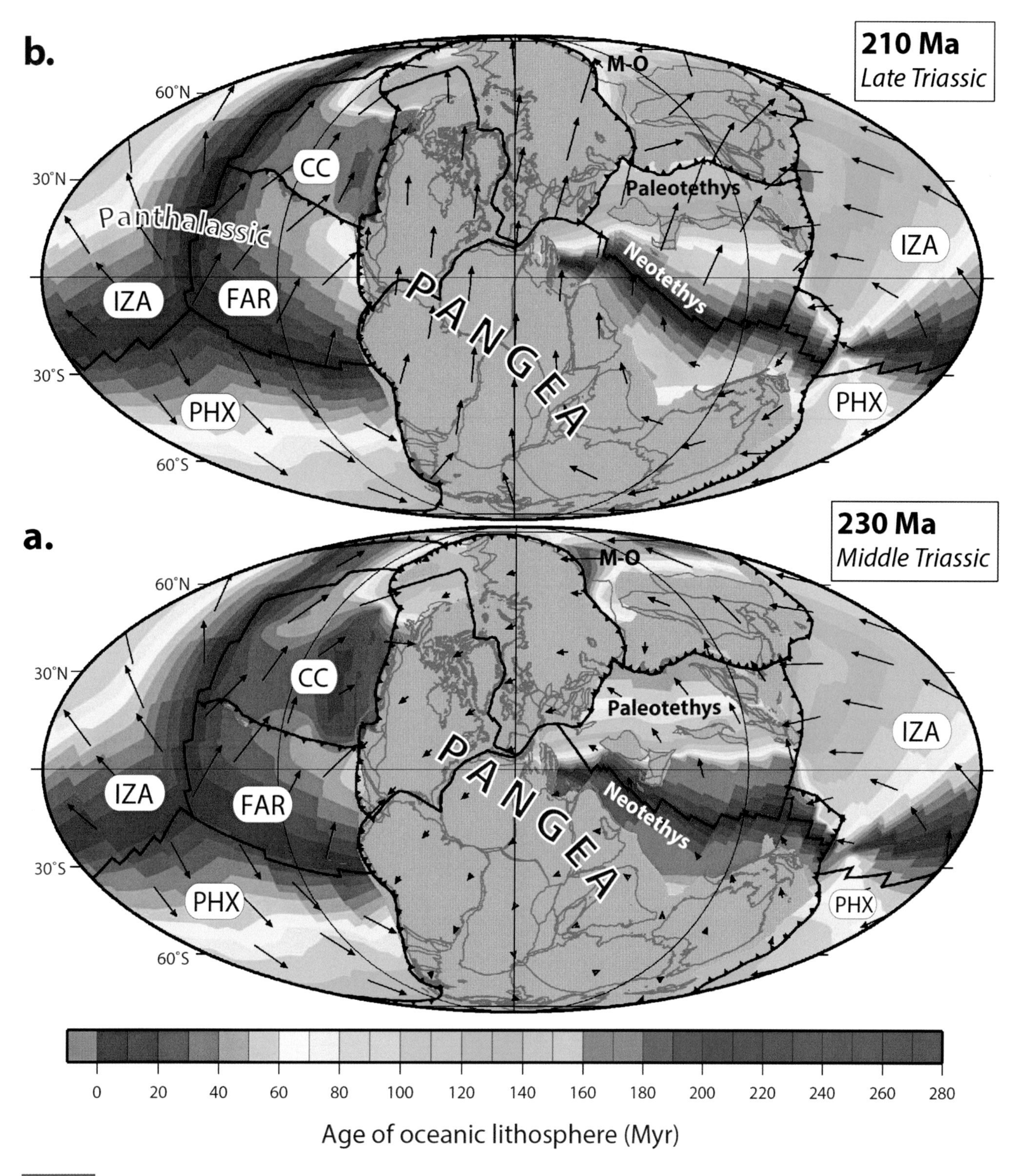

Fig. 11.1 Continental blocks (grey) and oceans (other colours) with plate velocity vectors and the ages of their oceanic lithosphere in the (a) Middle Triassic (230 Ma: Carnian) and (b) Late Triassic (210 Ma: Late Norian). CC, Cache Creek Oceanic Plate; FAR, Farallon Plate (including the continental Alexander and Wrangellia terranes near the plate boundary to CC); IZA, Izanagi Plate; M-O, Mongol–Okhotsk Ocean; PHX, Phoenix Oceanic Plate. EARTHBYTE mantle frame (see Chapter 2).

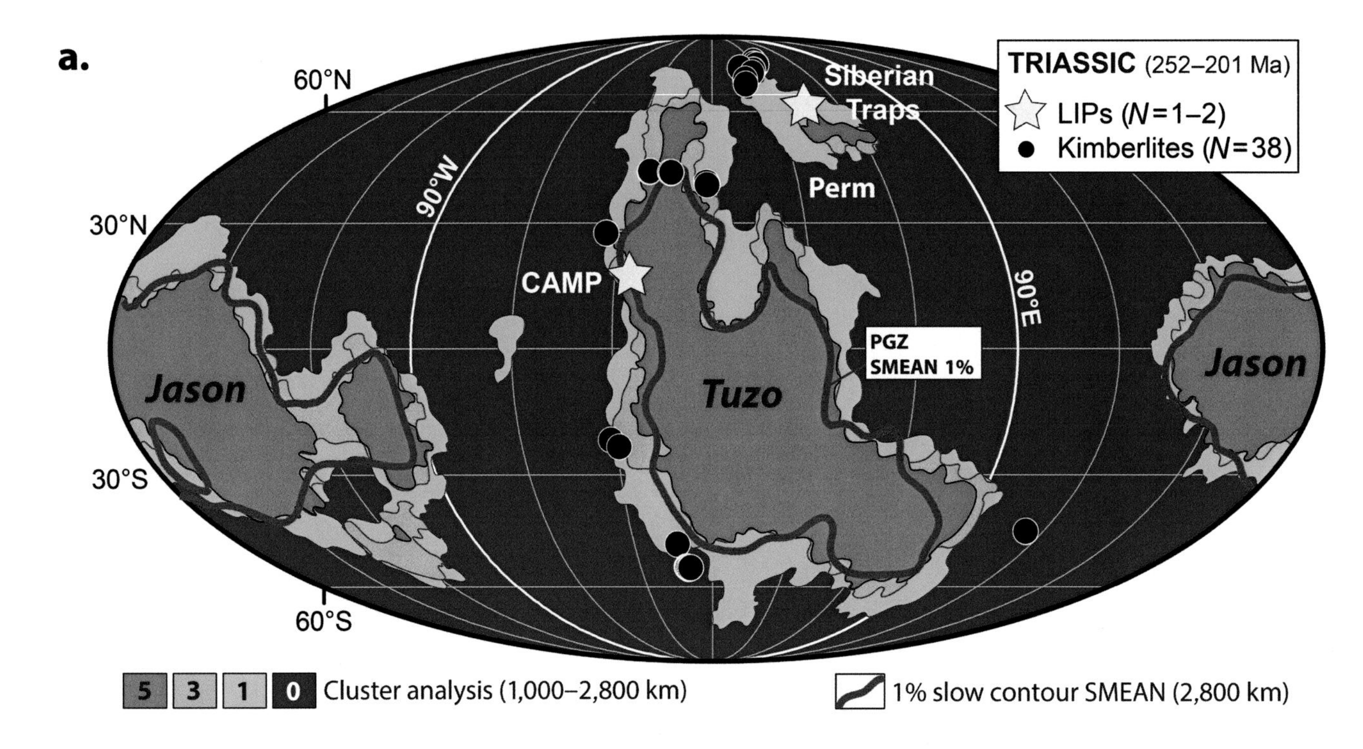

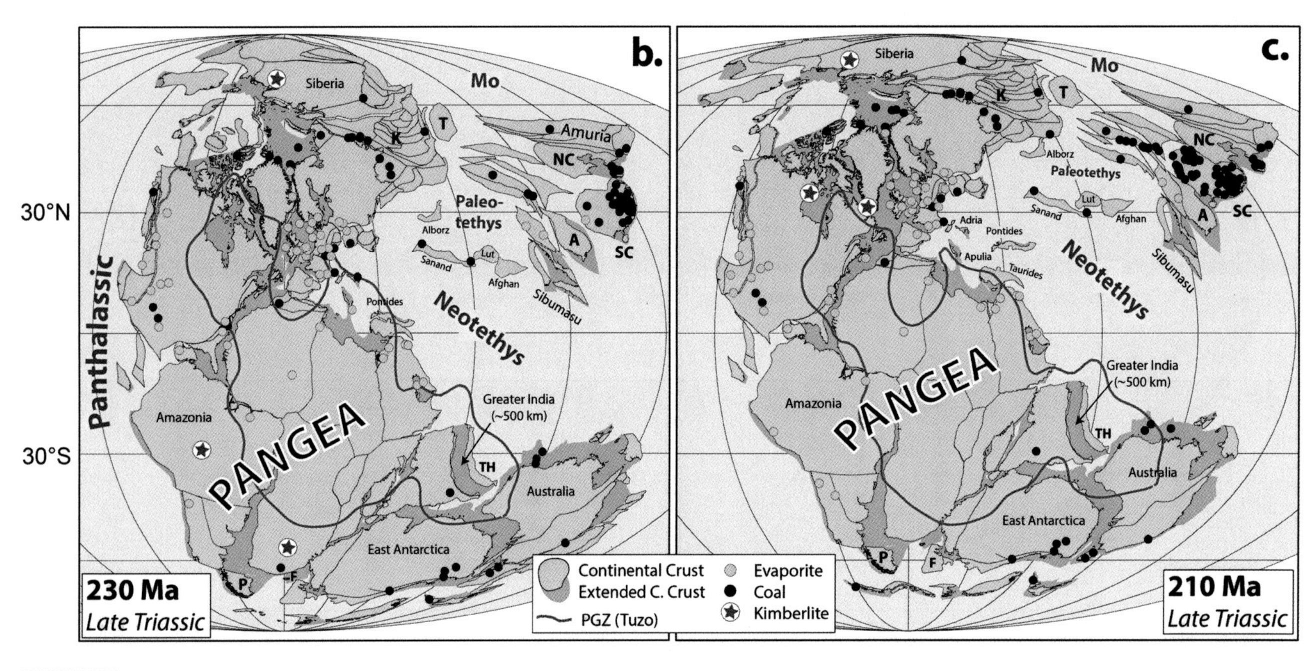

Fig. 11.2 (a) CEED mantle frame reconstruction of the estimated eruption centres (yellow stars) for the Siberian Traps and the Central Atlantic Magmatic Province (CAMP) together with 38 Triassic kimberlites (black dots), and draped on seismic voting-map contours in the lower mantle (Lekic et al., 2012). Areas 5, 3, and 1 define the LLSVPs and 0 (blue) denotes faster regions in the lower mantle. The 1% slow SMEAN contour is shown for comparison (see Chapter 2 for further details). CAMP and most kimberlite locations erupted near or over the Tuzo Plume Generation Zone (PGZ), whilst the Siberian Traps LIP (252 Ma) and Siberian kimberlites (231–215 Ma) appear to be related to the Perm Anomaly. Outline Earth geography (major crustal units) at (b) 230 Ma (Carnian), and (c) 210 Ma (Late Norian). A, Annamia; F, Falkland; K, Kazakh terranes; Mo, Mongol–Okhotsk Ocean; NC, North China; P, Patagonia; SC, South China; T, Tarim; TH, Tethyan Himalaya. CEED palaeomagnetic frame and the location of continents and plate geometries may differ in details from Fig. 11.1 (e.g. in South America and East Asia).

km), and from the Grand Banks of eastern Canada in the north-west and the margin of the Tethys Ocean in the south-east (over 3,500 km). Those extensions were responsible for great fluctuations in the local sea levels and also for the largely north–south-trending graben systems of western and central Europe, which evolved and deepened steadily throughout the Triassic (Ziegler, 1990).

In north-eastern Pangea, the Siberian Traps LIP continued to erupt throughout the Early Triassic and into the Middle Triassic in the Tunguska Basin, with the lavas interbedded with continental volcanic–sedimentary deposits, although only Early Triassic volcanics are known from the Lena–Khatanga basins in the north-east of the Siberian Craton. In Taimyr, sills and dykes are dated to the Carnian at 229–227 Ma (Walderhaug et al., 2005) and are thus considerably younger than the main body of the Siberian Trap flood basalts.

South-western Pangea. In south-western South America, the pre-Andean tectonic cycle reached its climax, with two rift stages, the first peaking in the Late Permian to Late Anisian, and the second in the Norian to Early Jurassic (Sinemurian). The two are separated from each other by a Ladinian to Carnian 'intermediate stage' of silicic–volcanic and volcanoclastic intercalations, which apparently preceded the later rifts in Chile (Fig. 11.6). Between those intercalations, there are thick sedimentary deposits reflecting the subsidence of various basins, with thick breccia deposits at most of their bases representing transgression over different Palaeozoic units. There were Late Triassic (232–220 Ma: Carnian and Norian) granites intruded in Argentina near Buenos Aires and also in Patagonia. There was also tension across a large proportion of the supercontinent; for example, eastward of the pre-Andean margin rifting continued from the Triassic into the Jurassic and that facilitated substantial north-western-trending depocentres covering most of central and southern Argentina. Substantial basalts were extruded within the Triassic of the Parnaíba Basin in north-eastern Brazil; eruptions which continued on into the Early Jurassic (Moullade & Nairn, 1983). The deposits of the later rifting in Chile include marine sediments as well as non-marine sediments reflecting large lakes, some of which extended south-eastwards into Argentina (Moreno & Gibbons, 2007).

Most of the African area was very stable and well above sea level, and was thus dry land with few undoubted Triassic rocks known, although some may represent lakes there (Fig. 11.6). In contrast, the Arabian Plate persisted as a relatively peneplained ENE-sloping passive and partly marine marginal platform in the Early Middle Triassic, as reflected in the facies patterns in the Arabian Gulf region. The sediments there were accentuated by a Ladinian subsidence event, and there was a sea-level low stand in the Early Carnian at about 235 Ma (Schandelmeier & Reynolds, 1997).

South-eastern Pangea. The chief constituents of eastern Gondwana were India, Antarctica, and Australasia, even though that considerable area had been reduced by the departure of the Cimmerian terranes at the north-eastern Gondwana margin in the Permian, but those three modern continents remained united throughout the Triassic (Fig. 11.6). However, as the Triassic progressed, incipient rifting between Madagascar (Indian Plate) and eastern Africa (Somalia) continued, with progressive marine incursions between them from the Neotethys to their north, as well as the elevation of eastern Africa into substantial mountains (Veevers, 2004). Nevertheless, in the same areas which have Lower and Middle Triassic marine sediments, they are followed by Late Triassic continental deposits.

Eastern Asia. Amuria and North China had amalgamated along the Solonker Suture during the Late Palaeozoic (Fig. 10.1c). During the Middle to Late Triassic, that combined continent accreted to Annamia (Indochina), and South China subsequently became attached to that by now substantial Eastern Asian continent in the latest Triassic or earliest Jurassic (Fig. 11.2c). In North China and surrounding areas, Late Triassic rocks unconformably overlie the Early and Middle Triassic with a widespread unconformity which reflects tectonic activity locally termed the Indosinian Orogeny, with nearly all the known Triassic rocks on that continent being non-marine and deposited either in intermontane basins or in lakes (Zhang et al., 2003). However, at the western margin of North China in the South Qilian area (Unit 456), the Early Triassic and Anisian rocks are all of marine origin below the unconformity, and South Qilian might have remained submerged then. In the Tibetan area, the Late Triassic collision of the Qiantang Terrane (Unit 616) with the Tarim–South China continent resulted in a marine regression which continued into the Early Jurassic.

The Cimmerian Orogeny. In the Late Triassic, at about 205 Ma (Norian), there was the onset of the Cimmerian Orogeny, which at its western end was initially caused by the collision of the Iranian blocks with the active margin of the Turan Terrane east of the Black Sea (Berra & Angiolini, 2014), and subduction continued below the Iranian margin into the Jurassic. The consequent tectonic events, which affected the whole Eastern Mediterranean area, as well as much of the Middle East, were extremely complex and are hotly disputed. There was certainly tectonic activity, some violent, in the various units in Greece (including Adria, Rhodope, and Crete), as well as in the Turkish and other Middle Eastern terranes. For example, in the Balkans of south-eastern Europe, east of the Apulia promontory there was a domain of Middle to Late Triassic deformation, accompanied by the deposition of flysch, which lay to the south of an area which escaped tectonism at that time, although it had previously been affected by the Late Palaeozoic Variscan events (Stampfli & Borel, 2004). In the eastern Mediterranean, faulting and rifting took place in the Middle to Late Triassic, and transform faults separated a narrow continental margin in the east from a deeper-water basin to the west. A rift splay branched off to the north-east and created the Palmyra and Sinjar basins within a rift that underwent thermal subsidence during the Middle to Late Triassic and Jurassic (Ziegler, 2001).

It is generally assumed that the same Cimmerian Orogeny extended much further to the east, through Afghanistan to South China, and was the same orogenesis that included the tectonic activity caused by the closure of the Paleotethys at its eastern end between Sibumasu and Annamia, which became united at the very end of Triassic time along the Inthanon Suture Zone (Sone & Metcalfe, 2008). However, since the numerous Triassic rock outcrops in the many separate blocks stretching from southern Europe to the Far East are divided from each other by later rocks, often by many hundreds of kilometres, the identification of only a single Cimmerian orogenic event might be over-simplistic.

Internal Reorganisation of Laurussia. Following the Middle Palaeozoic formation of Laurussia and the later unification of Pangea, there were several episodes of intra-plate deformation (mostly rifting and basin formations) prior to the breakup and sea-floor spreading that took place at different times during Pangea fragmentation. In the Late Palaeozoic, north-east–south-west orientated rift basins (Fig. 11.3), probably guided by the older Caledonian Suture between Baltica and East Svalbard, developed along the west Barents Sea margin (Faleide et al., 2010). In the Late Triassic, Laurentia (including North America, Greenland, Ellesmere, and the High Arctic blocks) started to readjust its positon in relation to Baltica and the Barents Sea region. That was accompanied by sinistral strike-slip faulting between Greenland and west Svalbard in the Barents Sea of 120 km between 220 Ma (Norian) and 190 Ma (Pliensbachian) (Fig. 11.4), but further south between West Greenland and Norway the movements were more oblique (transtension). That Late Triassic reorganisation coincided with rifting and subsequent sea-floor spreading in the Central Atlantic and led to almost orthogonal convergence of about 265 km in the High Arctic. The Lomonosov Ridge as part of the Barents Sea region probably collided with a potential North American continental part of the present day Alpha Ridge (ARC in Figs. 11.3 and 11.4), but it is uncertain whether it was a simple fold and thrust belt (named the Lomonosov Fold Belt by Nikishin et al., 2002), which is the most likely, or was also involved in any subduction.

During the Late Triassic and Early Jurassic, the Barents Sea and North Siberian margins were dominated by uplift, folding, and thrusting that clearly post-dated the deformation of the Late Palaeozoic Uralian Fold Belt caused by the Kazakh terranes colliding with Baltica (Figs. 9.1 and 9.10) and the 251 Ma Siberian LIP intrusion. The Taimyr margin is a fold and thrust belt which must have resulted from the final convergence between Siberia and the Kara Plate (now part of the Barents Sea within Baltica). At the same time, Novaya Zemlya was thrust westward into the East Barents Sea, crustal deformation took place in Pai–Khoi, and gentle Late Triassic and Early Jurassic inversion occurred within the West Siberian Basins (Torsvik & Andersen, 2002). The fold and thrust belt associated with that convergence, the Byrranga Fold Belt (Nikishin et al., 1996), probably developed simultaneously with the Lomonosov Fold Belt as well as strike-slip faulting between Greenland and the Svalbard–West Barents margin. The Siberian Plate moved perhaps 150 km (Buiter & Torsvik, 2007) between 220 and 190 Ma (Fig. 11.3), and Siberia therefore did not become a fully integrated part of Laurussia until the Early Jurassic. After that we refer to Laurussia and Siberia as Laurasia, although large parts of East Asia (Fig. 12.1) did not join Laurasia before the Late Jurassic to Early Cretaceous at about 145 Ma.

Facies, Floras, and Faunas

Triassic Climates and Sediments. The Triassic was a time of great continental emergence due to a combination of widespread orogenesis and low sea levels, particularly at the start of the period (Lucas in Selley et al., 2005). The deposition of marine rocks was chiefly confined to the Tethys Ocean margins and also the two areas on either side of Pangea. The latter two, the circum-Pacific and the circum-Arctic, were both boundaries of the Panthalassic Ocean.

The Triassic climates varied considerably, with consequent latitudinal contractions and expansions of the various temperature zones; in particular, cooler conditions were reflected in the reduced ammonoid diversity of the higher latitudes in the northern hemisphere (the Boreal Realm) and in the southern hemisphere (the Himalayan Province) in the earliest Triassic at the start of the Induan. That was in contrast to the warmest temperatures and the largest numbers of ammonoids near the beginning of the Anisian only 4 Myr later, and there was sporadic fluctuation for the rest of the period (Zakharov et al., 2008). As can be deduced from

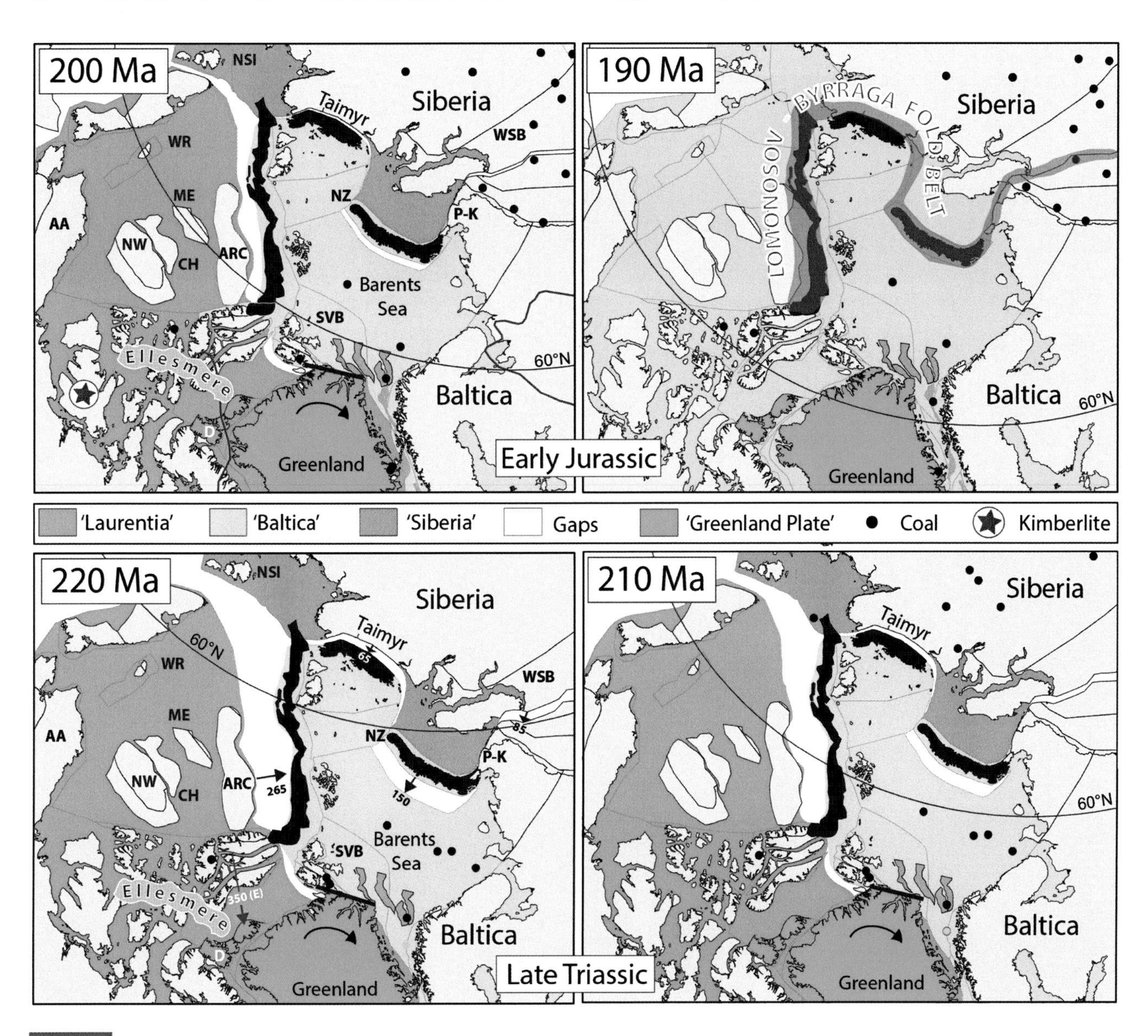

Fig. 11.3 Late Triassic–Early Jurassic development of the Arctic area at 220 Ma (Early Norian), 210 Ma (Late Norian), 200 Ma (Hettangian), and 190 Ma (Pliensbachian). Continental gaps (requiring younger convergence) between Laurentia, Siberia, and Baltica are shown as white regions. Black arrows with numbers denote the amount of convergence required. Note that the northern margin of Ellesmere is kept where it is today, which leaves a gap of about 350 km (blue arrow marked 350) which closed during the Eurekan Orogeny, which was caused by the Greenland Plate (Greenland, south-west Ellesmere, and Disco Island; shaded green) pushing northwards during the opening of the Labrador Sea at around 67 Ma. AA, Arctic Alaska; ARC, Alpha Ridge of potential continental origin (C. Gaina pers. comm., 2014); CH, Chukchi; ME, Mendeleev Ridge; NW, Northwind Ridge; P-K. Pai–Khoi; NSI, New Siberian Islands; NZ, Novaya Zemlya; SVB, Svalbard; WR, Wrangel Island; WSB, West Siberian Basin. The coals shown in the Barents Sea were probably deposited in mangrove swamps and other areas too low to be shown as land.

various factors, including the differing leaf morphologies of the plants found at many horizons in Europe and elsewhere, the Triassic climate varied between arid and semi-arid (Ziegler, 1990) and evaporites are found in both equatorial and subtropical latitudes (Fig. 11.2b,c). The polar forests and widespread reefs at lower latitudes reflect the mild climate of Carnian times (Figs. 11.5 and 11.6).

In the exposed continental land areas, particularly in northern Pangea, there were widespread and substantial desert conditions, which resulted in wadi, dune, and lacustrine deposits known as the New Red Sandstone over much of Europe and North America. Although that New Red sedimentation had started in the Permian, by far the larger bulk of it is of Triassic age. The falling sea levels and

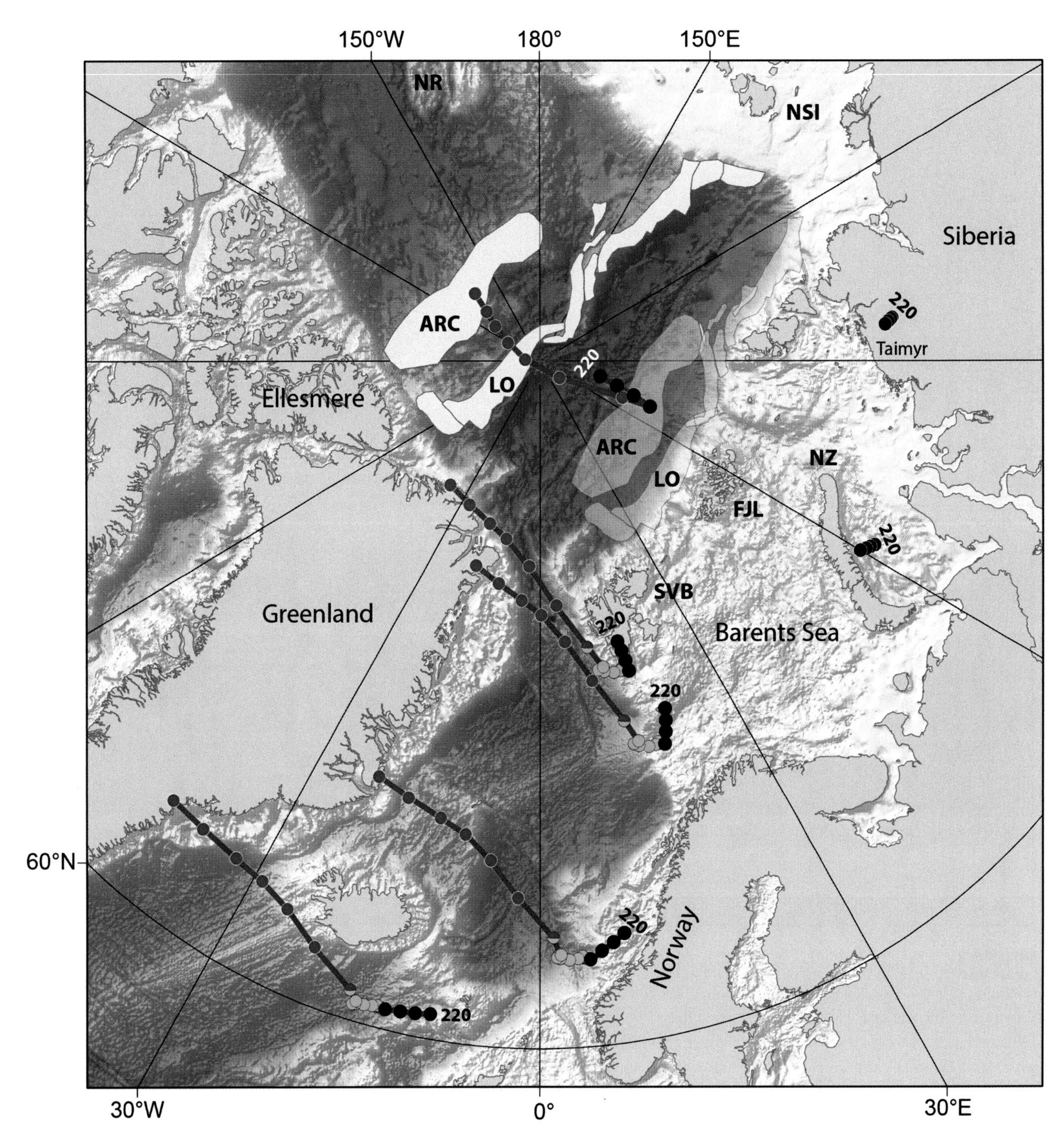

Fig. 11.4 Late Triassic–Early Jurassic plate adjustments in the north-east Atlantic and the Arctic. Dots and connecting lines from 220 Ma to the present (10 Myr intervals) show the trajectories of four distinct locations on Greenland, one in the High Arctic, and two in Siberia (Taimyr and Kara Sea) relative to a fixed Europe/Baltica. The former Laurentia was rotating clockwise in relation to Baltica/Siberia, causing orthogonal compression in the high Arctic, sinistral strike-slip faulting between north-east Greenland and Svalbard, but transtension/ extension further south between Greenland and Norway. At the same time, Siberia was indenting the Barents Sea in the east. Whether that was purely intra-plate deformation (folding and thrusting) within Laurussia (formed in the Silurian) or included subduction is uncertain. ARC, Alpha Ridge of potential continental origin (C. Gaina pers. comm., 2014); FJL, Franz Josef Land; LO, Lomonosov Rise; NR, Northwind Ridge; NSI, New Siberian Islands; NZ, Novaya Zemlya; SVB, Svalbard.

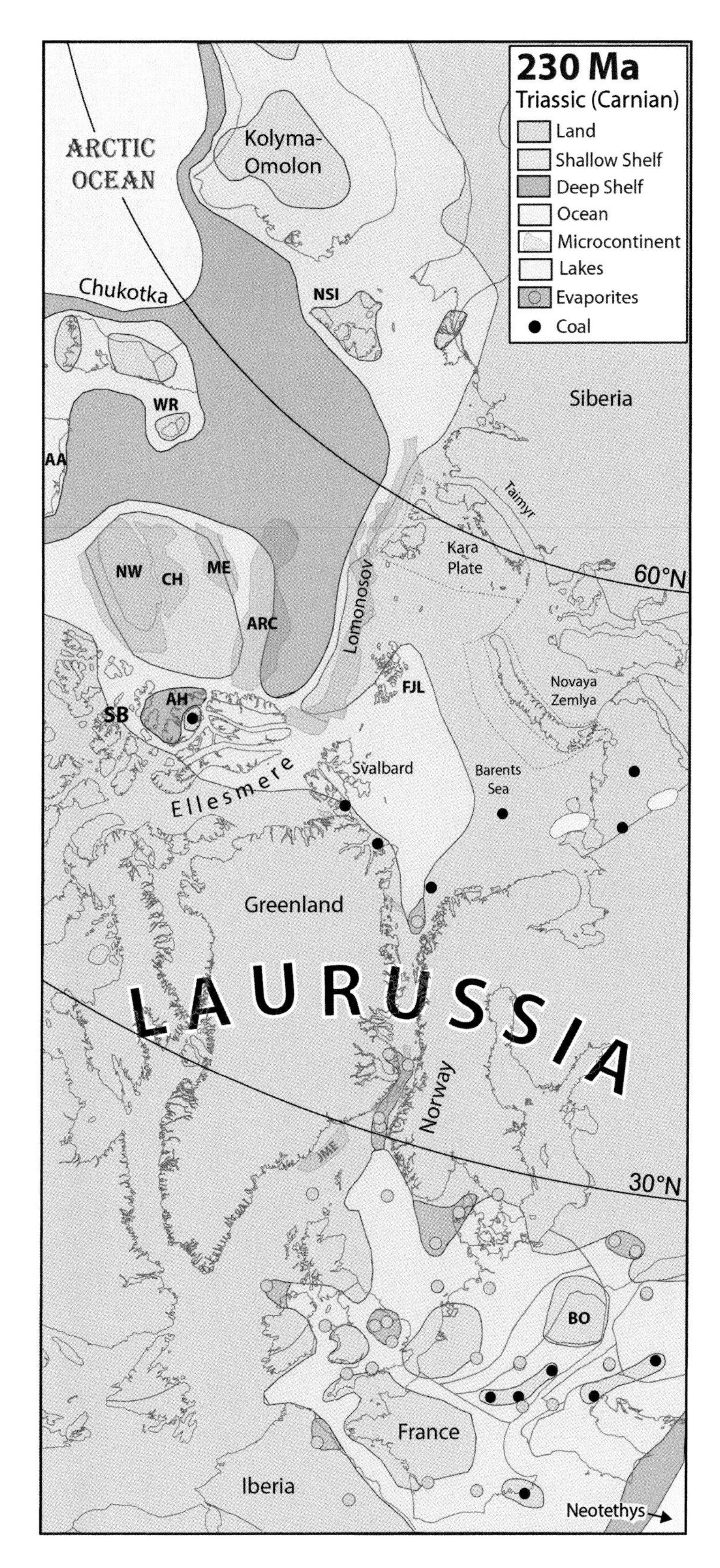

Fig. 11.5 Palaeogeography of parts of the Laurussian and Siberian sectors of central and north-western Pangea from the Arctic to the Neotethys oceans in the Early Late Triassic at about 230 Ma (Carnian), when there were substantial evaporites being deposited in much of Europe at the margins of the Zechstein Sea and its islands. AA, Arctic Alaska Terrane; AH, Axel Heiberg Islands; ARC, Alpha Ridge of potential continental origin; BO, Bohemian Massif; CH, Chukchi; FJL, Franz Josef Land; JME, Jan Mayen; ME, Mendeleev Ridge; NSI, New Siberian Islands: NW, Northwind Ridge; SB, Sverdrup Basin; WR, Wrangel Island. Facies taken from many sources, including Ziegler (1990).

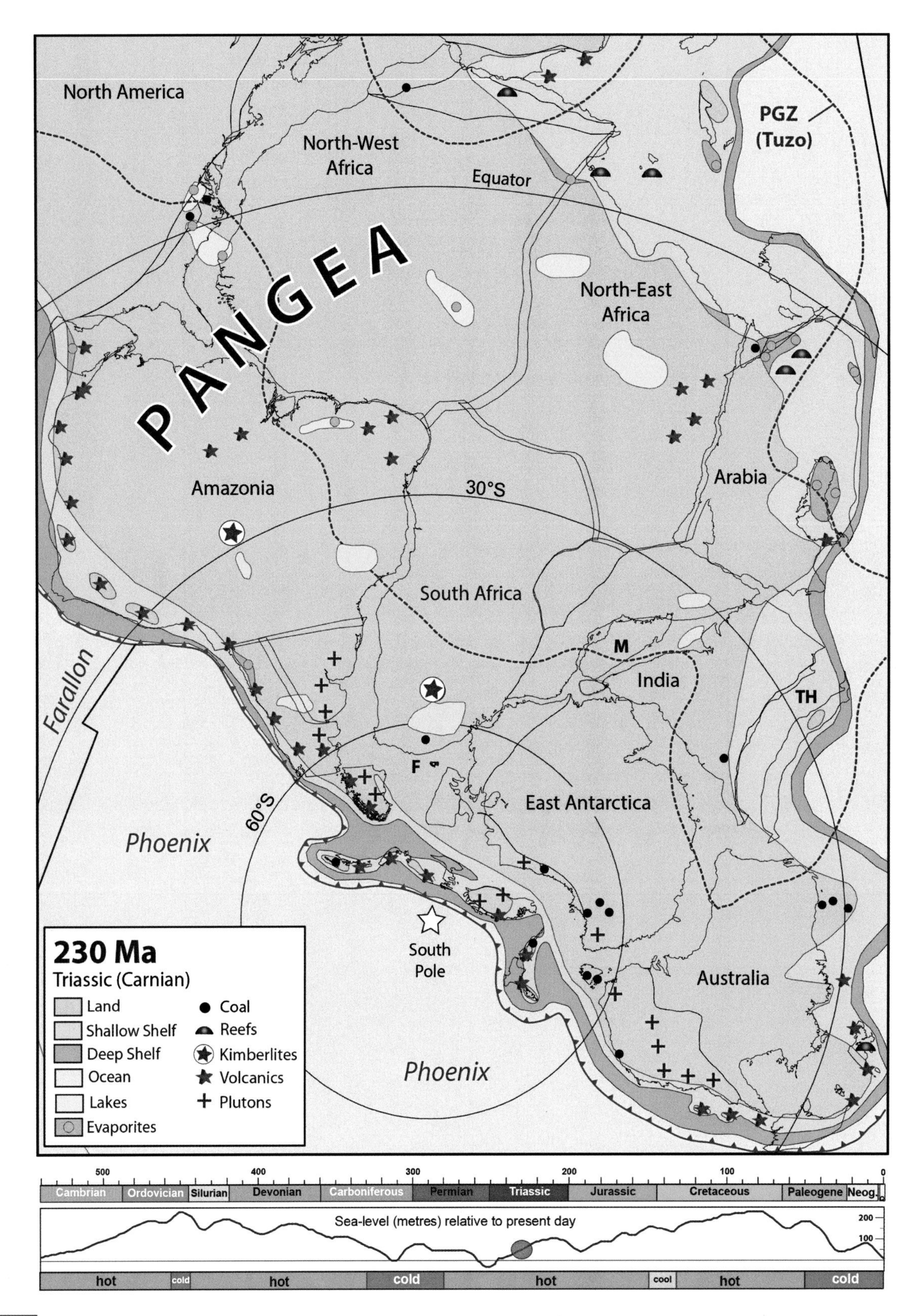

Fig. 11.6 Palaeogeography of the Gondwana sector of Pangea in the Early Late Triassic at about 230 Ma (Carnian). F, Falkland Islands; M, Madagascar; TH, Tethyan Himalaya. Bottom: Phanerozoic time scale and sea-level variations (Haq & Al-Qahtani, 2005: Haq & Shutter, 2008) and icehouse (cold) and greenhouse (hot) conditions.

contemporary orogeny and uplift caused rejuvenation of the Palaeozoic massifs as land areas, and in the Early Triassic (Scythian and Anisian) there was widespread generation of fluvial conglomerates collectively termed the Bunter Pebble Beds in northern Europe.

In the later Triassic (the Norian), sea level rose and flooded many parts of the cratons, particularly in northern Europe where much of it is termed the Zechstein Sea, which at times extended as far as Greenland and Svalbard. However, the transgressions and sea-level fluctuations were variable, and perhaps cyclical, and arid climates probably favoured the widespread dolomites and other evaporite rocks found over most of the Carnian and Norian carbonate platforms (Fig. 11.5). Those carbonates and evaporites were so thick and extensive that subsequent Tertiary tectonics caused many of them to mobilise into the substantial salt domes which concentrated many Jurassic source rocks into the hydrocarbon reservoirs of the North Sea and adjacent areas today.

Recovery from the End-Permian Extinctions. The effects of the end-Permian extinctions had been dramatic, with approximately 75% of species gone across all the different animal and plant groups. Among the marine benthos, hitherto dominant higher taxa, including fusulinid foraminifera, rugose corals, and trilobites, had vanished, and the total number of marine species known from the Early Triassic is less than one-third that in the Late Permian (Benton, 2008). However, that huge reduction gave the surviving stocks more living space and thus new opportunities for geographical expansion, which led in turn to progressive and substantial radiation into new taxa within various ecological niches in many different phyla. The molluscs dominated the Early Triassic shallow seas, with bivalves and gastropods benthic on the sea floor, and ammonite and belemnite cephalopods swimming. For example, only two ammonoid genera are known as occurring on both sides of the Permian–Triassic boundary, and one of those, *Ophiceras*, is thought to have diversified and evolved into over a hundred new genera by the end of the Early Triassic.

On the land, the dominant tetrapods changed dramatically from those in the Permian; for example, the dicynodont amphibian *Lystriosaurus* was a survivor which dominated the few known sites with Early Triassic faunas, and subsequently radiated quickly during the Middle and Late Triassic (Benton, 2005). However, one effect of the catastrophe was to reduce the amount of provincialisation, which is why no faunas are shown on our Triassic palaeogeographical maps. Another general result was that food chains became shorter and less aggressive: *Lystriosaurus* may have flourished simply because it was a herbivore which then had few predators.

The rates of recovery varied in different regions. For example, the high-diversity and complex latest Permian terrestrial ecosystems in Russia were not replaced until the Early Carnian, some 15 Myr after the start of the Triassic, whilst in South Africa recovery was much faster, with some vertebrate families almost back to their pre-extinction diversities within a million years (Benton, 2008).

The Development of Modern Reef Systems. After the slow regrowth of biohermal systems in the first part of the Triassic, reefs in the Late Triassic of northern Africa and north-western India reflected the geographical separation of the Tethyan Ocean region from the rest of the margins of Pangea, which mainly bordered the Panthalassic Ocean. That differentiation was because the Tethyan Ocean was a warmer-water domain driven by western boundary currents, whilst the Panthalassic reefs lived in transitional environments from a nutrient-rich cool-water realm in the temperate zone to a warm-water realm near the Equator, all driven by eastern boundary currents. The carbonates on the North American eastern margin of the Panthalassic reflected a latitudinal gradient from heterozoan (microbial and bivalve) patch-reef bioherms at higher latitudes to photozoan (coral-sponge) reefs at lower latitudes, which were progressively more substantial in size the closer they were to the Equator. Where it was cooler, there were only a few different, but nevertheless framework-building corals, and the microbes took over as the main bioherm builders, although bryozoans also played a part (Martindale et al., 2015).

The Origin of Mammals. Mammals evolved from reptiles in the Triassic, but were rare then, and most remained relatively small until the Cretaceous. There are many differences in both skeletons and soft parts between most reptiles and most mammals, and, although all mammals are warm-blooded, so were some reptiles. Thus it is difficult to differentiate between the two groups when fossil; although the possession of specialised milk-producing mammary glands (hence the name 'mammals') is unknown in reptiles, which makes a clear difference today. Mammals had been preceded by mammal-like reptiles such as the therapsids, some of which reached the size of bears and which were widespread and diverse in the Permian, but, like many other vertebrate groups, therapsids were much affected by the end-Permian extinctions, although they were not completely exterminated then. The earliest animals generally accepted as true mammals are known only from a few isolated fossil teeth found in the Middle Triassic of Britain and Germany. However, in the Late Triassic there were at least three well-defined genera, of which *Megazostrodon* is known from a complete skeleton from Lesotho in southern Africa, and related genera are also known from China and from Triassic fissure infillings in the Carboniferous Limestone of Britain. *Megazostrodon* was very shrew-like in size, proportions, and probably also its mode of life (Savage & Long, 1986).

Northern Pangea. Most of the Triassic rocks of western and central Europe consist of the continental to brackish-marine and desert red beds, pebbles, and playa deposits of the New Red Sandstone (Bunter Sandstone), followed by shallow-marine carbonates (the Muschelkalk) and fine-grained mudstones and evaporites (Keuper Marl) in the Zechstein Sea. These three layers form the 'Trias', from which the Triassic got its name. In addition to the effects of syndepositional tensional tectonics, there were frequent sea-level oscillations and an overall rise in sea level (Ziegler, 1990). That short-term cyclicity may have been caused by the alternate flooding and desiccation of silled rift-induced basins and broad-scale stress-induced low-relief lithospheric deformations. It has been suggested that an additional factor may have been the development of ice sheets at high latitudes in Siberia; however, the evidence for that seems weak. The cyclically rising sea levels caused the submergence of the North Sea–Ringkøbyn–Fyn highs, which led to a much enlarged North-west European Basin by the end of the Early Triassic. In today's higher latitude regions there were extensive shallow seas which combined to form the Arctic Ocean, and into which deltaic systems prograded from the Urals, the Fennoscandian High, the Canada–Greenland Shield area, and the Lomonosov High (Fig. 11.5). In the Sverdrup Basin of northern Canada the Lower Triassic clastic sediments are over 1,500 m thick, and consist of coarse conglomerates of fluvial and littoral origin at the basin margins, but there are finer sandstones and siltstones in the basin centre. There was a major transgression in the earliest Middle Triassic over all of Arctic North America, reaching its maximum in the Norian at about 220 Ma (Moullade & Nairn, 1983).

In north-eastern Pangea, in the Anabar and Olenek river basins and on the left bank of the Lena River of Siberia, there are successions representing the entire Triassic, with Early and Middle Triassic marine rocks including ceratitid ammonoids unconformably overlain by Carnian to Rhaetian non-marine deposits with bivalves and plants.

Southern Pangea. Although most of Africa was land with a few lake deposits, in northern Africa some marine rocks were deposited during transgressions over the craton, particularly in western Libya, where there are Ladinian and Carnian shallow-water marine sediments, including evaporites (Fig. 11.6). In southern Africa, the extensive Beaufort Series was deposited, which consists of non-marine sediments which host the remains of many large reptiles. Today's northern margin of India continued passive, with extensive Triassic sedimentary sequences, including many limestones with varied faunas, known there from Kashmir, Spiti (Tethyan Himalaya), and elsewhere. In Spiti, the earlier rocks are chiefly clastic, whilst in the later Triassic much thicker carbonates, many of them dolomites, abounded. However, although many of the later faunas are quite diverse, bioherms are small and sparse there (Moullade & Nairn, 1983).

There are very few coals known from rocks of Early and Middle Triassic ages. In contrast, by Late Triassic (Carnian) times, extensive coals were laid down in South Africa, India, Australia, New Zealand, and Antarctica (Veevers, 2004), as well as in the northern hemisphere wet belts (Fig. 11.2b,c). It is also noticeable from Fig. 11.6 that the coal swamps flourished right up to polar latitudes in Antarctica and elsewhere, which, together with the flourishing reefs at usually temperate latitudes in North Africa and the Middle East, reflected the mild global temperatures at the time.

Floral Realms. The Late Palaeozoic floral realms and provinces had been much affected by the end-Permian extinctions, but the scale and timing of those changes varied among regions and affected different plant groups (Rees, 2002). In particular the hitherto distinctive *Glossopteris* Flora was severely reduced and only lasted for part of the Triassic before becoming extinct. In contrast, one of the most important gymnosperm groups to re-emerge for the first time since the Carboniferous in the Triassic was the conifers, which evolved into today's six to eight dominant families, as well as the distinct and originally widespread Ceirolepidiaceae, which became extinct at the end of the Cretaceous. Many of the conifers were substantial trees, which came to dominate many forests, particularly in the temperate and higher latitudes. The Angaran and Gondwanan realms can be recognised in the northern and southern highest latitudes respectively, and there was a broad Euramerian Realm between them (Kenrick & Davis, 2004). However, the abundant Late Middle Triassic (about 240 Ma: Ladinian) floras of Argentina and Chile had much in common with those in the northern hemisphere (Rees, 2002).

As the Triassic progressed, the floras of the Euramerian Realm became more obviously divided into regions. For example, by the Late Triassic, a distinctive *Dictyophyllum–Clathropteris* Flora is found in North China, which includes the notable Yangchang Flora of 38 genera and a great many species, and which contrasts with the more temperate *Danaeopsis–Bernoullia* Flora in other parts of China (Zhang et al., 2003) and in the former Kazakh terranes (which by the Triassic were a sector of Pangea welded to Siberia), as well as further to the north (Fig. 11.2b,c).

Extinctions Later in the Triassic. The enormous CAMP intrusion, peaking at 201 Ma, and shown on Fig. 12.1, was the chief cause of another major biological extinction event, although not such a big one as at the end of the Permian (Wignall, 2007). CAMP polluted both sea water and the atmosphere for over a million years, starting with explosive rather than initially effusive lavas in North-West Africa (Dal Corso et al., 2014), which had a deleterious effect on the atmosphere, and thus badly affected many groups of animals (23% of marine and 22% of terrestrial families became extinct: Benton, 1995) and many plants. Far from the CAMP eruption sites, the floras of East Greenland, for example, show several different extinction events, some probably augmented by the wetter conditions resulting from the widespread eruptions (Mander et al., 2013).

12 Jurassic

Sauropod dinosaurs. The sauropods were large herbivorous dinosaurs which appeared in the Late Triassic, but were common and diverse by the Jurassic and continued up to the end-Cretaceous mass extinction of all the dinosaurs. The group includes the largest land animals to have ever lived, with some sauropods reaching over 40 metres in length, and weighing up to 100 tonnes. They lived in herds, and, as well as being land dwellers, some spent much of their time wallowing in water. Credit: Christian Darkin/Science Source.

The Jurassic is famous (not least through the film *Jurassic Park*) for the many exceptionally large and diverse dinosaur reptiles on the land, and ichthyosaurs and other swimming reptiles in the seas, as well as for the emergence of the earliest bird. During the Early Jurassic, Pangea breakup saw the opening of the Central Atlantic Ocean, and subsequently the opening of the Somali and Mozambique Basins which divided the former Gondwana into two major blocks. However, despite the Central Atlantic Ocean opening, all of the former northern Pangea from western North America to Siberia (termed Laurasia) remained united throughout the Jurassic.

The name Jurassic (or Jurassique as it was originally known) was coined by the Frenchman Alexander Brongniart in the 1820s to mark the rocks in the Jura Mountains at the border between France and Switzerland. Its base is now defined at the base of the *Psiloceras planorbis* ammonite Biozone, the earliest zone of the Hettangian Stage, in Somerset, England, which is dated at 201 Ma. The global distributions of the continents and oceans at 200 Ma (Hettangian), 180 Ma (Toarcian), and 160 Ma (Oxfordian) are shown in Fig. 12.1. The Jurassic period lasted for 56 Myr, from 201 to 145 Ma, and there are eleven formal divisions within it: in ascending order, the Hettangian, Sinemurian, Pliensbachian, Toarcian, Aalenian, Bajocian, Bathonian, Callovian, Oxfordian, Kimmeridgian, and Tithonian stages, all named after north-western European localities, many of which are very well exposed along the 'Jurassic Coast' World Heritage Site in Dorset, southern England.

Tectonics and Igneous Activity

Large Igneous Provinces and Kimberlites. It is notable that of the four LIPs of Jurassic age (Fig. 12.2a; Appendix 1) only one was intruded into continental crust, the Karroo–Ferrar LIP of southern Africa and East Antarctica at 183 Ma; the other three (Argo Margin, 155 Ma; Shatsky Rise, 147 Ma; Magellan Rise, 145 Ma) were all extruded over ocean floors in the Late Jurassic, and they are the oldest known *in situ* oceanic LIPs. This leads us to wonder about the total number of older oceanic LIPs which were intruded but which will remain forever unknown. The Jurassic LIPs erupted near vertically over the Tuzo (Karroo and Argo) and Jason (Shatsky and Magellan) plume generation zones (Fig. 12.2a). Kimberlites were more abundant than in the Triassic and are known from Laurasia (North America, Siberia), Australia, Africa (Botswana, Sierra Leone, Swaziland, South Africa), and South America (Brazil). Apart from a few anomalous kimberlites in Siberia (171–158 Ma), they all erupted above the western margin of Tuzo. The CAMP is reconstructed both in Fig. 11.2a and in Fig. 12.2a because its age straddles the Triassic–Jurassic boundary.

The Breakup of Pangea. Parts of Pangea had broken up long before the Palaeozoic had ended; for example, the Neotethys Ocean opened in the Permian (Fig. 10.1c), but most of Pangea remained coherent into the Jurassic. In the Early Jurassic at about 195 Ma (Sinemurian), the Central Atlantic opened between North America and Africa–South America, probably triggered by the CAMP. This opening occurred initially with slow sea-floor spreading rates, accompanied by ongoing rifting in the northern Atlantic and Caribbean areas (Labails et al., 2010; Gaina et al., 2013b). The opening of a Central Atlantic led to a definite break between North and South Pangea (Fig. 12.1), with the result that much of the Palaeozoic Gondwanan continent regained its independence for about 20 Myr before Gondwana split into West and East Gondwana (Fig. 12.1c) at around 170 Ma (Bajocian: Gaina et al., 2013b), some time after the intrusion of the Karoo LIP (183 Ma: Pliensbachian, Fig. 12.1b). It is interesting that fragments of the old Gondwana (e.g. Florida and the eastern sector of Avalonia) remained attached to the Laurussian side of the Atlantic Ocean rather than staying with their original neighbours.

Oceans. At the dawn of the Jurassic, sea-floor spreading in the Panthalassic is modelled as a three-plate system of the Izanagi, Farallon, and Phoenix plates, as well as a continually diminishing Cache Creek Plate (Fig. 12.1a). Along the Farallon–Cache Creek Ridge, a ridge jump transferred the Alexander and Wrangellia terranes to the Cache Creek Plate (Shephard et al., 2013).

The birth of the Pacific Plate at around 190 Ma (Pliensbachian) established a more complex spreading ridge system in the Panthalassic, with multiple triple junctions and spreading centres (Seton et al., 2012). Spreading along those ridges was initially moderate in the Early Jurassic ($<1°$/Myr) but with a later increase in Late Jurassic to Early Cretaceous times (Fig. 12.3c). However, velocity estimates and plate geometries should be treated with great caution, since the Panthalassic plates are largely theoretical inventions for those times (Fig. 12.3a). The Pacific Plate is modelled with a zero velocity from 190 to 140 Ma (Fig. 12.3) followed by a sharp acceleration, which is even more unprecedented for the Phoenix Plate.

Along the north-western Panthalassic Ocean margin, the Izanagi Plate interacted with the Mongol–Okhotsk Ocean Plate (originally a separate Palaeozoic ocean basin between Amuria and Siberia; Fig. 10.1), and the latter continued its progressive diminution by subduction along the margin of southern Siberia (Fig. 12.1), and possibly also along the Amuria-facing margin of East Asia (Van der Voo et al., 2015). The Mongol–Okhotsk Ocean Plate finally disappeared between the Late Jurassic (150 Ma: Tithonian) and Early Cretaceous (140 Ma: Berriasian).

In the Early Jurassic, a small remnant of the Paleotethys Ocean may still have been separated from the widening Neotethys Ocean by the string of Cimmerian terranes (Figs. 12.1 and 12.2b), including Sanand, Lut, Afghan, and Lhasa (Southern Tibet). Closure of the Paleotethys and accretion of the Cimmerian terranes to enlarge Laurasia occurred from the Middle Triassic (Şengör & Natal'in, 1996), but that closure was undoubtedly complex and probably involved younger openings and closures of back-arc basins in the process.

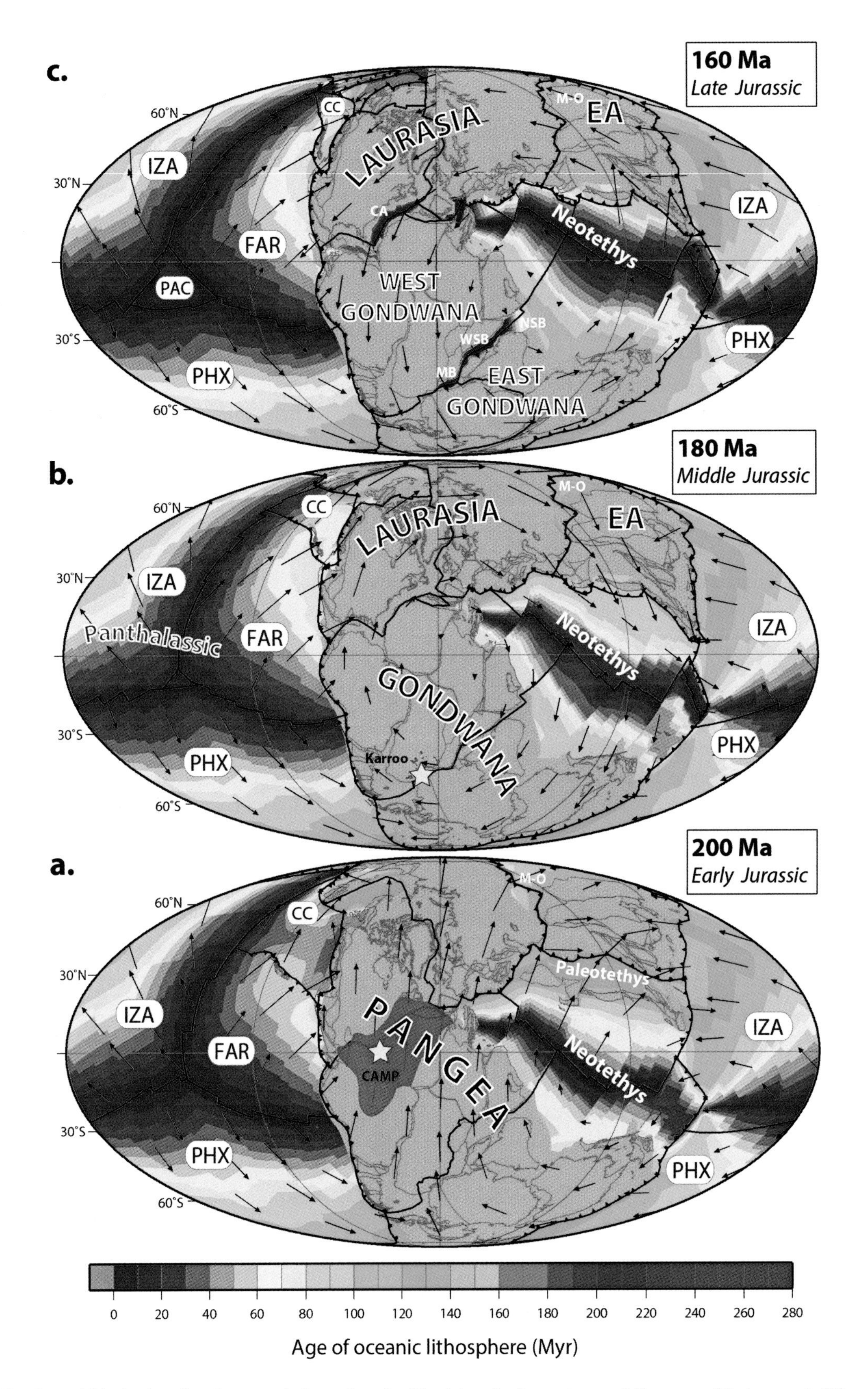

Fig. 12.1 Continental blocks (grey) and oceans (other colours) with plate velocity vectors and the ages of their oceanic lithosphere in the Jurassic: (a) Early (200 Ma: Hettangian); (b) Middle (180 Ma: Toarcian); and (c) Late (160 Ma: Oxfordian). Also shown are the reconstructed locations of the CAMP (Central Atlantic Magmatic Province) and Karroo LIPs (yellow stars are estimated eruption centres). CC, Cache Creek Plate; EA, Eastern Asia (Amuria, the China Bocks and former peri-Gondwanan fragments such as Sibumasu); FAR, Farallon Ocean Plate (including the continental Alexander and Wrangellia terranes near the plate boundary to CC; transferred to CC on the 180 and 160 Ma reconstructions due to ridge jump at around 185 Ma); IZA, Izanagi Ocean Plate; MB, Mozambique Basin; M-O, Mongol–Okhotsk Ocean; NSB, North Somali Basin; PAC, Pacific Ocean Plate; PHX, Phoenix Plate; WSB, West Somali Basin;. EARTHBYTE mantle frame (details in Chapter 2).

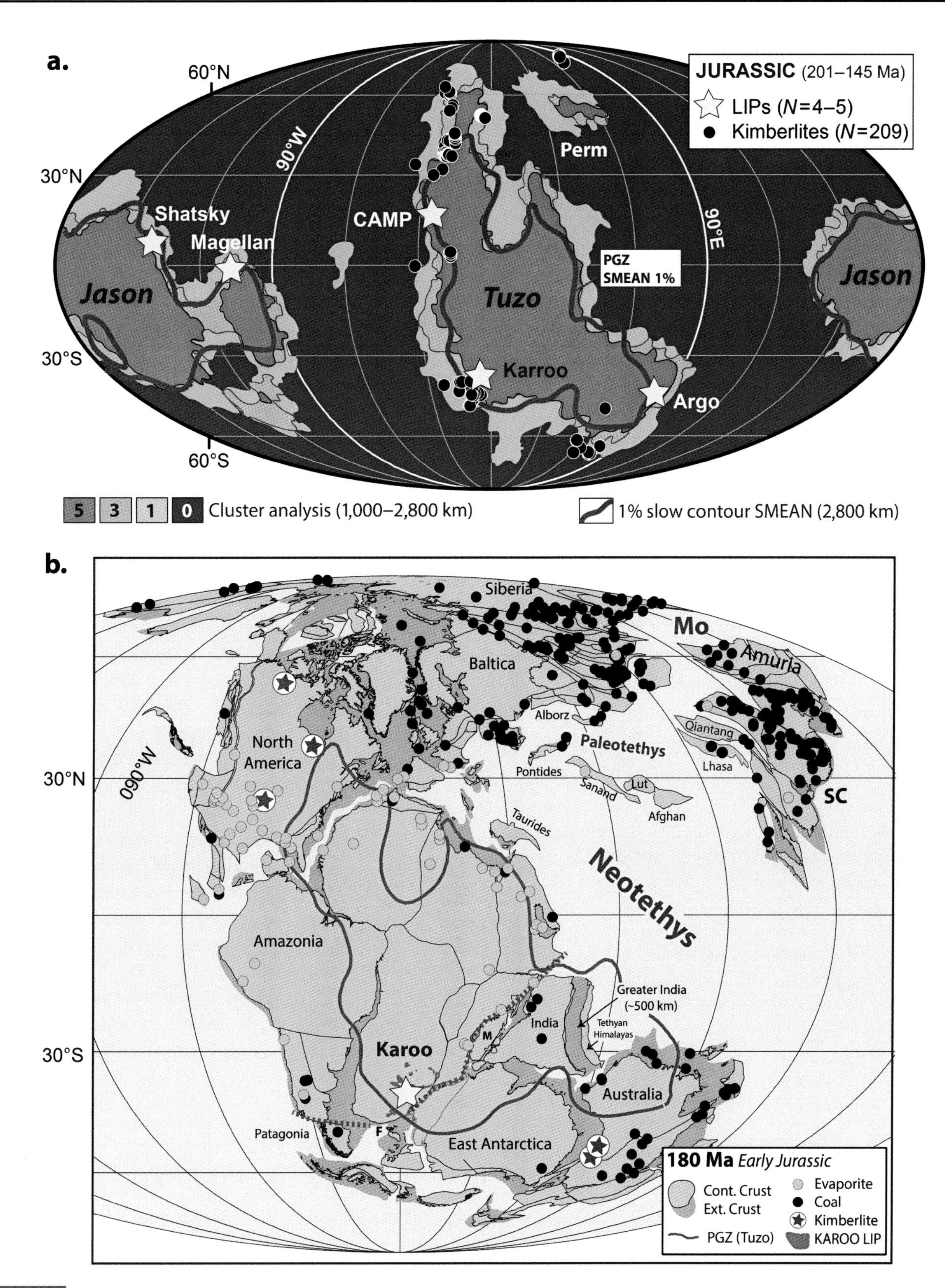

Fig. 12.2 (a) CEED mantle frame reconstruction of the estimated eruption centres (yellow stars) for the Central Atlantic Magmatic Province (CAMP; also shown in Fig. 11.2a), Karroo, Argo Margin, Shatsky, and Magellan Rise LIPs, together with 209 Jurassic kimberlites (black dots). These are draped on seismic voting-map contours in the lower mantle (Lekic et al., 2012). Contours 5, 3, and 1 define the

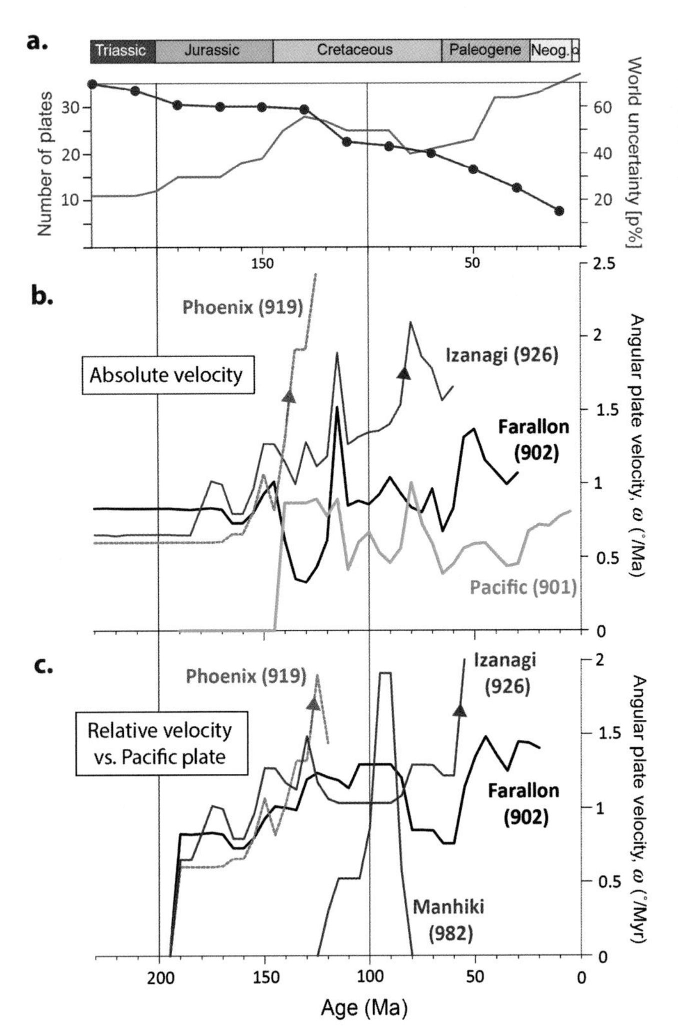

Fig. 12.3 (a) Estimated percentage of the Earth's lithosphere which has been lost through subduction over the past 230 Myr (red dots and line) and the number of plates (dynamic polygons: blue line), increasing from 11 in the Triassic to 37 today. (b) Absolute angular plate velocities for the Pacific, Farallon, Izanagi, and Phoenix plates (1°/Myr is about 10 cm/yr). Numbers in brackets are plate numbers. (c) Relative velocities (spreading rates) of the Phoenix, Manihiki, Farallon, and Izanagi plates versus the Pacific Plate. EARTHBYTE mantle frame (details in Chapter 2).

The opening of the Central Atlantic has often been assigned an age as late as 175 Ma, but sea-floor spreading probably started earlier, at around 195 Ma (Sinemurian), and that was attended by rifting in both the northern Atlantic and the Caribbean (Labails et al., 2010). The Central Atlantic Ridge became progressively connected with the Gulf of Mexico Ridge during the Late Jurassic.

The opening of the Somali and Mozambique Basins (Fig. 12.1c) at around 170 Ma (Bajocian), or slightly earlier (Aalenian) times, steered the breakup of Gondwana into West Gondwana (South America, Africa, and Arabia) and East Gondwana (East Antarctica, Australia, Madagascar, Seychelles, and India) (Gaina et al., 2013b). Early spreading in the Somali and Mozambique Basins probably connected with extensive dextral strike-slip faulting along the Agulhas–Falkland–Fracture Zone to the south-west, and its continuation into mainland South America, the Gastre Fault System (Fig. 12.4b,c).

Northern Pangea and Laurasia. The opening of the southern and central sectors of the Atlantic Ocean separated the northern and southern sectors of Pangea, after which the northern part is termed Laurasia. In north-western Laurasia, obduction of the Cache Creek Terrane, and accretion of the Stikinia Arc occurred along the western Laurentian margin between 175 and 172 Ma (Toarcian and Aalenian).

In north-eastern Siberia, the Verkhoyansk Fold Belt reflects the accretion of the substantial Kolyma–Omolon Microcontinent onto the main Siberian Craton, which acted as an indenter resulting in the Kolyma Orocline. Kolyma–Omolon is made up of ophiolites, olistostromes, and schistose units which were amalgamated during the Middle and Late Jurassic. That was followed near the end of the Jurassic by thrusting and strike-slip faulting and also by collision of the Alaskan and Siberian margins in the earliest Cretaceous, which resulted in further thrust and strike-slip deformation there (Oxman, 2003).

Southern Pangea and Gondwana. The opening of the Central Atlantic Ocean separated the northern and southern sectors of Pangea, and the southern part is usually termed

Fig. 12.2 (*cont.*) LLSVPs and 0 (blue) denotes faster regions in the lower mantle. The 1% slow SMEAN contour is shown for comparison (see Chapter 2 for further details). CAMP, Karroo and Argon LIPs and the majority of kimberlites erupted near or over the Tuzo Plume Generation Zone (PGZ), whilst Shatsky and Magellan are related to the Jason PGZ. (b) Outline Earth geography (major crustal units) at 180 Ma (Toarcian–Oxfordian) and the estimated eruption centre for the Karroo LIP (yellow star). The dotted line is the future breakup zone of Gondwana into West and East Gondwana (Fig. 12.1c), at about 170 Ma. The 'Greater India' area is added as a northern extension of India at this time (van Hinsbergen et al., 2012). North China and Tarim lay between South China and Amuria. CEED palaeomagnetic frame in which the location of continents and plate geometries may differ in detail from Fig. 12.1b. For example, in South America this reconstruction shows an offset of about 500 km between Patagonia and the immediate block to the north. F, Falkland Islands; M, Madagascar; Mo, Mongol–Okhotsk Ocean; PGZ, plume generation zone; M, Madagascar; SC, South China.

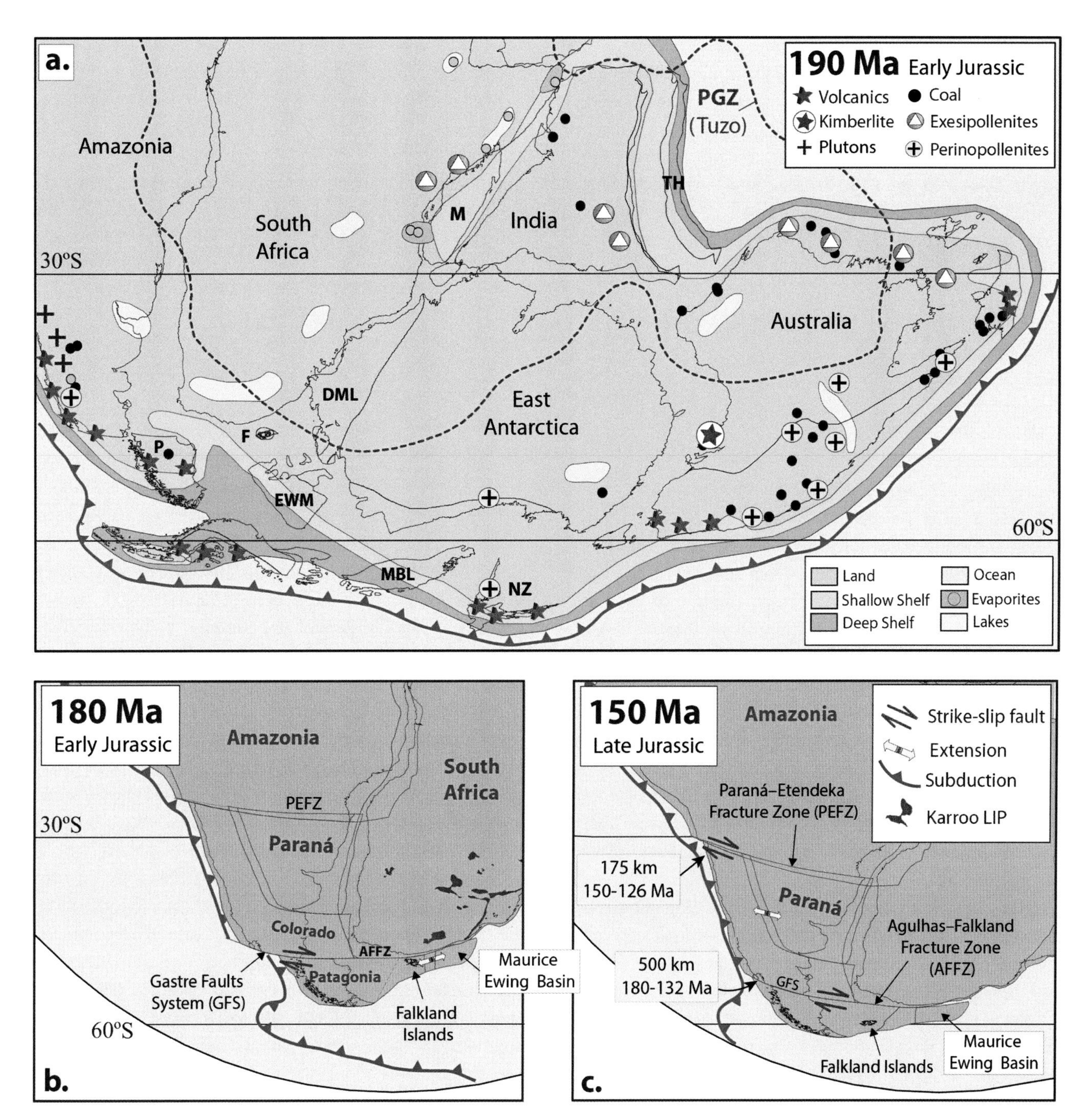

Fig. 12.4 (a) Hettangian to Middle Toarcian palaeogeography of central and south-eastern Gondwana on a 190 Ma reconstruction. The Cheirolepidiacean Phase spores and pollen were the *Exesipollenites* Association on the Panthalassic Ocean margin and the *Perinopollenites* Association on the Tethyan margin. DML, Dronning Maud Land; EWM, Ellsworth–Whitmore Mountains; F, Falkland Islands; M, Madagascar; NZ, New Zealand; P, Patagonia; PGZ, plume generation zone; TH, Tethyan Himalaya. Microfloral sites taken from Grant-Mackie et al. (in Wright et al., 2000). (b, c) South America and South Africa reconstructed at 180 and 150 Ma. In this model (Torsvik et al., 2008a), Patagonia moved with respect to Colorado between 180 (Toarcian) and 160 Ma (Oxfordian) whilst the Falkland Islands were being detached and rotated away from South Africa. Colorado and Patagonia later moved with respect to Paraná until ~132 Ma (Early Cretaceous), and dextral movements between Paraná and Amazonia (PEFZ) ceased at around 126 Ma. After that time, South America became a single rigid plate. Note that the then adjacent Antarctic blocks (Fig. 12.4a) are omitted here for simplicity.

Gondwana, although the constituent parts at its margins are not precisely identical to those of the pre-Permian Gondwanan continent, which nevertheless included most of the same area (Wegener, 1915).

In south-western South America, the pre-Andean tectonic cycle continued from the Triassic into the earliest Jurassic (Fig. 12.4a), but the late Early Jurassic saw the start of the true Andean tectonic cycle at about 180 Ma in the intrusion of a substantial granodiorite into earlier Jurassic basalts in northern Chile, a process which has continued until the present day (Moreno & Gibbons, 2007). Thick latest Triassic to Upper Jurassic volcanic rocks in central-west Argentina and Chile were probably associated with the accretion of an island arc there, after which there were andesite flows intercalated with continental sediments of Kimmeridgian age of about 155 Ma (Moullade & Nairn, 1983).

Rifting between Madagascar–India and East Africa enabled marine transgressions from the Neotethys Ocean (then at Gondwana's north) to reach as far as Madagascar and East Africa (Fig. 12.4a). From about 183 Ma (Early Toarcian), there was the widespread eruption of the Karroo Volcanics and contemporary rifting in south-eastern Africa, which eventually resulted in the separation of Gondwana at around 170 Ma (Bajocian) and the opening of the Somali and Mozambique basins (Figs. 12.1c and 13.1). The Karroo LIP and related dyke swarms in South Africa are also found in the Falklands and East Antarctica/Transantarctic Mountains (locally termed Ferrar), and partly coincided with the more protracted Chon Aike rhyolite volcanism seen in South America and parts of West Antarctica (Torsvik et al., 2008a).

Before the Early Jurassic, the Falklands (Fig. 12.4b) were located near the southern tip of Africa, as indicated by the presence of Karroo dykes and by the correlations between the basement, Palaeozoic stratigraphy, and structural trends (Torsvik et al., 2008a). Between the Early and Late Jurassic, the Falklands rotated and were also displaced about 500 km westward (together with the Patagonia block) along the Agulhas–Falkland Fracture Zone and the Gastre Fault System in South America (compare Fig. 12.4b and c).

Eastern Asia. The continental blocks which today make up eastern Asia were never parts of the Pangea supercontinent, and, as outlined in the previous chapter, many of them, including North and South China, had amalgamated to form a substantial separate continent. As the Jurassic progressed, the Mongol–Okhotsk Ocean closed steadily (Figs. 12.1 and 12.2) and was finally subducted near the Jurassic–Cretaceous boundary time at about 145 Ma (Van der Voo et al., 2015). The southern sector of that closing ocean was chiefly adjacent to Tarim and Amuria (Mongolia), and included the accreted arcs between them, which are identified as the Ala Shan Composite Terrane (Unit 451) in the Palaeozoic.

The northern sector of the Mongol–Okhotsk Ocean closure was chiefly caused by the interaction between the Siberian Craton and the hitherto separate Kolyma–Omolon, Microcontinent. In the Palaeozoic, the Omulevka Microcontinent was a key element in what was to enlarge to become Kolyma–Omolon, which lay not far from the Siberian Craton. During the later Triassic, rifting had developed in the Olmyakon Basin between Siberia and Kolyma–Omolon which evolved into a spreading ridge in the Early Jurassic. That spreading ceased in the Late Jurassic, and the oceanic area progressively closed, eventually with much nappe emplacement. In addition, off the other side of the Kolyma–Omolon Microcontinent to its north-east, the Alazeya Volcanic Arc was active, together with a back-arc basin in which subduction started in the Middle Jurassic at about 165 Ma (Callovian), and in which further nappes were emplaced in the opposite direction to those in the Olmyakon Basin, together with volcanism in the Uyandina–Yasachnyi Volcanic Belt in the same area (Oxman, 2003). That all resulted in very substantial and complex tectonics along the Verkhoyansk–Kolyma Belt, which stretches from the Sea of Okhotsk in the north-western Pacific to the Laptev Sea in the Arctic Ocean, and today includes the boundary within north-east Siberia between the Eurasian and North America plates.

The Cimmerian Orogeny. 'Cimmerian' has been used in two different ways by a variety of authors. It is chiefly used in this book for the elongate string of terranes which lay to the north of the rifting and spreading centre associated with the opening of the Neotethys Ocean in the Permian, an ocean which had originally lain between Gondwana to its south and the Cimmerian Terrane Group to its north. The Cimmerian terranes extend through the Middle and Far East to include the Lut, Karakorum, Afghan, Tibetan, and Sibumasu terranes, and many had Precambrian and Palaeozoic cores. The Neotethys progressively narrowed as the Cimmerian Terrane Group docked obliquely along the southern Laurasian continental margin between the Late Permian and the Jurassic–Cretaceous boundary time, as initially noted in the previous two chapters. However, the name is also used for the Cimmerian Orogeny, which is the collisional deformation seen within the Cimmerian belt (see Chapter 11). That orogenic activity had begun in the Triassic, and continued into the Jurassic, with subduction continuing below the Iranian margin of Gondwana.

Facies, Floras, and Faunas

Sea-Level Changes and Marine Transgressions. The beginning of the Jurassic saw major rises in sea level, with the result that, for example, nearly all of Europe, which had been a large area of land with sporadic deposition of non-marine sediments during the Late Triassic, started to become flooded in the latest Triassic (Rhaetian) at about 202 Ma. Within little more than a million years, during the Early Hettangian, the region had become transformed into an archipelago of islands of various sizes within shelf seas, a situation which continued for most of the Jurassic (Fig. 12.5). However, the surrounding lands were never very mountainous, so that nearly all the European Jurassic sediments are fine-grained. Thus, in the classic area of southern England, the rocks are largely shales, often with a high carbon content, in the Early Jurassic (the Lias), followed upwards by carbonates in the Middle Jurassic (the Inferior Oolite, of largely Bajocian age, which is overlain by the Bathonian Great Oolite). There was a comparable transgression in western and central Siberia, much of which was also flooded.

From the latest Triassic (Rhaetian) onwards, progressive rifting occurred between eastern Africa and Madagascar. As the Jurassic progressed, there was a marine transgression from the beginning of the Toarcian at about 180 Ma which progressively covered the Horn of Africa (Somalia and Eritrea), which had become completely covered by seas by the Kimmeridgian (155 Ma). In Kenya there is a full marine succession from the Toarcian upwards into the Cretaceous. Those seas progressed southwards, also in the Toarcian, as a relatively narrow gulf which reached as far as Madagascar, which is half covered by rocks of marine origin (Fig. 12.4a). However, no Early Jurassic marine rocks are known in either Antarctica or Australia, and the earliest are Bajocian (about 170 Ma) in the Perth Basin (M.K. Howarth in Cocks, 1981), which suggests that both those continents were largely land at the time.

At the Arctic margin of North America, there was a complex history of transgressions and regressions during the Early and Middle Jurassic from the Hettangian to the Early Oxfordian (200 to 160 Ma), mostly caused by tectonic control rather than eustatic sea-level changes. Those fluctuations were followed by a major transgression in the Middle Oxfordian which persisted until the earliest Cretaceous (Moullade & Nairn, 1983).

Oxygen Excursions. Reflected in the sediments, the Jurassic saw substantial fluctuations in global climate. One of the most important was the Toarcian Oceanic Anoxic Event at about 183 Ma, when there was widespread bottom-water anoxia, a negative carbon isotope excursion of 5–7‰, and various biotic extinctions (Jenkyns, 2010). Data from fossil wood indicates that carbon which was unusually isotopically light was present in the atmosphere during the event. Possible causes include outgassing from the Karroo–Ferrar LIP volcanism in Africa and East Antarctica, which might have disrupted ocean-water dynamics, leading to the disassociation of methane hydrates, perhaps aided by astronomical forcing. However, for whatever reason, extensive organic shales with up to 18% organic carbon were deposited not only in Europe (where the event was first noted), but also elsewhere, including Argentina, where the Nequén Basin was an extensional trough with one end linked to the open Panthalassic Ocean, and in which there is a thick Triassic and Jurassic succession with light-carbon-rich black shales in the Toarcian. At higher latitudes there were fewer changes; for example, the spore–pollen assemblages of the Hettangian to Middle Toarcian Cheirolepidiacean Phase were divided between the *Exesipollenites* Association on the southern margins of Gondwana and the *Perinopollenites* Association on the northern (Tethyan) margin (Fig. 12.4); and, although there was a faunal turnover in the Middle Toarcian, the succeeding assemblages had very similar distributions (Grant-Mackie et al. in Wright et al., 2000).

Non-marine Faunas and Floras in Eastern Asia. The distinctive Late Jurassic plants and animals which are widely found in eastern Asia today, particularly in North China (Hebei Province), were first noted in the 1920s by Alfred Grabau, who termed them the Jehol Fauna. The early Jehol Fauna consists of a wide variety of plants and palynomorphs, as well as many invertebrates, notably ostracods, bivalves, and insects, and vertebrates, mostly fish, and is probably of Kimmeridgian age. The middle fauna is comparable and is known from tuffaceous bands in thick predominantly volcanic rocks. However, the upper Jehol Fauna, which occurs in lacustrine deposits which span the Jurassic–Cretaceous boundary in age, is famous not only for its diverse plants and invertebrate faunas such as conchostracans, but also for the diverse fauna of relatively advanced birds such as *Cathayornis* and *Confuciusornis*, which had smaller claws and other more evolved features than previously known among European predecessors such as *Archaeopteryx* (see below); and the birds are associated in the same deposits with the remains of small feathered dinosaurs such as *Compsognathus*.

Jurassic Extinctions. Both at the Triassic–Jurassic boundary at 201 Ma and also some 20 Myr later, in the Toarcian, there were two faunal extinction events which, although nothing like as severe and comprehensive as the end-Permian and end-Cretaceous extinctions, were nevertheless quite substantial. For example, in the Toarcian of Yorkshire, England, some 85% of the numerous bivalve mollusc species became extinct when the local environment suddenly reacted to the Toarcian Oceanic Anoxic Event described above. However, the end-Triassic event may have been

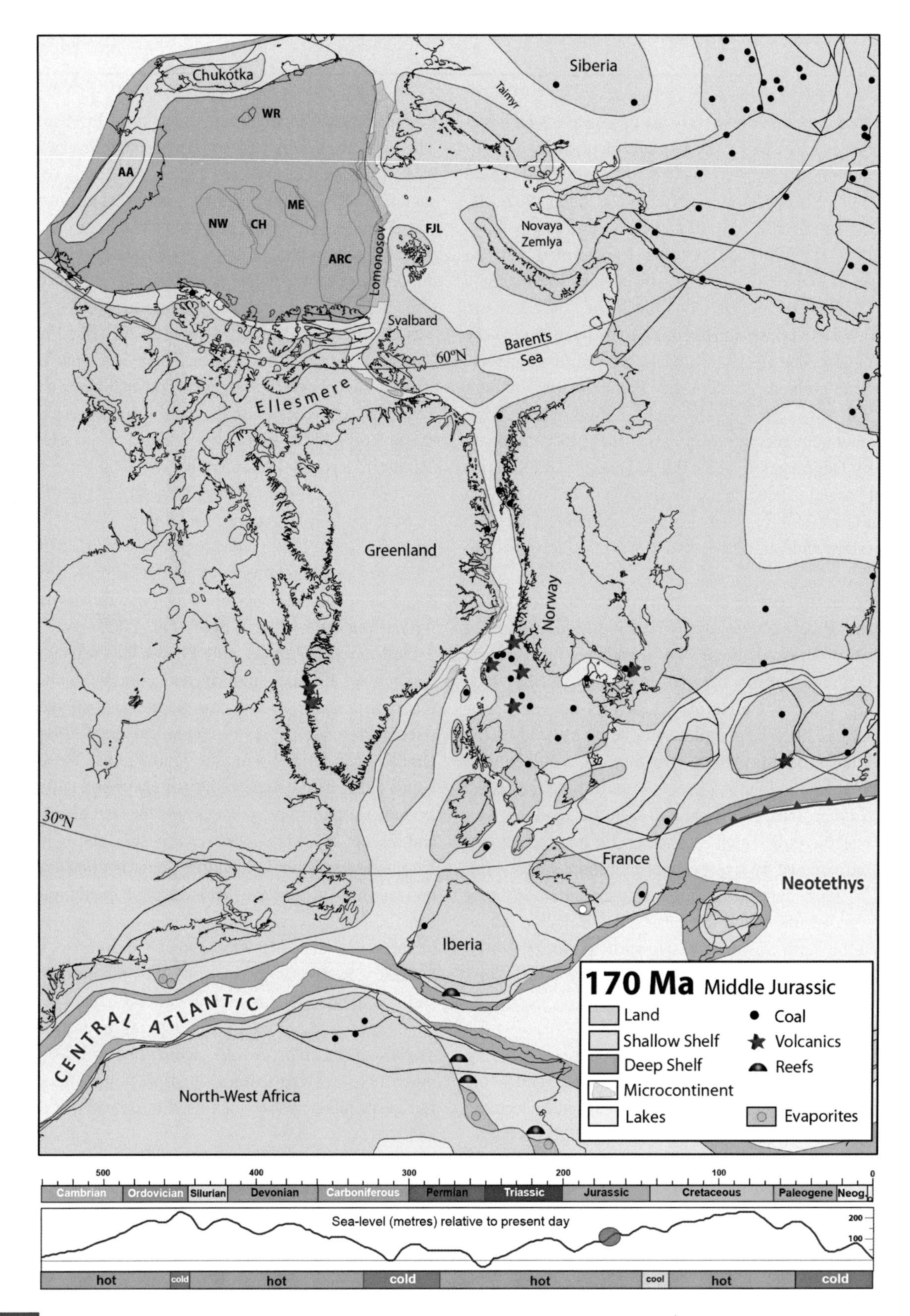

Fig. 12.5 Palaeogeography of central Eurasia, the Arctic Ocean area and north-western Africa at 170 Ma (Bajocian) showing the many small islands in Europe, the opening of the Central Atlantic Ocean, and the widening Neotethys Ocean. Most of the Arctic microcontinents coloured grey are today submerged within the Arctic Ocean. AA, Arctic Alaska; ARC, Alpha Ridge potential continental block; CH, Chukchi; FJL, Franz Josef Land; ME, Mendeleev Ridge; NW, Northwind Ridge; WR, Wrangel Island. Bottom: Phanerozoic time scale and sea-level variations (Haq & Al-Qahtani, 2005: Haq & Shutter, 2008) and icehouse (cold) and greenhouse (hot) conditions.

enhanced locally by the blocking of the extensive seaway and subsequent increased salinities and deposition of evaporites in north-western Europe (Wignall & Bond, 2008), as well as by the possible global climatic changes triggered by the CAMP LIP eruptions in Africa and South America. Conodonts also made their final appearance just before that event (although there had been far fewer of them in the Triassic than in their Palaeozoic heyday), as did two hitherto important brachiopod orders, the spiriferids and the athyrids.

Terrestrial Vertebrates and Dinosaurs. The most well-known Jurassic vertebrates are the dinosaurs, which included virtually all the large terrestrial animals from the end of the Late Triassic (Norian) to the end of the Cretaceous (see also the next chapter). Dinosaurs were reptiles within the Archosauria (which have three openings termed fenestrae for muscles in their skulls and also laid eggs with calcite shells), which had distinctive foot and ankle bones. Their comparable relatives and ancestors had all become extinct by the end of the Triassic, leaving the dinosaurs to radiate rapidly during the Jurassic. Most of the earlier genera were bipedal, but a minority became quadrupedal and to resist predation some developed substantial dermal bones ('armour plating') to compensate for their relative lack of speed. Most were herbivores, but a minority, such as the famous *Tyrannosaurus*, developed stronger jaws and larger teeth to become ferocious predators. *Tyrannosaurus* was largely solitary but many of the smaller predators hunted in packs. Because dinosaurs lived almost exclusively between about 45° north and south of the Equator (Rees et al., 2004), virtually all were cold-blooded, like most modern reptiles.

Although the dinosaurs were the largest vertebrates, they were not alone: other reptiles such as crocodiles and turtles thrived, and mammals also radiated. Mammals had evolved from reptiles in the Middle Triassic, but were rare then, and most species remained relatively small until the Cretaceous. However, although their habitats expanded in the Jurassic to include shallow freshwater marshes, like modern water voles, they did not extend to the oceans or the air: the marine mammals (such as whales and dolphins) and the bats did not evolve until the Paleogene (Savage & Long, 1986).

Marine Faunas. The shallow seas and oceans were alive with swimming reptiles and molluscs. The former included the shorter-necked ichthyosaurs and the longer-necked plesiosaurs, some of which reached lengths of over 30 m, and which were close relatives of the dinosaurs which had taken to the seas. They radiated quickly in the Latest Triassic and Early Jurassic from much smaller eel-like Early Triassic ancestors. As well as those two dominant groups, the fish, including sharks (some of which also reached very substantial sizes), were also varied and abundant, and occupied several levels in the predation hierarchy.

Invertebrate faunas were dominated by the molluscs, with varied bivalves and gastropods on the sea floors, and the cephalopods (chiefly ammonites and belemnites) which swam. The ammonites continued their rapid radiation from the Triassic and evolved so fast that numerous successive ammonite biozones have been found to be the best tools for stratigraphical correlation during all of the Jurassic. Although the brachiopods never regained their Palaeozoic diversity, two abundant groups which radiated extensively and have survived until today were the rhynchonellides, which were largely ribbed, and the terebratulides, which were largely smooth, but both had functional pedicles which enabled them to live above the sea floor, and they were usually attached to rocks, to seaweed, or to each other; however, apart from the burrowing lingulides, all the brachiopods were epifaunal (living on or above the sea floor). But burrowing into the sea floors (infaunal habitats) offered much better evolutionary opportunities than before. As well as the less common lingulide brachiopods, there were abundant and diverse bivalve molluscs and a variety of arthropods which burrowed. Irregular echinoids made their first appearance in the Early Jurassic and rapidly evolved as infaunal deposit feeders. The latter had much diversified by the Middle Jurassic so that both sand dollars and other irregulars had evolved into extremely efficient bulk sediment feeders (A.B. Smith in Selley et al., 2005).

The detailed ecology of the benthic faunas underwent substantial development as the Jurassic progressed. Many hitherto shallow burrowers, including worms and bivalve molluscs, went even deeper within the substrate, and many other molluscan groups, such as oysters, developed cementing for attachment to a variety of organic or inorganic substrates. Others, for example the scallops, even learnt to swim so as to avoid the increasing numbers and variety of predators. However, in contrast to the other mollusc groups, the largely epifaunal gastropods did not radiate so substantially during the Jurassic.

The Origin of Birds. The group of flying reptiles called pterosaurs had existed since the Late Triassic, but their flight was supported by a membrane stretching from an extended finger to the sides of their bodies, and, although the pterosaurs survived for a considerable time (they were among the many victims of the Late Cretaceous extinction event), they were not closely related to the birds, whose wings were an evolutionary development of the entire forelimb (Benton, 2005).

Fine-grained latest Jurassic (Tithonian) limestones deposited in a hypersaline lagoon at Solnhofen in Bavaria,

Germany, from about 150 Ma, have been extensively quarried for centuries because of their superb quality as lithographic stones for printing plates, and they have a great variety of fossils within them which are exceptionally preserved. After assiduous collecting for more than 150 years, among the many different vertebrate and invertebrate species recognised from Solnhofen, eight skeletal specimens (three complete) and a single feather are known of an extraordinary animal named *Archaeopteryx.* It was about the same size as a turkey, and is extraordinary because it has characters which are typical of both small dinosaurs and also birds. The long bony tail, sharp-clawed forelimbs, primitive pelvis, and teeth are all similar to those of therapod dinosaurs; but, unlike dinosaurs, *Archaeopteryx* has clavicles fused to form a furcular (wishbone), a uniquely avian character, as well as small feathers surrounding its body and many much longer feathers on its tail and forelimbs, both clearly similar to those on the wings of birds. Perhaps those small downy feathers initially evolved to serve as insulation and thus indicate that the animal might have been warm-blooded. Because of that mixture of its characters, *Archaeopteryx* has become one of the most-quoted examples used to vindicate Darwin's theory of evolution objectively, since it is one of the few important 'missing links' to be found directly as fossils. Since the discovery of *Archaeopteryx*, many other primitive birds have been found and described from the younger Mesozoic rocks, particularly in China (see above), as avian radiation proceeded quickly during the Latest Jurassic (Tithonian) and Cretaceous.

Floras. The lack of substantial ice caps signified a relatively equable global climate, which allowed much of Gondwana to be colonised by the widespread *Dictyophyllum* Flora, which included a few taxa which were less common in Europe, and also the gingkos and their relatives, which are less common in Gondwana. However, unlike today, there was no tropical rainforest, since the equatorial regions were semi-arid and had the least vegetation. The low latitudes were either desert or seasonally wet, the middle latitudes were mostly warm temperate and had the greatest diversity of plants, and the higher latitudes were cool and temperate: limited precipitation was the main restriction on plant growth near the Equator. The mid-latitude flora was dominated by conifers, cycads, pteridosperms, and extinct relatives: the common deciduous trees of today were very much a minority, but the landscape was savannah-like (rather than mainly dense stands of trees), through which the herbivorous dinosaurs could roam freely, rather like elephants today. The floras in polar latitudes were of lower diversity and were dominated by large-leafed conifers and gingkophytes, which were probably deciduous, as well as ferns (Rees et al., 2004; Kenrick and Davis, 2004).

13 Cretaceous

The Fish Clay is a thin grey-to-black marl between carbonates and marks the Cretaceous–Paleogene boundary at Stevns Klint in eastern Denmark. One layer in the Fish Clay has an anomalously high iridium concentration, which is due to a Late Cretaceous asteroid impact in the Caribbean (Alvarez et al., 1980). Credit: Holly Stein/CEED.

The Cretaceous saw the highest sea levels in the Phanerozoic, with the result that large parts of the continents were submerged under shallow seas (Fig. 13.10). As the long Cretaceous epoch progressed, much of the general location and outlines of continents and oceans steadily became more recognisable to a modern geographer. In particular, a much higher proportion of old Cretaceous ocean floor is still preserved today when compared with the preceding periods.

'Cretaceous' means 'chalk-bearing' and the name was coined in the 1820s by the Frenchman d'Omalius d'Halloy to denote the distinctive fine white limestones (the Chalk) which principally outcrop in southern England, Belgium, and France, including the White Cliffs of Dover, England. The base of the Cretaceous is not yet formally defined, but is often taken at the base of the *Berriasella jacobi* ammonite Biozone (Berriasian Stage) in France, which is dated at 145 Ma. The global distribution of the continents and oceans at different times within the period is shown in Figs. 13.1 and 13.2 at 130 Ma (Hauterivian), 110 Ma (Albian), 90 Ma (Turonian), and 70 Ma (Maastrichtian). The base of the Aptian is taken at about 121 Ma, rather than the date of 125 Ma given by some authors. The system lasted for 80 Myr, from 145 to 65 Ma, the longest within the Phanerozoic, and is also the one with the most stages, which are, in stratigraphic ascending order, the Berriasian, Valanginian, Hauterivian, Barremian, Aptian, Albian, Cenomanian, Turonian, Coniacian, Santonian, Campanian, and Maastrichtian, all named from various areas in Europe. The first three stages are often grouped as the Neocomian.

Tectonics and Igneous Activity

Large Igneous Provinces and Kimberlites. Notable characteristics of the Cretaceous are the exceptionally high number of LIPs (Figs. 13.3 and 13.4; Appendix 1) and also about 75% of all known Mesozoic–Cenozoic kimberlites (Fig. 14.2b), which were intruded into many areas. Nearly all of the kimberlites originated above the margin of Tuzo PGZ, except those in North America and Canada (Figs. 13.3a and 13.4a). As in the Jurassic, the majority of the LIPs are situated on the ocean floor and only five are found on continental lithosphere: the Deccan Traps (~65 Ma) in India, the Madagascar LIP (~87 Ma), the Rajhmahal Traps (~118 Ma) in India, the Bunbury Basalts of south-west Australia (~132 Ma), and the Paraná–Etendeka flood basalts in South America (~134 Ma). All Cretaceous oceanic LIPs erupted right above the plume generation zones of Tuzo and Jason (Nauru, Ontong Java, Nui, and Hess). There is also a High Arctic LIP (HALIP), originally considered to include only Upper Cretaceous volcanics, on Axel Heiberg Island in the Canadian Arctic (Tarduno et al., 1998), but now extended to be grouped with magmatism in many other places in the High Arctic (Fig. 13.8). Because radiometric ages for HALIP show a wide range (130–80 Ma) and a plume centre is not readily defined, HALIP is not shown in our reconstructions of LIPs. The Cretaceous Caribbean LIP (CLIP) is also not included in our analysis because of large plate reconstruction uncertainties and an unknown plume centre. If HALIP and CLIP were sourced by deep plumes, they would probably relate to the northern margin of Tuzo and the eastern margin of Jason respectively.

Pangea Breakup. The southern hemisphere continents became steadily fragmented as the Cretaceous progressed (Figs. 13.1 and 13.2). In the Early Cretaceous, this fragmentation included the sea-floor spreading between South America and Africa from around 130 Ma and widening of the West Somali Basin between East Africa and Madagascar–India until about 120 Ma. Sea-floor spreading had begun in the Mid–Late Jurassic between Australia and Greater India and it probably progressed southwards between the Indian and Antarctic plates (Early Cretaceous) and between the Indian and Australian plates in the Mid–Late Cretaceous (Figs. 13.2 and 13.9). In the High Arctic (Gaina et al., 2013a), short-lived sea-floor spreading in the Amerasian Basin took place between 141 Ma (Berriasian) and 125 Ma (Barremian). Australia finally separated from Antarctica in the Late Cretaceous at around 85 Ma (Santonian).

Panthalassic Ocean. Near the Jurassic–Cretaceous boundary (145 Ma), Panthalassic Ocean sea-floor spreading between the four main oceanic plates, Pacific, Farallon, Izanagi, and Phoenix, led to the enlargement of the Pacific Plate and gradual subduction of its neighbours along adjacent active margins (Fig. 13.1a). One consequence of that development was the subduction of the Cache Creek Plate beneath north-west North America (Shephard et al., 2013). The Early to Middle Cretaceous marked a significant increase in both absolute plate velocities and sea-floor spreading rates in the Panthalassic Ocean (Fig. 12.3b,c), often termed the 'Middle Cretaceous sea-floor spreading pulse' (e.g. by Seton et al., 2009). The eruption of the Ontong Java–Nui mega-LIP (Ontong Java, Manihiki, and Hikurangi Plateaus) started at around 123 Ma and probably led to the breakup of the Phoenix Plate into four plates, Hikurangi, Manihiki, Chasca, and Catequil (Fig. 13.5a) at around 120 Ma (Aptian). After the terminations of the Hikurangi–Manihiki–Chasca–Catequil plates at around 85 Ma, the Pacific Plate became the dominant plate in the Panthalassic Ocean (Fig. 13.5c), and it is from that time that the overarching name is considered to be the Pacific Ocean. A major event in the Pacific Ocean was the inception of the Kula Plate at about 83 Ma (Campanian) from pieces of the Izanagi, Farallon, and Pacific plates. Spreading continued along the Pacific–Izanagi Ridge after the establishment of the Kula–Pacific Ridge to the east, connected via a large offset transform fault (Fig. 13.5d). The Pacific–Izanagi Ridge rapidly approached the East Asian margin during the Late Cretaceous and was very near to it by the Early Paleocene (Figs. 14.1a and 14.4).

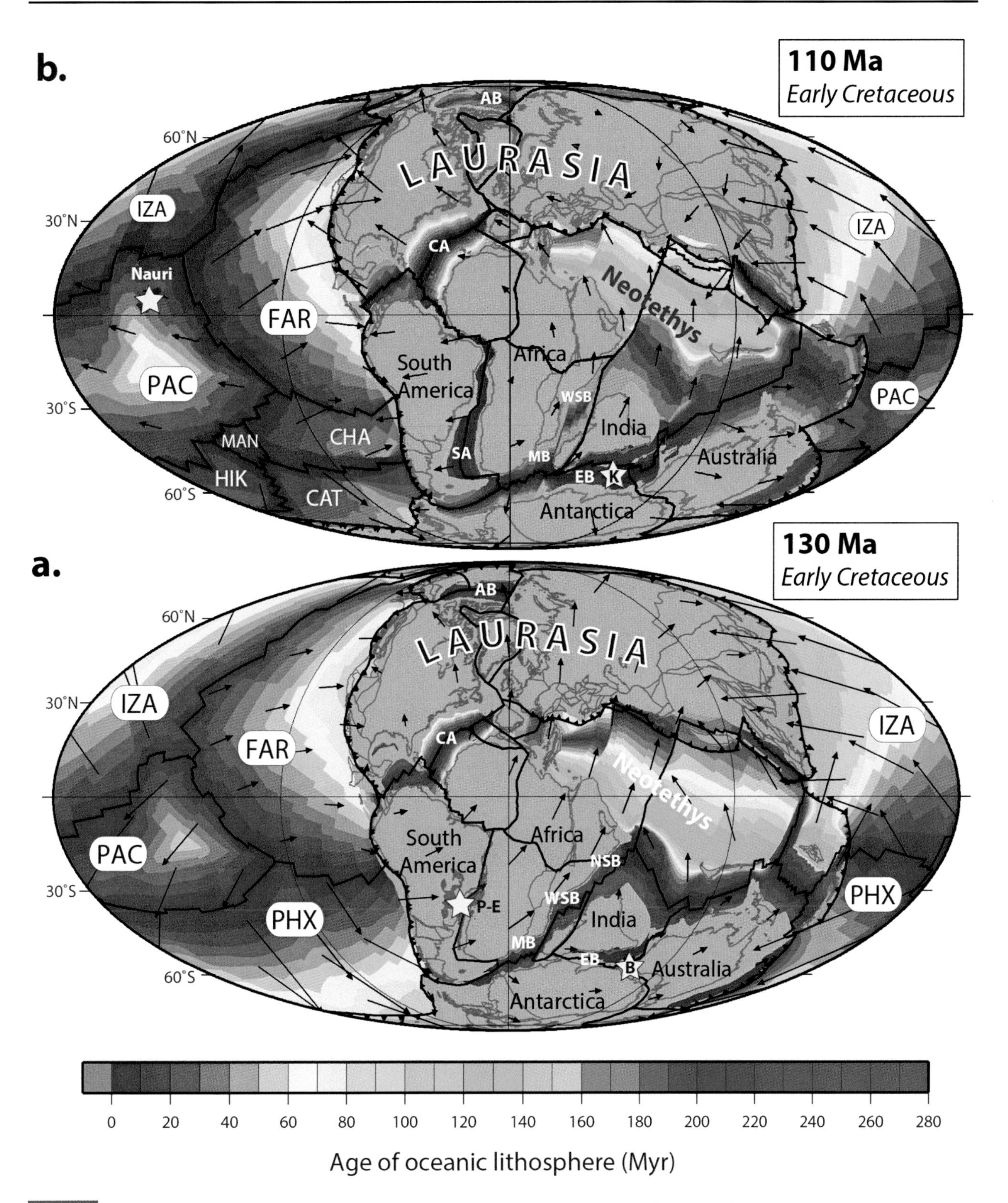

Fig. 13.1 Continental blocks (grey) and oceans (other colours) with plate velocity vectors and the ages of their oceanic lithosphere in the Early Cretaceous at around 130 Ma (Hauterivian) and 110 Ma (Campanian). Also shown are the reconstructed locations of the Paraná–Etendeka (P-E at 134 Ma), Bunbury (B at 132 Ma), and Southern Kerguelen (K at 114 Ma) LIPs (yellow stars are estimated eruption centres). AB, Amerasia Basin; CA, Central Atlantic; CAT, Cateqil Plate; CHA, Chasca Plate; EB, Enderby Basin; FAR, Farallon Ocean Plate; HIK, Hikurangi Plate; IZA, Izanagi Ocean Plate; MAN, Manihiki Plate; MB, Mozambique Basin; NSB, North Somali Basin; PAC, Pacific Ocean Plate; PHX, Phoenix Plate; WSB, West Somali Basin. Generated from the EARTHBYTE mantle frame (see Chapter 2).

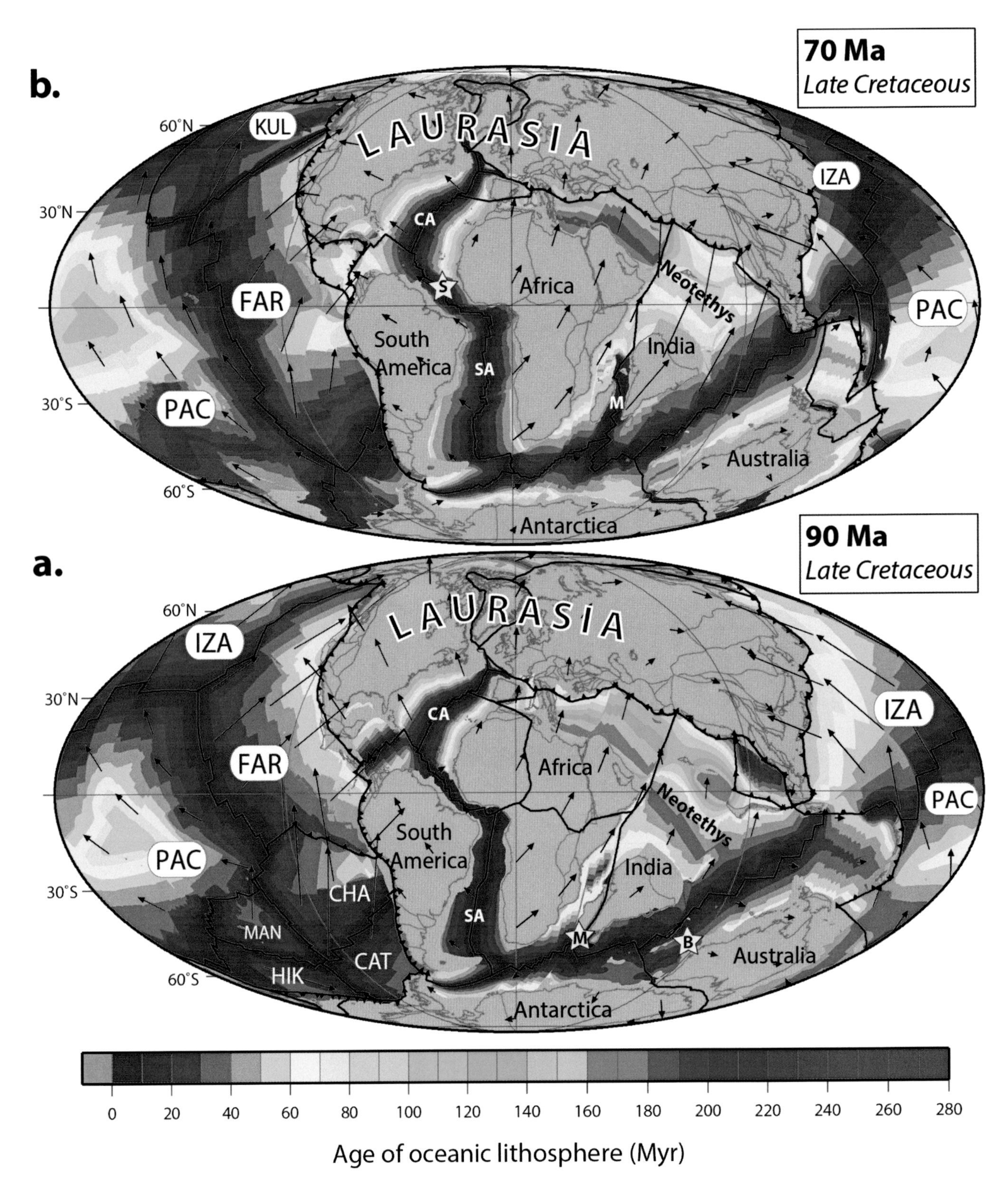

Fig. 13.2 Continental blocks (grey) and oceans (other colours) with plate velocity vectors and the ages of their oceanic lithosphere in the Late Cretaceous at around 90 Ma (Turonian) and 70 Ma (Maastrichtian). Also shown are the reconstructed locations of the Madagascar (M, 87 Ma), Broken Ridge (B, 95 Ma), and Sierra Leone (S, 73 Ma) LIPs (yellow stars are estimated eruption centres). CA, Central Atlantic; CAT, Cateqil Plate; CHA, Chasca Plate; ESB, East Somali Basin; FAR, Farallon Ocean Plate; IZA, Izanagi Ocean Plate; KUL, Kula Ocean Plate; NSB, North Somali Basin; HIK, Hikurangi Plate; M, Mascarene Basin; MAN, Manihiki Plate; PAC, Pacific Ocean Plate; PHX, Phoenix Plate; SA, South Atlantic. Generated from the EARTHBYTE mantle frame (see Chapter 2).

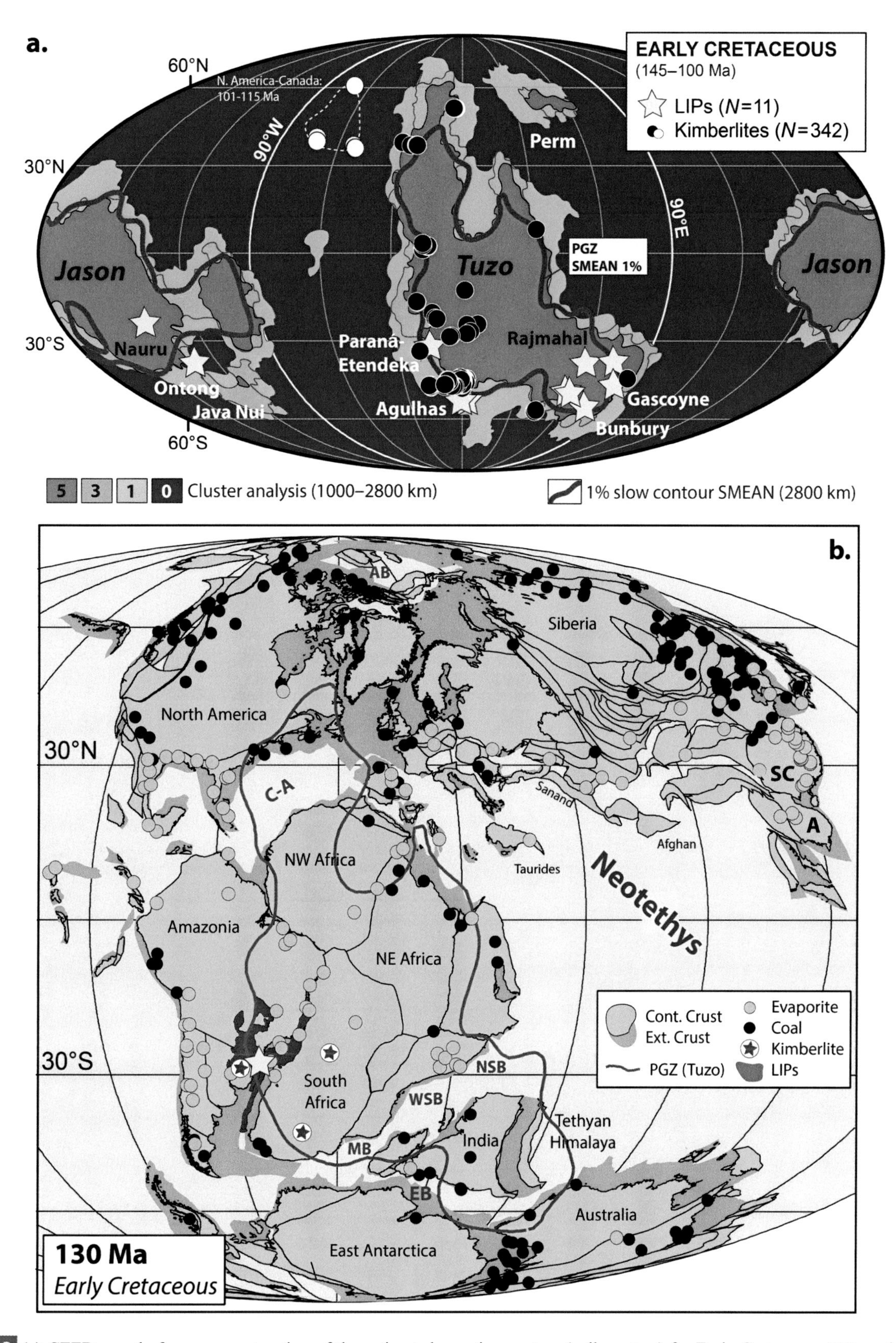

Fig. 13.3 (a) CEED mantle frame reconstruction of the estimated eruption centres (yellow stars) for Early Cretaceous LIPs and 342 kimberlites (black and white dots). These are draped on seismic voting-map contours in the lower mantle (Lekic et al., 2012). Areas 5, 3, and 1 define the LLSVPs and 0 (blue) denotes faster regions in the lower mantle (see Fig. 2.16c for detailed description). The 1% slow SMEAN contour is shown for comparison (see Chapter 2). (b) Outline Earth geography (major crustal units) at 130 Ma (Hauterivian) and the estimated eruption centre for the Paraná–Etendeka LIP (yellow star). CEED palaeomagnetic frame and the location of continents and plate boundaries may differ in details from Fig. 13.1a. A, Annamia; AB, Amerasian Basin; MB, Mozambique Basin; C-A, Central Atlantic; EB, Enderby Basin; NSB, North Somalian Basin; PGZ, plume generation zone; SC, South China; WSB, West Somalian Basin.

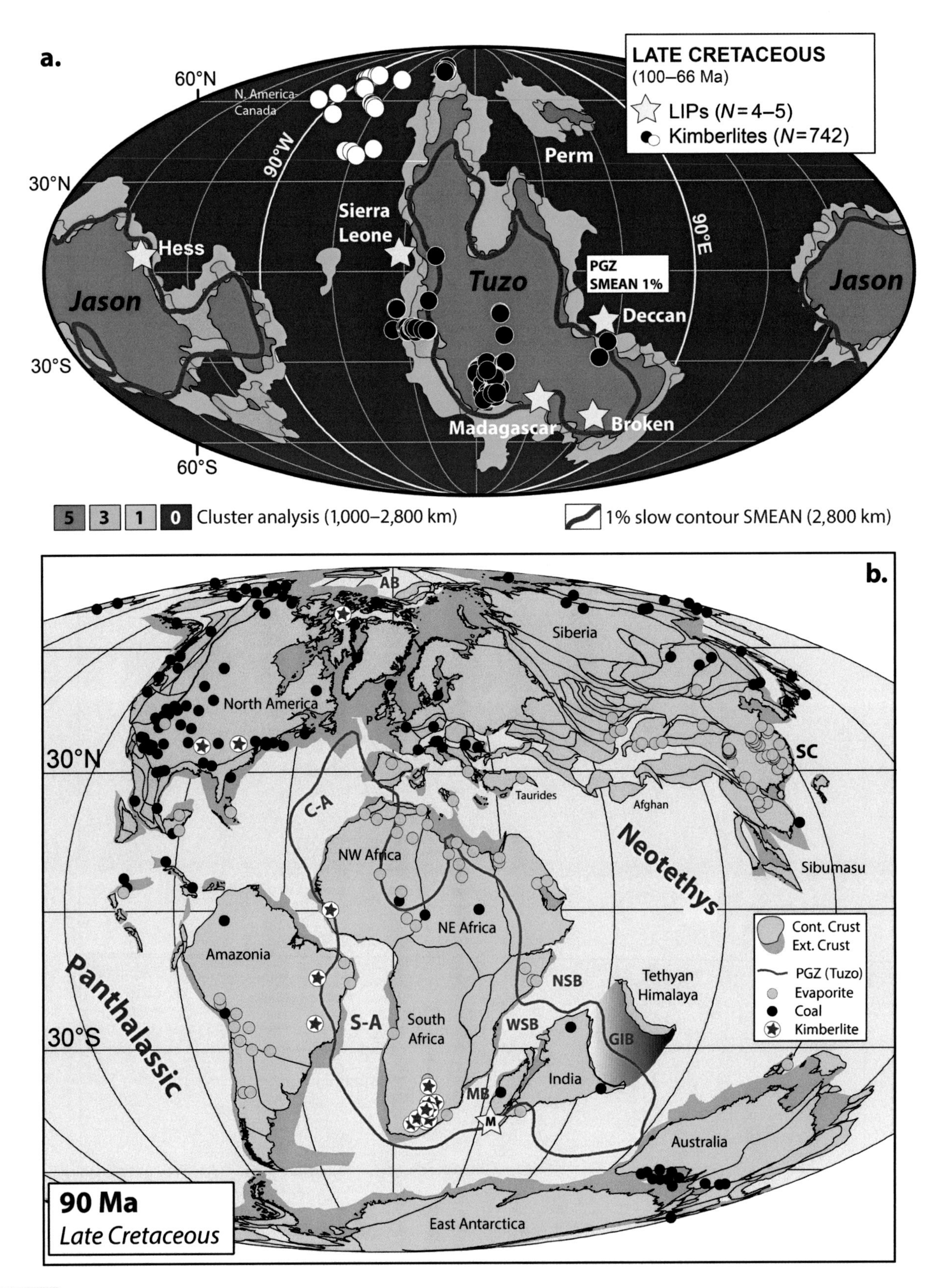

Fig. 13.4 (a) CEED mantle frame reconstruction of the estimated eruption centres (yellow stars) for Late Cretaceous LIPs and 742 kimberlites (black and white dots). These are draped on seismic voting-map contours in the lower mantle (Lekic et al., 2012). Areas 5, 3, and 1 define the LLSVPs and 0 (blue) denotes faster regions in the lower mantle. The 1% slow SMEAN contour is shown for comparison (see Chapter 2 for further details). (b) Outline Earth geography (major crustal units) at 90 Ma (Turonian) and the estimated eruption centre for the Madagascar (M) LIP (yellow star). CEED palaeomagnetic frame and the location of continents and plate boundaries may differ in detail from Fig. 13.2a. AB, Amerasian Basin; MB, Mozambique Basin; C-A, Central Atlantic; GIB, Greater India Basin; NSB, North Somalian Basin; PGZ, plume generation zone; S-A, South Atlantic; SC, South China; WSB, West Somalian Basin.

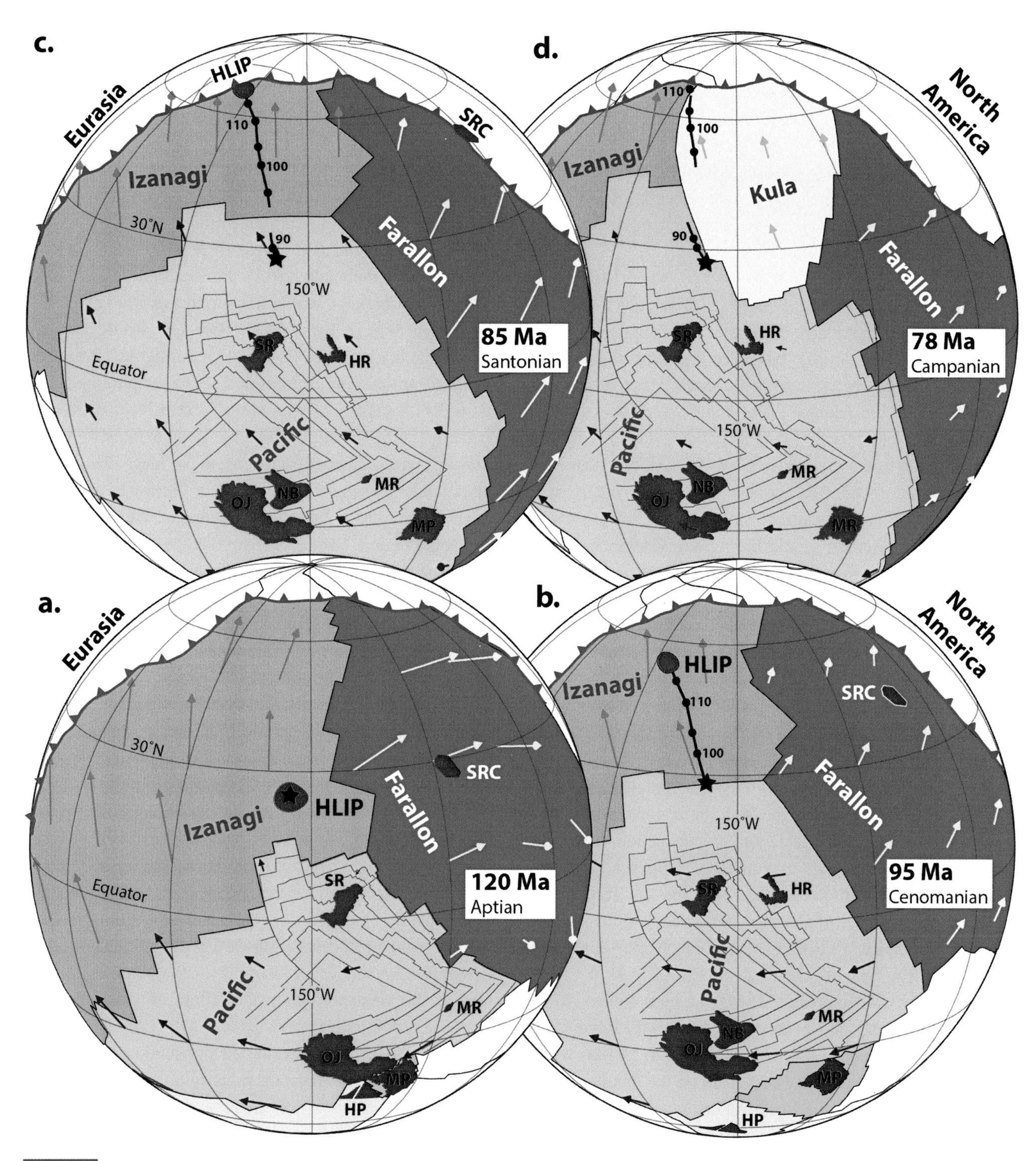

Fig. 13.5 Panthalassic reconstructions from (a) Aptian (120 Ma), (b) Cenomanian (95 Ma), (c) Santonian (85 Ma), and (d) Campanian (78 Ma) times in a CEED mantle frame. Also shown are the estimated surface location of the Hawaii Plume (black star) and plate velocity vectors for the Pacific, Izanagi, Farallon, and Kula plates: the plume track (black dots) is shown at 5 Myr intervals. Plate polygons are those of Shephard et al. (2014). Annotated LIPs: HLIP, Hawaii LIP (here assumed a starting age of 120 Ma); HP, Hikurangi Plateau (123 Ma); HR, Hess Rise (99 Ma); MP, Manihiki Plateau (123 Ma); MR, Magellan Rise (145 Ma); NB, Nauru Basin (111 Ma); OJ, Ontong Java (123 Ma); SR, Shatsky Rise (147 Ma); SRC, Shatsky Rise Conjugate (postulated), which started to subduct at about 85 Ma (Santonian) along the North American margin.

Aptian Hawaii LIP? Like some other hotspots, Hawaii may also have been linked to a starting LIP, now long subducted; but where, and at what depths in the mantle, we should look for such a Hawaii LIP (HLIP) is uncertain. Examples of LIPs being recycled, through subduction, into the Earth's mantle include the ongoing subduction of Ontong Java and Hikurangi Plateaus today. Subduction of LIPs and seamounts has been proposed to explain flat subduction and the absence of arc magmatism.

With a theoretical starting age of 120 Ma, HLIP would have erupted on the Izanagi Plate (Fig. 13.5a), and a chain of volcanic islands could have been constructed on that plate up until 95 Ma (Fig. 13.5b); but after that time the Hawaii Plume would be under the Pacific Plate. Parts of the old plume trail on the Izanagi Plate would have been trapped on the Kula Plate after 83 Ma (Campanian) (Fig. 13.5d) due to a relocation of the mid-ocean ridge. Most of that plate was subducted before 65 Ma (earliest Paleocene), but a part, including possible remnants of the Hawaiian hotspot track, may still be preserved in the Bering Sea (Steinberger and Gaina, 2007).

HLIP would have stayed on the Izanagi Plate until about 85 Ma when it reached the trench. If subducted, we would expect to find the subducted HLIP and adjacent lithosphere in tomographic models (as faster than average mantle velocities) at high northerly latitudes (70° N) and at depths of around 1,000 km, depending on the slab sinking speed and subduction angle.

The compositional buoyancy of a LIP should resist subduction and may prohibit the slab from sinking into the mantle (accretion), but accretion or subduction of a LIP could initially be reflected in reduced convergence velocities at the trench. Unfortunately, plate velocities for Izanagi are ambiguous (modelled under a 'symmetric sea-floor spreading' assumption). Collision along the North American margin with a twin of the Shatsky Rise LIP (Liu et al., 2010; Sigloch & Mihalynuk, 2013) has been linked to the Laramide Orogeny and basement uplifts in the Late Cretaceous to Early Eocene from around 85–80 to 55 Ma (B3 in Fig. 13.6c). A Shatsky Rise conjugate on the Farallon Plate would have arrived at the North American margin at around 85 Ma, at the same time as a 120 Ma HLIP would have reached the trench (Fig. 13.5c). The Farallon Plate is much better constrained than Izanagi because of magnetic anomalies preserved on the conjugate Pacific Plate. A Late Cretaceous deceleration of Farallon occurred at around 88 Ma (Fig. 13.6e), which may be linked to a compositionally, buoyant LIP (the Shatsky Rise conjugate) arriving at the North American margin. However, that deceleration is model driven and interpolated since there are no magnetic anomalies in that region from about 120 to 84 Ma (Aptian to Santonian). Collision with a twin of the Shatsky Rise LIP is modelled in a setting with eastward subduction of the Farallon Plate beneath accreted magmatic arcs and North America in the Late Cretaceous (Fig. 13.6c,d). Hildebrand (2014), however, argued that North America was the lower plate during the Laramide and older events, with westward subduction beneath a Cordilleran Ribbon Continent (named Ruby), which would add severe difficulties to the story of Sigloch and Mihalynuk (2013). For example, the Farallon Plate, at least in parts, would not subduct as in Fig. 13.5 and therefore a twin of the Shatsky Rise LIP would not have reached the margin of North America during the Cretaceous.

Intra-Panthalassic Subduction. Plate tectonic modelling of the Panthalassic realm is almost exclusively carried out assuming intra-oceanic spreading centres/transforms and 'Andean type' subduction zones at or near the circum-Panthalassic margins (i.e. along the Eurasian and North American margins in Fig. 13.5). Subducted slabs, interpreted from 'blue regions' of faster-than-average seismic velocities in the mantle (Fig. 2.18), have been identified at various depths beneath the Panthalassic (van der Meer et al., 2010, 2012; Sigloch & Mihalynuk, 2013). The large number of imaged slabs appears to demonstrate considerable intra-oceanic subduction, and subduction polarity shifts have also been postulated along the North American Margin. Figure 13.6 shows an example of complex models which can be derived from a combination of slab imaging, geological interpretations, near-vertical slab sinking, and plate modelling. In the Early Cretaceous, at around 140 Ma (Fig. 13.6a), the offshore margin of western North America is littered with intra-oceanic subduction, and associated volcanic arcs were mostly accreted to the North American margin by the Early Eocene at 50 Ma. Large-scale Late Cretaceous to Paleocene northward terrane translation (e.g. Wrangellia–Alexander) is also part of the complex tectonic story of the western margin Cordillera of North America (Johnston, 2008; Hildebrand, 2014).

Tethys and Indian Oceans. At the south-eastern end of the Tethys domain, a landward ridge-jump propagated southward to open the Gascoyne, Cuvier, and Perth Abyssal Plains between Greater India and Australia. The mid-ocean ridge that led to separation of Greater India from Australia–Antarctica started at around 136 Ma (Valanginian) northwest of Australia and reached the southern tip of India at 130–126 Ma (Barremian). Sea-floor spreading in the Enderby Basin (between East Antarctica and India, Fig. 13.1) was abandoned in the Aptian, when a ridge jump transferred the Elan Bank and South Kerguelen Plateau to the Antarctic Plate (Gibbons et al., 2013). Initiation of sea-floor

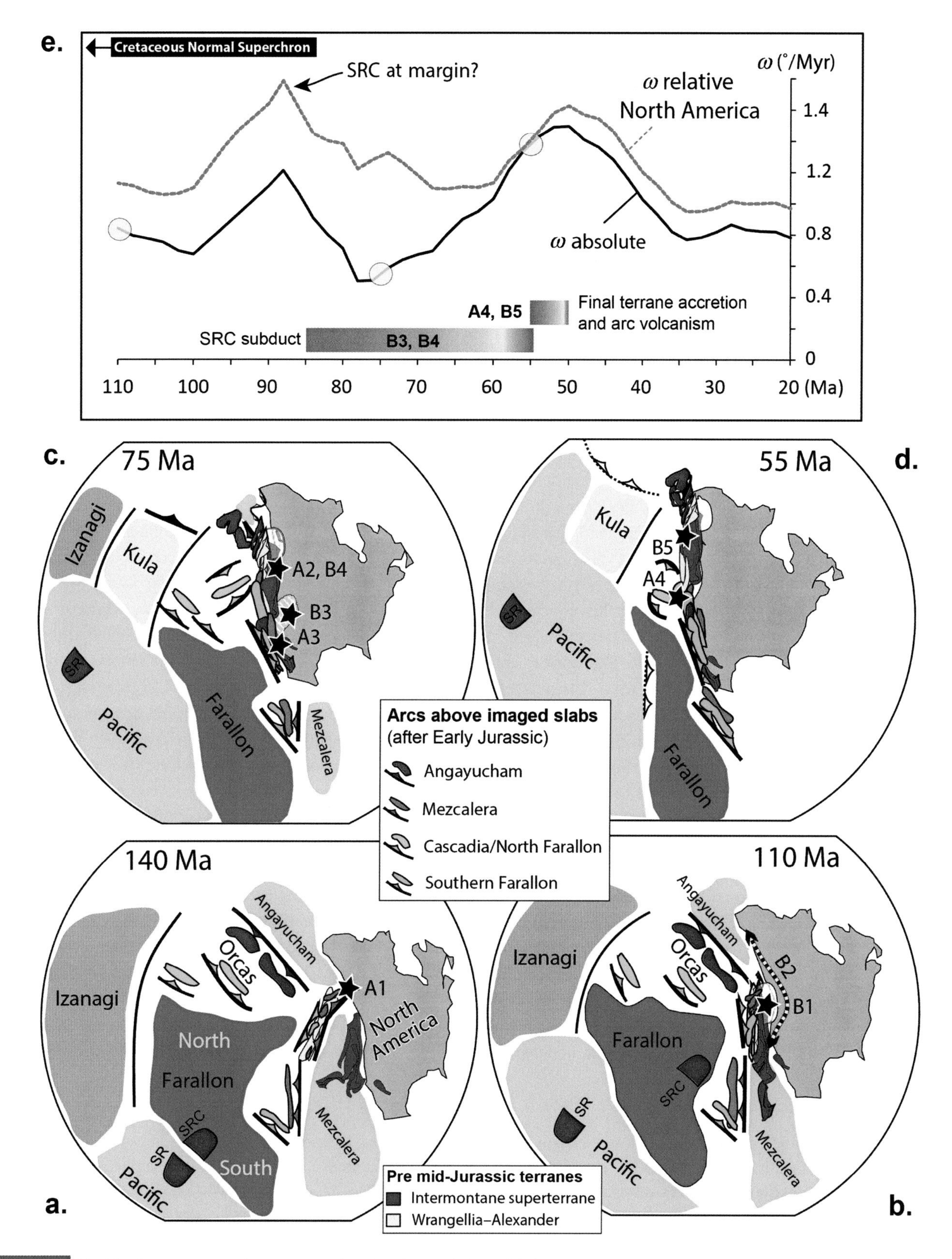

Fig. 13.6 (a)–(d) Plate reconstruction cartoons showing the evolution of inferred trench (based on imaged slabs) and terrane geometries along the western margin of North America from Early Cretaceous (140 Ma) to Early Eocene (55 Ma) times. SR, Shatsky Rise; SRC, Shatsky Rise Conjugate. Annotated black stars denote interpreted tectonic events as follows: A1, initiation of Rocky Mountain deformation

spreading in the Enderby Basin accommodated strike-slip motion between India and Madagascar, which was connected with the West Somali Basin Spreading Ridge until about 120 Ma (Aptian). It is unclear whether magmatic activity could have assisted the separation of India and Australia/East Antarctica, but the Wallaby Plateau LIP offshore from western Australia erupted in the Early Aptian at 123 Ma (Olierook et al., 2015) and further to the west the Maud Rise LIP erupted off Dronning Maud Land/East Antarctica a little earlier at around 125 Ma, whereas the first sign of volcanic activity registered on the Kerguelen Plateau was 118 Ma (Middle Aptian). An older event at about 130 Ma (Hauterivian) formed the Bunbury basaltic province in south-west Australia.

Parts of Australia (North, South, and West Australian cratons) and Antarctica (Mawson Craton) had stayed together since the Late Mesoproterozoic, but Cretaceous rifting (pre-drift extension) eventually established sea-floor spreading, with Australia drifting slowly away from Antarctica (Fig. 13.9b) from around 85 Ma (Santonian) (the oldest magnetic anomaly is Chron 34: 83.5 Ma). That coincided with the opening of the Tasman Sea in the south-western Pacific, when the Lord Howe Rise and smaller continental fragments were detached from the East Australian margin.

The Indian subcontinent, together with Madagascar and the Seychelles, had started to break away from Africa during the Jurassic, and their conjugate margins were completely separate before the end of the Jurassic. Sea-floor spreading in the Enderby Basin led to a sinistral shear-zone between India and Madagascar, with India displaced northwards along the eastern margin of Madagascar. That caused convergence and perhaps subduction between north-western India and Somalia/Arabia (Gaina et al., 2015). Between 100 and 90 Ma (Turonian to Santonian), those displacements were reversed. Soon afterwards, a Late Cretaceous LIP event occurred from 91 Ma (Turonian) to 84 Ma (Santonian), which covered most of Madagascar with flood basalts and opened the Mascarene Basin (Fig. 13.7). That early opening led to the counter-clockwise rotation of India relative to Arabia and several hundred kilometres of convergence were accommodated by Late Cretaceous Santonian to Early Campanian subduction (Gaina et al., 2015).

The opening of the Mascarene Basin has been traditionally modelled as a rather symmetrical opening between Madagascar and India–Seychelles, but the Indian Ocean (like the north-east Atlantic) could have been littered with small continental fragments (Fig. 13.7), due to plate–plume interactions and complicated ridge jumps. India, along with the Seychelles, the Laxmi Ridge (now offshore from western India), and several other continental fragments (part of the Mauritian Microcontinent), first separated from Madagascar, whilst others (including Mauritius, the Cargados Carajos, and the Nazareth Banks) were probably parts of Madagascar at the initial breakup (Torsvik et al., 2013). During the opening of the Mascarene Basin, which lasted from 83.5 to 61 Ma (Campanian to Early Paleocene), three major ridge jumps took place at 80, 73.6, and 70 Ma, and Mauritius and other potential continental fragments of Mauritia were gradually transferred to the Indian Plate (Fig. 13.7); after 70 Ma (Maastrichtian) all the Mauritian fragments were integral parts of the Indian Plate until the Late Paleocene (Fig. 14.7).

Offshore from what was to become India's Himalayan margin, there was a chain of land areas, which may have been more than one microcontinent, and included the Tethyan (Tibetan) Himalaya (Fig. 13.9) which moved northwards relative to cratonic India between 118 Ma (Aptian) and 68 Ma (Maastrichtian). At least one oceanic basin, the Greater India Basin, must have formed between the Tethyan Himalaya and cratonic India by the Late Cretaceous, but the reality could have been much more complex, with many basins alternating with extended continental fragments (van Hinsbergen et al., 2012).

Atlantic and Arctic Oceans. In the southern South Atlantic, sea-floor spreading started in the Early Cretaceous at about 130 Ma (Hauterivian), shortly after a peak in magmatism, which included the intrusion and eruption of the Paraná–Etendeka LIP at 134 Ma (Fig. 13.1a) which affected large parts of Brazil and to a lesser extent Namibia. Sea-floor spreading had propagated northward to the central segment of the Ocean at around 112 Ma (north of the Florianopolis

Fig. 13.6 (*cont.*) (160–155 Ma) and Franciscan subduction complex/South Farallon (165–155 Ma); B1, Omenica magmatic belts (124–90 Ma); B2, margin-wide deformation: Sevier and Canadian Rocky Mountains from about 125 Ma (Barremian); A2, Carmacks volcanic episode (72–69 Ma); A3, Sonora volcanism: Tarahumara ignimbrite province (85 Ma); B3, subduction of Shatsky Rise Conjugate, Laramide Orogeny and basement uplift (85–55 Ma); B4, northward movements of Wrangellia–Alexander, intermontane, and Angayucham terranes along the North American margin (85–55 Ma); A4, terminal terrane accretions: Siletzia, Pacific Rim (55–50 Ma); B5, explosive termination of Coast Mountain arc volcanism (55–50 Ma). © Sigloch & Mihalynuk, 2013. (e) Absolute velocity (ω) and relative velocity (in relation to North America) for the Farallon Plate from 110 to 20 Ma, calculated from a CEED mantle frame. Some of the marked tectonic events in (a)–(d) are indicated. Velocities are averaged over a 10 Myr window.

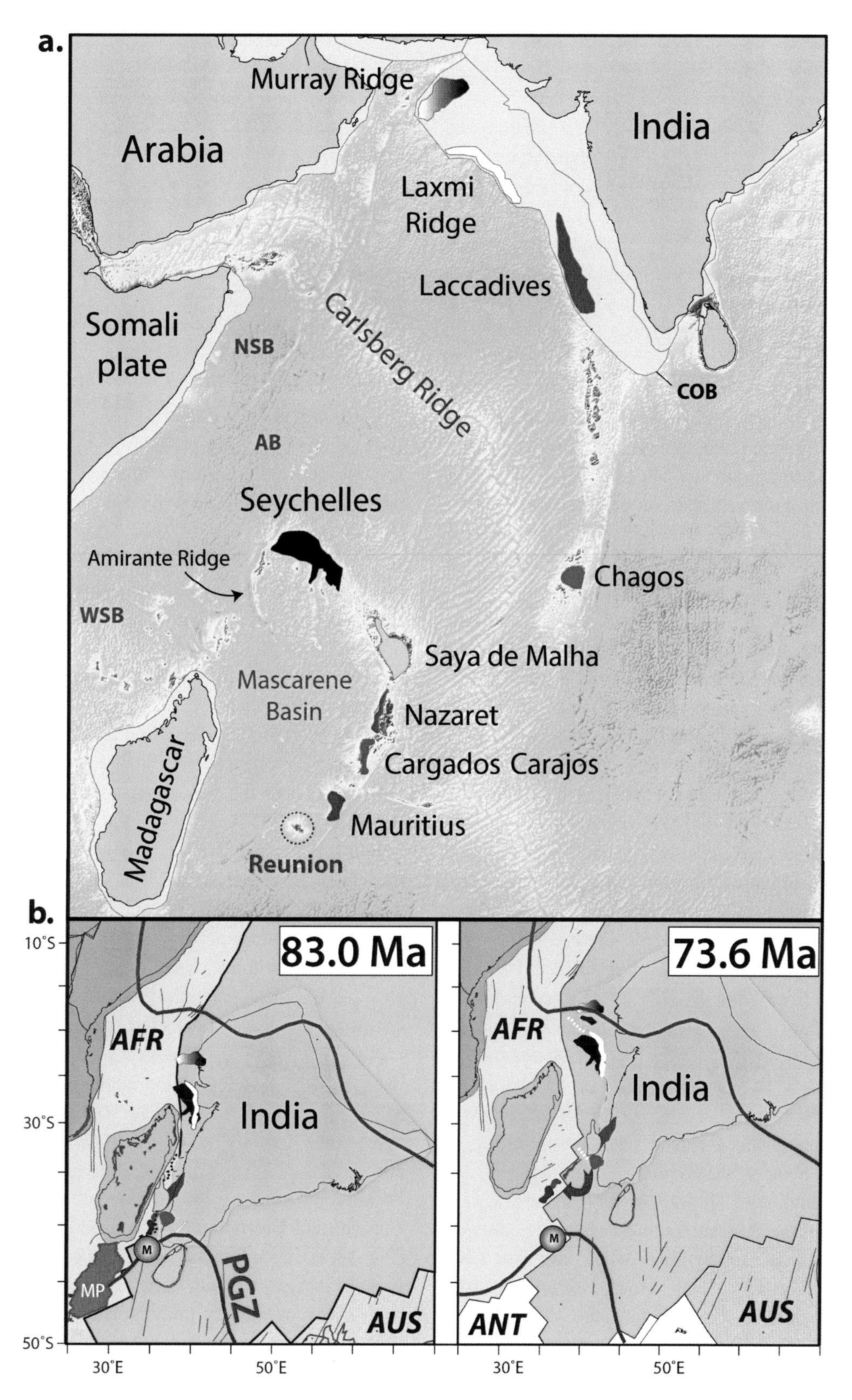

Fig. 13.7 (a) Present-day locations of postulated continental fragments in the Indian Ocean draped on ETOPO1 bathymetry. AB, Amirante Basin; COB, Continent–Ocean Boundary; NSB, North Somali Basin; WSB, West Somali Basin. (b) Late Cretaceous plate reconstructions (CEED mantle frame) with surface location for the Marion Hotspot (M). During the early opening of the Mascarene Basin (83.5–70 Ma), some Mauritian fragments (e.g. Mauritius and the Cargados Carajos) were attached to Madagascar but relocated to the Indian Plate through south-west-propagating ridge jumps (dotted white lines in diagram for 73.6 Ma indicate extinct spreading ridge 80 Myr old). All Mauritian fragments were part of the Indian Plate after 70 Ma until the Late Paleocene at about 56 Ma. AFR, African Plate; ANT, Antarctic Plate; AUS, Australian Plate; MP, Madagascar Plateau.

Fracture Zone), and by the Middle Albian (~100 Ma) sea-floor spreading to the north of the Niger segment, at which time the South Atlantic became fully connected with the Central Atlantic (Figs. 13.2 and 13.9).

In the Central Atlantic, sea-floor spreading propagated north-eastwards to the Iberia–Newfoundland margins at around 125 Ma (Barremian) (Figs. 13.1 and 13.8). Spreading between Iberia and Newfoundland was connected to a rift zone adjacent to the Porcupine and Rockall Plateaus, and continued northwards to the Labrador Sea (between Greenland and North America) and between Greenland and Eurasia. Breakup between Porcupine and North America (Fig. 13.10) occurred after 105 Ma (Albian). The detailed position of Iberia with respect to its neighbouring plates is debated, but the Iberian Plate may have acted independently from about 121 to 83 Ma (Aptian to Campanian), and its plate boundaries probably accommodated Albian and Aptian counter-clockwise rotation, which opened the Bay of Biscay and produced a diverse system of strike-slip, transpression, and transtension north of Africa (Vissers & Meijer, 2012).

Initiation of sea-floor spreading in the Labrador Sea is also much debated. Rifting and volcanic activity between North America and Greenland had started in the Late Jurassic to Early Cretaceous, but sea-floor spreading in Labrador probably did not start before 70–67 Ma (Campanian), and marked the end of the stable Laurentian continental landmass.

The north-east Atlantic (between Greenland and the British Isles and Norway) was characterised by widespread Late Jurassic to Early Cretaceous and Late Cretaceous to Paleocene rift basins, but sea-floor spreading there did not start before the Early Eocene. The Early Eocene also marked the opening of the Eurasia Basin in the High Arctic, but other (difficult to identify) episodes of sea-floor spreading may have occurred in the Amerasia Basin (Fig. 13.8) from the Late Cretaceous to the Paleocene, as the North American and Eurasian Plate experienced extension to accommodate the opening of the Labrador Sea (Gaina et al., 2013a).

Africa, Arabia, and India. Major Cretaceous rifts developed within the African continent which acted as diffuse plate boundaries subdividing Africa into five major tectonic blocks (Fig. 3.4): South Africa, North-East Africa, North-West Africa, Somalia, and the Lake Victoria Block (Torsvik et al., 2009; Gaina et al., 2013b). The last two were attached to the South African block during the Cretaceous, whilst predominantly extension took place between North-West Africa, North-East Africa, and South Africa from 130 to 120 Ma (Early Cretaceous) to the Late Cretaceous at about 84 Ma (Hauterivian to Santonian).

There is a widespread Upper Cretaceous (Cenomanian to Middle Turonian) unconformity on the Arabian Plate which was probably caused by the onset of Semail ophiolite obduction along the south-eastern margin of the plate. The ophiolite suite that is best preserved today is in the Musandam Peninsula of Oman (Fig. 13.9). It is dated to 96–95 Ma (Cenomanian) and was emplaced south-westwards onto the continental margin (Glennie, 2006). Ophiolite obduction probably began at around 96 Ma and ended at about 70 Ma (Maastrichtian), with deposition of shallow-marine limestones which continued until the Late Eocene at around 35 Ma (Searle et al., 2014).

Ophiolites are also found in south-eastern Oman but they have different ages and probably formed near a plate boundary between eastern Arabia and north-west India. One of those ophiolites, the Late Jurassic age (150 Ma) Masirah Ophiolite (MO in Fig. 13.9), was obducted at the very end of the Cretaceous at 65 Ma (Gaina et al., 2015). The Masirah Ophiolite and others found in Pakistan and Afghanistan were probably parts of Jurassic–Cretaceous oceanic crust (North and West Somali Basins) which formed between India and Arabia after Gondwana breakup. Oblique convergence and north-westward intra-oceanic subduction between Arabia and India, starting at around 84 Ma (Santonian), probably led to ophiolite obduction onto south-east Arabia and the north-western part of the Indian Plate (Gaina et al., 2015).

South America. The western margin of South America is the classic example of an Andean-type subduction zone, which is characterised by subduction of oceanic plates beneath a continental margin (Fig. 13.9a) and mountain range uplift without continent–continent collision (Kay et al., 2005). The Andes Mountains stretch for some 8,000 km from Venezuela to the southernmost tip of South America (Tierra del Fuego) and have a complex deformational and magmatic history. Subduction along the western margin has been active since at least the Jurassic, with the Farallon and Phoenix plates originally being subducted beneath the northern and southern parts of South America. From about 120 to 85 Ma (Aptian to Santonian), three oceanic plates (Farallon, Chazca, and Catequil) interacted with the Andean margin (Fig. 13.9a). Late Jurassic and Cretaceous deformation at the western margin of South America was dominated by extension, and subsequently predominantly contraction from the Late Cretaceous to the Recent (Maloney et al., 2013).

As mentioned in Chapter 2, there is no consensus on the localisation and amount of intra-plate deformation in the South American continent. For example, Torsvik et al. (2009; Fig. 12.4b,c) and Moulin et al. (2010) have rather mobilistic views, whilst, for example, Heine et al. (2013; South America model portrayed in Fig. 13.1a) invoke less dramatic adjustments. Torsvik et al. (2009) divided South

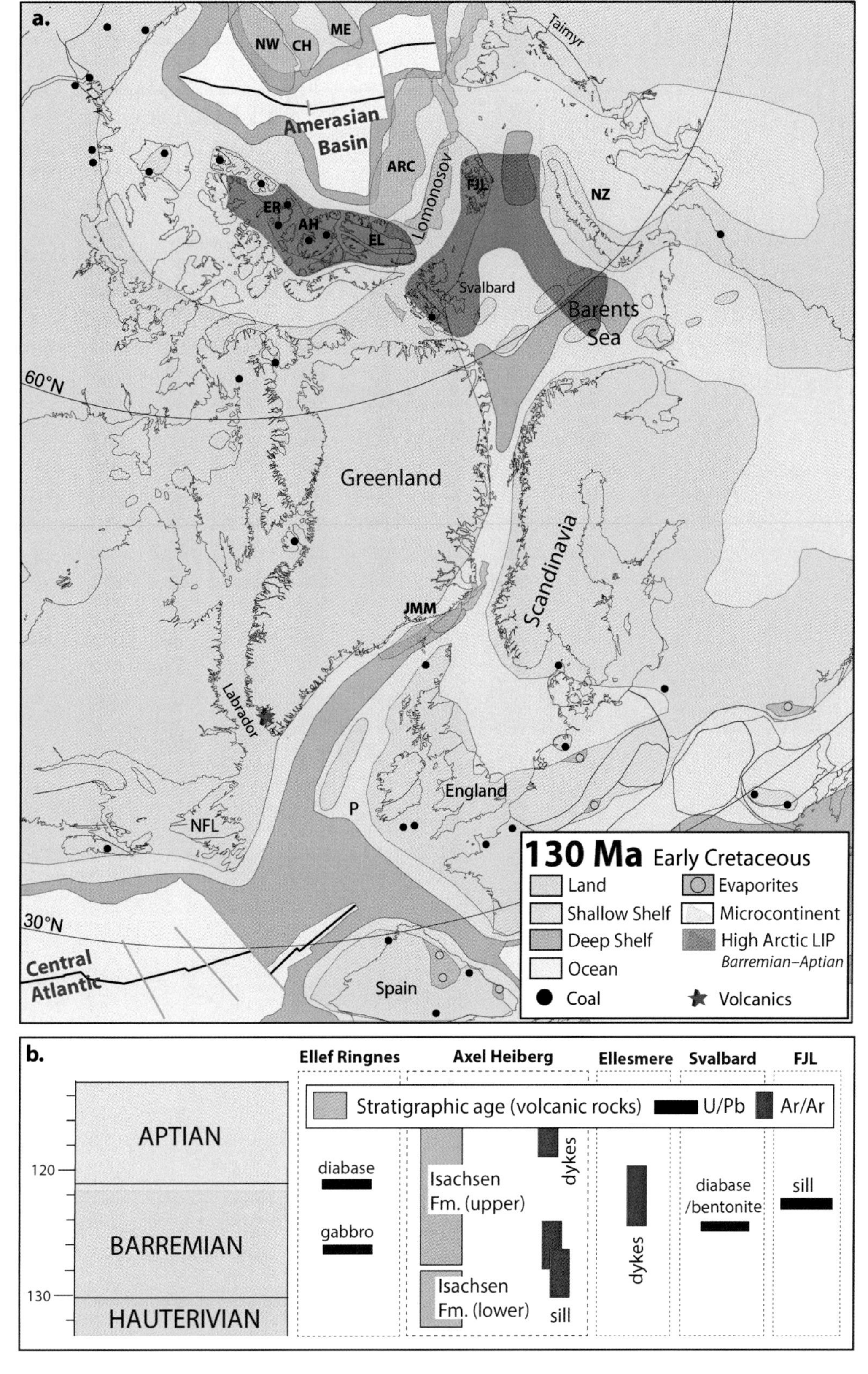

Fig. 13.8 (a) Palaeogeography of western Europe, north-eastern North America, and the Arctic area in the Early Cretaceous at 130 Ma (Hauterivian), showing the contemporary lands, seas, and oceans and the extent of the slightly later (Barremian to Aptian: about 121 Ma) High Arctic Large Igneous Province. The microcontinental blocks now submerged under the Arctic Ocean are shown in grey: it is uncertain what the sea depths above them were at that time. AH, Axel Heiberg Island; ARC, Alpha Ridge of potential continental origin; CH, Chukchi; EL, Ellesmere Island; ER, Ellef Ringnes; FJL, Franz Josef Land; JMM, Jan Meyen Microcontinent; ME, Mendeleev Ridge; NFL, Newfoundland; NW, Northwind; NZ, Novaya Zemlya; P, Porcupine Bank. (b) Compilation of stratigraphic and isotopic ages (U/Pb and Ar/Ar) for the initial volcanism in Ellef Ringnes, Axel Heiberg, Ellesmere, Svalbard, and Franz Josef Land. Data from many sources, including Corfu et al. (2013) and Evenchick et al. (2015).

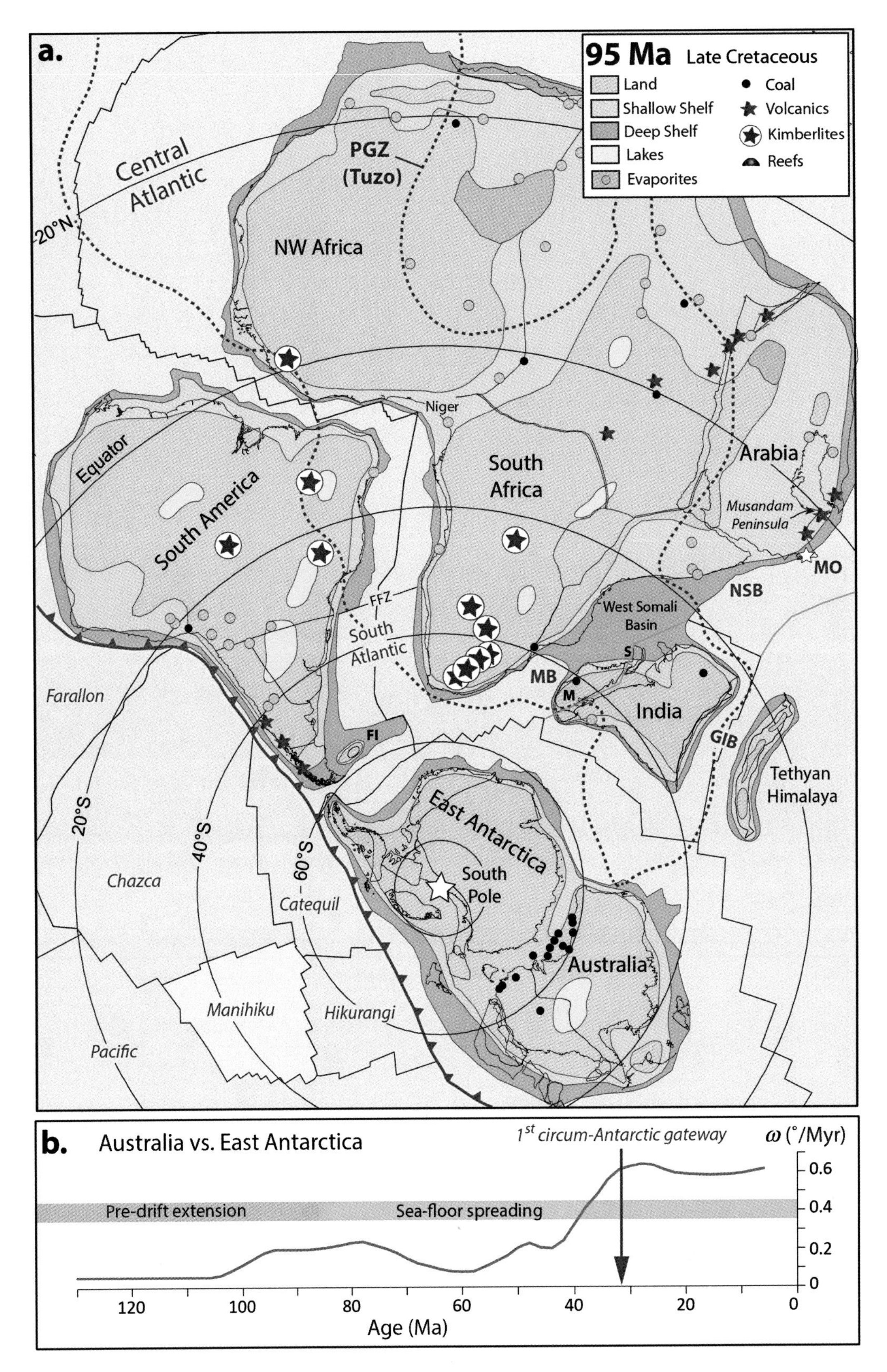

Fig. 13.9 (a) Late Cretaceous palaeogeography of Africa, India, and the southern hemisphere at 95 Ma (Cenomanian), a time when sea level was at its highest and many of the old cratonic areas were flooded, such as in the Trans-African Seaway. The oceanic plates in today's Pacific area are synthetic. FI, Falkland Islands; FFZ, Florianopolis Fracture Zones; GIB, Greater India Basin; M, Madagascar; MB, Mozambique Basin, MO, Masirah Ophiolite (150 Ma) obducted at the end of the Cretaceous (65 Ma); NSB, North Somali Basin; PGZ, plume generation zone. (b) Relative plate velocity between Australia and East Antarctica. Velocities are averaged over a 10 Myr window and plotted at 2 Myr intervals.

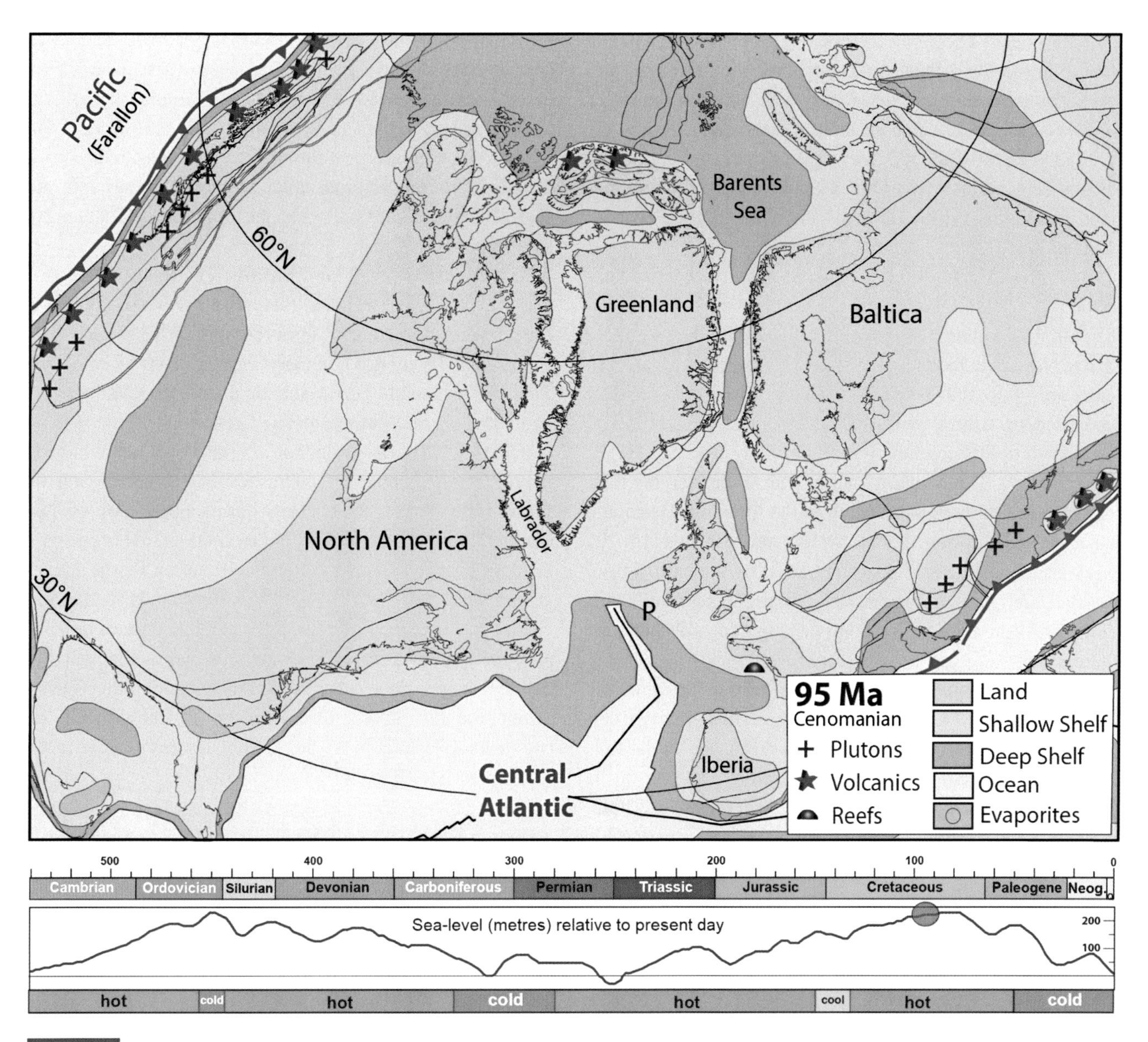

Fig. 13.10 Late Cretaceous palaeogeography of North America and Europe, at 95 Ma (Cenomanian), a time when sea level was at its highest and many of the old cratonic areas were flooded, such as in the seaway running north–south across North America. Data from many sources, including Ziegler (1990). P, Porcupine bank.

America into four main blocks: Amazonia, Paraná, Colorado, and Patagonia. A very crucial boundary in this model is that between Amazonia and Paraná, where a dextral transtensional fault zone (the Paraná–Etendeka Fault Zone: Figs. 2.15b and 12.4b,c) was active between 150 and 126 Ma (Late Jurassic to Barremian). That rift boundary was important not only in reducing the pre-drift extension on the Brazilian (Santos) and the conjugate African margin to acceptable values, but also in providing a zone of lithospheric weakness which has allowed upside-down drainage of the Tristan Plume to cover vast areas of Brazil with the Paraná–Etendeka flood basalts. The Paraná–Etendeka LIP was intruded at about 134 Ma (Valanginian) and occupies virtually all of the 1,200,000 km^2 of the Paraná Basin; there are also many related dykes beyond the borders of the main basin.

Laurasia. After the opening of the Central Atlantic Ocean, a large united continent to its north which stretched from western North America to eastern Siberia was formed, termed Laurasia. The eastern Asian blocks, including the united North and South China continents, joined Laurasia at about the Jurassic–Cretaceous boundary time (145 Ma), and thus Laurasia was at its largest only during the Cretaceous (Figs. 13.1 and 13.2). It was subsequently split into its two modern halves by the opening of the North Atlantic Ocean in the Early Paleogene. However, the latter was preceded by a buildup of extensional stress between the major sectors of Laurasia which eventually facilitated rifting in the Labrador–Baffin Bay and Norwegian–Greenland areas, and crustal separation between the Grand Banks of Newfoundland and Iberia had started by the Early Cretaceous at about 125 Ma (Aptian).

In western North America, igneous activity changed substantially with the onset of the Laramide Orogeny at about 85 Ma (Santonian), when intrusion within the immense Sierra Nevada batholith ended, and the magmatism migrated eastwards into the Rocky Mountains, where the orogeny consisted of rifting, with the development of basins and associated uplifts continuing until the Early Eocene (see the next chapter). The northern Interior and Arctic margins were transgressed through subsidence in the Sverdrup Basin and adjacent areas during the late Aptian and Albian (115–105 Ma). That subsidence was accompanied by the extrusion of basalts and the intrusion of gabbro dykes and sills (Fig. 14.8): activity which continued sporadically until the Turonian.

The Mediterranean Area and the Initial Phase of the Alpine Orogeny. Although most of the Africa–Eurasia collisional events occurred in the Tertiary, in the Cretaceous there was a considerable amount of tectonic activity in the Alpine area of southern Europe (Froitzheim et al., 2008; Schmid et al., 2008; Gaina et al., 2013b). Intra-oceanic subduction in the eastern Alpine realm and eastern Mediterranean region had already started in the Middle Jurassic and culminated in ophiolite emplacement over both the African (Adriatic) and the Eurasian units in the Early Cretaceous. That was followed by intra-continental subduction in the eastern Alps, where tectonic shortening is documented in the Valanginian at about 137 Ma in the Carpathians and Dinarides, which resulted in nappe formation which had formed a pile over 10 km thick before the end of the Cretaceous, and which in turn resulted in metamorphism as high as eclogite grade during the Turonian at about 92 Ma. After the Turonian, oceanic subduction of the Piemonte–Ligurian (or Alpine Tethys) Ocean occurred in the Alps and Carpathians, and the previous continental nappe stack underwent exhumation and resultant cooling, which lasted from about 90 Ma (Turonian) to 60 Ma (Paleocene). To the southeast, in Greece and Turkey, northward oceanic subduction was active throughout the Cretaceous. Contraction in the western Mediterranean region was focused in the Pyrenees and was associated with the rotation of Iberia versus Eurasia. After about 85 Ma (Santonian), convergence was accommodated throughout the Mediterranean region, with Europe as the down-going plate in the Alps and Carpathians, and Africa (Adria) as the down-going plate everywhere else. In the eastern Mediterranean region, a phase of intra-oceanic subduction started in the Cenomanian at around 95 Ma, leading to ophiolite emplacement throughout the Late Cretaceous in Turkey, Syria, and Cyprus, which was synchronous with, and also probably connected with, the ophiolite obduction history of northern Oman.

The Deccan Traps Flood Basalts. The Deccan Traps LIP was intruded into the Indian Plate as it progressively moved over the Réunion Hotspot for a few million years during the Cretaceous–Paleocene boundary interval from 67 to 63 Ma. That LIP resulted in major climatic disturbance, and was one of the two prime causes of the Cretaceous–Tertiary (K-T) boundary biological extinction events (see below). More than 2 million cubic kilometres of lava erupted onto the Indian subcontinent, the majority during a period of less than a million years (Courtillot & Renne, 2003). Offshore areas in Western India were also affected by Réunion Plume activity at that time, and Silhouette and North Islands in the Seychelles are remnants of an alkaline plutonic–volcanic complex (63.5–63 Ma: Early Paleocene) which overlapped with the final stages of the cataclysmic volcanism in India (Owen-Smith et al., 2013). The Deccan LIP event subsequently led to the final separation of the Seychelles from India in the Late Paleocene (Fig. 14.6).

Facies, Floras, and Faunas

Climate. Since it was such a long period, it is no surprise that Cretaceous climates varied considerably. For most of the period it was much warmer than average, including some of the warmest climates in the whole Phanerozoic, with a temperature peak in the Turonian at about 90 Ma, and thus lush and varied vegetation flourished over all of the Arctic and Antarctic land areas. In contrast, although there is no evidence for Cretaceous glaciation, the Earth subsequently became significantly cooler near the end of the Campanian at about 70 Ma, to reach a Cretaceous minimum at the K-T boundary at the very end of the period at 65 Ma, although the temperatures rose again to become more equable during the subsequent Paleocene. In addition to the effects of the Deccan Traps and the Mexican meteorite impact, that cooling must also have played at least some part in the end-Cretaceous terminal extinction event (see below).

Sea Levels. The Cretaceous saw the highest sea levels in the whole Phanerozoic, which peaked at about 95 Ma (Late Cenomanian). That can be explained by a combination of very different factors, including the absence of polar ice caps, the high temperatures (which increased the volume of the sea water), the larger than average sizes of the mid-ocean ridges, younger than average age of the Cretaceous sea floor, and faster sea-floor spreading (e.g. Seton et al., 2009). As a consequence of the sea-level rise, the total land area was much reduced, with substantial transgressions over the margins of all the cratons; and only some 18% of the Earth's area was above sea level by comparison with 28% today. For example, North America was divided into three separate emergent land areas by the seaways which stretched across the craton from north to south and eastwards from the centre (Fig. 13.10). Greenland was another separate large island to the east of the three North American emergent domains. In south-eastern North America and the Mexico area, the whole area was progressively flooded, enabling free faunal communication between the Pacific and the Atlantic, and there were many bioherms which included agglomerations of large rudist molluscs. However, paradoxically and in contrast, at the Arctic margin of North America, tectonic activity actually caused regression during the Cenomanian (Moullade & Nairn, 1983).

The Cenomanian sea-level rise also resulted in another seaway in north-western Africa, running from the Tethys in the north, southwards across the Sahara, to what is now the Bay of Guinea in the eastern Atlantic (Fig. 13.9). In contrast to much of the rest of the world, Antarctica and Australia appear to have been above sea level for most of the Cretaceous, even when sea level was at its highest during the Cenomanian (Fig. 13.9), although marginal marine basins surrounded them, as well as island arcs.

As mentioned in Chapter 4, the Tuzo LLSVP (Fig. 2.2) was located beneath the heart of Pangea (Fig. 10.1) near the Permian–Triassic boundary, and that concentration of continents above Tuzo, which was a region of upwelling in the mantle (high dynamic topography), contributed to the all-time low sea levels then (Fig. 16.2). The subsequent dispersal of Pangea, starting in the Early Jurassic but accelerating during the Cretaceous, with continents drifting in the direction of regions of negative dynamic topography, may alone have led to a global eustatic sea-level rise of the order of 50–100 m (Conrad et al., 2014). That magnitude is equivalent to other important mechanisms for sea-level change such as oceanic crust production rates and glaciations.

The Chalk Seas. As the name 'Cretaceous' suggests, the system is best known for the widespread very white and fine-grained and often pure limestones collectively known as 'the Chalk', which are chiefly made up of the calcitic shells of dead coccolithophores, which are microscopic nannoplankton. Coccoliths live in the clear waters of the modern tropics at depths from 50 to 200 m, and, although coccolith blooms are still relatively common in the warmer oceans of today, because of the high Cretaceous temperatures they have never since been as abundant, and post-Cretaceous true chalks are uncommon. Chalky sediments blanketed large areas of the floors of the shallow seas which covered much of Europe and North America in the second half of the Cretaceous, and progressive local folding and the deepening of the grabens such as the one in the Central North Sea enabled the chalk to reach a thickness of over 2,000 m in a few places (Ziegler, 1990). Most of the bands seen, which are often picked out by lines of flint in the Chalk, represent seasonal variation, and have been linked to orbital changes. However, sedimentation was not as even as appears at first glance in many outcrops, and there is local evidence of slumping and sliding as well as much bioturbation caused by burrowing animals, largely arthropods. Because the Chalk substrates were soft and appear to have remained relatively uncompacted for some time after their initial deposition, the benthic faunas that lived in them tended to have specialised modes of life; for example the abundant large bivalve *Inoceramus* had one valve which was much inflated so as to elevate the other above the sea floor, and other molluscs developed spectacular spines to keep them stable in the soupy sediment.

The high global sea levels meant that many of the previous land areas were flooded (Figs. 13.8–13.11), and as a result the input of clastic sediments into the chalk seas was low. However, some were deposited and thus, although the rock appears to be a limestone, the proportion of fine clastic mudstone in the Chalk can be as high as 30% in some beds. Although chalk deposition persisted until the Maastrichtian in a few places such as Denmark, the youngest chalks at most sites, for example in south-eastern England, are of Campanian age, and were laid down more than 10 Myr before the end of the Cretaceous (Mortimore, 2011).

Interbedded with the chalk at many horizons, sometimes as continuous layers and sometimes as discrete nodules, there are quantities of cherts colloquially known as 'flints'. The flints formed from siliceous oozes which became concentrated on the shallow sea beds, and which often entombed dead seashells, which are sometimes consequently silicified and exquisitely preserved as fossils. Well-known chalk fossils include *Micraster*, an echinoderm sea urchin which originally lived burrowed within the soft chalky sediments, specimens of which are commonly found isolated as flint silica moulds of their interiors in soils overlying chalk rocks. The original calcium carbonate shell has been leached away by modern weathering, and the remaining fossils are fancifully termed 'shepherd's crowns' in folklore.

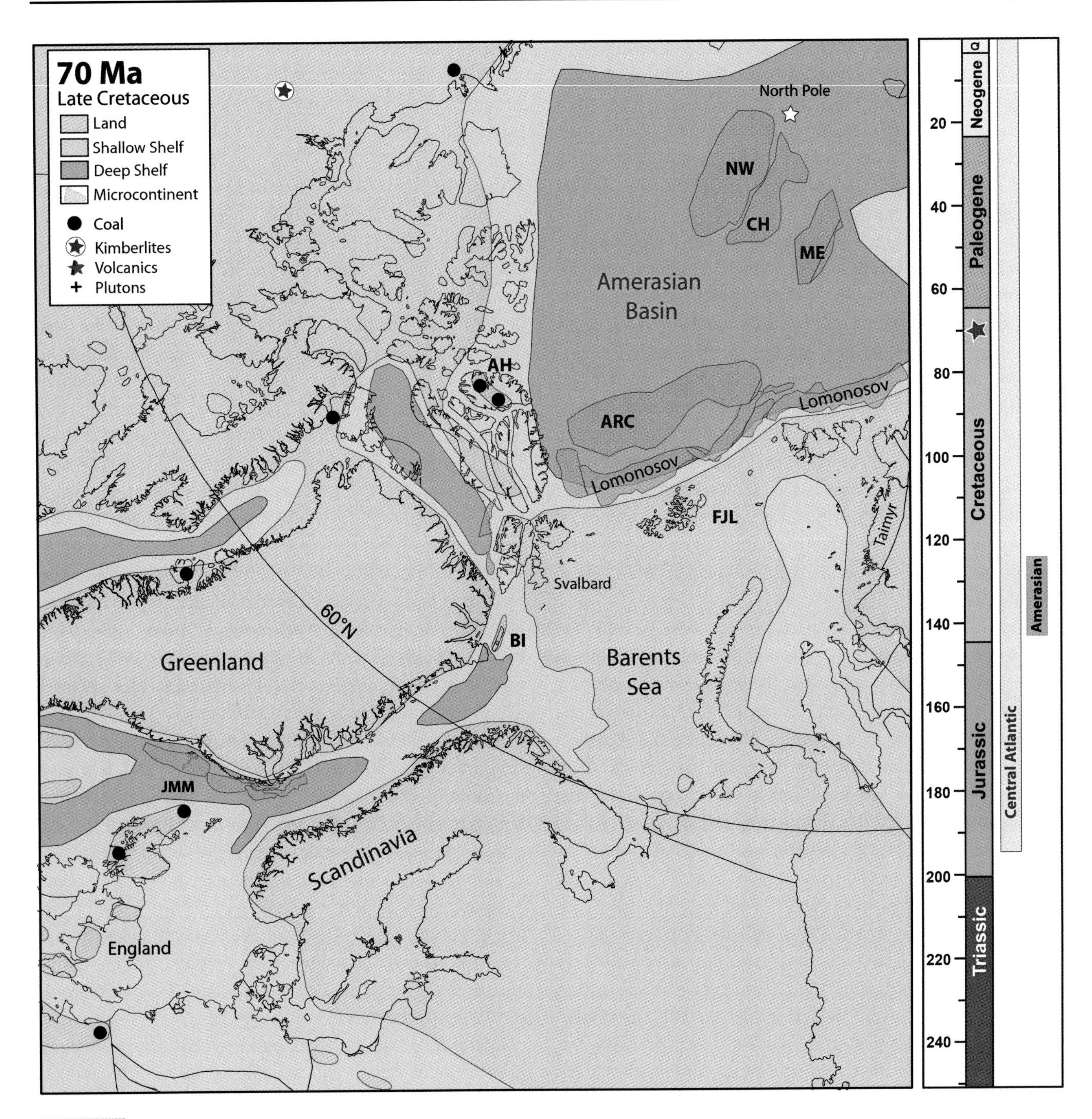

Fig. 13.11 Palaeogeography of Western Europe, north-eastern North America, and the Arctic area in the latest Cretaceous at 70 Ma (Maastrichtian), showing the contemporary lands and seas (there were no true oceans in that area). The microcontinental blocks now submerged are shown in grey: it is uncertain what the sea depths above them were at the time. AH, Axel Heiberg Island; ARC, Alpha Ridge of potential continental block; BI, Bear Island (Björnøya); CH, Chukchi; FJL, Franz Josef Land; JMM, Jan Mayen Microcontinent; ME, Mendeleev Ridge; NW, Northwind Ridge.

The Non-marine Sediments of Southern England. Over a substantial area of southern and south-eastern England, non-marine Lower Cretaceous sediments were deposited from the Late Berriasian (about 140 Ma) to the Aptian (120 Ma), and they are collectively termed the Wealden. There are two major facies groups: largely oxidised clastic rocks of various grain sizes indicating distant meanderplain to proximal braidplain river deposits and prodeltaic fan settings; and mudstone-dominated successions which were deposited in lakes, lagoons, and on mudflats of fluctuating but mostly low salinities (Radley & Allen, 2012). The area varied between submergence and emergence, and, in the latter, muds with desiccation cracks and animal footprints are often seen.

The Wealden is famous for the animals and plants which have been found fossil within it for many centuries; these include the first bones ever recognised from anywhere as a separate group of extinct reptiles which were named 'dinosaurs' (terrible lizards) by Richard Owen in the 1840s. Those bones, which had been found in Sussex and recognised as different from modern reptiles by Gideon Mantell, a local medical doctor, were of *Iguanodon*, a fairly large (up to 8 m), common and herbivorous erect bipedal dinosaur which lived in substantial herds. Some of the Wealden rocks are packed with remains of freshwater molluscs, including the small snail *Viviparus* and both large and small bivalves, as well as diverse freshwater ostracod arthropods. The remains of fossil forests with large tree trunks are often preserved, as well as leaves and seeds, and also varied and abundant insects, including dragonflies. Above the largely non-marine Wealden, there are conformable marine rocks which reflect the subsequent rise in global sea levels.

North-East Laurasia and China. The Mongol–Okhotsk Ocean between Siberia and the formerly independent combined Amuria–North China continent closed at about the Jurassic–Cretaceous boundary time at 145 Ma. Most of that region remained land after the closure, and there were numerous large and small lakes in the Amuria–North China area, including the very big Quingyang Lake in Shaanxi, Gansu, and Ningxia provinces, which covered an area of over 130,000 square kilometres (Zhang et al., 2003). The freshwater sediments laid down in the lakes contain abundant fossils, including conchostracans, ostrocods, bivalves, insects, fish, and plants. There were substantial mountains at the eastern margin of South China, including many active volcanoes, which were often separated by intermontane basins, many of which contained smaller freshwater lakes with different freshwater animals and plants from those in North China.

In addition to the lakes, there were many swamps in the Early Cretaceous (Neocomian), resulting in extensive coal deposits in north-eastern China, eastern Mongolia, and Transbaikalia, and at many sites in north-eastern Siberia. After a hiatus, coarse-grained continental red beds were laid down unconformably over Neocomian rocks in the Late Cretaceous, particularly in north-western China, but also in the former South China. In the latter (especially in Guangdong Province) abundant dinosaur eggs have been found fossil; these are exported for sale in street markets all over the world.

North-west Laurasia, Including Europe. Although still united as parts of the former Pangea, many parts of the North American and European cratons became flooded, particularly during the unusually high sea-level stands prevalent during much of the Cretaceous, with the result that the land areas were smaller and more divided than hitherto (Fig. 13.11). The Chalk seas (see above) covered much of the continental shelf during the latter part of Cretaceous time, and there is little evidence that much of the surrounding land was mountainous, except where active orogenesis was occurring in the mountains today forming the western margin of North America. Since the ambient temperatures were also higher than normal, the flora radiated (see below) and was clearly luxuriant at many sites.

In the north-western Arctic margin, thick sediments were deposited, for example in the deepening Sverdrup Basin, but, owing to a combination of variable tectonic activity (including orogeny in the Brooks Range of northern Alaska) and high sediment input, the various basins there were divided into several discrete depositional areas, with small horsts and local grabens, and the rocks in the median part of the basin were pierced by halokinetic deposits (Moullade & Nairn, 1983).

Africa, Arabia, India, and South America. As noted above, Africa and South America were united within West Gondwana at the start of the Cretaceous, but, between 130 and 100 Ma, they split from the south northwards towards the Equator to initiate the South Atlantic Ocean (Fig. 13.1). In northern and eastern Africa there was a passive margin on the southern margin of the Neotethys Ocean, and some flexing of the Arabian Plate facilitated intra-shelf basins with sediments, mostly on an extensive carbonate platform, including the oil-bearing Sirt Basin of Libya; and local unconformities, such as the hiatus during the Late Valanginian to Early Hauterivian in the Kuwait Basin. However, there were also interbedded clastic deposits derived from a delta inland to the west, and in the subsurface there were substantial Aptian and Albian halokinetic movements in the Precambrian evaporites, causing salt domes which were important in concentrating the abundant Upper Ordovician and Lower Silurian hydrocarbon source rocks there into oil and gas reservoirs of global industrial importance. The wide areas of flooded craton during the Cenomanian sea-level high stand at 95 Ma are shown in Fig. 13.9. Sediments deposited within a compressive foreland basin setting were deposited along the eastern margin of the Arabian Plate during the Middle Turonian, together with the ophiolitic activity noted above.

In South America there were many basins, some of which had occasional marine incursions, but many of which housed extensive lakes, with extensive freshwater sediments in many places; for example, in the Neuquén Basin of northern Chile there are over 6,000 m of Cretaceous sediments varying from marine neritic, through sublittoral and littoral, deltaic, and freshwater to continental (Moullade & Nairn, 1983).

In India, the northern (subsequently Himalayan) margin was passive, with a mixture of shallow-marine sediments, whose distribution was largely controlled by the global eustatic changes, with the result that a higher proportion of the craton was flooded during the Late Cretaceous, with its peak in the Cenomanian, reaching as far south as the Coromandel coast of south-east India (Fig. 13.8), where marine sedimentation started in the Late Albian and continued without break into the Early Paleocene. The only known sedimentary Cretaceous rocks in Antarctica are in the Antarctic Peninsula (Francis et al., 2008).

Radiations of Flowering Plants and Terrestrial Animals. Although they are first known from the Lower Cretaceous, it was during the Late Cretaceous that angiosperms (flowering plants) became dominant for the first time over the gymnosperms (for example, living conifers and cycads, as well as numerous extinct pteridosperms) which had dominated the Triassic and Jurassic floras. Prior to then, a typical Middle Cretaceous forest, such as an Aptian one in Antarctica, consisted of three assemblages: a conifer and fern assemblage with mature conifers of mainly araucarian (monkey-puzzle tree) type with an understorey of ferns; a mixed conifer, fern, and cycad assemblage with araucarian conifers and gingko trees; and a disturbance flora, growing in back-swamp areas of a braided floodplain, of liverworts, shrubs, ferns, and just a few angiosperms (Francis et al., 2008). In contrast, a Late Cretaceous (Coniacean to Campanian) forest from Antarctica was characterised by a much greater variety of gymnosperms, which are closely comparable to those in living families. Angiosperms also flourished, partly because of their propensity for insect pollination, and also because their seeds have greater protection than those in other plant groups.

By the end of the Cretaceous, from 50% to 80% of land flora genera were flowering plants, and therefore formed the most varied vegetation, particularly at lower latitudes. However, although flowering plants were 61% of the species present, they represented, for example, only 12% of the area of the vegetation cover on a land surface in Wyoming, USA, which was exceptionally preserved since it was covered by a volcanic ash deposit (Friis et al., 2011). The rise in the abundance and diversity of the flowering plants had a direct effect on the development of new ecologies and the behaviour of arthropods such as bees and other insects which developed new mutual symbioses, so that the latter pollinated the plants, a radiation which has continued on with increasing complexity up until the present day.

As in the Jurassic, the larger terrestrial animals continued to be dominated by the dinosaurs right up until the end of the Cretaceous, and there were many new genera and families among them. However, and much less conspicuously, the Cretaceous also saw further development in mammal evolution. An exceptional phosphatised specimen of the small shrew-sized *Spinolestes*, found in rocks of Barremian age (125 Ma) in Spain, is the oldest mammal in which skin, hair, and internal organs, including the liver and lungs, have been preserved, all similar to those in modern mammals, as well as its dermal armour plates made of keratin, like those of modern armadillos.

Radiation in Marine Animals. Although ammonite cephalopods had dominated the seas in the Triassic and Jurassic, it was not until Cretaceous times that some of their simple symmetrical spiral shells evolved into a surprising variety of asymmetrical forms of very different shapes, most of which are termed heteromorphs, in some of which the shell spirals give the misleading impression of having uncoiled during life. Like their relative the pearly *Nautilus* of today, most ammonites lived in the oceans and used their shells as buoyancy mechanisms so that they could live at some depth during the day and rise to near the surface during the night in order to track the comparable movements of the plankton upon which they fed. However, a minority of ammonite genera appear to have existed only as benthic browsers on the ocean floor over a wide depth spectrum, with those in deeper-water continental shelves and ocean floors having much thinner shells than their shallower-water relatives, which needed more robust shells to protect them from predators.

Invertebrate animal groups less conspicuous than the ammonites also flourished. For example, the Bryozoa, which are colonial animals and most of which have skeletons of calcite which make them occasionally rock-forming in abundance, radiated rapidly during Cretaceous times. Of those, the most dominant today are the cheilostomes, of which over 1,000 genera and 4,000 species are known, but which evolved only during Cretaceous times at about 155 Ma (P.D. Taylor in Selley et al., 2005). Bryozoa have a wider temperature tolerance (and thus a greater latitudinal range) than the more noticeable corals, which are also colonial and often rock-forming. Both groups, together with calcareous algae, crinoids, stromatoporoids, and other biota with calcitic skeletons, form the main framework for reefs and other biohermal mounds up to the present day, although the bryozoan bioherms do not reach the large sizes of the coral reefs.

End-Cretaceous Extinctions. A combination of the toxicity caused by the Deccan Traps LIP and the effects of the fall of a large meteorite at Chicxulub, near the Yucatán Peninsula in the Gulf of Mexico, at the end of the Cretaceous at about 65 Ma caused rapid and devastating changes in both climate and sea composition, which were the immediate causes of the extinction of much of the world's biota, both on land and in the seas and oceans. However, the three previous sharp eustatic falls in sea level which had occurred during the last stage of the Cretaceous, the Maastrichtian, constituted a relevant background factor which had progressively increased the survival problems for many groups, and meant that the number of genera had already declined over the few million years before 65 Ma. In addition, as outlined above, during the final few million years of the Cretaceous, the Earth's climate had become progressively colder from a maximum at between 90 and 70 Ma, which must also have been a factor.

The most famous extinction was of the large dinosaurian reptiles on land and the related ichthyosaurs at sea, although it is now agreed that the continued existence of the birds (which had evolved from dinosaurs in the Jurassic) means that the dinosaurian clade did not really end in disaster at 65 Ma. The only marine reptiles to survive into the Paleocene were turtles and some specialised crocodilians. Other important groups, such as the ammonite cephalopods and the larger planktonic foraminifera in the invertebrates, also became extinct. However, both the ammonites and the non-avian flying reptiles (pterosaurs) had declined steadily since the Campanian, and there are few records of Maastrichtian forms.

Although most of the major phyla survived the extinctions, which were much less drastic than the Permian–Triassic turnover, many hitherto important groups within them did not (MacLeod et al., 1997). Those included the large oyster-like bivalves called inoceramids, which are the most common bivalves found in the Chalk; the bizarre bivalves named rudists, whose conical shells took the apparent forms of large simple corals, the largest of which was over half a metre high and which had formed the major constituents of many bioherms in warmer Cretaceous seas; and the distinctive belemnite cephalopods, whose conical buoyancy chambers are often preserved as fossils resembling bullets in Jurassic and Cretaceous rocks. In many groups the extinction produced varied results; for example, prior to the end of the Cretaceous there were roughly equal proportions of regular and irregular echinoids (sea urchins), and whilst most of the irregular forms survived, only a few regular families persisted into the Neogene, thus skewing the overall abundance in favour of the irregulars. It is notable that most of those irregulars are infaunal burrowers and suspension feeders, in contrast to the regular echinoids, which are largely epifaunal deposit feeders.

In contrast, most fish and reptiles such as lizards and snakes, as well as the majority of plants, appear to have been far less affected at the K-T boundary than the other biota noted above, although the sharp drop in global temperatures at 65 Ma, which took some time to rise again in the succeeding Paleocene, caused many substantial changes in ecology and habitats in both marine and terrestrial biotas (see below).

14 Paleogene

Spider and cricket trapped in a piece of Oligocene tree resin (amber) from the Baltic.

The Cretaceous–Tertiary (K-T) boundary event is one of the best-known milestones in geology, as described in the previous chapter: in addition to the many extinctions at that time, the global temperature dropped sharply and remained low for most of the Paleocene. Modern mountain belts, including the Himalaya, Alps, Cordillera, and Andes, were all initiated or much enhanced during the period. The mountains which they have produced have been the subjects of interests ranging from art to warfare over many centuries, and they have all attracted much attention from geologists, who have struggled to unravel their histories, often with vehement debate, since the birth of the subject over two hundred years ago.

The Paleogene is divided into the Paleocene (base at 65 Ma), Eocene (base at 56 Ma), and Oligocene (base at 34 Ma), which together form the first half of the Tertiary Era of earlier authors, and which are grouped together within a single chapter here. The Paleocene was originally thought of as the earlier part of the Eocene, and was not identified as a separate time period until the 1870s, whilst the terms Eocene (meaning 'Dawn of the Present' from the Greek *eos* 'dawn' and *kainos* 'recent') and Oligocene had both been coined by the famous British geologist Charles Lyell in the 1830s. The base of the Paleocene is defined by an iridium anomaly at El Kef, Tunisia, which reflects ejecta derived from the Caribbean Chicxulub meteorite impact at 65 Ma. The global distributions of the continents and oceans are shown in Fig. 14.1 at 60 Ma (Paleocene: Selandian) and 40 Ma (Eocene: Bartonian).

Tectonics and Igneous Activity

Large Igneous Provinces and Kimberlites. There were three continental Paleogene LIPs and no Oceanic LIPs at that time (Fig. 14.2). The Deccan Traps erupted near the Cretaceous–Paleogene boundary, the North Atlantic Igneous Province became active at 62 Ma, and the East African LIP erupted at 31 Ma (Oligocene). These LIPs were all erupted near vertically above the Tuzo margin. There are 125 kimberlites and they are found in Africa (Namibia, South Africa, Tanzania), India, the USA, and Canada. The North American kimberlites are anomalous in that they erupted above cold regions in the deep mantle.

The Himalayan Orogeny. The Himalayan Mountains include the world's highest mountain range, whose history has been vividly described in the book by Searle (2013). The Himalaya extend from the Karakorum Mountains in the west to north-western Myanmar (Burma) in the east from about 70° to 95° E in longitude, and represent the thrusting of Asia to the north over the Indian Plate to the south.

From the Early Eocene at around 52 Ma, the eastern sector of the Neotethys Ocean closed along the Indus–Yarling Suture Zone between the Tethyan (Tibetan) Himalaya and the Lhasa Terrane of Tibet. South of that suture lies the Himalaya, which is made up from a jumble of metasedimentary rocks that were scraped off the now-subducted Indian continental crust and mantle lithosphere as the collision progressed. Following its accelerating departure from Madagascar at around 83 Ma, with peak Paleocene velocities of around 18 cm/yr (Fig. 14.3), India has continued to push northwards until the present day, indenting into Asia and raising the southern Asian margin, particularly in the Tibet Plateau.

The India–Asia collision is commonly assumed to have happened in the Early Eocene when the Indian Plate was witnessing a major deceleration from 18 cm/yr to about 5 cm/yr (Fig. 14.3e), but all previous plate reconstructions put India too far south of Asia at that time. This has resulted in plentiful models where India is extended northwards (Greater India) in order to be adjacent to Asia by the Early Eocene. In a simple static model, Greater India must have included several thousand kilometres that must have undergone later crustal shortening; however, those convergence estimates are much larger than values documented in the geological record of Asia and the Himalaya.

Van Hinsbergen et al. (2012) maintained that the Himalayan Orogeny is best divided into two phases (Fig. 14.3). The first phase was a collision between India and a microcontinent to the north of India during the Paleogene at about 50 Ma (Early Eocene): a microcontinent that was subsequently fractured, but many fragments of it are still preserved in the Tethyan (Tibetan) Himalaya area. That collision was followed by subduction of a largely oceanic Greater India Basin in the Greater Himalayan area. The second phase was the more substantial India–Asia collision with thicker Indian continental lithosphere, which did not start until near the Oligocene–Miocene boundary time at 25 to 20 Ma, and which has continued until the present day. However, the second phase, the 'hard collision' (Fig. 14.3d,e), was not associated with any plate velocity changes for India, which has been rather stable at about 5 cm/yr since the Middle Eocene.

The Alpine Orogeny. In contrast to the Himalayas, the Alps of southern Europe are the result of a complex series of movements between the African Plate and Eurasian Plate, but, because there has been a considerable amount of lateral movement between the two plates combined with the more direct collision, the Alps form an arcuate structure and understanding them must be integrated with the complex history of the Mediterranean Sea to their south. It is over-

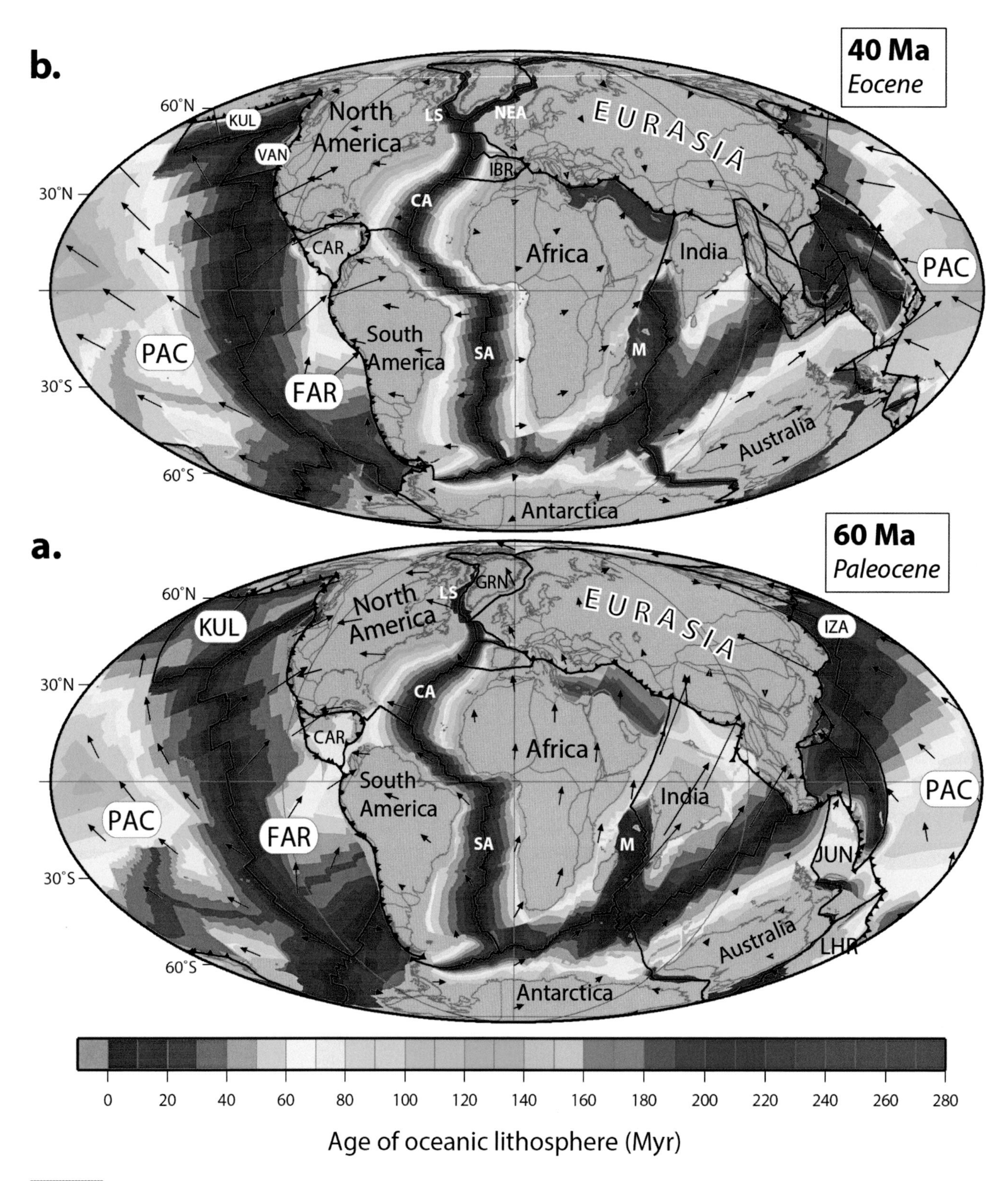

Fig. 14.1 Continental blocks (grey) and oceans (other colours) with plate velocity vectors and the ages of their oceanic lithosphere in the Paleocene (60 Ma: Selandian) and Eocene (40 Ma: Bartonian). CA, Central Atlantic; CAR, Caribbean Plate; FAR, Farallon Ocean Plate; GRN, Greenland; IBR, Iberia; IZA, Izanagi Ocean Plate; JUN, Junction Plate; LHR, Lord Howe Rise; LS, Labrador Sea; NEA, North-East Atlantic; PAC, Pacific Ocean Plate; KUL, Kula Plate; M, Mascarene Basin; SA, South Atlantic; VAN, Vancouver Plate. Generated from EARTHBYTE mantle frame (see Chapter 2).

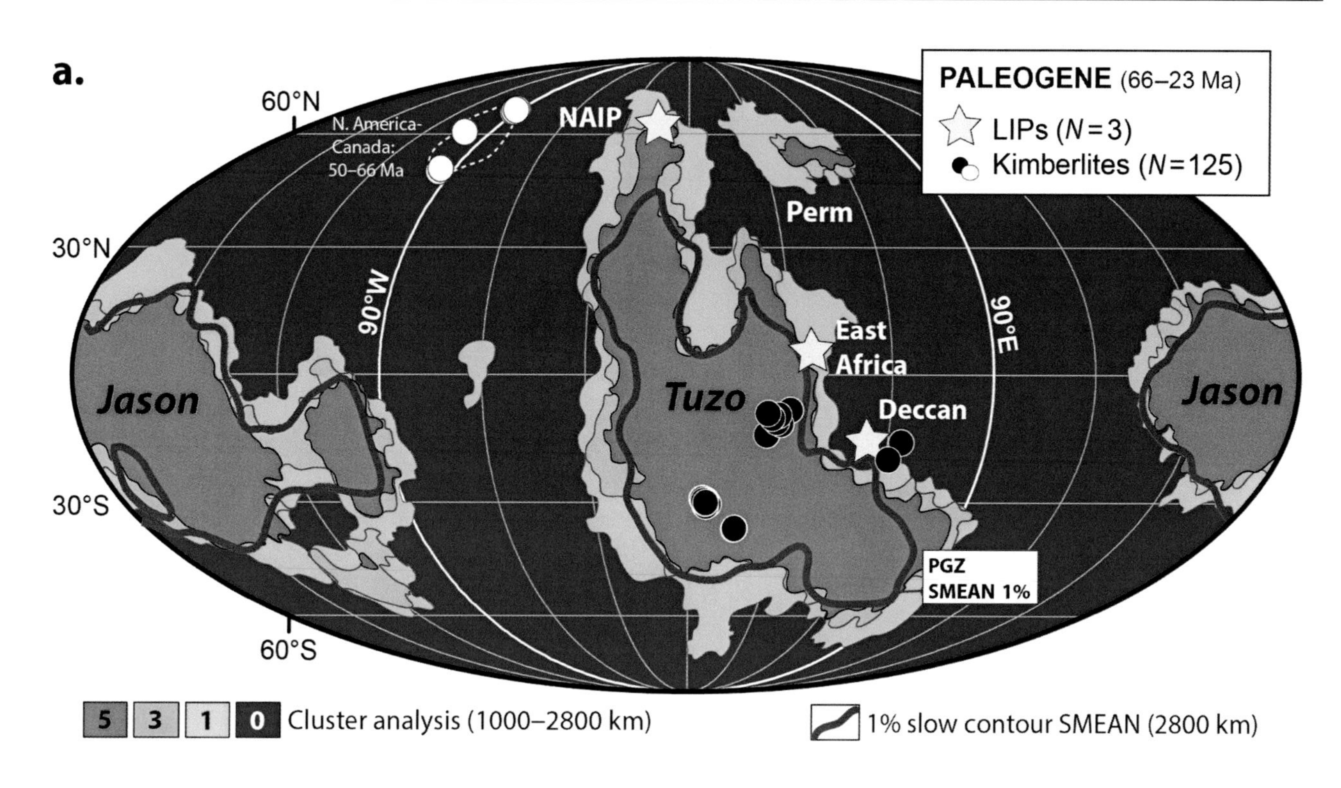

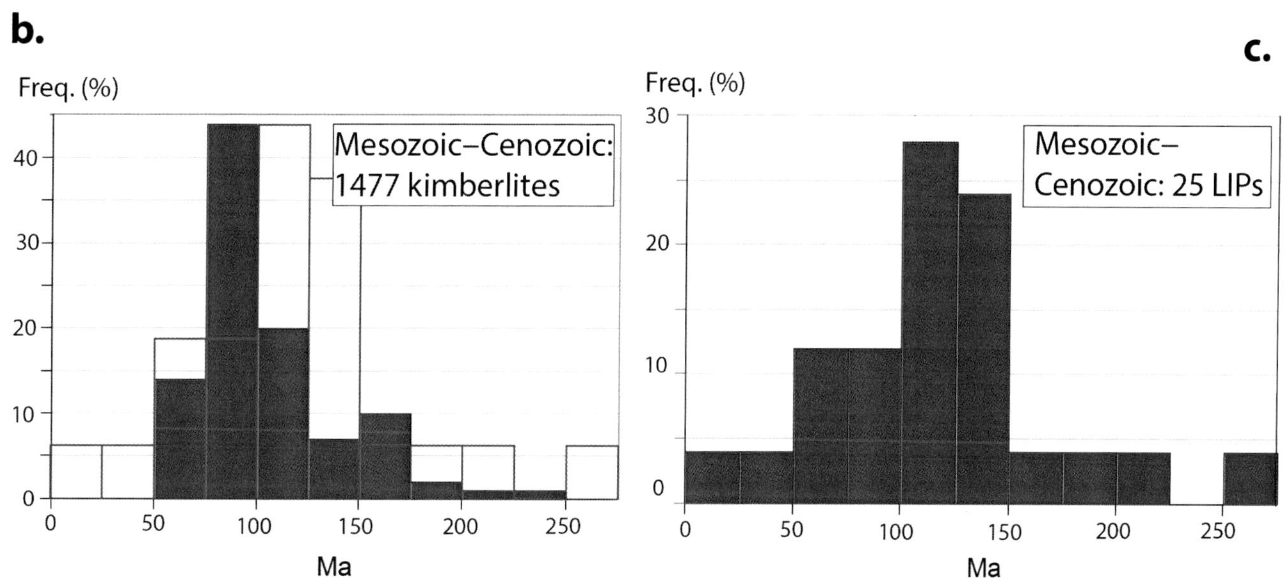

Fig. 14.2 (a) CEED mantle frame reconstruction of the estimated eruption centres (yellow stars) for the Deccan and East Africa LIPs, and 125 Paleogene kimberlites (black and white dots). These are draped on seismic voting-map contours in the lower mantle (Lekic et al., 2012). Areas 5, 3, and 1 define the LLSVPs and 0 (blue) denotes faster regions in the lower mantle (see Fig. 2.16c for detailed description). The 1% slow SMEAN contour is shown for comparison (see Chapter 2 for further details). Histogram (frequency) of (b) kimberlites and (c) LIPs for Mesozoic–Cenozoic times; 75% of all those kimberlites erupted in the Cretaceous with a near 50% peak between 100 and 75 Ma (25 Myr bins). In Fig. 14.2b we have also scaled the LIP frequency to the highest kimberlite peak (black lines). Note that kimberlites often coincide with LIPs or are somewhat younger than the main LIP events. LIPs show a peak between 125 and 100 Ma.

simple to regard the Mediterranean as a simple remnant of a single original Tethys Ocean. The Mediterranean region can be roughly subdivided into a western 'Alpine Tethyan' realm, genetically linked to the opening of the Central Atlantic Ocean, which separated the Adriatic promontory of Africa from Europe and ran from south-eastern Iberia to the Tornquist Line in the Carpathian realm, and an eastern realm that followed the Paleotethys–Neotethys logic. There,

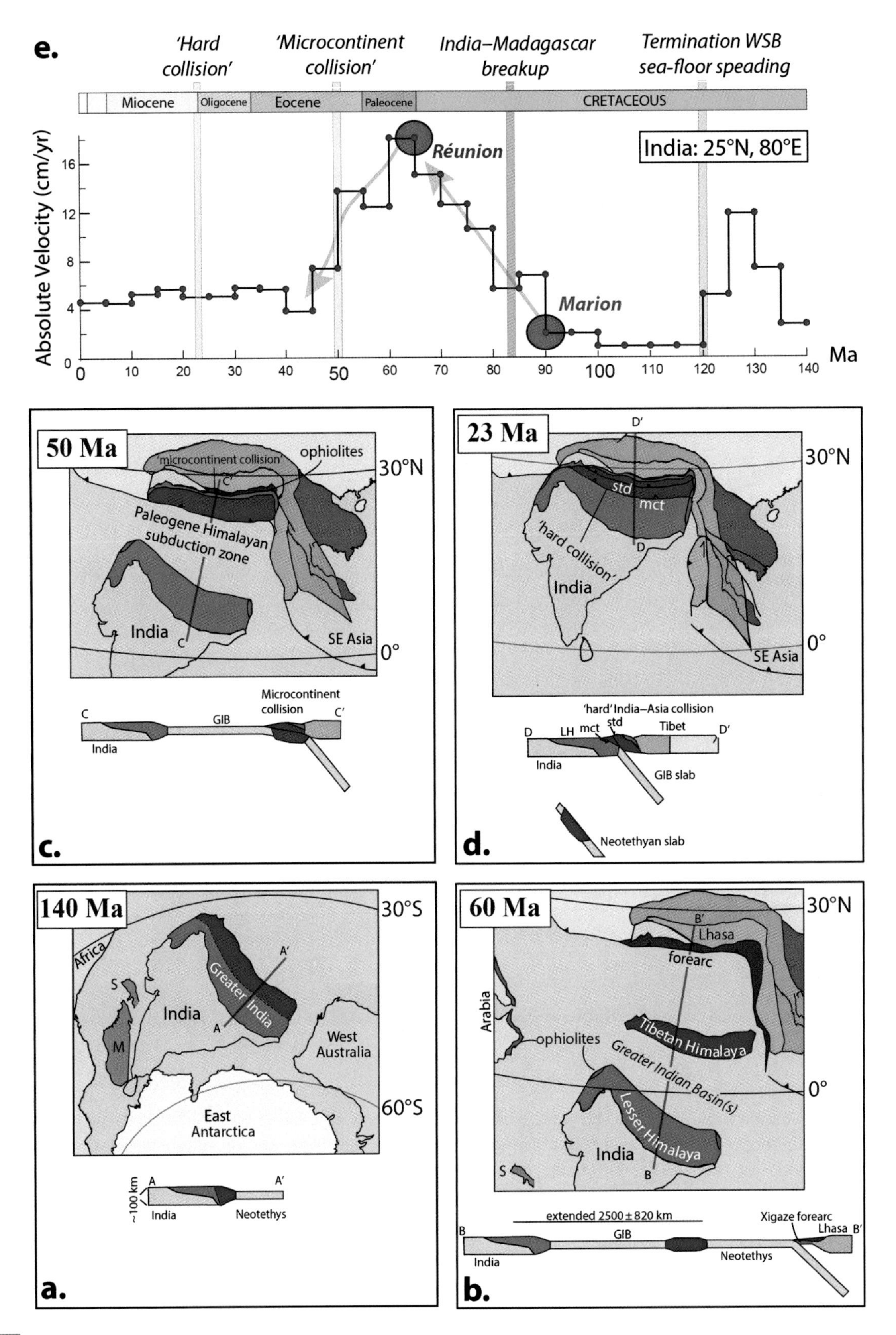

Fig. 14.3 Northward movements of India and surrounding areas from (a) 140 Ma (Early Cretaceous) through (b) 60 Ma (Paleocene) and (c) 50 Ma (Early Eocene) to (d) the end of the Paleogene at 23 Ma (Oligocene–Miocene boundary), together with vertical sketches along the section lines shown on each map. M, Madagascar; S, Seychelles. Simplified and updated from van Hinsbergen (2012 and pers. comm.).
(e) Absolute motion of a location in India (CEED mantle frame) from 140 Ma to present time. The eruption times for the Madagascar LIP (linked to the Marion Plume) and the Deccan Traps (linked to the Reunion Plume) are indicated by large red circles. WSB, West Somali Basin.

a Triassic Neotethys opened through rifting of 'Cimmerian' blocks, including movement of the Sakarya continent away from Gondwana, consuming the Paleotethys Ocean to its north. The Alps formed at the junction between those two systems.

The southern Eurasian margin was passive from the earliest Cretaceous onwards until the Eocene at around 45 Ma, when the convergence of Africa and Europe caused compression at the Helvetic margin of Eurasia, causing its partial inversion. That was followed by Middle Eocene to Miocene flexural subsidence of the European Plate under the load of the advancing tectonic wedge (Cavazza et al., 2004). After that there was a complex series of interactions, which varied along the lengthy Alpine–Carpathian Chain, and resulted in all the aspects of a major orogeny, including sub-horizontal nappe emplacements and the development of narrow basins which became filled with either flysch or molasse, depending on the availability of adjacent bodies of water, which could be either marine or lacustrine. In the western Mediterranean region, northward subduction became associated with volcanic arcs and subduction metamorphism in Eocene time. Arrival of the Adriatic promontory in the northward-dipping subduction zone of Greece and Turkey since latest Cretaceous time led to intra-continental subduction and nappe stacking, which remains active today.

Pacific Ocean. East-dipping subduction continued along the Middle America margin bordering the Pacific Ocean. The Pacific–Izanagi Ridge was subducted under the East Asian margin at around 55 Ma (Early Eocene), leading to the end of the Izanagi Plate (Fig. 14.4b,c), and followed by a change in spreading direction along the Kula–Pacific spreading ridge. After 55 Ma, the eastern Pacific was dominated by the rupture of the Farallon Plate close to the Pioneer Fracture Zone, forming the Vancouver Plate outboard of North America at about 52 Ma (Eocene: Ypresian) (Fig. 14.4c). Further south, spreading continued along the Pacific–Farallon, Pacific–Antarctic, and Farallon–Antarctic Ridges.

Spreading between the Kula–Pacific and Kula–Farallon plates continued until 40 Ma (Late Eocene), but then ceased, leading to the already large Pacific Plate being further augmented by the smaller Kula Plate. The intersection of the Murray Fracture Zone (and probably also the Mendocino Triple Junction) with the North American subduction zone in the Oligocene led to substantial strike-slip movement and the establishment of the San Andreas Fault and the Basin and Range extensional province in western North America and also corresponded to the establishment of the Juan De Fuca Plate (~37 Ma) at the expense of the Vancouver Plate.

A Santonian Hawaii LIP? In Chapter 13 we considered a hypothetical Hawaii LIP (HLIP) erupting in the Early Aptian (120 Ma) on the Izanagi Plate, which was later subducted at around 85 Ma (Santonian). Here we debate a scenario where HLIP was much younger (say 85 Ma) and originally impinging on the Pacific Plate and leading to a major plate reorganisation and the birth of the Kula Plate shortly thereafter.

Initiation of the Kula Plate in the Cretaceous (83–79 Ma) is interesting because the oldest dated fragments of the Emperor Chain (81–76 Ma; Detroit ODP Sites 884 & 1203) are coeval with the birth of the Kula Plate. If a HLIP assisted plate reorganisation, and initiated the Kula Plate, it could have been relocated from the Pacific to the Kula Plate (Fig. 14.4a). Although the western margin of the Kula Plate is generally modelled as a transform boundary, a minor segment could have been initiated as a spreading centre (paralleling the southern Kula Plate spreading axis) which later evolved into a pure transform margin. Depending on the age and size of HLIP, we predict it to have reached the trench between 55 and 50 Ma, and we should expect to locate the HLIP tomographically beneath the Central Bering Sea at depths of 600 km or more, depending on the slab sinking speed and subduction angle. It is interesting to note that a sharp decrease in plate velocities for both the Pacific Plate and the faster moving but coupled Kula Plate is recorded between 55 and 48 Ma (Early Eocene). That is the expected time for a hypothetical and compositionally buoyant HLIP arriving at the trench, obstructing plate motions, and shortly after 47 Ma the Pacific (including Kula) Plate radically changed its northerly plate motion (Fig. 14.4b,c) (paralleling the Emperor seamount chain) to a WNW course (Fig. 14.4d, e), which has been the same ever since, thus creating the Hawaii area seamounts in its wake.

The Emperor–Hawaiian Chain. The chain of volcanic islands and seamounts stretches over almost 6,000 km, from the Detroit Seamount (81–76 Ma) adjacent to the Aleutian Trench in the north-west Pacific to the active submarine volcano Loihi near the Island of Hawaii. A pronounced 60° bend in the Emperor–Hawaii Chain between 47.9 Ma (Kimmei Seamount, Fig. 2.5b) and 46.7 Ma (Daikakuji Guyot) has long been a mystery. Originally, and since the novel paper by Wilson (1963), the bend was interpreted as caused by a rapid change in the movement of the Pacific Plate from a northward (Emperor Chain) to a north-westerly (Hawaiian Chain) direction. The Emperor–Hawaiian Chain was explained by the Pacific Plate drifting over a stationary mantle plume (Hawaii). The Hawaii Hotspot is currently located vertically above the northern margin of the Jason LLSVP (Figs. 2.2 and 15.4).

Explaining the 47 Myr Emperor–Hawaii Bend by only a change in the direction of the Pacific Plate has turned out to be problematic. Pacific and African fixed hotspot reference

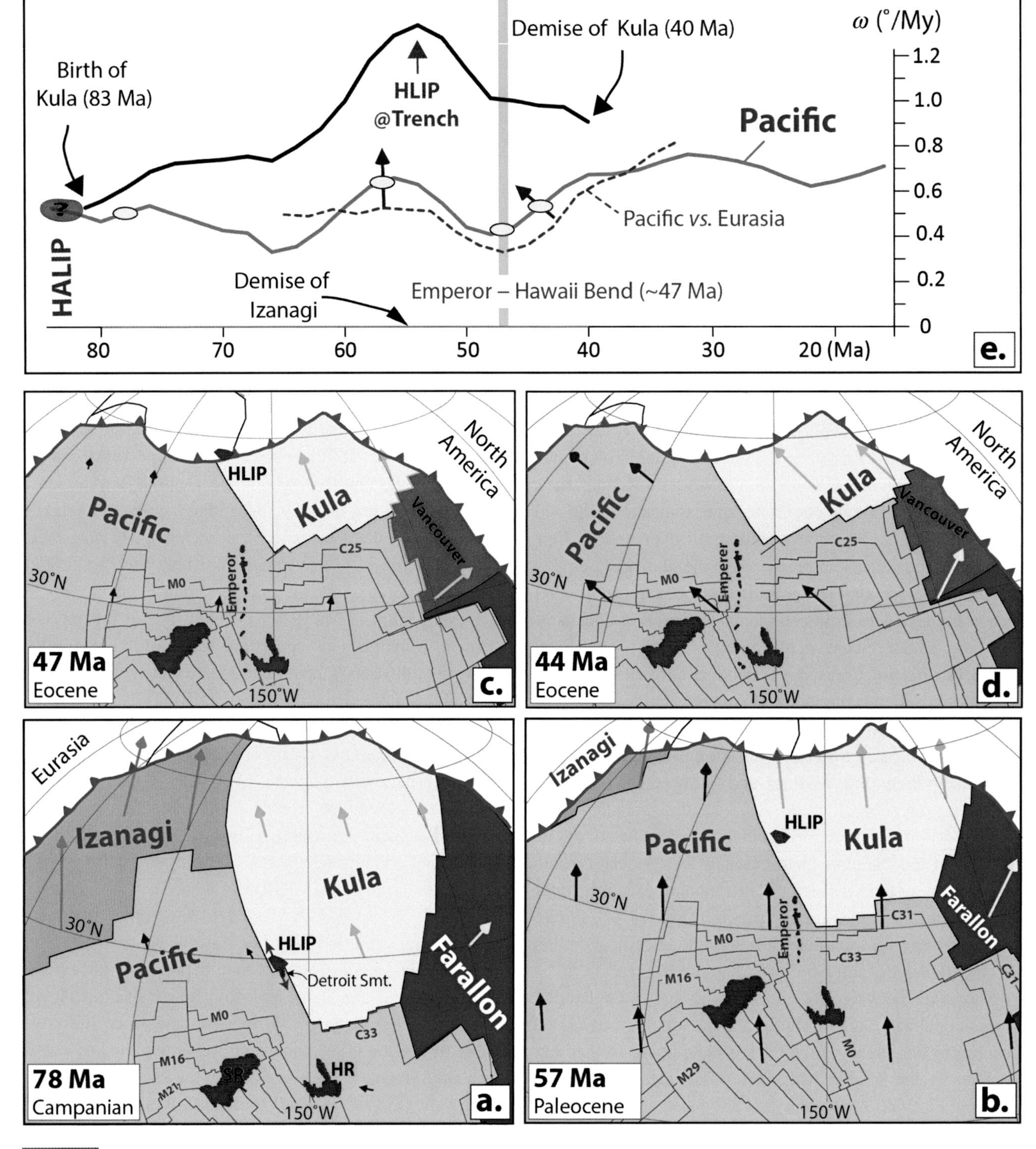

Fig. 14.4 (a)–(d) North Pacific reconstructions from Campanian (78 Ma) to Eocene (44 Ma) times in a CEED mantle frame with plate velocity vectors for the Pacific, Izanagi (before 55 Ma), Farallon, Kula (before 39 Ma), and Vancouver (after 52 Ma) plates. Dynamic self-closing plate polygons are similar to those of Shephard et al. (2014). Annotated LIPs: HLIP, Hawaii LIP (tentatively assumed a starting age of 85 Ma); HR, Hess Rise (99 Ma); Shatsky Rise (147 Ma). We also show identified magnetic anomalies and some important anomalies are annotated (M29 at 168 Ma, M21 at 147.7 Ma, M16 at 139.6 Ma, MO at 120.4 Ma, C33 at 79 Ma, C31 at 67.7 Ma, C25 at 55.9 Ma; time scale Gee & Kent, 2007). (e) Absolute velocities for the Pacific (85–15 Ma) and Kula (79–40 Ma) plates, averaged over a 10 Myr

frames do not agree with each other (Fig. 2.5a), and the bend is hardly visible if the Hawaii track is predicted from an African hotspot reference frame assuming fixed hotspots and using a plate chain (model A), which links the Pacific and Africa via Marie Byrd Land and East Antarctica (Fig. 14.5b).

The way to resolve this issue was elegantly outlined in Steinberger et al. (2004). They replaced a fixed hotspot reference frame with a global moving hotspot frame (Chapter 2) and that produced a much better fit with the Emperor–Hawaii Chain (Fig. 2.5a). But the calculated path was still much to the west of the Emperor Chain. Their initial attempt to mimic the trend of the Emperor–Hawaii Chain used conventional plate circuits linking the Pacific to the rest of the world via Marie Byrd Land and East Antarctica (model A). An alternative plate chain model (model B in Fig. 14.5b), however, linking Africa and the Pacific Plate via Lord Howe Rise, Australia, and East Antarctica (before 43.8 Ma) has proved much more realistic (red thick line in Fig. 14.5a). In fact, the shape of the bend can be reproduced quite well with plate chain model B, even without Hawaiian hotspot motion (white dotted line in Fig. 14.5a). However, hotspot locations before the time of the bend are predicted to lie too far south compared with the actual age progression along the Emperor track. One may therefore simply argue that the Emperor–Hawaii hotspot track including its bend can be explained by a change in Pacific Plate motion that can be appropriately modelled with plate chain model B and southward advection of the Hawaii Plume in a convecting mantle.

A moving hotspot reference frame (Model B plate chain) captures the Emperor–Hawaii Chain quite well (Fig. 14.5a). But is there any independent evidence for the modelled hotspot motion of Hawaii which has been dominated by southward motion (8°) for the past 80 Myr (Fig. 14.5c,d)? That evidence comes from palaeomagnetism. If the Hawaii Hotspot has remained above a fixed mantle plume, then the palaeolatitude recorded from volcanoes and seamounts (irrespective of their age) should be the same as the present latitude for Hawaii (19.4° N). However, the palaeolatitudes for the Emperor seamounts show a change from north to south with decreasing age. Paleocene–Eocene seamounts (Suiko, Nintoku, and Koko) indicate southward motion of the Hawaii Hotspot of 8–3° and statistically overlap with the modelled hotspot motion of Hawaii (Fig. 14.5c). The large offset (14°) for Detroit (81–75 Ma) is not captured in any hotspot models; added plume advection has been suggested, but the Hawaii Plume interacting with a spreading ridge (Tarduno et al., 2009) is perhaps a more attractive explanation for this large Late Cretaceous (Campanian) offset. Plumes modify plate boundaries, new plates can be created, spreading ridges can jump towards plumes (Figs. 14.6 and 14.7), but plumes can also temporarily be captured by ridges (upside-down drainage; Sleep, 1997). Unlike the younger seamounts in the Emperor Chain (e.g. Suiko), the Detroit Seamount has a mid-ocean ridge basalt (MORB) geochemical signature. A Kula–Pacific spreading ridge therefore probably lay north of the Detroit Seamount during its formation (Fig. 14.4a), and the Hawaii plume, which was originally located to the south, was captured (pinned) by a more northerly ridge (Tarduno et al., 2009). This is plate tectonically possible since the plate geometries and boundary types between Kula, Pacific, and Izanagi are open to the scientist's imagination.

Late Cretaceous ridge–plume interaction influenced only the construction of the northernmost seamounts of the Emperor Chain (e.g. Detroit Seamount), and the 47 Ma Emperor–Hawaii Bend can be explained by a combination of Pacific Plate motion change and plume advection. There is some misapprehension in the literature that the southward movement of the Hawaii Plume must have ceased after the bend, but advection may have occurred well into the Miocene. The model shown in Fig. 14.5 demonstrates that the geometry (including the bend), ages, and palaeolatitudes of the Hawaiian–Emperor Chain fit very well for the past 65 Myr and are within reasonable error margins for the past 80 Myr.

Why a Bend in the Emperor–Hawaiian Chain at 47 Ma? This is still a mystery, partly because the two key oceanic plates, Izanagi and Kula, are mostly long gone. Izanagi is the least well constrained plate and the orientation and obliquity of the Izanagi–Pacific ridge versus the Eurasian margin vary from model to model. Conversely, the projected

Fig. 14.4 (*cont.*) window and shown in 2 Myr steps. The two plates are coupled and show acceleration of different magnitudes from 66 Ma to 56–54 Ma followed by a sharp de-acceleration at 48–47 Ma, coeval with the Emperor–Hawaii Bend. Also shown is the relative velocity of the Pacific Plate versus Eurasia at around the Emperor–Hawaii Bend time (red dashed line). Yellow shaded ellipses on Pacific Plate velocity curve denote reconstruction times in (a)–(d), and the Pacific velocity vectors are shown at 57 and 44 Ma, and mirror the change from a northerly to north-west plate direction before and after the bend.

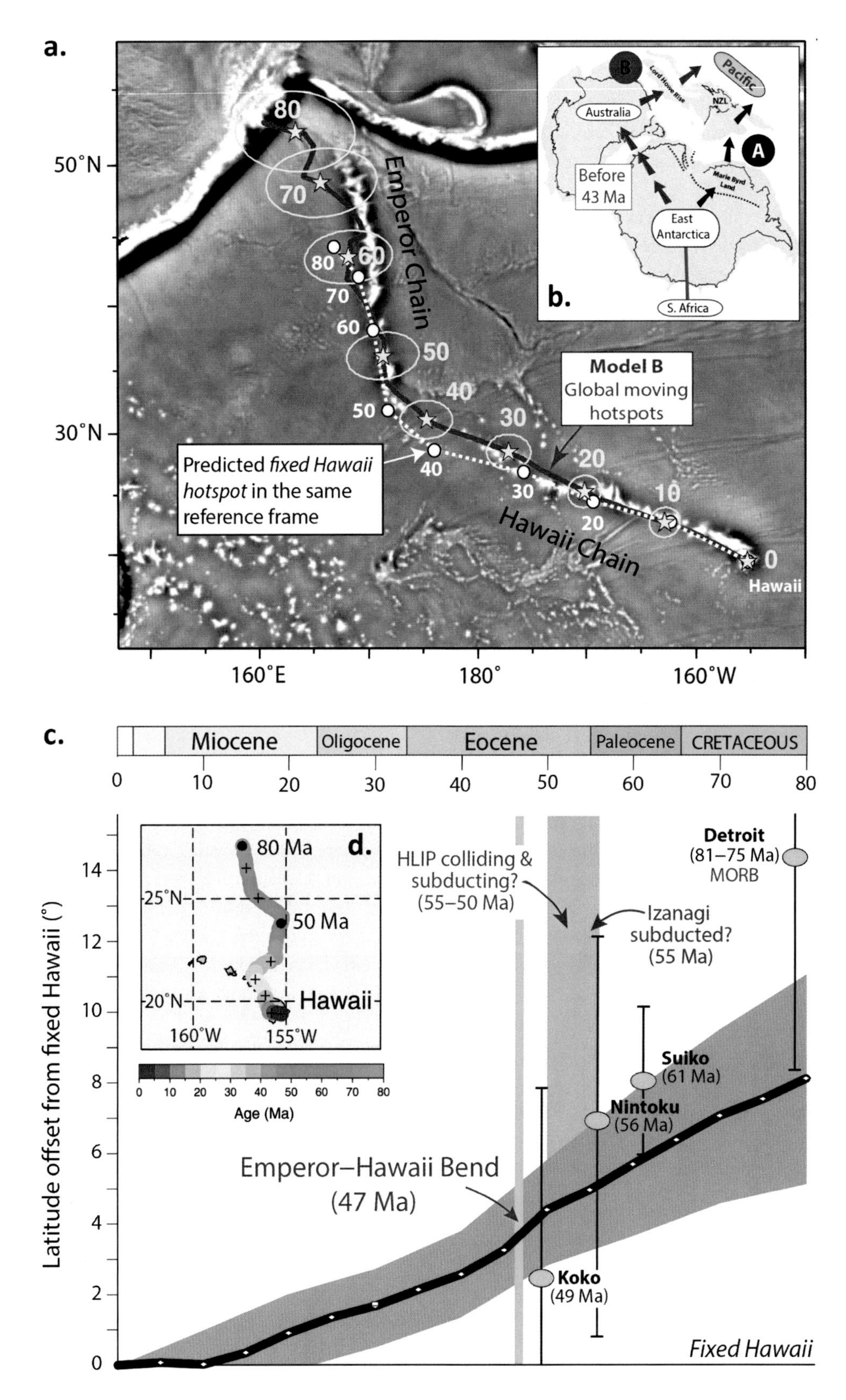

Fig. 14.5 (a) The computed track (red thick line; yellow stars with uncertainty ellipses in 10 Myr intervals) for a global moving hotspot reference frame with model B plate circuits (Fig. 2.6 and Doubrovine et al., 2012) captures the Emperor–Hawaii Chain for the past 65 Myr quite well. The track of an assumed fixed Hawaii Hotspot in the same reference frame as the red line (calculated in 10 Myr

conjugate margin of the synthetic Kula Plate before the bend is partly preserved on the Pacific Plate (Chron 33–25: 79–55.9 Ma; Chron 34 has also been identified by some authors), and therefore the orientation and location of the Kula–Pacific ridge (Fig. 14.4a–d) is known with at least some confidence.

Seton et al. (2015) argued that the arrival and subduction of a conjectured Izanagi–Pacific ridge along the Eurasian trenches at around 55–50 Ma led to a 'cascade of events that culminated in a reorganisation of Pacific mantle flow, triggering tectonic changes around the Pacific Ocean basin'. They emphasised that the large Pacific Plate changed from being one predominantly surrounded by spreading ridges in the Paleocene (Fig. 14.1a) to a plate with subduction systems developed along its western boundary, and therefore increased slab pull along the Eurasian margin from the Eocene onwards. Both the Pacific absolute velocity and the relative velocity versus Eurasia (almost stationary) slow down before the bend, followed by a steady increase in velocities thereafter (Fig. 14.4e). The bend is associated with a change from a northerly to a north-westerly motion of the Pacific Plate (Fig. 14.4), and it is likely that attempts to subduct the HLIP or seamounts older than Detroit along the northern margin of the Kula Plate (coupled to the Pacific) could obstruct northward movements of the Pacific Plate. A marked velocity decrease for the Kula Plate is indeed seen after 55 Ma (Fig. 14.4e). Therefore, competing forces associated with subduction of a potential Aptian HLIP or seamounts to the north and Izanagi–Pacific ridge subduction to the west combined with slow (~0.1°/Myr) but steady southward movement (Fig. 14.5c) of the Hawaii plume led to the spectacular Emperor–Hawaii bend discussed above.

Tethyan and Indian Oceans, and the Red Sea. In the Paleocene, before 55 Ma, subduction was occurring along the Tethyan subduction zone, consuming crust that had formed during previous Neotethys Ocean spreading. Near the beginning of the Eocene, the northern tip of Greater India started to collide with Eurasia and gradually led to the formation of the Himalayas.

The Afar LIP in north-eastern Africa had started to erupt from about 31 Ma (Early Oligocene), and extrusion of the plateau basalts there continued until 26 Ma (Late Oligocene). That LIP was indubitably linked with the subsequent Neogene Red Sea opening (Chapter 15), since there are also Early Miocene tholeiitic dyke swarms striking parallel to the Red Sea axis (Almalki et al., 2015). The initial rifting led to the sea-floor spreading which divides the African and Arabian plates. Extension along the East Africa rifts also led to the formation of the Neogene Somalia Plate from about 11 Ma.

Following peak Deccan magmatism at around 66 Ma, sea-floor spreading was initiated between the Laxmi Ridge and the Seychelles at 63–62 Ma, but sea-floor spreading was still ongoing in the Mascarene Basin. After 61 Ma (Paleocene), there was a radical change in the Indian Ocean architecture: the Réunion Plume was now located beneath the south-west margin of India (Fig. 14.6), which assisted a major north-east ridge jump that led to the termination of sea-floor spreading in the Mascarene Basin shortly after Chron 27. By 56 Ma (Early Eocene), Mauritian fragments (except Laccadives)

Fig. 14.5 (*cont.*) intervals from Doubrovine et al., 2012) is shown as large open white circles and dotted lines. The fixed Hawaii hotspot track reproduces the bend very well, and therefore the entire bend can be explained by a change in Pacific Plate motion with model B plate circuits. The fixed Hawaii hotspot track, however, plots too much to the south, compared with measured radiometric ages along the track, which is the effect of the southward hotspot drift. Gravity anomaly map in background indicates actual track geometry. (b) Two relative plate circuit models between Indo-Atlantic (Africa) and Pacific hotspots before the Middle Eocene (Chron 20 at 43 Ma). After Chron 20, models A and B follow the same plate motion chain through East Antarctica and Marie Byrd Land. Originally dubbed models 1 and 2 in Steinberger et al. (2004). NZL, New Zealand. (c) Palaeomagnetically derived latitudes with 95% error bars for seamounts along the Emperor Chain (Detroit, Suiko, Nintoku, and Koko; Tarduno et al., 2003; Doubrovine & Tarduno, 2004). Latitudes are shown as latitude offsets from zero (observed latitude minus latitude of Hawaii) and compared with latitudinal estimates of plume advection (thick black line with grey band corresponding to uncertainty ellipses; Doubrovine et al., 2012). For a system with fixed plumes (and no true polar wander) all the latitudes should be on the zero line, i.e. at the same latitude as Hawaii today. (d) The modelled surface motion (colour coded) of the Hawaii Hotspot back to 80 Ma, which is dominated by southward motion that is clearly reflected by the palaeomagnetic data. For this model, the assumed starting age of the plume is 120 Ma (Early Cretaceous), the same as the assumed HLIP age in Chapter 13. Resulting southward motion is quite dependent on this age and tends to be less with a later starting age assumed. Some of the Emperor seamounts (Detroit, Suiko, and Nintoku) indicate more advection of the Hawaii plume conduit than that estimated from numerical modelling but the latitudes are clearly within errors, except perhaps for Detroit. Detroit, however, has a mid-ocean ridge basalt (MORB) geochemical signature, and this oldest known seamount in the Emperor Chain may represent volcanism at the Kula–Pacific spreading ridge a few hundred km to the north of the hotspot, due to plume–ridge interaction (see text and Fig. 14.4a).

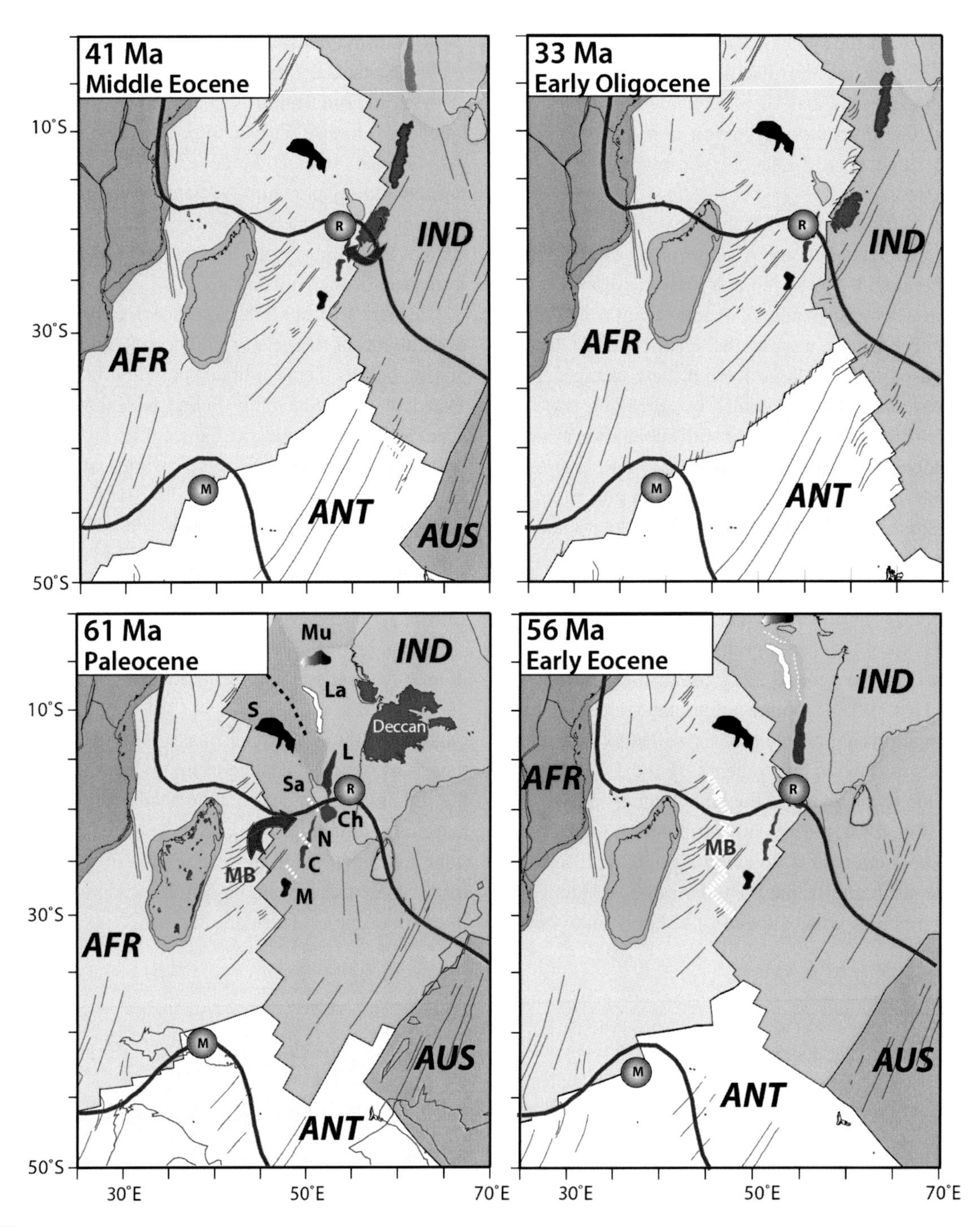

Fig. 14.6 The evolution of the Indian Ocean area (Torsvik et al., 2013) from 61 Ma and 56 Ma (Paleogene) to 41 Ma (Eocene–Bartonian) and 33 Ma (Eocene–Oligocene boundary). The predicted locations of the Marion (M) and Réunion (R) hotspots are shown in magenta and the thick red lines are the plume generation zones (PGZs) at the core–mantle boundary. Extinct spreading ridges between various microcontinents are shown as dashed white lines. Ridge jumps propagated south-west while the Marion Plume was in the proximity of the active plate boundary. Red arrows indicate direction of plate boundary relocation towards closest hotspot. Oceanic floor fabric and direction of spreading between major tectonic plates are depicted by interpreted fracture zones. Major plate boundaries are shown as thick black lines. Outlines of major volcanic plateaus and provinces are shown in magenta. AFR, Africa Plate; ANT, Antarctica Plate; AUS, Australia Plate; C, Cargados Carajos continental fragment; Ch, Chagos continental fragment; IND, India Plate; L, Laccadives continental fragment; La, Laxmi Ridge continental fragment; M, Mauritius continental fragment; Mu, Murray Ridge continental fragment; MB, Mascarene Basin; N, Nazareth continental fragment; S, Seychelles Microcontinent; Sa, Saya de Malha Bank.

and the Seychelles became part of the African Plate. Thereafter, the Réunion Plume was beneath the slowly moving African Plate. An important ridge jump occurred at about 41 Ma (Late Eocene) when the Réunion Plume was located at the position of Saya de Malha/Nazareth. That ridge jump led to the separation of Chagos from other Mauritian elements, and Chagos once again became part of the Indian Plate (Fig. 14.6).

Atlantic Ocean. Sea-floor spreading propagated into the north-east Atlantic (between Greenland–Eurasia margin), forming a triple junction between North America, Greenland, and Eurasia (Fig. 14.1). The North Atlantic Igneous Province (NAIP) LIP linked to the Iceland Plume, which was located beneath East Greenland throughout the Paleogene, erupted from about 63 Ma with two major peaks: at 62 and 55 Ma. North-east Atlantic sea-floor spreading started at around 54 Ma (Early Eocene) with spreading along the Reykjanes, Ægir (now-extinct), and Mohn's Ridges (Fig. 14.7b). The Jan Mayen Microcontinent (JMM) in the Norwegian–Greenland Sea had been divided into several segments (e.g. Gaina & Ball, 2009; Peron-Pinvidic et al., 2012), and Torsvik et al. (2015) also proposed that a continental fragment now beneath south-east Iceland is a south-westward extension of JMM. JMM fragments fringed the East Greenland margin until 52 Ma (Early Eocene) when the Reykjanes Ridge propagated northwards, and detached the most southerly fragment of JMM from East Greenland by 47 Ma (Fig. 14.7b). Two ridge jumps occurred before 33 Ma (Oligocene) (Fig. 14.7c) and a continuous plate boundary was established west of JMM at around 27 Ma (Oligocene) by a ridge jump in the direction of the Iceland Plume (Fig. 14.7d). The Ægir Ridge became extinct from that time and JMM became part of the Eurasian Plate (Fig. 14.7e).

Spreading in the Eurasian Basin in the Arctic also started at around 54 Ma (Early Eocene) along the Gakkel Ridge. This ridge probably connected to the Baffin Bay ridge axis through the Nares Strait (Wegener Fault) and Mohn's Ridge to the south through major strike-slip faults, with compression between Greenland and Svalbard (Fig. 14.8). Following the concept of Oakey and Damaske (2006), we keep south-west Ellesmere (including Devon Island) semi-locked to the Greenland Plate so as to minimise strike-slip motion and deformation along the Wegener Fault during the opening of the Labrador Sea and Baffin Bay. The other Ellesmerian terranes to the north are kept semi-locked to North America, and, because the Greenland Plate was moving northwards during the opening of the Labrador Sea, this caused compression in between them (grey shaded area in Fig. 14.8 labelled Eurekan). Eurekan deformation continued until about 33 Ma (Early Oligocene) when Labrador–Baffin Bay sea-floor spreading gradually ceased, and Greenland once again became attached to North America.

Laurasia and North America. Except for pre-drift extension and rift basin formations, Laurasia remained tectonically united until the Atlantic Ocean floor spreading between Norway and Greenland finally started in the Early Eocene, although a shelf sea had separated the North America and Eurasia landmasses before then. The Western Interior Seaway, which had extended across North America during the Cretaceous (Fig. 13.10), was still present as remnants in the Paleocene, but had completely disappeared before the start of the Eocene.

During the Paleogene, mountainous uplift and associated foreland basins developed in the Canadian Arctic Archipelago in the Eurekan Orogeny and also in the western Barents Sea near Svalbard (Smelror et al., 2009). The Eurekan Orogeny, as discussed above, affected the area between north-western Greenland and Ellesmere Island (Fig. 14.8), and started as extensional structures with grabens and half-grabens in the Cretaceous, followed by compression possibly associated with strike-slip deformation and south-west splays of the Nares Fault system in the Paleogene (Spencer et al., 2011).

The rising mountains over much of north-western North America and in the hinterlands of Asia led to the development of very large river systems which discharged northwards into the surroundings of the expanding Arctic Ocean. Chief among them were and are the Mackenzie River in Canada and the Lena River in Russia, both of which have Paleogene to Recent sediments over 15 km thick in their respective deltas.

Africa, Antarctica, and Australia. There is a substantial rift system extending for 3,000 by 750 km (comparable in area to the East African Rift systems of today) within West Antarctica from the Ross Sea to the Bellinghausen Sea at the southern end of the Antarctic Peninsula. The rift runs along the margin of the Transantarctic Mountains, which have been rising since Late Cretaceous times, and is characterised by the intrusion of mostly alkali volcanic rocks ranging in age from Oligocene or earlier to the present day.

Slow rifting of Australia from Antarctica, which had begun in the Jurassic, led to ultra-slow sea-floor spreading in the Mid–Late Cretaceous. Shallow water passages were probably formed south of Tasmania in the Eocene, but deep oceanic gateways were probably in place after the Early Oligocene as sea floor formed between Tasmania and Antarctica, and between the Antarctic Peninsula and Antarctica (the Drake Passage) (Fig. 13.9a). With the last land bridges between Gondwana continents broken, the Great Southern

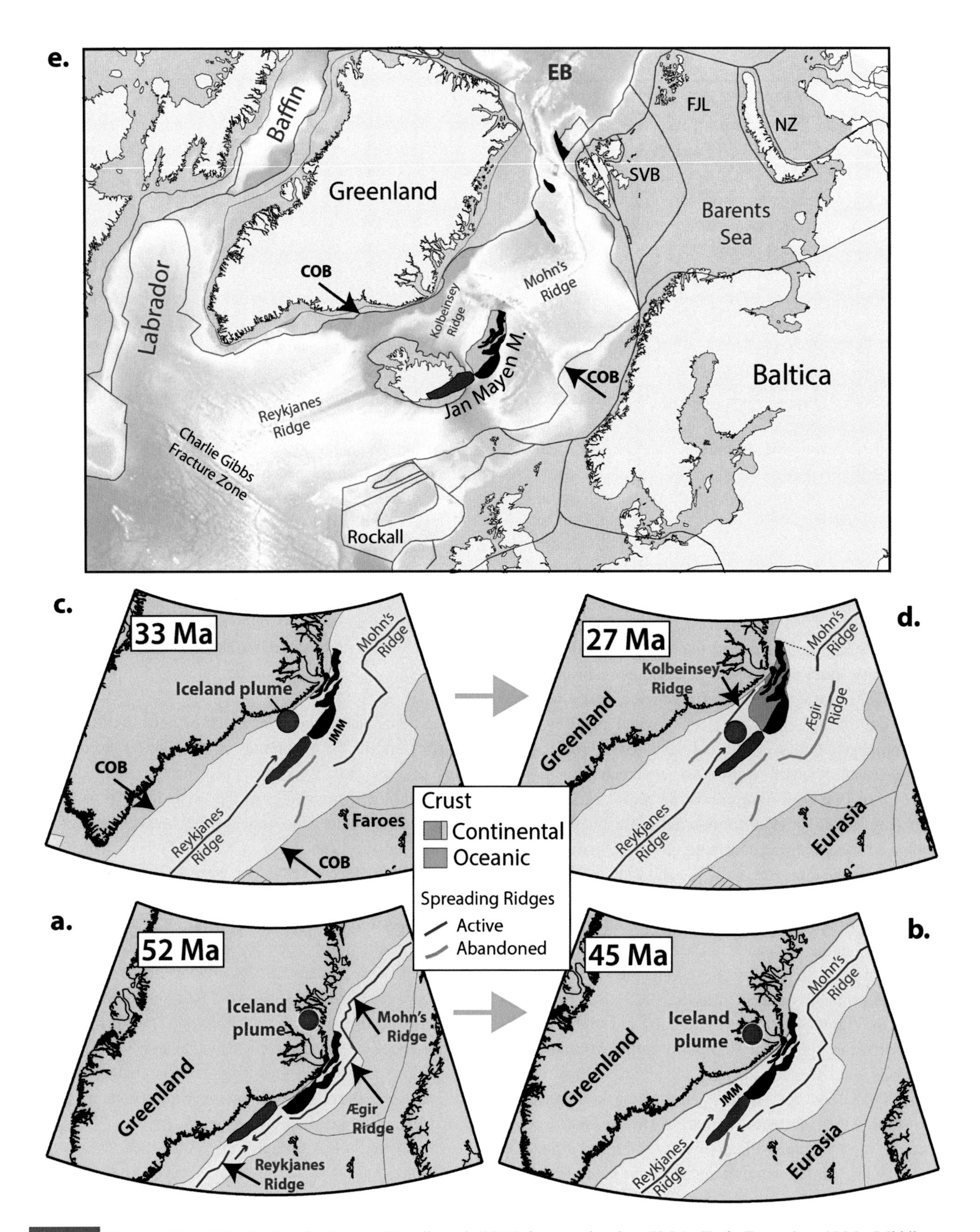

Fig. 14.7 The evolution of the North Atlantic area (Torsvik et al., 2015) from pre-breakup 52 Ma (Early Eocene) to 45 Ma (Middle Eocene), 33 Ma (Eocene–Oligocene boundary), 27 Ma (Middle Oligocene), and present-day panel. The computed Iceland plume location (red-filled circle) is shown together with a CEED absolute plate reconstruction. Major continental entities are shown in yellows (a–d) or grey (e) and tectonic boundaries within these regions are indicated as thin dark grey lines. Main tectonic blocks of the Jan Mayen Microcontinent (JMM) are shown in black. Active mid-ocean ridges are shown as thick magenta lines, whilst abandoned ridges are shown as green lines. Light brown (d) or grey (e) area around the main JMM blocks indicates extended continental crust. The southward extension of JMM beneath Iceland proposed by Torsvik et al. (2015) is shown in dark brown. COB, transition from continental to oceanic crust; EB, Eurasian Basin; FJL, Franz Josef Land; NZ, Novaya Zemlya; SVB, Svalbard.

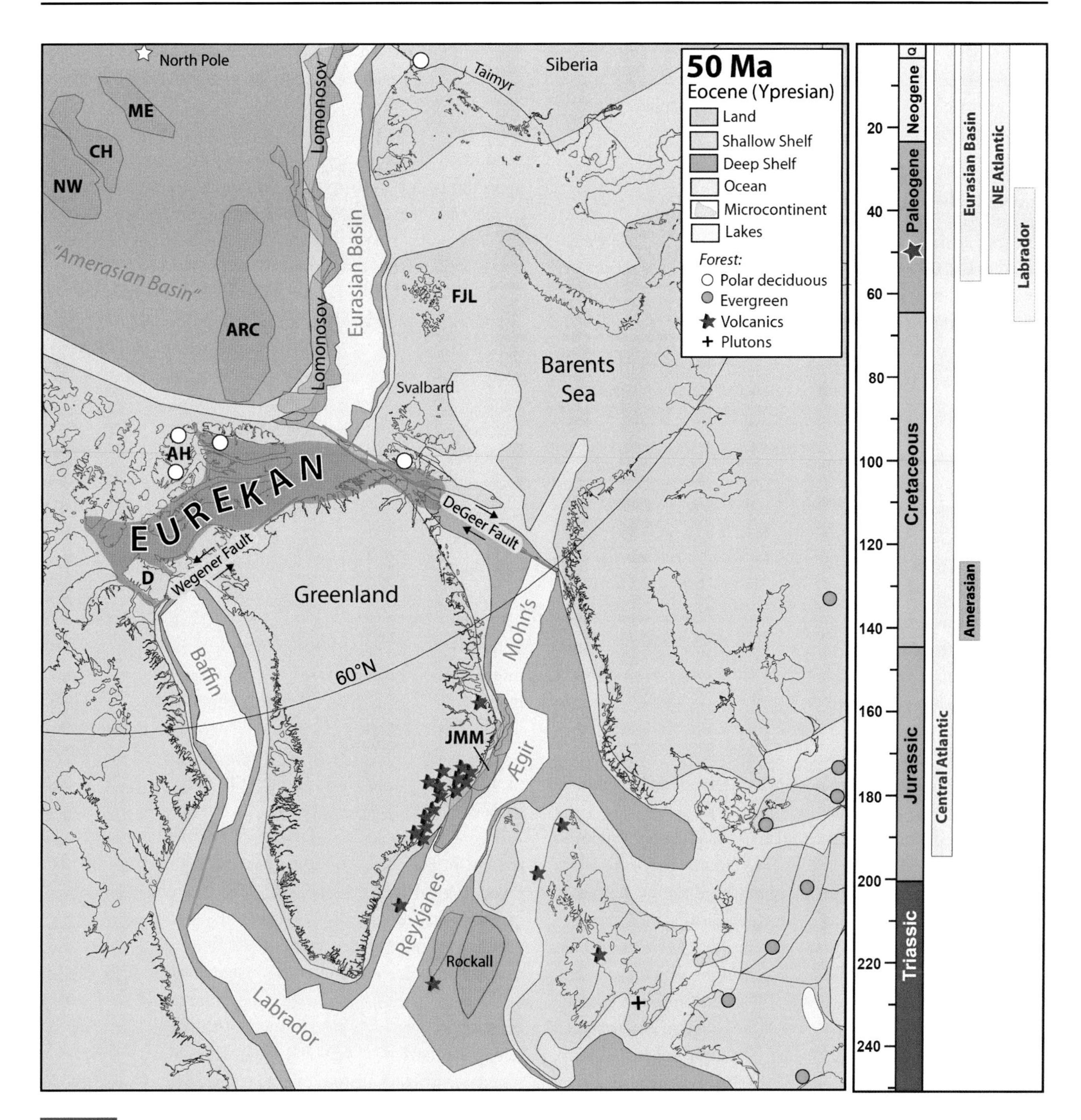

Fig. 14.8 The palaeogeography of north-western Europe, north-eastern North America, and parts of north-western Siberia and the Arctic area at 50 Ma (Early Eocene: Ypresian), showing the spreading arms of the North Atlantic and Arctic oceans and including the area of the Eurekan Orogeny in north-east Canada and Greenland. Green lines are transform faults, and the microcontinental blocks now submerged are shown in grey: it is uncertain what the sea depths above them were at that time. Also shown are the sites where polar deciduous forests and the more temperate evergreen forests have been found. AH, Axel Heiberg Island; ARC, Alpha Ridge; CH, Chukchi; D, Devon Island; FJL, Franz Josef Land; JMM, Jan Mayen Microcontinent; ME, Mendeleev Ridge; NW, Northwind Ridge. Floral data sites from Collinson & Hooker (2003). Time scale on right shows periods of ocean spreading in various basins.

Ocean was born and the Antarctic Circumpolar Current established. This deep-water circulation continues today, and has a huge effect on the southern hemisphere weather climate patterns.

Facies, Floras, and Faunas

Climate. Global temperatures increased in the Late Paleocene and peaked over the Paleocene–Eocene boundary (Fig. 16.3), an event known as the Paleocene–Eocene Thermal Maximum (PETM). That resulted in a sharp negative excursion in the oxygen, carbon, and nitrogen isotope ratios followed by a significant increase in phosphorus concentrations, whose effects have been documented in North America, France, Spain, Uzbekistan, and Egypt (Khozyem et al., 2013). That temperature maximum led to increased humidity, reflected by increased kaolinite content in the sediments, and the associated anoxia was probably tied to a bloom of nitrogen-fixing cyanobacteria in the oceans. The PETM appears to have triggered the important turnover in calcareous nannoplankton, when the genera that had radiated and were dominant in the Paleocene became largely extinct, and the survivors are essentially the modern-aspect flora which had previously been of low diversity (Aubrey et al., 1998). Although warmer temperatures continued (with some fluctuation) until the Middle Eocene (Bartonian: 40 Ma), they subsequently declined steadily, so that by the Miocene (see next chapter) there was a more marked temperature gradient from Equator to poles than previously.

Because of fluctuating sea levels, a land bridge intermittently developed across the Bering Straits between North America and Eurasia, thus allowing non-marine faunas to cross between the two continents sporadically.

There is no evidence for any ice caps at the poles during the earlier part of the Paleogene, including all of the Paleocene. However, there was steady cooling during the Eocene, with the first evidence of glaciation recorded in Antarctica in the Late Eocene at about 37 Ma. That Late Eocene cooling, with the coldest interval in the Early Oligocene at about 34 Ma, was echoed in the changes in vegetation and a mammalian faunal turnover noted below. The eventual Plio-Pleistocene Ice Age is described in the next chapter.

Recovery from the End-Cretaceous Extinction. As noted in the previous chapter, the Cretaceous–Tertiary (K-T) boundary event was caused by the combination of the catastrophic meteorite impact in Mexico, the outpourings of the Deccan Traps LIP in India, and probably also the sea-level falls during the preceding Maastrichtian, all of which caused a marked deterioration in the global climate. The drop in sea levels reached a threshold in the Paleocene (Fig. 16.2), draining the extensive Late Cretaceous marine carbonate platforms and much increasing the exposed overall land area of the planet. However, as before in the Permian–Triassic event and others, the removal of that quantity of varied biota from many different ecological niches provided an excellent opportunity for the survivors to colonise new habitats and evolve, and the extinction does not appear to have strongly affected some major groups such as most of the plants, corals, and marine arthropods (MacLeod et al., 1997).

There was a marked general low in both the diversity and abundance of nearly all groups for a million or more years at the beginning of the Paleocene before radiation proceeded rapidly; in addition, there was an overall marked scarcity of radiolarian and associated biosiliceous oozes on the ocean floor at the same time, which underlines the reduction in overall oceanic biological productivity. However, evidence of rich radiolarian assemblages in sediments across the K-T boundary in New Zealand suggests enhanced oceanic upwelling caused by climatic cooling in that high-latitude region. True analysis is complex, since most of the earliest-known Paleocene deposits are found in deep sea basins; for example, most of those deposits contain relatively unspecialised assemblages of planktonic foraminifera which nevertheless have few genera in common with those in latest Cretaceous (Maastrichtian) strata. The larger benthic foraminifera are almost unknown from Lower Paleogene rocks, but some Cretaceous genera reappeared in the Middle Paleocene, and rapidly diversified in the Late Paleocene.

The New Vegetation. Although flowering plants were present in the Mesozoic, they became dominant only in the Paleogene. In the forests, birches, beeches, elms, dicotyledon groups (such as *Magnolia* and laurels), and monocotyledon groups (including palm trees) were abundant and were distributed in chiefly latitude-dominated provinces (Collinson & Hooker, 2003). Pinoceae (firs, pines, etc.) trees also became progressively more abundant and diverse in the more temperate latitudes; but at higher latitudes the Paleogene forests were dominated by polar broad-leaved deciduous forests, including beeches. Such forests have since become completely extinct in the Arctic, but remnants of the forest which had previously covered most of Antarctica, characterised by the Southern Beech (*Nothofagus*), still remain in many fragments of the former Gondwana, such as New Zealand, Australia, New Caledonia, New Guinea, and South America (Argentina and Chile), and provide links with the previously abundant *Nothofagus* fossils found in the Paleogene and Early Neogene rocks of Antarctica and elsewhere. Grasses radiated in the Paleocene, but the areas of prairies and savannahs only became very extensive in the Neogene.

Radiations in the Animal Kingdom. Rapid radiation of planktonic foraminifera had started a very short time before the end of the Cretaceous but proceeded at an even greater pace in the Paleocene, so that there was a virtually complete replacement of Late Cretaceous species by Paleogene ones in an interval of less than 1.5 Myr, with the result that foraminifera are by far the best fossils for documenting the successive biozones so useful for marine stratigraphical correlation, particularly in the boreholes of the oil industry. Insects also radiated dramatically and probably numbered well towards the millions of species known today by the end of the Paleogene, although it is interesting that of the 37 orders of insects, only 3 newly evolved in the Eocene, and more than 25 of the orders have Palaeozoic representatives (E.A. Jarzembowski in Selley et al., 2005).

Reptiles such as turtles and crocodiles also flourished, as did their cousins the birds, with the first penguin-like swimming birds recorded from the Eocene. On the sea floor, the marine benthos also changed rapidly, with a great diversity of new epifaunal gastropods evolving, particularly in the warmer waters of the tropics and especially in reef habitats. In more temperate and deeper waters, echinoderms such as starfish proliferated, the latter on top of the sea floor. The marine infauna was dominated by bivalve molluscs, irregular echinoids, arthropods, and worms (the latter not often preserved as fossils), but all continued to radiate more as ever deeper-water new ecological niches were progressively exploited.

The Rise of the Mammals. Although the earliest true mammals are known from the Triassic at about 210 Ma, they were tiny insectivores, much like living shrews. Some larger families and genera had evolved before the end of the Cretaceous, but it was not until the extinction of the non-avian dinosaurs at the end of the Cretaceous that the mammals evolved quickly to progressively exploit the vacant and varied previous dinosaur habitats. Thus, in contrast to the Mesozoic, when there were relatively few larger mammals, they soon became the dominant group of vertebrates in the Paleogene and have remained so until the present day. Mammals had separated into the Eutheria (most of them) and the Marsupalia, or pouched mammals (notably kangaroos), before the end of the Cretaceous. The relative isolation of Australia after Pangea breakup enabled progressive radiation of the marsupials there without competition from other mammal groups, although marsupials are also known from the Eocene of Antarctica (Savage & Long, 1986).

During the Paleocene and Eocene the extensive mammal radiation expanded into swimming forms such as whales and dolphins, some of which eventually became the largest animals on Earth, and also into the air (for example, bats). Like the dinosaurs which had preceded them, most early mammals were herbivores, whilst a minority were carnivores; and although large mammals such as elephants are the most noticed today, the groups of smaller sizes, particularly the rodents, are the most diverse and abundant.

The temperature drop (Fig. 16.3) across the Eocene–Oligocene boundary (noted in 'climate' above), was echoed in the changes in vegetation from Eocene dense forests to Oligocene more open country, which was in turn matched by much mammalian faunal turnover and radiation. For example, on the Chinese–Mongolian plateau (which would have been most influenced by the temperature drop because of its height above sea level), the dominance of the perissodactyls of the Eocene was abruptly replaced by a dramatic rise in rodent and lagomorph faunas in the Early Eocene, many of which included the first appearances of modern common families, as may be most easily assessed by the analysis of their increasingly complex teeth, which are very commonly preserved as fossils (Meng & McKenna, 1998). Sizes were also affected by a variety of factors, both external (such as climatic fluctuations) and from direct competition with other animals; for example, it is not obvious why the Eocene saw some of the largest land mammals ever known, whilst the Oligocene mammal fauna was mostly much smaller.

15 Neogene and Quaternary

Svalbard glacier and sea ice. Credit: Morgan Jones/CEED.

This period, which takes us up to the present day, has been dominated by very varied tectonic events, including the mountain building in the Alps, the Himalaya, and the western Americas (Cordillera and Andes), all of which have not yet come to an end. The climate was dramatically changed by the Plio-Pleistocene glaciation, which also appears to be continuing today.

The Neogene is divided into the Miocene (base at 23 Ma) and Pliocene (base at 5 Ma), both names coined by the famous British geologist Charles Lyell in the 1830s. The succeeding Quaternary is divided into the Pleistocene (base at 2.6 Ma), a name also coined by Lyell, and the Holocene for approximately the most recent 100,000 years. The global distribution of the continents and oceans is shown in Fig. 15.1 at 20 Ma, Early Miocene, and Fig. 15.2, today. Because modern plates, continents, and terranes, and their positions and biota are in general well known, we have done little original work on them and thus this chapter is relatively short to make our voyage through time complete.

Tectonics and Igneous Activity

Large Igneous Provinces and Hotspots. There is only one Neogene LIP, the Columbia River Basalt in north-west North America, which erupted in a back-arc setting between the Cascade Volcanic Arc and the Rocky Mountains at 16.7 Ma (Early Miocene). Volcanism there was most active in the Middle Miocene but continued until 5.5 Ma, near the end of the Miocene. The basalts were initially erupted in the southern Oregon Plateau, but spread northwards, inundating

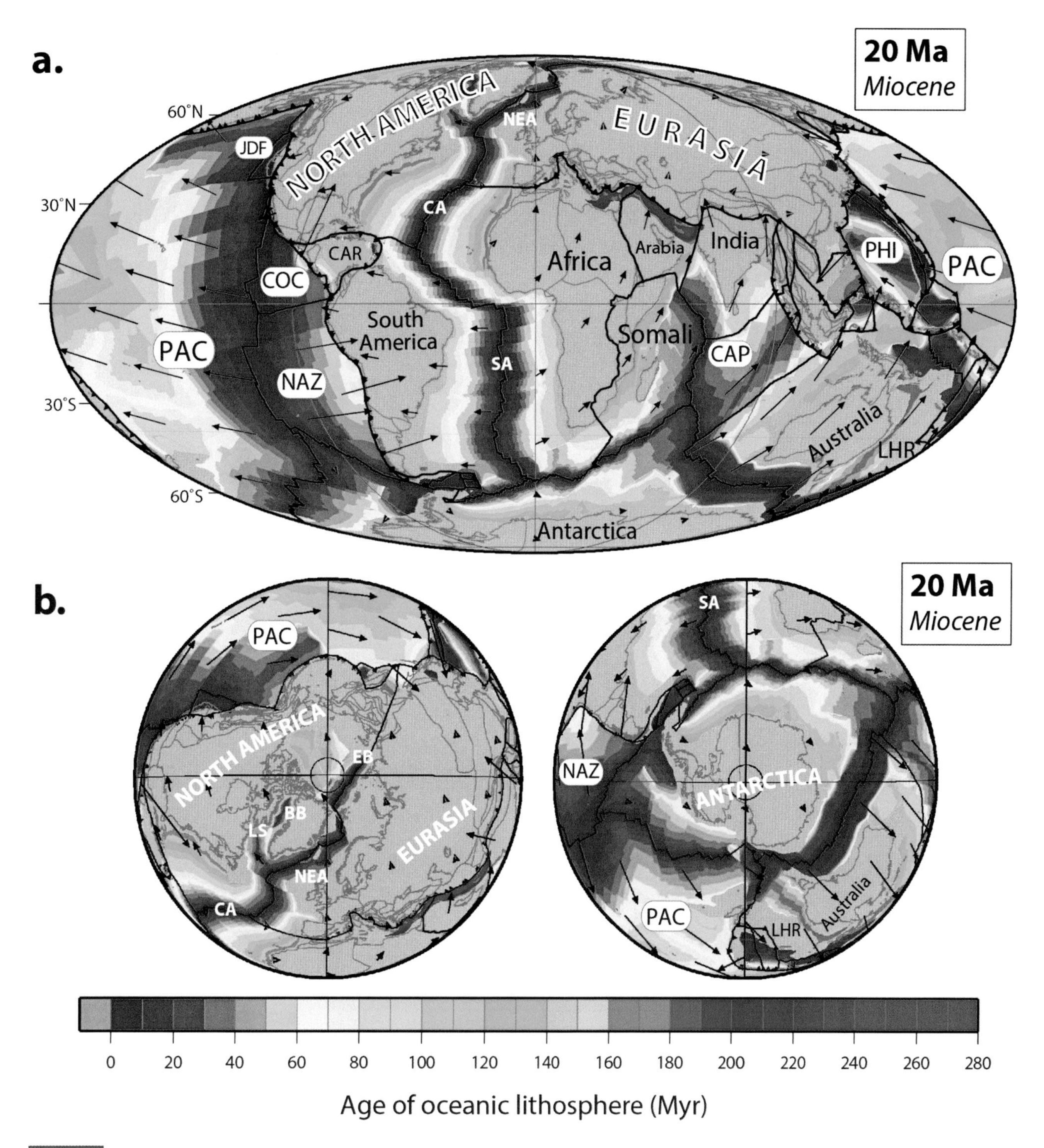

Fig. 15.1 Mollweide (a) and polar projected (b) continental blocks (grey) and oceans (other colours) with plate velocity vectors and the ages of their oceanic lithosphere at 20 Ma (Miocene: Burdigalian). BB, Baffin Bay; CA, Central Atlantic; CAP, Capricorn Plate; CAR, Caribbean Plate; COC, Cocos Plate; EB, Eurasian Basin; JDF, Juan de Fuca Plate; LHR, Lord Howe Rise; NAZ, Nazca Plate; NEA, North-East Atlantic; LS, Labrador Sea; PAC, Pacific Ocean Plate; PHI, Philippine Plate; SA, South Atlantic. EARTHBYTE mantle frame (see details in Chapter 2).

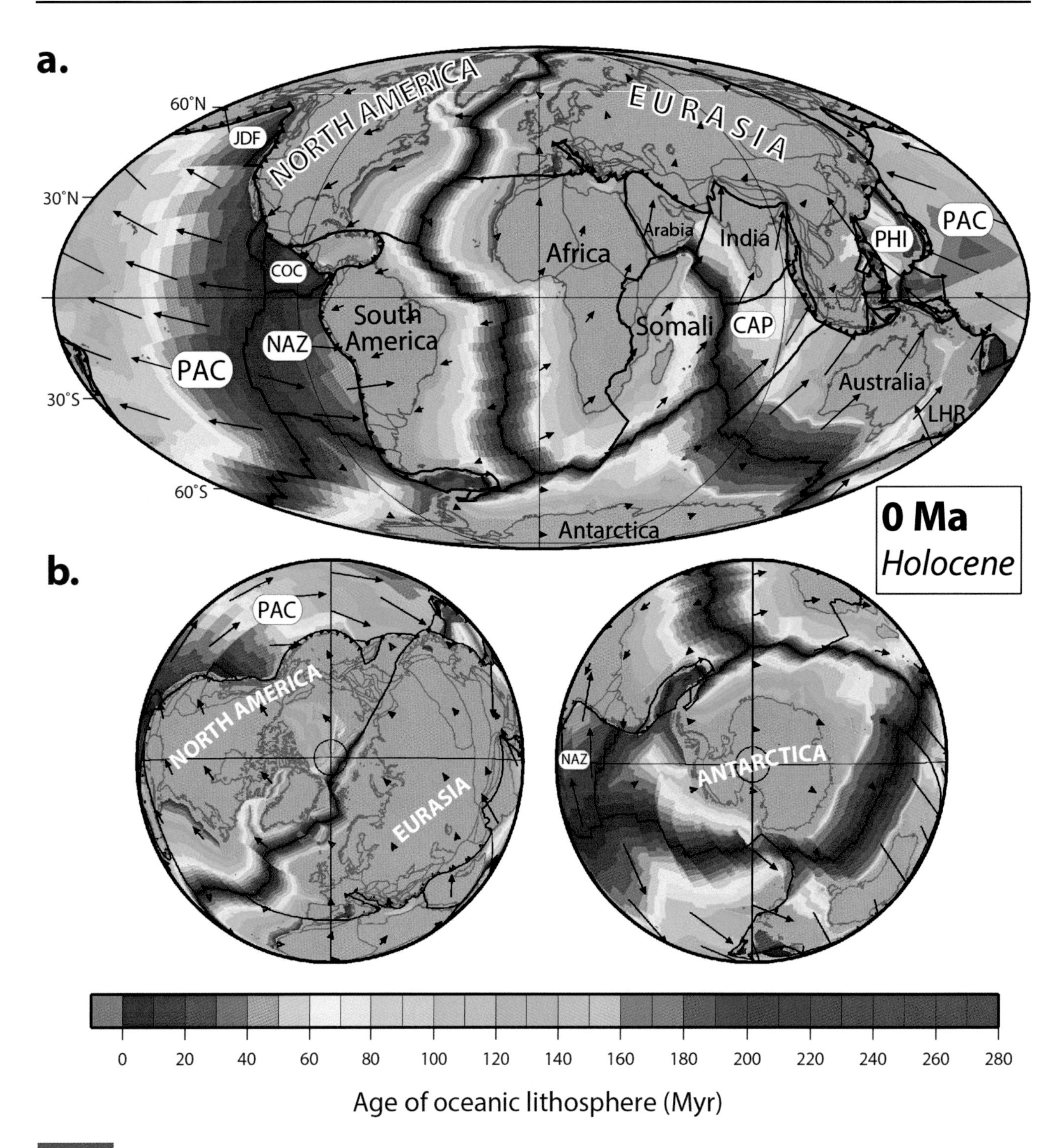

Fig. 15.2 Mollweide (a) and polar projected (b) continental blocks (grey) and oceans (other colours) with plate velocity vectors and the ages of their oceanic lithosphere today (Holocene). BB, Baffin Bay; COC, Cocos Plate; NAZ, Nazca Plate; PAC, Pacific Ocean Plate. EARTHBYTE mantle frame (see details in Chapter 2).

the Columbia River basin and finally spreading into the offshore fore-arc basin, eventually totalling 210,000 km^2 in area, and covering much of Oregon, Washington, and adjacent states (Reidel et al., 2013).

There are few Neogene kimberlites, but about 50 hotspots that were active during the Neogene, and most of them started much earlier. For example, the Tristan Hotspot in the South Atlantic has been active since the Paraná–Etendeka

LIP first erupted in the Early Cretaceous at 134 Ma. Many hotspot catalogues have been compiled in recent years, but it has long been suspected that many of the hotspots included in these compilations do not have a deep plume origin. For example, Courtillot et al. (2003) considered that only 7 out of 49 hotspots were sourced from plumes at the core–mantle boundary. Those were named 'primary' hotspots, and included Afar, Easter Island, Iceland, Hawaii, Louisville, Réunion, and Tristan, which are all located above or near the edges of Tuzo and Jason (Fig. 15.3c). Courtillot et al. (2003) also identified 'secondary' plumes, originating from the base of the transition zone on the tops of Tuzo and Jason, and a third type of superficial ('Andersonian') hotspots linked to lithosphere tensile stresses and decompression melting. More recently, French and Romanowicz (2015) identified 20 primary or clearly resolved plumes in the Earth's mantle from seismic tomography (Fig 15.3a,c). The pattern of primary/clearly resolved hotspots is similar to that for reconstructed LIPs for the past 300 Myr (Fig. 15.3d).

The majority of LIPs and kimberlites appear to be sourced by plumes from the plume generation zones at the core–mantle boundary (CMB), but, based on global tomographic models, there are exceptions such as the youngest and smallest LIP on Earth, the Columbia River Basalt, and Cretaceous–Tertiary kimberlites in north-west North America. Additionally, no hotspots in this region or in nearby offshore areas (Bowie, Raton, Cobb, Yellowstone, Guadalupe, and Socorro in Fig. 15.3c) were imaged to be situated above low-velocity regions (Fig. 15.3a) at the CMB (French & Romanowicz, 2015). The Columbia River Basalt is also unusual, when compared with other LIPs, in that it erupted in a back-arc setting near a convergent plate margin (Fig. 15.3b), and alternatives to a deep plume origin include back-arc extension, lithospheric delamination, and slab tearing (Liu & Stegman, 2012). It is interesting that practically all the anomalous kimberlites (not erupted above Tuzo or Jason) in North America and Canada are Late Cretaceous to Paleogene in age (115–50 Ma); an age which corresponds to the Sevier (125–105 Ma: Hildebrand, 2014) and Laramide (85–50 Ma) events along the western margin of North America (Fig. 13.6b–d).

Oceans. Spreading in the high Arctic (Eurasian Basin) and the South, Central, and North Atlantic Ocean continued though the Neogene (Fig. 15.1). In the Caribbean, the Cayman Trough continued to expand and develop, and the Chortis Block moved over the promontory in the Yucatán area. Subduction of Atlantic oceanic lithosphere along the Lesser Antilles Trench, which has been active since the Eocene (Boschman et al., 2014), connects to the Mid-Atlantic Ridge along the Researcher Ridge and Royal Trough.

In the eastern Pacific, a further rupture of the Farallon Plate occurred in the earliest Miocene at around 23 Ma, leading to the establishment of the Cocos and Nazca plates (Fig. 15.1) and initiation of the East Pacific Rise (spreading ridges running from the northern end of the Gulf of California southwards to Antarctica), the Galápagos Spreading Centre (between the Nazca and Cocos plates), and the Chile Ridge (between the Nazca and Antarctic plates). The Juan de Fuca Plate formed in the Late Eocene at around 37 Ma (Figs. 15.1 and 15.3b), and is limited at its southern end by the Mendocino Fracture Zone as well as subducting along the Cascadia Subduction Zone.

What is today the Mediterranean Sea was previously a junction area of the Neotethys running from the southern Alps, through the Aegean and Anatolia eastwards, and westwards to the eastern end of the Atlantic Ocean, which extended into the western and northern Mediterranean realm as a branch known as the Alpine Tethys, or Piemonte Ligurian Ocean. The western Alpine Tethys Ocean was an ocean basin between Iberia, Africa, and Adria in the west of the Mediterranean, and it was slowly closing throughout much of the Cenozoic, but was subducted largely during the Miocene, forming the Betic–Rif–Tell–Apennine thrust belt. This major oroclinal structure includes the Pre- and Sub-Betic Mountains of southern Spain, the Prerif of Morocco, the Tell Mountains of north-west Algeria, the Maghrebides of Sicily, and the Apennines of Italy. Those thrust belts are the remnants of Iberian, African, and Adriatic passive margins which were deformed by thin-skinned folding and thrusting. That led to the accumulation of a substantial accretionary wedge in the eastern Atlantic by Late Tortonian time at about 10 Ma, and consequently the closure of the Mediterranean connection to the Atlantic, which culminated in the massive sea-level drop and the accumulation of thick evaporites in isolated Mediterranean basins during the Late Miocene Messinian Salinity Crisis (see below), but that isolation ended when the arc was breached at the Straits of Gibraltar, and the Mediterranean was refilled within a very short time (Flecker et al., 2015).

Southward subduction of Europe below Adria, which had formed the Alps since the Cretaceous, largely came to a halt in the Neogene. In the western Alps, underthrusting slowly continued until 7 Ma (Messinian). In the eastern Alps, the polarity of the orogeny reversed and Adria started to underthrust the eastern Alps from about 10 Ma (Late Miocene), creating the Southern Alps fold–thrust belt. Southward and westward subduction of the eastern end of the Alpine Tethys, with Eurasia as the downgoing plate, continued in the Carpathian realm until Early Tortonian time at around 10 Ma (Ustaszewski et al., 2008). From the Dinarides to Greece and

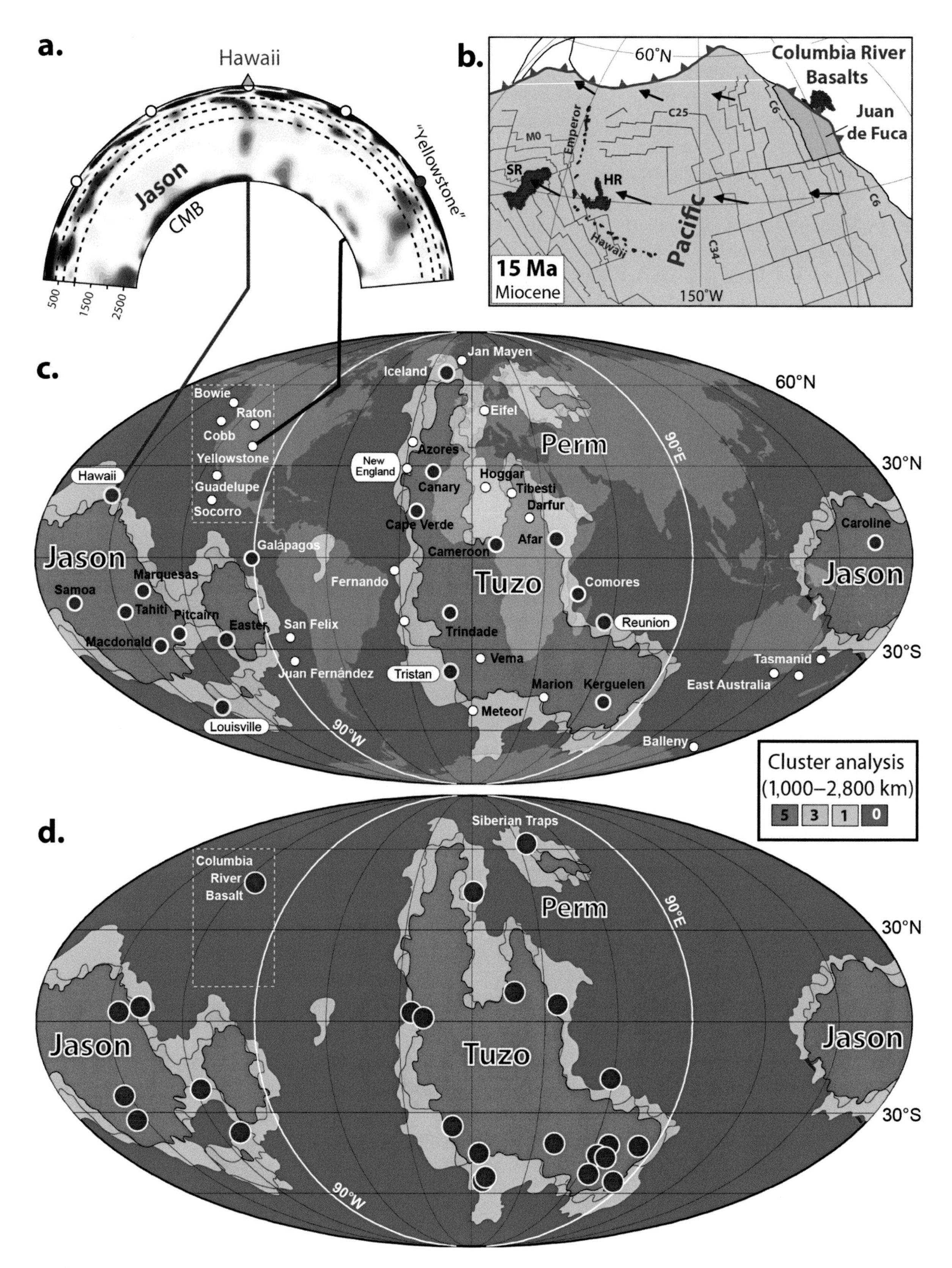

Fig. 15.3 (a) Two-dimensional cross-section of shear-wave velocity anomalies across the Hawaii and Yellowstone hotspots. A broad plume beneath Hawaii extends continuously from the core–mantle boundary (CMB) at the margin of Jason to the uppermost mantle. Conversely, negative anomalies (warm colours) are not seen in the lower mantle beneath the Yellowstone Hotspot

Turkey, Adria and Africa underwent slow subduction. From western Greece towards Cyprus, this included consumption of Triassic Neotethyan oceanic crust, which may form the oldest ocean floor that is still preserved in the modern ocean floors (Speranza et al., 2012).

Although plate convergence and subduction was active throughout the Mediterranean region, the most prominent tectonic feature was the formation of three major extensional back-arcs, normally interpreted as a result of rapid slab roll-back. The overriding Eurasian Plate in the western Mediterranean became host to two oceanic basins on either side of the Corsica–Sardinia block, as well as the Algerian basin between the Balearic Islands and northern Africa, all with Neogene back-arc ocean floor. The eastern Alps and units to the south which had collided with Adria in the Paleogene (Tisia and Dacia) became extended above the Carpathian subduction zone to form the Pannonian Basin (Faccenna et al., 2014). The Aegean extensional back-arc, which had started forming in the Eocene, accelerated in Early Miocene time, leading to the modern orocline. Many of these features remain active today, as can be seen from the results of widespread GPS campaigns as well as active seismicity.

India–Asia Collision and South-East Asia. The initial India–Asia collision (Fig. 14.3) occurred during the Early Eocene followed by subduction of a largely oceanic Greater India Basin (van Hinsbergen et al., 2012). The second phase was the more substantial India–Asia collision, which may have begun in Late Paleogene–Early Neogene times (25–20 Ma), marked by a sudden propagation into the Asian interior uplifting the Tien Shan mountain range and the Altai Mountains of Mongolia, uplift which is still continuing today. In many plate models, the Australian Plate actually includes the Indian Plate in the Late Paleogene, but from the Early Neogene at about 20 Ma (Miocene) that plate has been subdivided into three plates: India, Capricorn, and Australia (Fig. 15.1a).

The evolution of South-East Asia is complex and dominated by the opening of a series of back-arc basins due to subduction roll-back (Hall, 2012). Australia began to collide with South-East Asia in the Early Miocene when the Sula Spur collided with the Sulawesi Volcanic Arc (Fig. 15.4a). Current Australian subduction beneath Java began in the Middle Eocene (45 Ma), and subduction roll-back into the Banda Embayment began in the Miocene at about 15 Ma, which caused extension of the Sula Spur. Soon afterwards, the first stage of extension formed the North Banda Sea during the Miocene between 12 and 7 Ma (Fig. 15.4) and remnants of the Sula Spur were carried south-eastwards above the subduction hinge. Collision of the Banda Arc with Australia and Timor started in the latest Miocene and continues today.

Arabia–Eurasia Collision. The Western Tethyan Orogen, which includes the belt of mountains in southern Spain and northern Africa, through the Alps, Carpathians, and Anatolia, and continuing into the Zagros Mountains, resulted from the closure of various parts of the Tethys Ocean. The still actively deforming Zagros fold and thrust belt in Iran is the result of the consequent Arabia–Eurasia collision, which probably started in Late Paleogene or Neogene times (Fig. 15.5). This is an interesting collision for many reasons, but perhaps its most startling feature is how the Eurasian blocks involved in it had once drifted off Arabia during the opening of the Neotethys Ocean in the Early Permian (Fig. 10.1) and later rejoined the same region. The formerly peri-Gondwanan Eurasian blocks involved in the Arabia–Eurasia collision included Sanand and Lut (now in Iran) and the Taurides and Pontides (Fig. 15.5a). The two latter blocks now form the Anatolian Plate, which includes all of Turkey.

Estimates for the Arabia–Eurasia collision vary from about 35 Ma to 20 Ma (Fig. 15.5e). Absolute velocities for Africa and Arabia declined in the Early Paleogene, followed by increased velocities after 45 Ma (Middle Eocene), peaking at 4 cm/yr at about 30 Ma (Oligocene). Conversely, convergence velocities between Arabia and Eurasia were

Fig. 15.3 (*cont.*) (French and Romanowicz, 2015), which have been linked to the Columbia River Basalt. (b) North-east Pacific reconstructions in the Miocene (15 Ma) in a CEED mantle frame with plate velocity vectors for the Pacific Plate. Dynamic self-closing plate polygons are those of Shephard et al. (2014). We show the location of three LIPs, the Columbia River Basalt (15 Ma) erupted on the North American Plate but close to the Juan de Fuca Trench, and the Hess (HR, 99 Ma) and Shatsky (SR, 147 Ma) Rises on the Pacific Plate. Also shown is the outline of the Emperor–Hawaii Chain and identified magnetic anomalies (MO, 120.4 Ma; C34, 83.5 Ma; C25, 55.9 Ma; C6, 20.1 Ma). (c) Distribution of hotspots (Steinberger, 2000) draped on seismic voting-map contours in the lower mantle (Lekic et al., 2012). Contours 5, 3, and 1 define the LLSVPs and 0 (blue) denotes faster regions in the lower mantle. Many hotspots appear to overlie regions of slower than average shear-wave velocities (notably those associated with Tuzo) but there are clear exceptions (e.g. Yellowstone). Twenty hotspots thought to be sourced by deep plumes from the core–mantle boundary (primary and clearly resolved plumes in French and Romanowicz, 2015) are shown as large white or yellow circles with red filling (also identified by Courtillot et al., 2003). Others of unknown origin are shown as smaller circles with white filling. (d) Reconstructed LIPs for the past 300 Myr (CEED mantle frame). Similar to Fig. 2.16c but Emeishan LIP not shown here because that LIP was calibrated in longitude to fall at the margin of Jason.

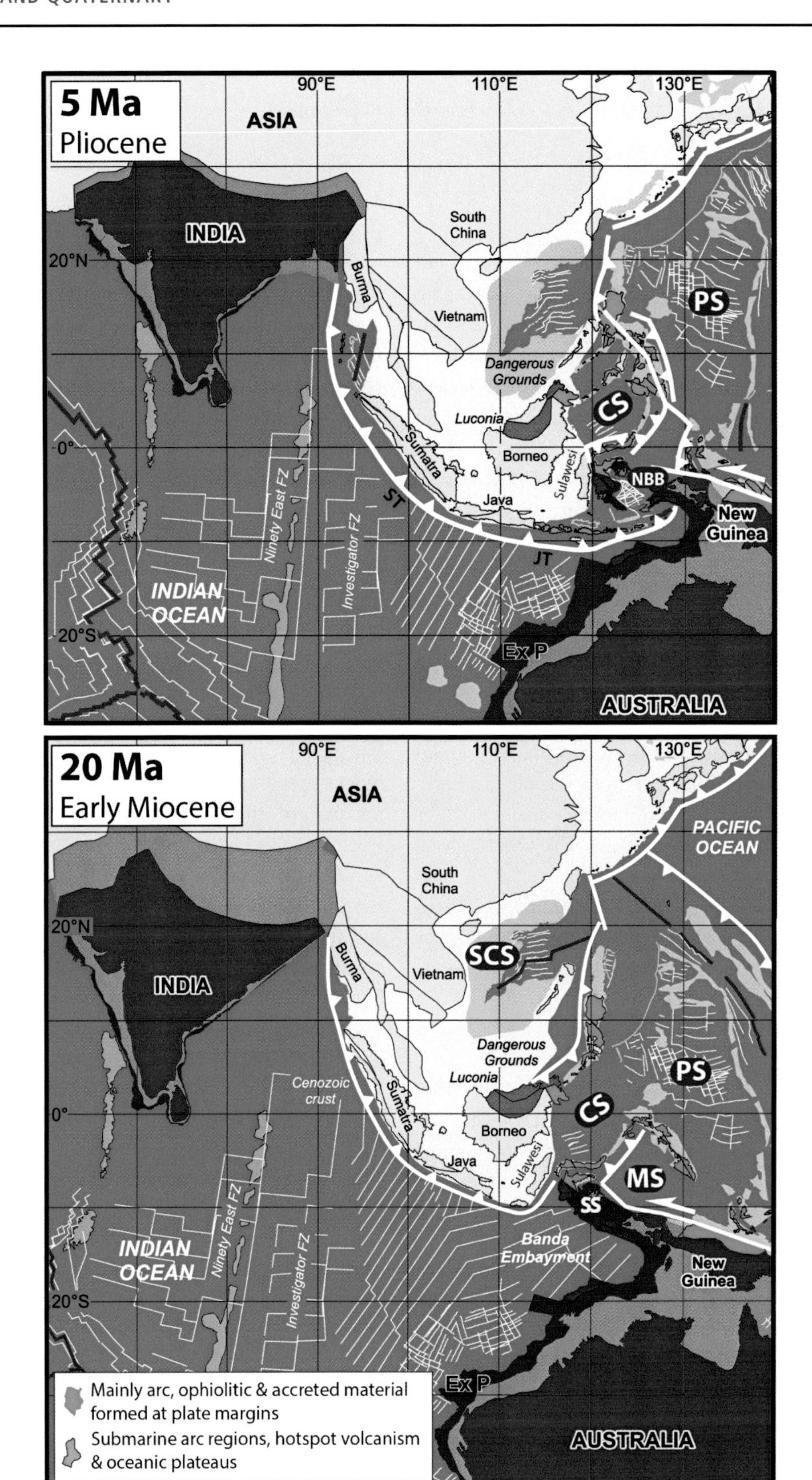

Fig. 15.4 Early Miocene and Pliocene reconstructions of South-East Asia and adjacent areas, showing the progressive collision of India with Asia, and some of the many movements between Australasia and the South-East Asian promontory of Indonesia (Hall, 2012). CS, Celebes Sea; Ex P, Exmouth Plateau; FZ, fault zone; JT, Java Trench; MS, Molucca Sea; NBB, North Banda Basin; PS, Philippine Sea; SCS, South China Sea; SS, Sula Spur; ST, Sunda Trench.

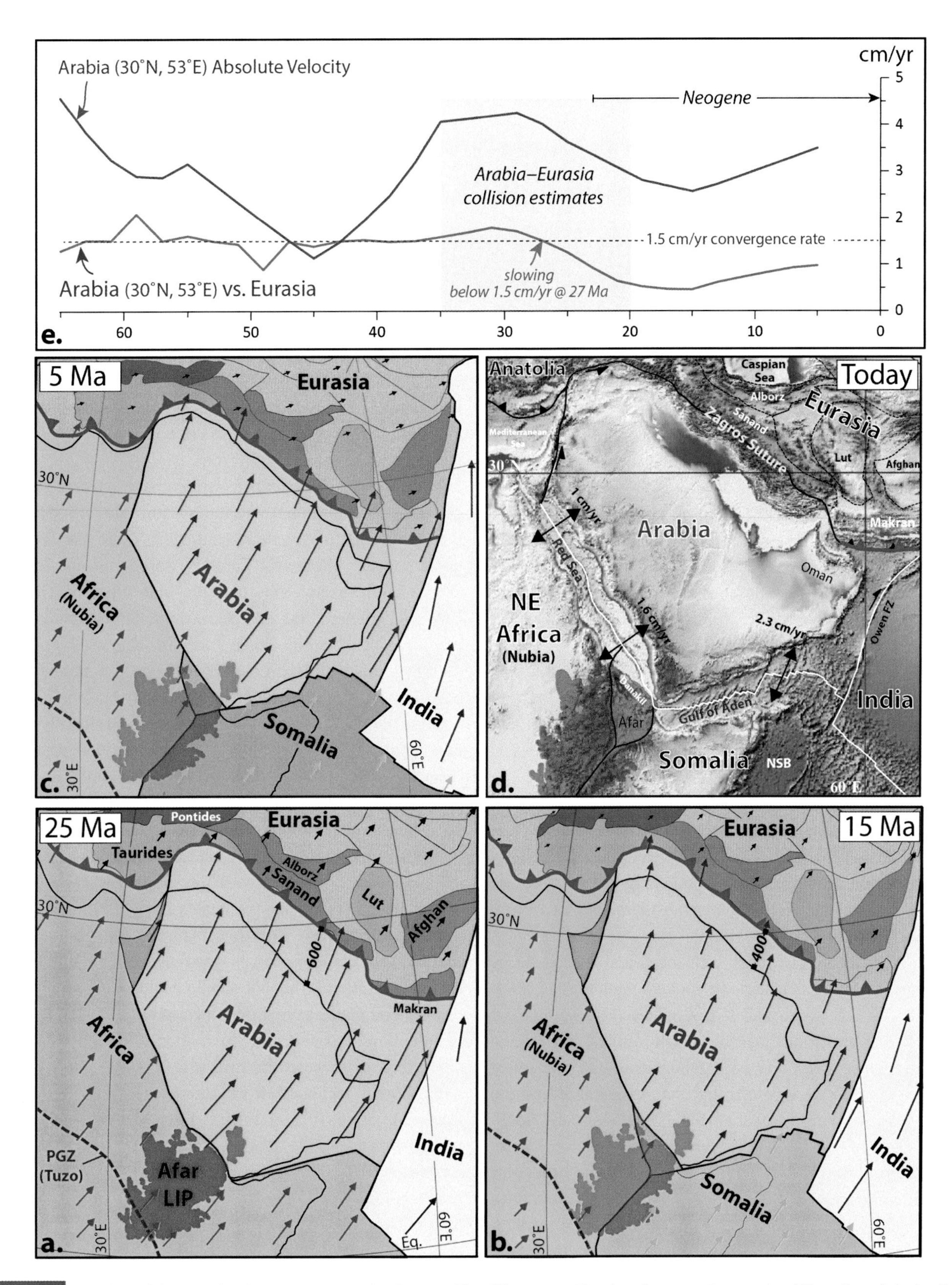

Fig. 15.5 (a)–(d) Arabia–Eurasia plate convergence for the past 25 million years. Numbers between the margin of Eurasia and the leading edge of Arabia at 25 Ma (Late Oligocene) and 15 Ma (Middle Miocene) denote gaps (required convergence) in kilometres; these are maximum convergence estimates since internal shortening between the Eurasian blocks has not been accounted for (see McQuarrie & van Hinsbergen, 2013). At 5 Ma the required convergence is 150 km (3 cm/yr). CEED mantle frame with absolute plate velocity vectors. (e) Absolute velocity for Arabia and relative velocity between Arabia and Eurasia calculated for a location at the leading edge of Arabia (30° N and 53° E). Convergence rates are about 1.5 cm/yr but there is a notably deceleration below this value at around 27 Ma (Late Oligocene) and within the range of estimated collision ages for Arabia and Eurasia.

steady at around 1.5 cm/yr for most of the Paleogene, but dropped below 1 cm/yr at around 27 Ma (Late Oligocene). That is the Arabia–Eurasia collision time estimated by McQuarrie and van Hinsbergen (2013), and that time is also near the first phase suggested for sea-floor spreading in the Red Sea.

The Afar LIP erupted from a plume over the margin of Tuzo (Fig. 15.5a) in north-eastern Africa at around 31 Ma (Oligocene) and may have been instrumental in triggering an early but short-lived 2 Myr phase of sea-floor spreading in the Red Sea, which according to Almalki et al. (2014) started in the Late Oligocene at around 26 Ma, but finished shortly afterwards because of the Arabia–Eurasia collision. That first phase of sea-floor spreading was followed by a phase of Neogene lithospheric extension and mafic dyke activity to about 6 Ma (latest Miocene), and overlapped with sea-floor spreading in the Gulf of Aden, which had probably started at around 18 Ma (Early Miocene) or perhaps a little earlier (Leroy et al., 2012). The onset of a second and still ongoing phase of sea-floor spreading started in the Pliocene at about 5 Ma.

The Western Americas. Western North America has witnessed a long and complex tectonic history, which comprised subduction, terrane accretion, strike-slip faulting, and shortening from Jurassic to Paleogene times (the Cordilleran, Sevier, and Laramide events). Conversely, the Neogene is known for extensional collapse and Basin and Range extension, Cascadia arc volcanism (subduction of the Juan de Fuca Plate), LIP volcanism (Columbia River Basalts), the Yellowstone Hotspot, and strike-slip motion along the San Andreas Fault and associated fault systems, most famously in the San Francisco area.

The Andes has been much influenced and shaped by its Neogene history, which started with the Farallon Plate being replaced by the Nazca and Cocos plates (Fig. 15.6a,b) at about 23 Ma (earliest Miocene), which resulted in a change in convergence direction: the Farallon Plate had previously been moving north-eastwards, but that was replaced by the Nazca Plate moving eastwards (Fig. 15.6) whilst the South American Plate maintained its slow westerly course. Therefore, the Andean margin changed from oblique to near orthogonal convergence and the Nazca Plate also underwent a small increase in absolute speed and convergence rates with South America, peaking at about 13 cm/yr in the Early Miocene (Fig. 15.6e).

The modern Andes is commonly considered as resulting from Nazca Plate subduction, and the Neogene change to more orthogonal convergence has been linked with major uplift, which culminated in the Late Miocene and Pliocene. Conversely, the deformation, uplift, and magmatic history of the southern Andes (south of 47° S) has been linked to the northward propagation of the Chile Triple Junction (Kay et al., 2005), which is being subducted beneath the South American Plate (Fig. 15.6c,d). But far-field stresses must have been important in the development of the modern Andes: most subduction models show that trenches tend to retreat, therefore favouring back-arc extension (and not mountain building) unless external stresses force the upper plate (in this case South America) to move towards the subduction zone faster than the trench retreat (Husson et al., 2012). South America has indeed had a history of being driven westwards faster than the trench retreat, and the westward drift of South America is about 5° during the Neogene up to today (Fig. 15.6e).

Unlike many orogenic belts, which have active island arcs offshore, the substantial volcanoes within the orogen are all within the South American continent. There are four volcanic zones, Northern, Central, Southern, and Austral (Fig. 15.6d), separated by large flat-slab segments (Bucaramanga, Peruvian, and Pampean), and the Patagonian volcanic gap at the Chile Triple Junction (Ramos & Folguera, 2009). A distinctive difference between the northern and southern Andes is that the former is chiefly built on previous oceanic lithosphere, whilst the southern part is largely underlain by fragments of continental basement that previously formed parts of Gondwana and peri-Gondwana, as well as previously accreted remains of Mesozoic and earlier island arcs. The combination of subduction on the western side and over-thrusting on the cratonic side has led to an exceptional 13 km tectonic relief between the bottom of the adjacent trench and the top of the mountains (Armijo et al., 2015).

Net Lithosphere Rotation. Estimating net lithosphere rotation (NR) is important in geodynamic modelling, and a basic assumption is that NR should be zero unless individual lithospheric plates have different couplings to the underlying mantle flow. For the past 30 million years NR has been characterised by westward drift (Fig. 15.7a) and NR values have slowly increased from about 0.1°/Myr in the Late Eocene to about 0.15–0.2°/Myr (Fig. 15.7b), depending on which absolute plate motion frame is used. The current westward drift is primarily driven by the large and fast-moving Pacific Plate that has controlled about 40% of the total NR over the past 5 million years. The model of Torsvik et al. (2010) yields the lowest NR values, and for the past 10 million years NR has been about 1.5 cm/yr for an equatorial site (Fig. 15.7a).

Going back in time, a minor NR peak in the Paleocene at around 60 Ma (Fig. 15.7b) has been noticed in many studies, and this peak of about 0.3°/Myr has been linked to the initial India–Asia collision, or the terminal subduction (55 Ma) of

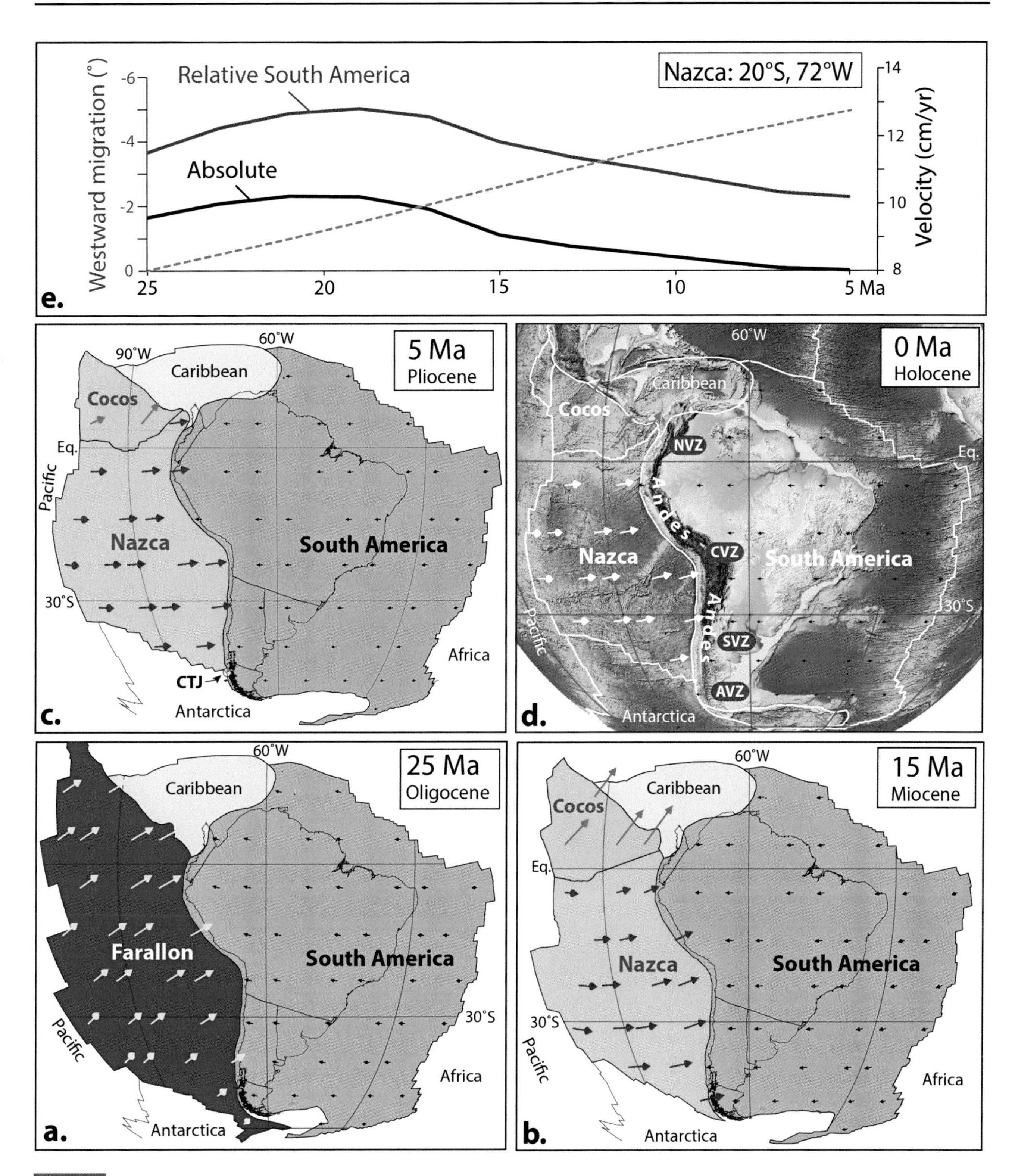

Fig. 15.6 (a)–(d) Farallon and Nazca plate interaction with South America over the past 25 million years. CEED mantle frame with absolute plate velocity vectors. (e) Absolute velocity for Nazca and relative velocity between Nazca and South America calculated for a location at the leading edge of Nazca (20° S and 72° W). In an absolute reference frame, the Nazca trench and South America are migrating westward at an almost linear rate (stippled blue line), amounting to about 5° for the past 25 million years. AVZ, Austral Volcanic Zone; CTJ, Chile Triple Junction; CVZ, Central Volcanic Zone; NVC, Northern Volcanic Zone; SVZ, Southern Volcanic Zone.

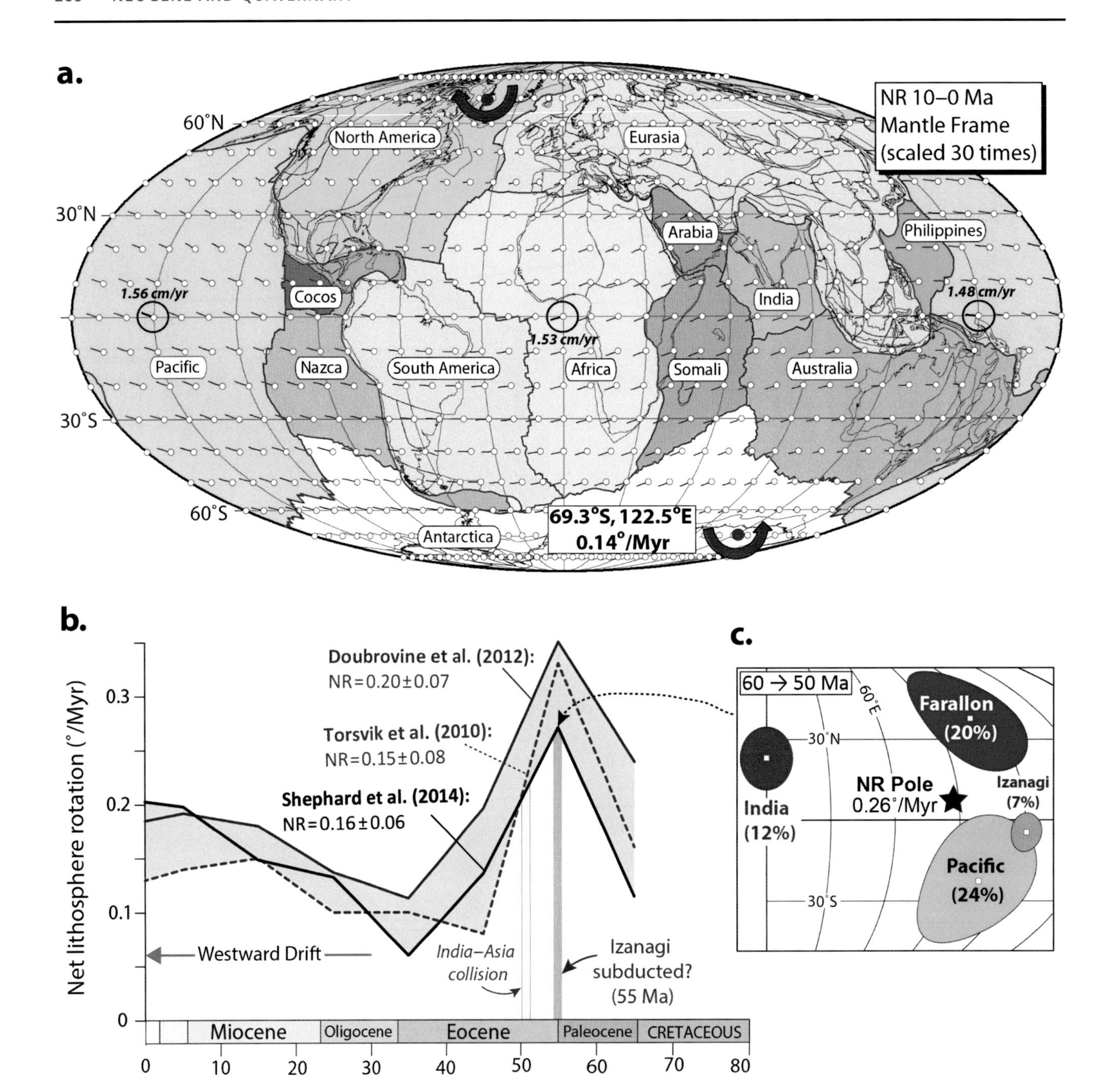

Fig. 15.7 (a) Net rotation (NR) velocity field from 10 Ma (Late Miocene) to the present (Torsvik et al., 2010). The counter-clockwise NR pole at high southerly latitudes (69.3° S, 122.5° E) results in westward drift, which has been calculated at three equatorial locations (1.48–1.56 cm/yr; vectors with black circles). The NR velocity field is draped on a simplified present-day plate polygon model (Mollweide projection). (b) Estimated NR for the past 70 Myr (calculated over a 10 Myr window) from three different studies; mean NR ranges from about 0.2°/Myr (Doubrovine et al., 2012) to 0.15°/Myr (Torsvik et al., 2010) but there is a notable high in the Early Eocene at 55 Ma (calculated between 6 and 50 Ma) and lows at 45–35 Ma (Middle and Late Eocene). (c) Total NR pole and NR poles for the Pacific, Farallon, Izanagi, and Indian plates at 55 Ma (calculated from Shephard et al., 2014, model). Sizes of colour shaded ovals are scaled to their contribution to total NR (e.g. the Pacific Plate contributed 24% of the total NR at that time).

the Izanagi Plate (Izanagi–Pacific ridge). The latter plates, however, controlled only 12% (India) and 7% (Izanagi) of the total torque between 60 and 50 Ma (Paleocene and Early Eocene), and the Pacific (24%) and Farallon (20%) plates were the big torque players. Between 60 and 50 Ma, the NR pole was close to the Pacific, Izanagi, and Farallon rotation poles, whilst the India pole was far away. During the rest of the Eocene, between 50 and 40 Ma, NR was reduced because the Pacific Plate changed its course from north to north-west after 47 Ma, as seen in the Emperor–Hawaii Bend, whilst the Farallon Plate maintained its north-easterly course.

The world uncertainty (a measure of how much oceanic crust is inferred in a reconstruction that has been subducted afterwards) quickly approaches 50% when going back to the Cretaceous, and NR spikes before the Paleogene are probably all artefacts reflecting how we organise the Panthalassic or Neotethyan oceanic plates and assign their velocities. As an example, there is a large NR spike in many models (e.g. Seton et al., 2012) in the Early Cretaceous shortly after 140 Ma (Fig. 15.8c), which is exactly when the Pacific Plate is assigned a velocity for the first time (it had become progressively larger without any velocity from the Early Jurassic at 190 to 140 Ma) (Fig. 15.8a) in these models, but, even more importantly, a doubling in the velocity (Fig. 12.3b) for the Phoenix Plate in the Early Cretaceous, which alone contributes 29% to the total NR (Fig. 15.8b), causes an abrupt increase of NR.

Facies, Floras, and Faunas

Climate and Sea Levels. Throughout most of this period, global temperatures have progressively decreased, although with various fluctuations. During the Middle and Late Miocene, climatic differences between low and high latitudes increased significantly, reflecting the start of the polar cooling described below, and those changes obviously restricted the distributions of animals and plants with limited habitat tolerance. Between the polar and tropical regions, at middle to higher latitudes, there was an expansion of drier and more open environments with grasslands which suited many species (such as horses and buffalo), but the ranges of many other animals dwindled as the forests became reduced in area.

Global sea level progressively fell during the Miocene (Fig. 16.2). For example, during the Middle Miocene, comparable calcareous nannoplankton indicate that the Mediterranean was connected to the Tethys and north-eastern Indian oceans through the widening Red Sea, although the shallowness of those seas is demonstrated by the thick evaporites in all three areas. However, further shallowing led to the Messinian Salinity Crisis, as the Alpine Tethys Ocean dwindled in the Mediterranean area. In the latest stage of the Miocene, the Messinian at about 6 Ma, thick deposits of halite and anhydrite formed beneath the western Mediterranean, indicating that the area was completely cut off from the open oceans, and became a series of large progressively desiccated lakes of very high salinity, which resulted in erosion at the basin margins and the extensive deposition of non-marine sediments there. That crisis included a local drop in sea level of at least 60 m (it may have been much more), and continued until the Straits of Gibraltar were breached soon afterwards.

The Plio-Pleistocene Glaciation. Although the Earth's temperature has gradually dwindled since the Early Eocene (Fig. 16.3), it was during the Pliocene, at about 2.7 Ma, that the first phase started of the remarkable Plio-Pleistocene glaciation, which has probably not yet ended today. Although a hundred years ago it was concluded that there were fewer than ten glacial episodes within it, through the analysis of complete cores in both deep-sea marine sediments and in the extensive ice caps in the higher latitudes, it is now known that there have been at least 17 glacial periods alternating with a corresponding number of warmer periods termed 'interglacials', including the present. The existence of ice sheets has a profound effect on both atmospheric and oceanic circulations, with climatic belts being pushed by up to 2,000 km towards the Equator, resulting in a marked narrowing of the Equatorial Zone during the glacial intervals. In addition, the desert areas become much more extensive during the glacial periods, even in areas some distance away from the ice caps, whilst in the interglacials the deserts are much reduced in size, enabling the enlargement of many areas fringing the erstwhile deserts which can support many more diverse faunal and floral communities and habitats. For example, during the postglacial optimum between 9,000 and 6,000 years ago, the Sahara Desert was much less extensive than today, and its surrounding lakes (for example, Lake Chad) were much larger. Animals such as giraffe and hippopotamus are known, not only from the Sahara, but from all across northern Africa, as well as in the more temperate regions such as England.

The Development of Modern Faunal Provinces. This is not a book on modern faunas and their distributions, but some of them are helpful in interpreting past geography. An example is Wallace's Line, which separates Australasia from most of the East Indies today, and which demarcates the land regions to its south-east whose larger faunas are dominated by marsupials (kangaroos and others) in contrast to the lands to the north-west, which have pantropical faunas dominated by other mammals (elephants, tigers, and others). That line is chiefly the result of the large but dwindling distance which previously separated the Australasian sector of Gondwana from the varied East Asian continents described in earlier chapters, and which allowed separate evolution within the two regions. However, the reality is additionally much more complex. For example, of the 13 primate species (monkeys etc.) found in Borneo today, only three have crossed Wallace's Line during the Pleistocene, and that is because the tropical forests to which they are restricted have been much reduced at various times (caused by the varying

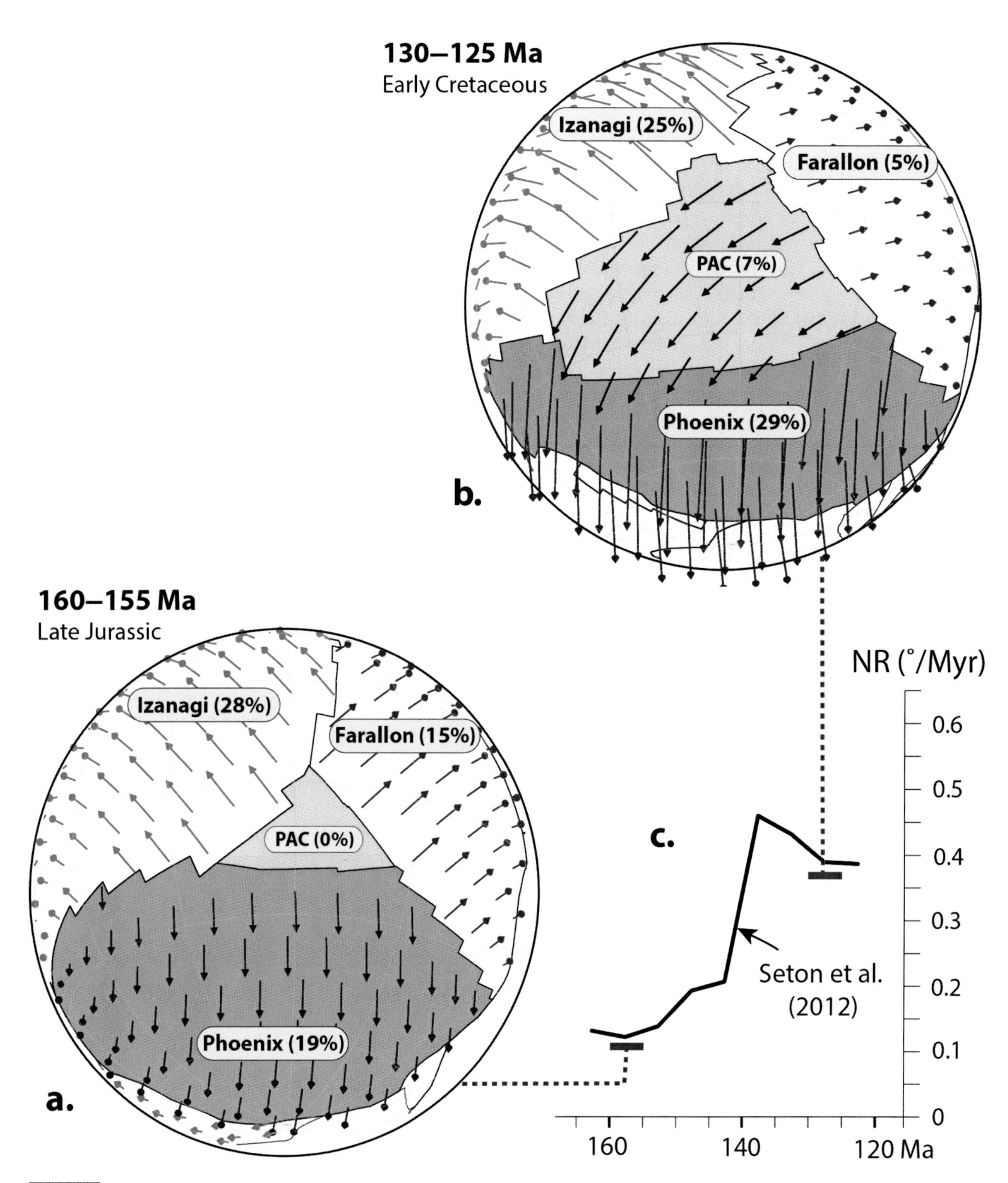

Fig. 15.8 (a, b) Izanagi, Farallon, Pacific, and Phoenix velocity fields at 160–155 Ma (Late Jurassic) and 130–125 Ma (Early Cretaceous) in the Seton et al. (2012) absolute plate motion model. (c) Variation in net lithosphere rotation (NR) in Late Jurassic–Early Cretaceous. The rapid increase in NR is caused by the accelerating Phoenix Plate combined with the emergence of the Pacific Plate. The percentage contribution to total NR is listed for each plate. In the Early Cretaceous four plates alone contributed 66% of the total NR.

intensities of the Plio-Pleistocene glacial maxima), thus inhibiting the monkeys from spreading (D. Brandon-Jones in Hall & Holloway, 1998), rather than through the existence of any oceanic barriers to migration. That sort of effect is impossible to decipher in, for example, the Carboniferous, when detailed correlation is on too large a scale, and also the rocks and fossil records are patchy.

Most of the birds flew across Wallace's Line many years ago, and thus the Australasian parrots are not very different from their South American cousins. However, those that had adapted to a flightless running life could not cross the line, which explains the differences between the African ostriches and the Australian emus and their relatives. Comparably, insects, such as the flies which are so irritating to humans in Australasia, consist of genera which are largely cosmopolitan.

Planktonic foraminifera were obviously able to be transported freely through the ocean currents, and thus their distributions were related to temperature and (more broadly) latitude, even though they cannot actively swim (but can travel up and down the water column by altering their buoyancy), and have no separate faunal provinces. In contrast, the larger foraminifera are benthos on the sea floors and were confined to the shelves and thus various faunal provinces can be recognised; for example, in the Late Miocene there were an American Province, a Mediterranean Province, and an Indo-West Pacific Province. Some notable molluscs were also limited to the same Indo-West Pacific Province, including the planktonic cephalopod *Nautilus*, and the large bivalve *Tridacna*, the giant clam (C.G. Adams in Cocks, 1981). The Indo-West Pacific Province also hosts the highest zooxanthellate coral biodiversity in the world today, but that high biodiversity has only been located there since the Miocene: and thus the province cannot be recognised in the corals in the Paleogene (M.E.J. Wilson & B.R. Rosen in Hall & Holloway, 1998). The Red Sea also hosts a diverse coral fauna, but the oldest sediments within it are Oligocene, and thus that rich fauna has largely developed during the Neogene and Quaternary (Almalki et al., 2015).

Links between America and the Rest of the World. With the widening Atlantic Ocean, North America has had progressively more tenuous connections with the north-west of Eurasia, and the western North American margin has also mostly been widely separate from the north-east of Eurasia, apart from the fluctuating land bridge between Siberia and Alaska in the Bering Straits region. The relationships between North and South America have also fluctuated as the Isthmus of Panama and the Caribbean microplates performed their complex dance through time. In the Early Miocene at about 15 Ma, the isthmus consisted of a series of disconnected islands, and it was not until the Pliocene, at about 3 Ma, that there was a complete land bridge between the two major continents. It was then that what has been termed the Great American Biotic Interchange (GABI) began, which had previously been deemed to be the ending of the relative isolation of South America, and had begun during the progressive breakup of Pangea in the Cretaceous (Fig. 14.2). However, the GABI event did not affect many of the floras (whose seeds are often windblown or dispersed by birds).

Although it was previously thought that seals (which swim in salt water) and bats (which flew) were the only mammals which lived in South America from before 10 Ma (Late Miocene), there are now known to be at least two well-documented sites, in La Venta, Colombia, and the Fitzcarrald Arch in Peru, which are earlier. In the latter, which is of late Middle Miocene age (about 12 Ma), a freshwater lacustrine tidal basin has preserved 24 terrestrial and two aquatic mammal taxa as fossils which represent both the latest species of some extinct groups and also the earliest forms of some of the modern South American mammalian fauna, such as large sloths (Tejada-Lara et al., 2015). Most of the other mammal groups appear to have crossed the Isthmus of Panama soon afterwards, and before the GABI event, which is now identified as being less significant than previously suggested (Cody et al., 2010).

The Rise of Humans. Humans (the family Hominidae) are placental mammals and are within the Primates, an order which also includes creatures such as lemurs and lorises. The higher primates are mostly monkeys, and include gorillas, chimpanzees, and hominids. From genetic data, gorillas appear to have diverged from the others in Miocene times at about 10 Ma, and hominids from chimpanzees in the latest Miocene at about 6.5 Ma. Hominids are distinguished from the other apes by the development of bipedalism, and, apart from our own genus *Homo*, are all known only from southern and eastern Africa. Other genera closely related to humans also existed in that area; as well as the oldest *Homo*, the species *H. habilis* which has been found in deposits in Tanzania dated from 2.4 to 1.6 Ma. *Homo habilis* had a brain size of about 600 cm^3, and evolved into several other species, including the Neanderthals (*H. neanderthalensis*), and our own *H. sapiens*, which has an average adult brain size of about 1,300 cm^3. Several groups of those later humanoids (but not *H. sapiens*) migrated out from Africa at various times from about 2 Ma onwards, even as far as China (Peking Man) and Indonesia (Java Man). However, modern humans, *Homo sapiens*, which did not evolve until about 300 thousand years (ka) ago, did not leave Africa for the

Middle East until about 100 ka, although they subsequently spread relatively quickly into southern Europe (Stringer, 2002). The oldest true *Homo sapiens* fossil remains in China and Australia date from about 50 ka, but humans did not reach America, almost certainly via one of the ephemeral Siberian land bridges during a low glacially induced sea-level stand, until 13 ka, after which they spread relatively quickly and reached southernmost South America by 11 ka. Humans were all nomadic hunter-gatherers until the advent of agriculture, which necessitated permanent settlements, the oldest of which date from about 10 ka and are in Turkey and Syria.

16 Climates Past and Present

The first lava flows of the Siberian Traps north of Norilsk (Talnakh) erupted onto continental deposits of sand and coal. The end-Permian mass extinction is linked to the Siberian Traps. Credit: Henrik Svensen/CEED.

Will it rain next week? Is global warming real? Have we come to the end of the Ice Age, or are we living in an interglacial period? The weather and climate are perennial topics, not only of conversation, but also of worry and economic concern. Much is known about how climate has changed over the past hundred years or so, but only geologists can add the deeper time perspective which we feel is essential to our collective better understanding of this very complex topic, which involves the interaction of so many variables. That requires a seamless integration of many branches of geology, ranging through stratigraphy, sedimentology, palaeontology, isotope analysis, age dating, tectonics, and many more disciplines.

Many scientists argue that an increase in global surface temperatures over the past decades has been driven largely by human-induced (anthropogenic) emissions of greenhouse

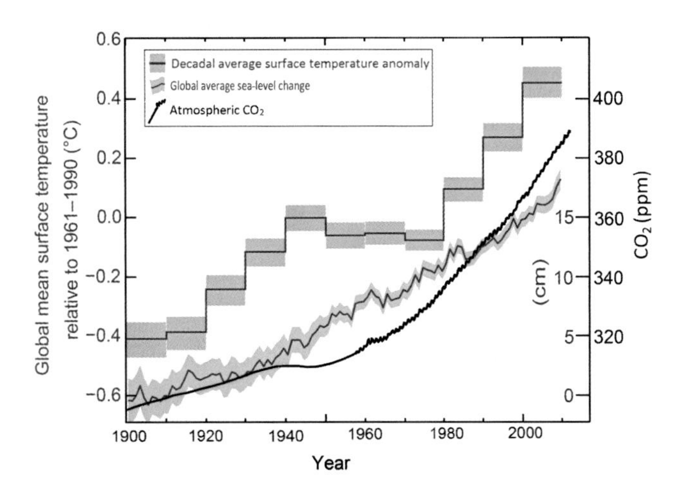

Fig. 16.1 Decadal average surface temperature anomaly, changes in atmospheric pCO_2, and global average sea level. Data from IPCC (2013).

gases with carbon dioxide concentrations having increased by about 40% since pre-industrial times (Fig. 16.1). The Earth's surface has become successively warmer, the Greenland and Antarctic ice sheets and glaciers worldwide have lost mass, and the global mean sea level has risen by about 20 cm over the past century.

An interesting question is whether geologists who interpret palaeoclimate millions of years in the past can contribute something of value about the causes of recent climate changes (global warming). Some may argue 'probably not much', except that we can point to obvious facts: sea level was at times much higher than today (by several hundred metres); temperatures were higher (by more than 10 °C); and atmospheric pCO_2 has been more than ten times higher than 'normal'. These 'highs' are the signature of the greenhouse climate (warm and humid conditions) which has dominated the Phanerozoic climate history (Fig. 16.2). Those climate changes are mostly related to rather slow geological processes such as plate tectonics (including continent–ocean distribution, mountain building, topography, weathering, and more), true polar wander (rotation of the Earth's lithosphere and mantle with respect to the spin axis), and flow in the mantle generating dynamic topography. Much depends on our ability to date rocks precisely so that the rates of change of geological processes can be accurately assessed (often limited to uncertainties of millions of years). In addition, estimates of atmospheric pCO_2 and temperatures in the deep past are based on various proxies and modelling approaches. It is also important to remember that the Earth's rotation has slowed steadily during its history, largely due to the ceaseless friction of the tides as they ebb and flow. Thus a day during the early Earth was about two to three hours, increasing to about twenty hours by the Late Precambrian, and a year in the Early Cambrian consisted of about 420 days (Fig. 16.4), which will cumulatively have made the ancient climates less similar than those of today.

Our planet has also undergone rapid climate–environmental changes and five major mass extinctions are known in the Phanerozoic. These are known as the *big five* and include the End Ordovician, the Late Devonian, the End Permian (the largest), the End Triassic, and the End Cretaceous (K-Pg) extinction events (Fig. 16.2). In addition there have been many smaller events, as well as a sixth ongoing (Holocene) mass extinction probably caused by human activity. The causes for mass extinctions are debated, but many consider large igneous provinces as an important cause.

Some Factors Affecting the Climate

Factors commonly invoked to force or drive climate change include the amount of energy from the Sun, orbital forcing, the magnetic field strength, and plate tectonics. True polar wander can also be considered as an external driver for climate changes, and catastrophic volcanism (LIPs) can clearly force rapid climate changes and mass extinctions. These five external drivers influence the climate system and lead to measurable climate responses such as changes in land surface, vegetation, ocean, ice, atmosphere, and life.

Changes in the Sun's Energy Output. Astronomical data suggest that the Sun steadily increases its heat output by up to 1% per 100 million years, and thus in its early days might have emitted as little as 45% of its current output. However, the atmosphere was probably much denser in Earth's early days than now and much of it was composed of carbon dioxide, and perhaps also methane, and thus surface temperatures might not have been so very different from today. It is also worth remembering that probably 30–40% of the Earth's surface heat is generated by radioactive decay in the Earth's interior, which would make the variation in direct solar radiation less significant.

Changes in the Earth's Orbit. Due to mutual gravitational interaction between the Sun, other planets and their satellites, the Earth's orbit varies from near circular to elliptical over a cycle of about 110 thousand years (ka), which means that the distance from the Sun and hence solar radiation and climates are affected. The Earth's axis of rotation (axial tilt) also varies from 21.8° to 24.4° in cycles of about 40 ka, as does its wobble, which causes a precession of the equinoxes and has a cycle of about 22 ka, and the three together are termed Milankovitch cycles. Such cycles have certainly affected climate changes and are deemed to have been the chief triggers underlying the changes between the

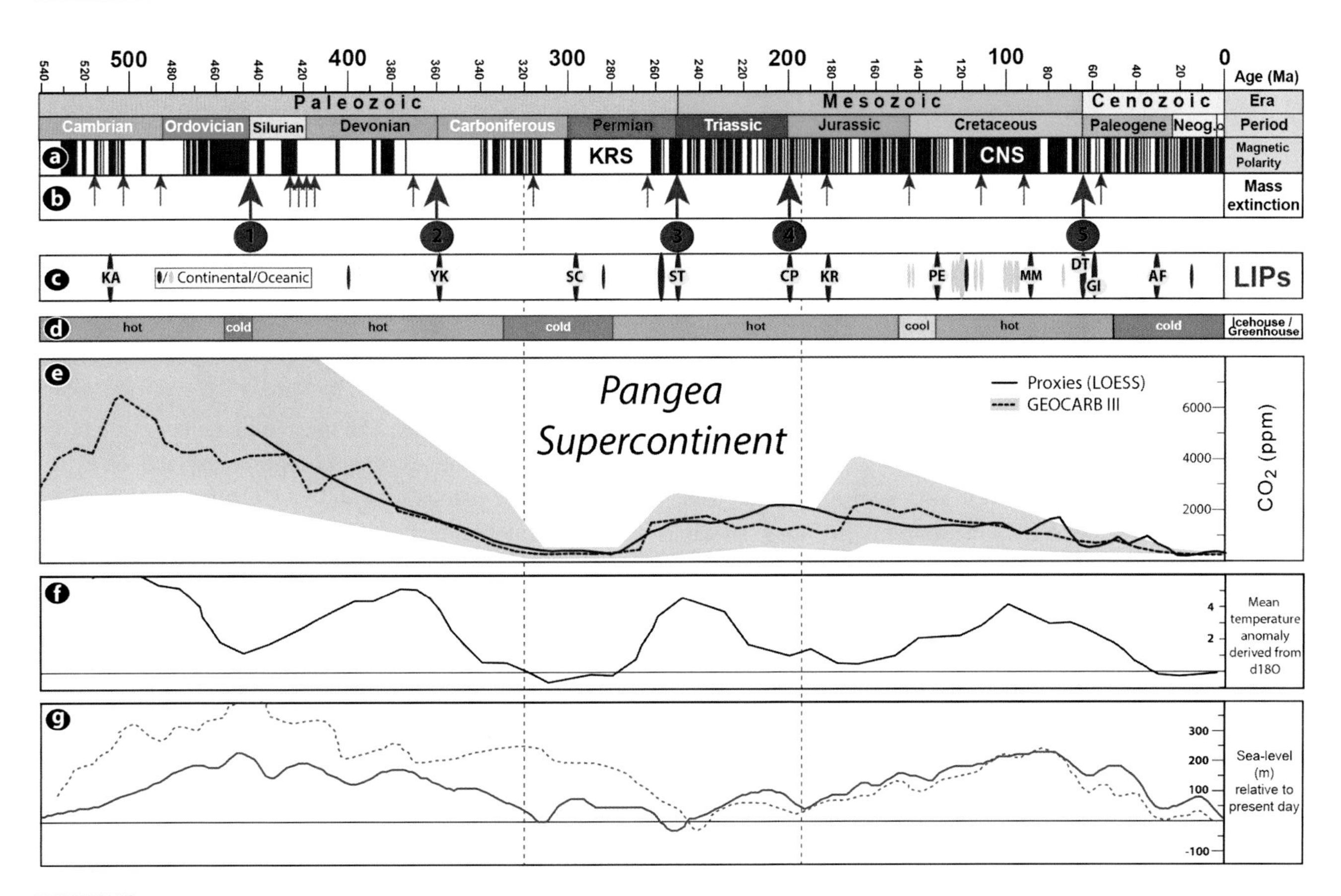

Fig. 16.2 Phanerozoic time scale and (a) magnetic polarity (Eide & Torsvik, 1996; Gee & Kent, 2007), (b) extinction events (five major and an impact scenario commonly invoked to explain the Cretaceous–Paleogene event), (c) LIP events (listed in Appendix 1), (d) icehouse (cold) vs. greenhouse (hot) conditions, (e) atmospheric pCO_2 (Royer, 2006), (f) mean temperature anomaly (Royer et al., 2007; Veizer et al., 2000; Came et al., 2007), and (g) global sea-level variations (red dashed line after Hallam, 1988; black line after Haq & Al-Qahtani, 2005; Haq & Shutter, 2008). AF, Afar LIP; CNS, Cretaceous Normal Superchron; CP, Central Atlantic Magmatic Province; DT, Deccan Traps; GI, North Atlantic Igneous Province (NAIP); KA, Kalkarindji LIP; KR, Karroo LIP; KRS, Kiaman Reverse Superchron; PE, Paraná–Etendeka LIP; SC, Skagerrak Centred LIP; ST, Siberian Traps; YK, Yakutsk LIP.

glacial and interglacial periods during the current Ice Age. In particular, the timing of the equinoxes directly affects the climates at different latitudes.

Changes in the Earth's Magnetic Field. Rarely invoked to impact the climate system, but abrupt geomagnetic field variations, perhaps resulting in enhanced cosmic-ray induced nucleation of clouds, have been argued by Courtillot et al. (2007) to correlate with certain climatic variations. On a similar note, Knudsen and Riisager (2009) argue for a correlation between the Earth's dipole moment and the amount of precipitation (based on speleothem $\delta^{18}O$ proxy record) in the tropics during Holocene times.

Plate Tectonics. The greenhouse climate that dominated the Phanerozoic climate history is commonly associated with continental dispersal, high sea-floor production, high sea level, and carbon dioxide production. Dispersal and assembly of landmasses that open and close oceanic gateways can profoundly affect ocean circulations, climate, and distributions of fauna and flora. Formation of supercontinents also reduces the total continental shelf areas and leads to arid conditions with extreme seasonal variations.

The oceanic 'carbon source–sink' processes depend on the configuration of oceanic areas through time, and subduction processes contribute to changes in global geochemical cycles through sediment recycling and related volcanism. In the spreading-rate hypothesis (Berner et al., 1983), commonly dubbed BLAG (after the names of the authors), CO_2 is removed from the atmosphere by chemical weathering, then deposited in the oceans, subducted, and eventually returned to the atmosphere through volcanic activity. In essence, fast sea-floor spreading leads to rapid CO_2 input in the atmosphere and a greenhouse climate. But that is compensated by higher chemical weathering on land, and thus increased CO_2 removal and reduced warming. Conversely, slow spreading is associated with slow CO_2 input (icehouse climate), decreased chemical weathering on land, and therefore reduced CO_2.

The BLAG hypothesis of carbon cycling provides long-term climate stability, but imbalances between CO_2 atmosphere input and carbon weathering and burial may drive climate changes over tens of millions of years (Ruddiman, 2014). Another proposal explaining how plate tectonics may control atmospheric CO_2 is the uplift weathering hypothesis (Raymo et al., 1988). In this hypothesis mountain building and uplift accelerates weathering, removing more CO_2 from the atmosphere, and therefore cools the global climate.

True Polar Wander (TPW). This is also an important, albeit relatively unexplored, mechanism that may slowly impact the climate system. During TPW events, when the entire Earth's surface is rotating slowly relative to the spin axis, some areas become warmer (moving away from the poles), others colder, whilst regions close to the pole of rotation should experience only minor changes. Although TPW rates may appear slow, about 10 cm/yr (1°/Myr) today, the magnitude is comparable to, and can be even higher (perhaps up to 3°/Myr) than, the velocity of individual continents ('continental drift'), but it affects all plates in chorus. TPW may have played a subsidiary role in both the end-Ordovician and the Plio-Pleistocene northern hemisphere glaciations.

Large Igneous Provinces. Abrupt changes in the atmospheric concentration of greenhouse gases have occurred throughout Earth's history, and many of these climatic and environmental perturbations show a causal relationship with LIP eruptions (Fig. 16.2). LIPs may have caused or contributed to four of the 'big five' biotic extinction events in the Phanerozoic: in order, the End Ordovician, the End Devonian (Yakutsk LIP), the End Permian (Siberian Traps LIP), the End Triassic (Central Atlantic Magmatic Province CAMP LIP) and the end Cretaceous (K-Pg: Deccan Trap LIP). The K-Pg mass extinction event, however, is unique because it is coincident with the Chicxulub bolide impact. The oldest Phanerozoic extinction, and the third in importance, at the end of the Ordovician (Hirnantian), is not linked with any known LIP eruption, but it must not be forgotten that from the Cretaceous onwards most LIPs were emplaced on the ocean floors, and no Palaeozoic oceanic LIPs are known due to subsequent subduction.

LIPs represent huge volumes of basalts, perhaps erupted over a million years or less, and are likely to have strong climate impact through the release of vast amounts of carbon dioxide, methane, and acid compounds into the atmosphere, leading to ocean acidification and global warming. Over the past decade the scientific emphasis on the causes and triggers of past global warming has gradually shifted from gas hydrate dissociation and lava degassing to solid earth degassing caused by contact metamorphism in sedimentary basins (Svensen et al., 2004). A key factor here is the magma emplacement environment and the potential to produce poisonous gases: the two largest mass extinction events (end-Permian, end-Triassic) are both coeval with intrusions of sills into evaporite-rich sedimentary basins in Siberia (Siberian Traps) and in Brazil (CAMP).

How Past Climates Are Deciphered

Oxygen Isotope Values. The most important climatic record in the ocean is the oxygen isotope signal and $\delta^{18}O$ is a measure of the ratio of ^{18}O and ^{16}O. Oxygen isotopes documented from sediments and well-preserved fossils within them reflect the temperature of the sea water at the time of deposition, but $\delta^{18}O$ variations are also sensitive to continental ice sheet changes. A number of lows and highs are recorded in the Cenozoic (Fig. 16.3), which reflect global warming and cooling, and growth and decay of ice sheets. Negative excursions correspond to times of warmer temperature, and a 1‰ decrease corresponds to about a 4 °C temperature increase (Fig. 16.3).

Carbon Isotope Values. Carbon isotope ratios ($\delta^{13}C$) documented from sediments record the chemical composition of the contemporaneous sea water. However, bulk sediment sampling also reflects subsequent diagenetic alteration, and the best results are therefore obtained from well-preserved shells, such as oysters in the Mesozoic with low-magnesium calcite, a mineral which is relatively resistant to diagenesis (Korte et al., 2009). Positive carbon isotope excursions correspond to periods of low atmospheric carbon dioxide content, whereas negative excursions potentially correspond to times of higher atmospheric carbon dioxide (^{12}C enrichment). Two pronounced Early Eocene negative excursions correspond to the Eocene thermal maxima (ETM1 and ETM2 in Fig. 16.3).

Distributions of Coal, Peat, Limestones, and Evaporites. Coals, lignites, and peats indicate a prolonged excess of precipitation over evaporation at their deposition sites, as well as requiring relatively rapid burial. Today such conditions occur in both temperate and equatorial belts. The migrations of the many kinds of sedimentary belts towards or away from either the poles or the Equator are good indicators of progressive climate change. That is particularly true of the locations of reefs; however, while the abundances of most reef-building organisms such as corals increase with warmer temperatures, others (such as bryozoans) are capable of being the frameworks for bioherms at cooler latitudes. Coals, reefs, and glacial features are therefore shown in the

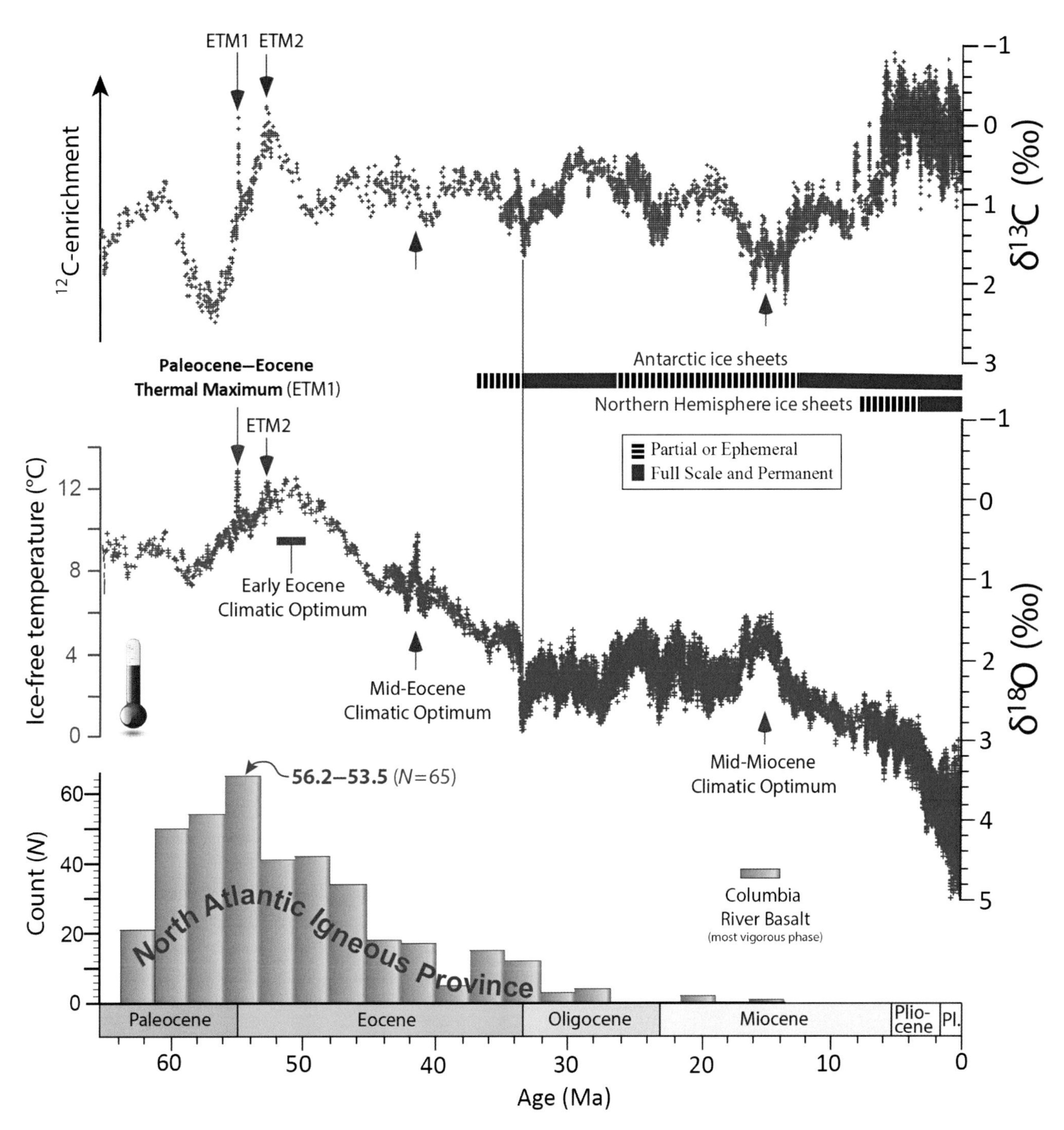

Fig. 16.3 Global deep-sea oxygen (Zachos et al., 2008) and carbon isotope (Zachos et al., 2001) records for the Cenozoic. The left-hand δ^{18}O temperature scale is calculated for an Earth with ice-free oceans and thus not applicable after the onset of large-scale glaciation in Antarctica at around 35 Ma. The Early Eocene Climatic Optimum (about 2 Myr long), the Middle Eocene Climatic Optimum, and the very short-lived Early Eocene hyperthermals (brief intervals of extreme global warmth and massive carbon addition) such as the Paleocene–Eocene Thermal Maximum (PETM), also known as the Eocene Thermal Maximum 1 (ETM1) and the Eocene Thermal Maximum 2 (ETM2) are indicated in the diagram. Note that the carbon isotope curve (δ^{13}C) is plotted so that lows appear as peaks (^{12}C-enrichment) for easier comparison with the oxygen isotope curve. The lower diagram is a frequency plot (*N*: number) for all isotope ages from the North Atlantic Igneous Province. The analysis is based on 330 published ages (only six U/pb ages), and the largest peak was between 56.2 and 53.5 Ma, which is contemporaneous with ETM1.

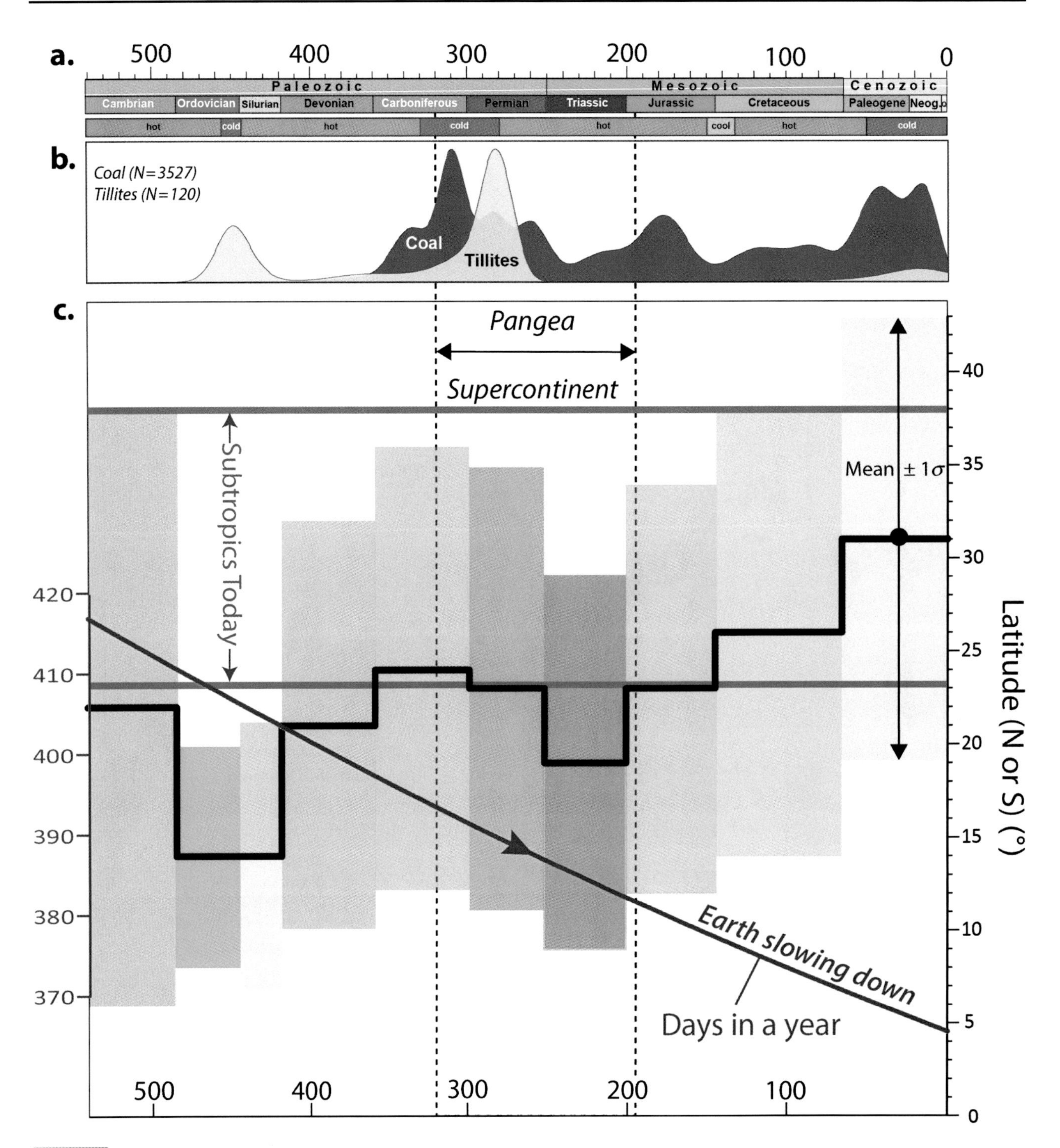

Fig. 16.4 (a) Phanerozoic time scale, icehouse (cold) vs. greenhouse (hot) conditions; (b) kernel density estimates for coal ($N = 3{,}527$) and tillites ($N = 120$); and (c) reconstructed latitude ($\pm 1\sigma$) for around 2,000 evaporite sites (averaged for the Cenozoic and older geological periods), and the approximate number of days in a year. Lithological data from Boucot et al. (2013).

many facies maps in the preceding chapters, as well as providing some independent check on the positioning and palaeolatitudes of some of the terrane units. The published sources for the sites shown are too varied to mention individually, and many are drawn from our previous papers, but a major source for the positions of coals, evaporites, and glacial features (Figs. 16.4 and 16.5) has been the compilation by Boucot et al. (2013).

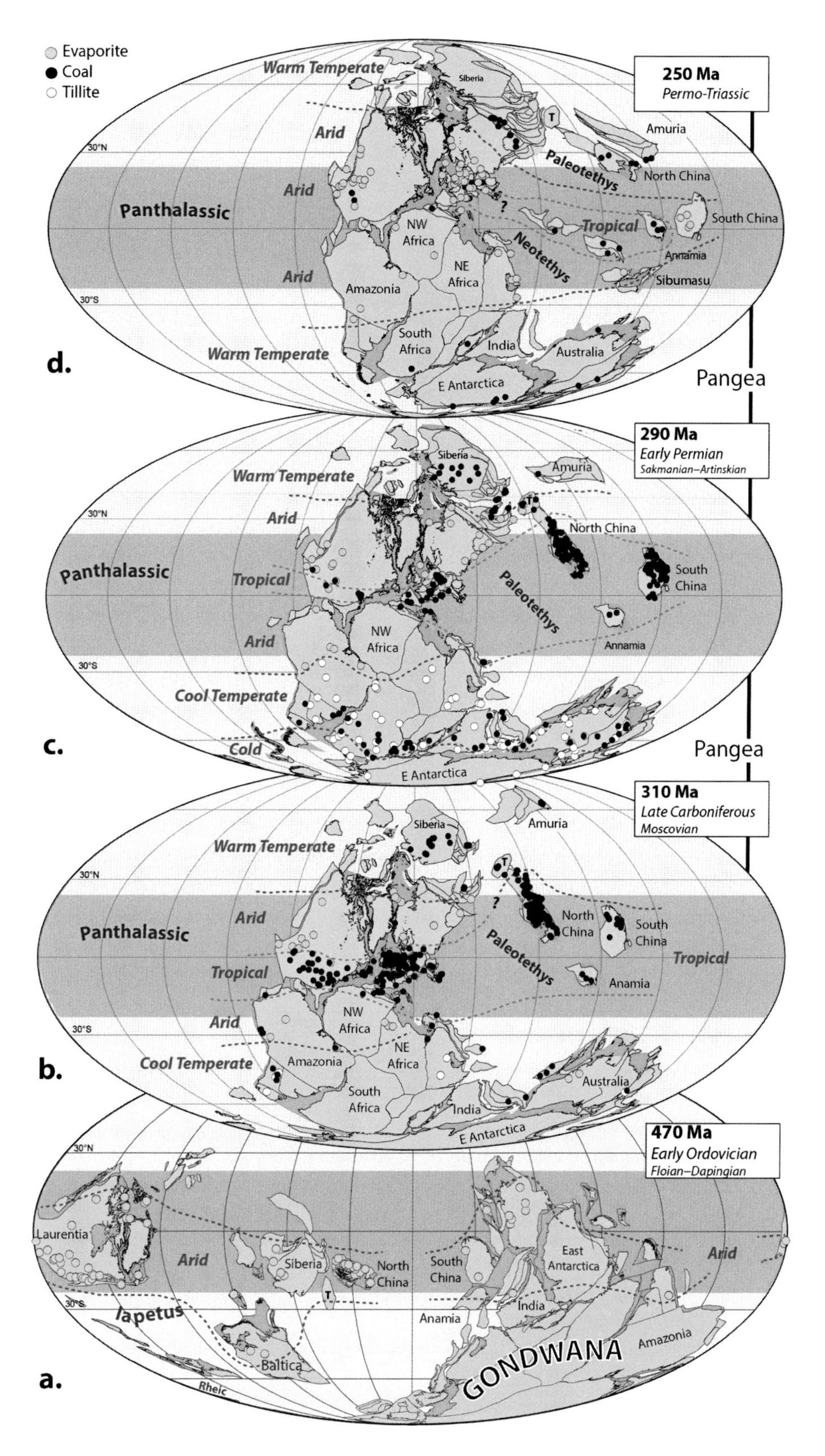

Fig. 16.5 Outline Earth geography in the (a) Early Ordovician (470 Ma), (b) Late Carboniferous (310 Ma), and (c) Early Permian (290 Ma), and (d) near the Permian–Triassic boundary (250 Ma). Also shown are occurrences of evaporite–coal–tillite and tentative outlines of climate gradients. Darker shaded area around the Equator is the present tropics. T, Tarim.

Coal occurrences show a pronounced peak (forest collapse) in the Late Carboniferous (~310 Ma), and smaller peaks in Late Permian, Middle Jurassic, and Cenozoic times (Fig. 16.4). Evaporites are distributed throughout the Phanerozoic but with minor frequency peaks in the Ordovician, Triassic, and Late Cretaceous. Reconstructed evaporite sites average to subtropical latitudes of around 31° (north or south) for the Cenozoic, but mean evaporite latitudes become slightly lower in the deeper past, and there are two minima in the Triassic (19° N/S) and Ordovician–Silurian time (14° N/S). Lower mean latitudes with time may partly reflect the slowing down of Earth. In palaeoclimate modelling, the subtropical high is placed at around 30° north or south, but a faster spinning Earth displaces the subtropics equatorwards, and Christiansen and Stouge (1999) estimated that the subtropical high was displaced 5° equatorwards in Ordovician times.

The two local minima in Triassic and Ordovician–Silurian times probably reflect arid periods with evaporites also found at or near the Equator (e.g. Fig. 11.2b). This is readily illustrated for the Early Ordovician, where an equatorial to low-latitude arid belt includes Laurentia, Siberia, North and South China, and Western Australia (Fig. 16.5a).

Glaciogenic Rocks. Glacial tills, which usually include distinctively angular and poorly sorted fragments, and striated pavements are the best direct indications of where glaciers have been. Dropstones of angular or rounded boulders in finer sediments are also indicative of a glacial origin, although they should be interpreted with caution, since some have travelled into much warmer latitudes before their formerly encasing icebergs melted. Peak tillite occurrences are found in Early Permian and Late Ordovician times (Fig. 16.4).

Ancient glacial deposits, today found in continents located at low latitudes, were instrumental in Wegener's work, because the only possible explanation (unless there were extreme climatic conditions) is continental drift with a possible contribution from true polar wander: the movement of all continents with respect to the poles.

Phanerozoic Climates

The dawn of the Phanerozoic is exceptional in many ways: most continents were located in the southern hemisphere (Figs. 5.1 and 16.5a), atmospheric pCO_2 was perhaps 10–15 times current levels, and a global sea-level rise with largely warm surface sea-water temperatures appear to have characterised the Cambrian and much of the Ordovician (Fig. 16.2). Within each Phanerozoic chapter above, there are some comments on the climate, as well as references to the literature not repeated here, but a brief summary with some additions is as follows.

Cambrian. Prior to the Late Neoproterozoic, there would have been little or no vegetation of any kind on the land. Even by the beginning of the Cambrian, the run-off after rain would have been relatively unimpeded, and erosion very much faster than today.

Gondwana was by far the largest continent in the Cambrian, but its craton was variably overlain by shallow shelf seas, which varied rapidly, due partly to eustatic changes in sea level (Fig. 5.5), but more substantially due to local tectonics in the different regions. Much of the area of the major continental cratons was submerged under shelf seas for long parts of the 54 Myr of the Cambrian and reflects global eustatic sea levels, which steadily rose through time, although relatively small sea-level changes caused numerous transgressions and regressions (Figs. 5.4, 5.6, and 5.8).

There were numerous separate land areas around and upon the Gondwana Craton, although the main continental land area stretched uninterrupted from the Cambrian South Pole to north of the Equator (Fig. 5.4). The total Cambrian rock thicknesses in the two extensive areas of Siberia are about the same, at between 1,500 and 2,000 m: whether that is coincidence or was caused by continent-wide (or perhaps eustatic) sea-level changes is uncertain. It seems probable that the unusually widespread distributions of the Late Cambrian olenid trilobite faunas in Baltica, Laurentia, and Siberia were more due to global low sea-water oxygenation, rather than that the trilobites lived at any great water depths, not least since there is no evidence of any local tectonic activity which would have led to the formation of deeper-water basins on those three large cratons.

The oldest known continental LIP in the Phanerozoic, the Kalkarindji LIP, was intruded into a substantial area in northwestern Australia at around 511 Ma. The Kalkarindji LIP erupted near the Equator (Figs. 5.1b and 5.4) and volatile release of carbon dioxide and sulphur dioxide is estimated to account for only 0.5% of Cambrian atmospheric conditions. Kalkarindji is therefore considered unlikely to have had a major global effect on the Cambrian environment (Marshall et al., 2016).

Ordovician. Temperatures and sea levels varied greatly as the Ordovician progressed (Fig. 16.2), as testified both by carbon and oxygen isotope curves and also by the radiations seen in the biota. They rose steadily during the first half of the period, and the highest global sea level (the second highest in the Phanerozoic), and probably also the warmest temperatures, were at about 455 Ma in the Middle Ordovician and slightly later. That resulted in substantial transgressions with underlying unconformities in many areas, as well as distributing the distinctive graptolite faunas of the *Nemagraptus gracilis* Biozone, which mark the base of the Sandbian Stage. However, global temperatures fluctuated several times during the subsequent Katian, including a pronounced global warming in the Early Katian termed the Boda Event. The final large change was at the end-Ordovician Hirnantian glaciation.

Gondwana was particularly affected by the global temperature changes during the Ordovician, since it spanned so many latitudes from the South Pole to north of the Equator (Fig. 16.6a). From the Floian to the Early Katian (479–453 Ma), Dabard et al. (2015) recognised a series of cycles from gamma-ray spectral logs in the Armorican Massif of France (then a sector of Gondwana) which they interpreted as due to eustatic sea-level fluctuations caused by third-order glacio-eustatic cycles, although no undoubted glaciogenic rocks are known from Gondwana during that period.

The Early Katian Boda Warming Event is reflected in the patch carbonates, some of which were bryozoan bioherms, seen in the higher latitudes of Gondwana in Morocco, the Iberian Peninsula, Sardinia, and France. Although those bioherms were formed under cooler water than the many larger reefs of lower latitudes in more tropical sites, they are notable since the sequences in which they occur otherwise consist entirely of clastic rocks. The Boda Event was followed less than 10 million years later by the Hirnantian glaciation at the end of the Ordovician (Fig. 16.6a), which is best seen in Gondwana. Although various oxygen and carbon isotope excursions are known in some earlier Ordovician rocks, there were no Ordovician glacial intervals supported by glaciogenic sediments until the end-Ordovician Hirnantian glaciation.

The Late Ordovician South Pole lay under North-West Africa, and thus the surrounding areas were covered by a substantial ice cap during the Latest Ordovician (Hirnantian) glaciation, which lasted for probably less than a million years. The glacial and periglacial deposits are extensive and impressive, and this was one of only three major glacial periods in the whole Phanerozoic (Fig. 16.6a). Partly for tectonic reasons and partly because water was locked up in the ice caps, global sea levels dropped in the Hirnantian (Fig. 16.2), with the result that there are Ordovician–Silurian unconformities in most places round the world. Where Hirnantian rocks are preserved, it can be seen that not only did the sea level drop, but also the oxygenated zone extended to greater sea depths than usual. That deeper oxygenation enabled some opportunistic benthos, such as the *Hirnantia* Fauna, to colonise deeper parts of the continental shelves than usual (Fig. 6.9). Although its development resulted from the glacial cooling, the locally abundant brachiopods of the *Hirnantia* Fauna stretched from high to low latitudes, and, rather than interpreting that fauna as 'cooler-water', it was one whose generally lower diversity reflects the limited proportion of benthic taxa which reacted quickly to the changing environments.

Although no Hirnantian glaciogenic rocks occur in Laurentia, since it was at low palaeolatitudes, the glaciation had far-field effects there: firstly, in the eustatic lowering of sea level (Fig. 16.2) which led to widespread unconformities on the craton between rocks of Ordovician and Silurian ages; and, secondly, in the breakdown of many of the relatively fragile marine ecosystems.

The Hirnantian glaciation led to widespread extinctions at the Ordovician–Silurian boundary interval, such as that of the previously characteristic eastern Laurentian Richmondian brachiopod fauna. Even though the glacial interval lasted for less than a million years, two separate extinction phases have been recognised.

Enigmatic Late Ordovician Cooling. The cooling event is paradoxical because of its apparent association with extremely high atmospheric carbon dioxide levels (Figs. 16.2 and 16.7a), but Lowry et al. (2014) and others argue that perhaps the temporal resolution of the proxy record is too coarse to capture short-term variations (<1 Myr) in deep time. Based on climate simulations and other evidence, Lowry et al. (2014) concluded that atmospheric pCO_2 was the prime control on the Palaeozoic continental-scale glaciations, and that geography and solar irradiance were of secondary importance. In their global simulations the pCO_2 threshold permitting Late Ordovician ice accumulations in Gondwana (Fig. 16.7b) was 560 ppm or less whilst high pCO_2 concentrations (≥ 1120 ppm) result in an ice-free Early–Middle Palaeozoic Earth (Fig. 16.7a). Although geography was deemed unimportant by Lowry et al. (2014), plate tectonics also influences the atmosphere composition (albeit slowly) through volcanic degassing of carbon dioxide at subduction zones and mid-ocean ridges, effects that they did not attempt to include in their climate simulations. Large volcanic eruptions are known to impact the climate, and most modelling exercises have focused on low-latitude (tropical) eruptions because they have been considered more important and can lead to global cooling. On the other hand, high-latitude eruptions have been considered hemispheric rather than global, but Pausata et al. (2015) argue that high-latitude eruption can also have a global effect on climate.

For Mesozoic–Cenozoic times, plate tectonic degassing estimates can be derived from reconstructions of ocean floor production, indirectly by sea-level inversion, or by reconstructing the subduction evolution through a combination of

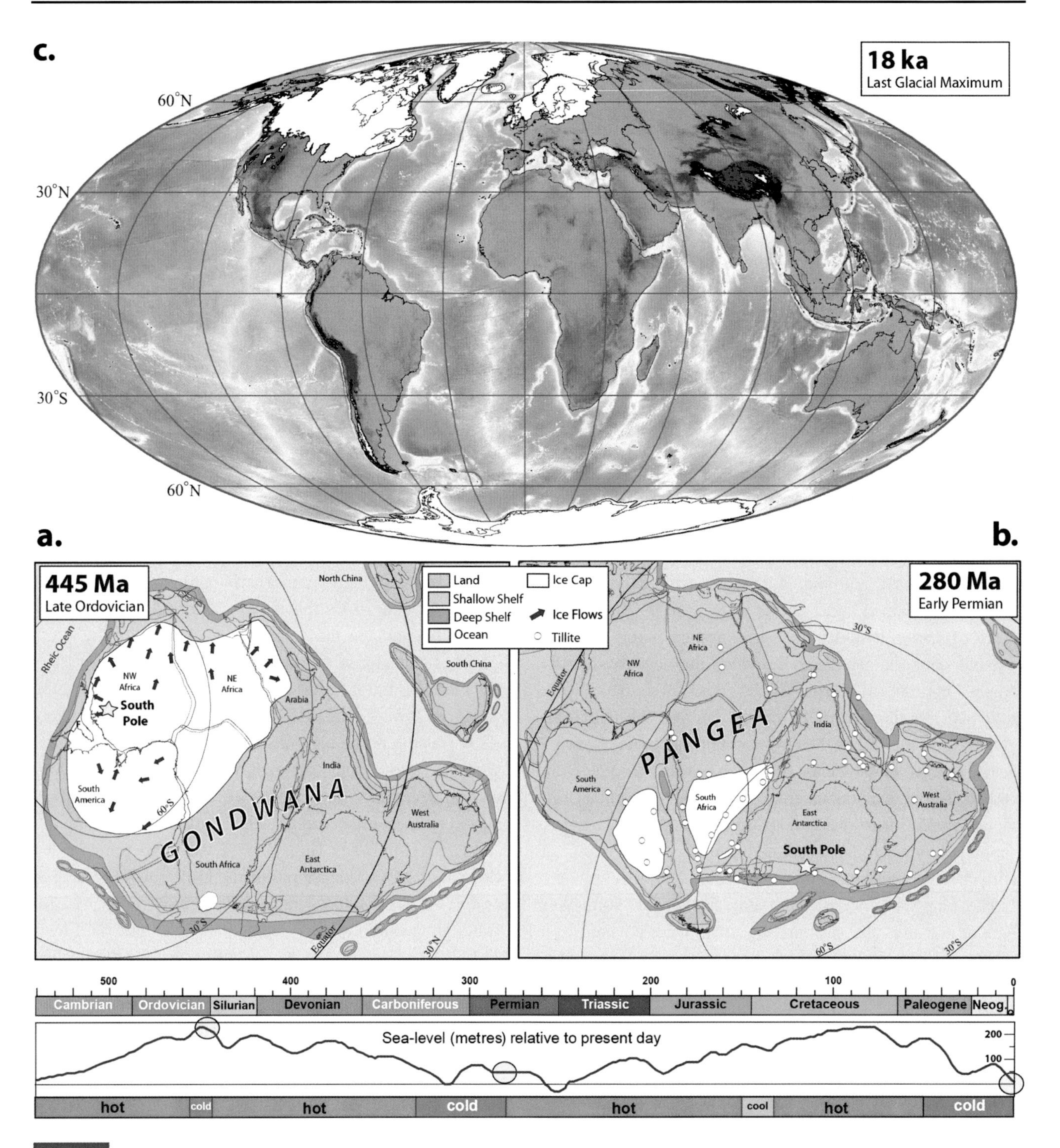

Fig. 16.6 (a) Reconstruction of the Late Ordovician glaciation, (b) Early Permian glaciation, and (c) the Last Glacial Maximum 18,000 years ago.

plate reconstructions and mantle tomography (van der Meer et al., 2014). These methods are not available for the Ordovician and therefore it is difficult to quantify plate tectonic degassing effects, but a few qualitative statements can be made. In the Cambrian (Figs. 5.3 and 5.4) and partly in the Early Ordovician (Figs. 6.2a and 6.3) there were massive subduction systems almost entirely surrounding Gondwana stretching from the Equator to the South Pole. Massive subduction and volcanic degassing of carbon dioxide near the South Pole changed dramatically in the Ordovician

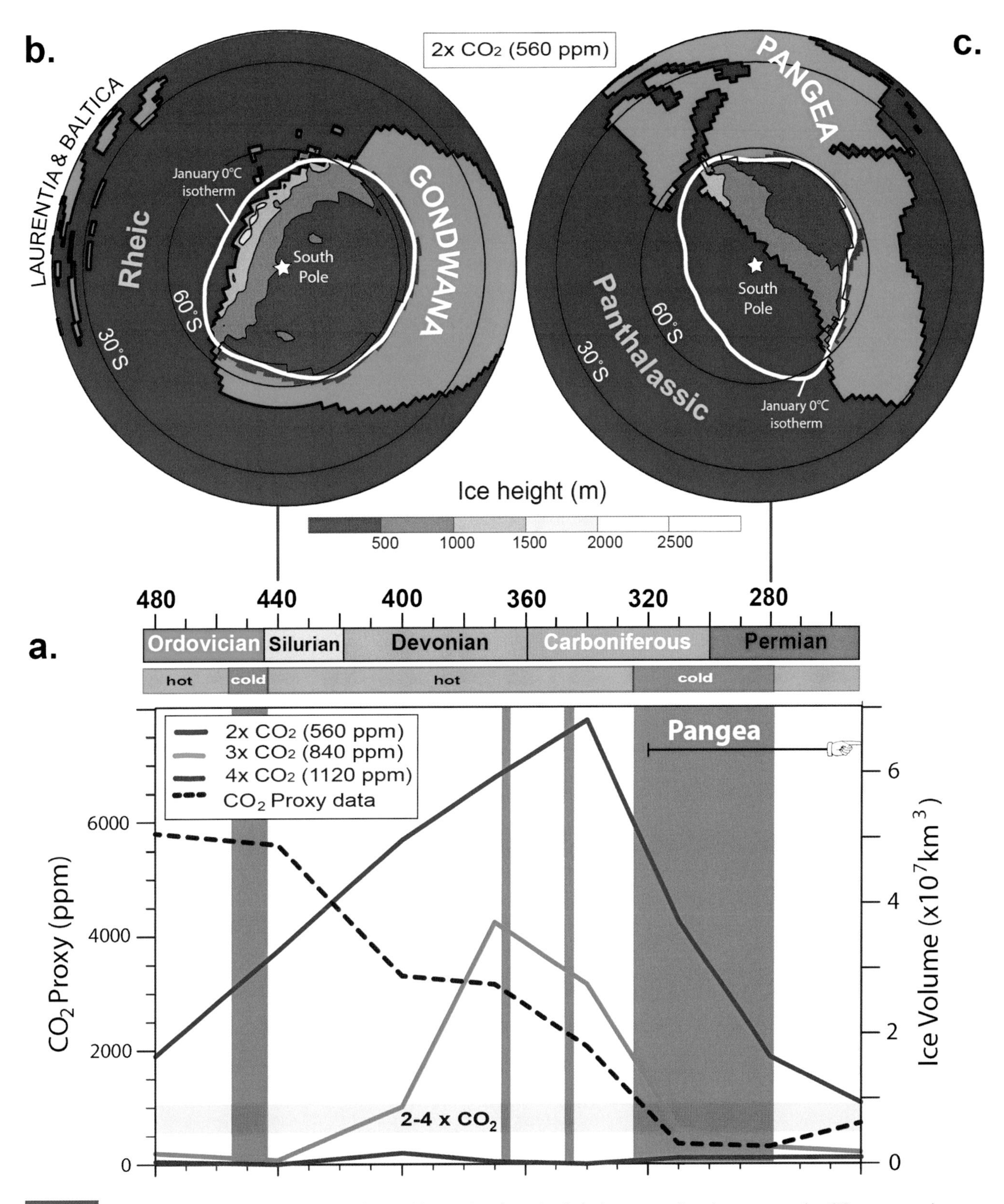

Fig. 16.7 (a) Palaeozoic ice volume estimates (after 5,000 years ice sheet simulation) at two to four times present-day CO_2 concentrations (560–1120 ppm) with constant solar luminosity (Lowry et al., 2014). Known glacial events include the Late Ordovician, Late Devonian–Early Carboniferous (minor), and Late Carboniferous–Early Permian glaciations. The dashed line is a proxy record of atmospheric CO_2 (similar to that in Fig. 16.2e). Note the high proxy values before the Late Carboniferous, which would inhibit glaciations, and the yellow horizontally shaded region shows CO_2 concentrations used in climate simulations. (b) and (c) Southern polar projected reconstructions with simulated Palaeozoic ice sheet heights at about 440 Ma and 280 Ma (2× CO_2 with southern hemisphere cold summer orbital configuration).

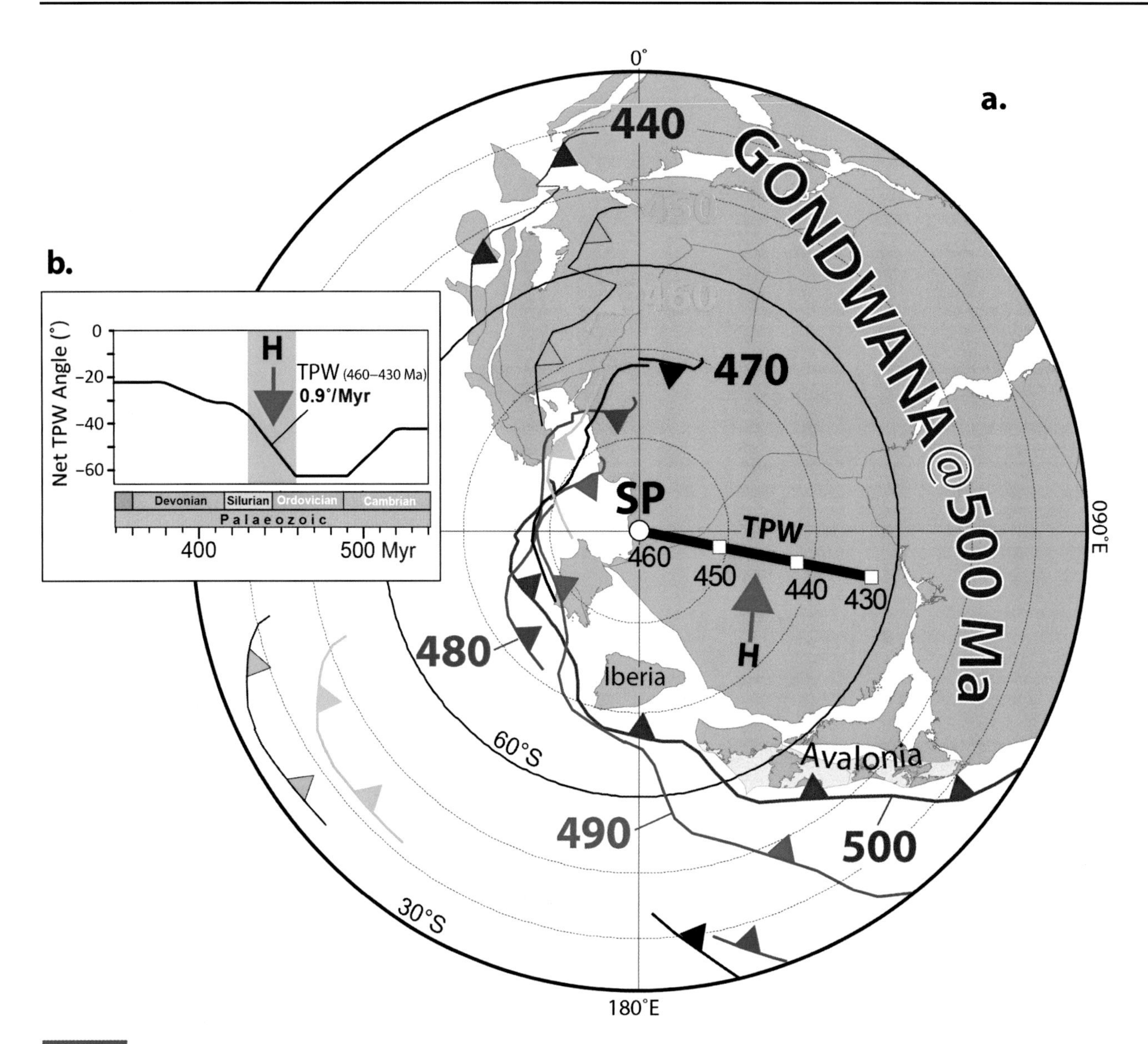

Fig. 16.8 (a) Reconstruction of Gondwana and peri-Gondwana subduction zones (red line and teeth) in the Late Cambrian when the South Pole was located in North Africa. Outboard Avalonia, Iberia, and other areas were adjacent to subduction at high latitudes at 500 Ma; the pattern of high-latitude subduction and high-latitude volcanism continued during the Ordovician but had shifted towards lower latitudes (<60° S) by the Early Silurian (~440 Ma). The subduction shift to lower latitudes explains the rapid true polar wander (TPW) between 460 Ma and 430 Ma which was centred around the time of the Hirnantian (H) glaciation (*c.* 445 Ma). (b) Net TPW in the Lower Palaeozoic shows the highest rates between 460 and 430 Ma (0.9°/Myr).

through the development of back-arcs which rifted off peri-Gondwana terranes (e.g. Avalonian terranes in Fig. 6.2), and, by the Mid–Late Ordovician, subduction zones had migrated to lower latitudes (Fig. 16.8) and there were fewer of them around Gondwana (Fig. 6.9).

During Late Ordovician to Early Silurian times, the Earth experienced one of the fastest periods of TPW (0.9°/Myr, Fig. 16.8) recorded for the Phanerozoic. Adding dense subducted material in the upper mantle at intermediate to high latitudes is the prime reason for TPW, and the maximum effect is achieved 30–40 Myr after subduction initiation, when slabs arrive in the mantle transition zone (Torsvik et al., 2014). Peri-Gondwanan subduction in Late Cambrian to Early Ordovician times (Fig. 16.8) therefore provides a simple explanation for high TPW rates between 460 and 430 Ma, at times when the Earth experienced both global warming (Boda Event, *c.* 453 Ma) and cooling (Hirnantian, *c.* 445 Ma). At the time of the Hirnantian glaciation, the south-polar region and that part of Gondwana had been dramatically transformed from a high-latitude volcanic

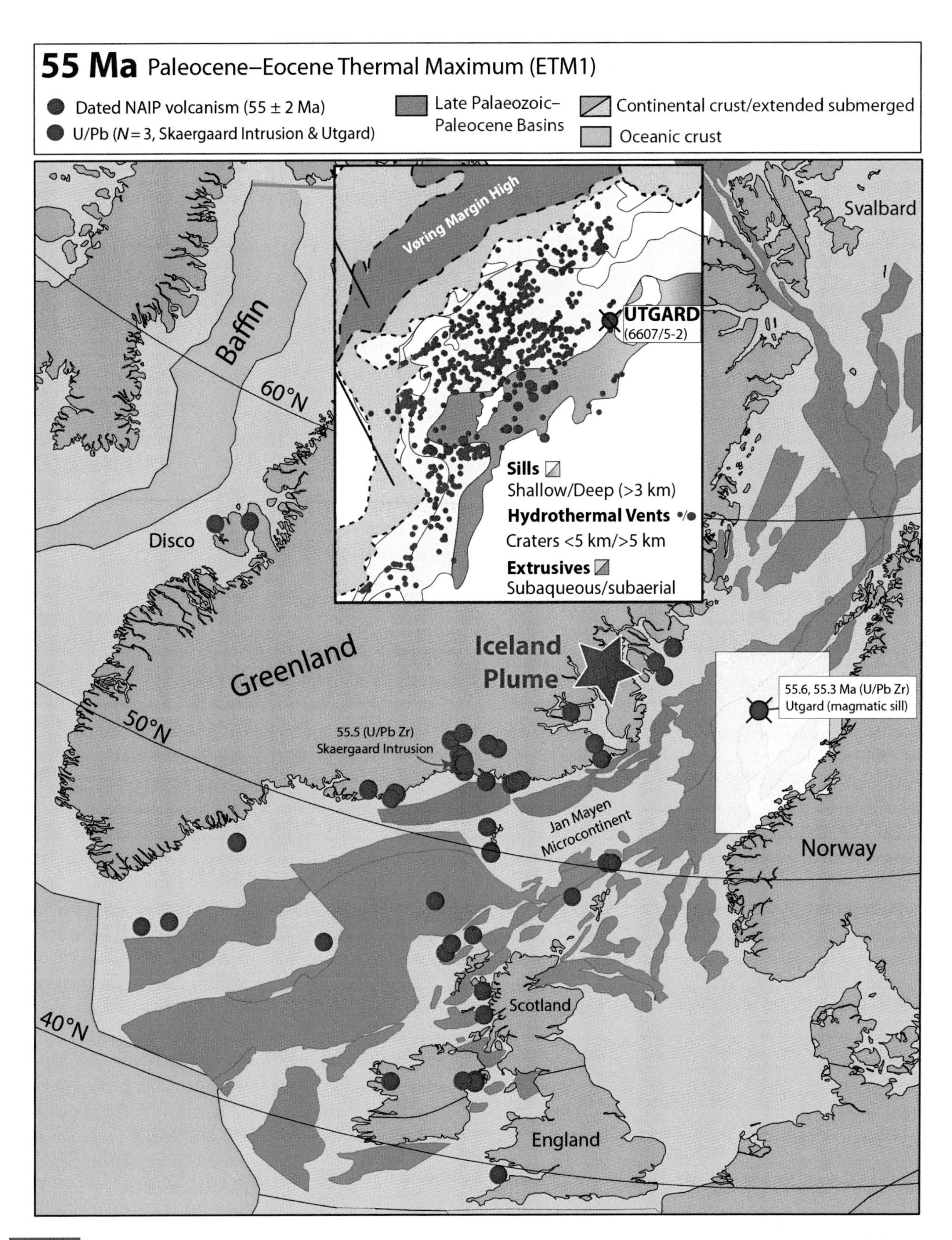

Fig. 16.9 Reconstruction of the North Atlantic at 55 Ma with the distribution of dated (57–53 Ma) onshore and offshore sample locations (red filled circles) for the North Atlantic Igneous Province, the location of the Iceland Plume with respect to Greenland (Torsvik et al., 2015), and rift basins developed from the Late Palaeozoic to the Paleocene (Faleide et al., 2010). The inset map demonstrates the extensive sill and hydrothermal vent complexes in the Vøring Basin off Norway (see white box in main map), and the location of the 6607/5-2 Utgard borehole where magmatic sills intruding organic-rich sediments are dated to 55.6 and 55.3 Ma (U/Pb zircon; Svensen et al., 2010). From a database of many hundred dated volcanics and intrusions there are only six U/Pb ages, ranging from 62.6 ± 0.6 Ma (Antrim lower basalt in Ireland) to 55.5 ± 0.1 Ma (Skaergaard Intrusion in East Greenland).

degassing scenario to a volcanic-free passive margin which faced a wide Rheic Ocean, but how this dramatic shift from high-latitude to low-latitude subduction and TPW could explain climate changes is at best speculative; it could be purely coincidental, and hence the end-Ordovician glaciation and extinction remains a mystery.

Silurian. The post-glacial slow warming was reflected in the global distributions and gradually expanding diversity of the Llandovery benthic faunas such as the brachiopods. Nevertheless, there are few reefs and bioherms of Llandovery age, and it was not until the Wenlock that they became substantial as the climate warmed. By the end of the Silurian, in Ludlow and Pridoli times, the climate had become even warmer, and there was substantial deposition of evaporites in equatorial areas such as Laurentia.

Devonian. There was a greenhouse global climate during most of the Devonian (Fig. 16.2), and thus the average temperatures were exceptionally high, although near the end of the period a progressive switch to a cooler environment led to significant changes. That cooling was probably the result of an eventually dramatic decrease in atmospheric carbon dioxide caused by its absorption by the massive increase of plants on the lands, notably through the advent of large trees. Following a low-stand near the Silurian–Devonian boundary, global eustatic sea levels rose substantially during Early Devonian and Middle Devonian times (Lochkovian to Givetian, from about 420 Ma) to a high-stand within the Late Devonian (Late Frasnian) at about 380 Ma. In particular, the largest development of reef systems known in Earth history occurred in the Devonian, and those reefs are estimated to have covered areas of perhaps as much as 5 million square kilometres, almost ten times the areas of comparable reef ecosystems found today.

There was clearly much differentiation in equatorial to polar temperatures during the Early Devonian. That is reflected in the development of provinces in the benthic faunas such as the brachiopods, which were largely developed at different palaeolatitudes. That provinciality rose to a peak in the Emsian and subsequently dwindled so that the provinces became progressively less obvious. The first evidence of Late Palaeozoic glaciation is some time after that, with the well-dated Late Famennian glaciogenic deposits in Bolivia and adjacent areas in that part of Gondwana which lay over the South Pole.

The sea-level rises during the Early Devonian led to transgressions over many Gondwanan cratonic areas, for example the substantial southward extension of the shallow sea over much of Gondwana and elsewhere.

The End Devonian extinction was one of the five major mass extinction events in Earth history and there is a causal relationship between this event and the multi-phase emplacement of the Yakutsk LIP (also known as the Viljuy Traps) at around 364 and 377 Ma (Ricci et al. 2013). The Yakutsk LIP intruded the Viljuy palaeo-rift in south-east Siberia and was emplaced at northerly latitudes of about 30°.

Carboniferous. Following the global change from a warmer to a cooler environment near the end of the Devonian (Fig. 16.2), there was a substantial but intermittent series of ice ages through much of the Carboniferous and into the earliest Permian (330–290 Ma). Although there are some rocks of glacial origin in the latest Devonian (Famennian, Fig. 8.3) and earliest Carboniferous of South America (Fig. 9.2), it was not until well into the Carboniferous (in Visean times) that glaciogenic rocks were deposited in more widespread areas in the Gondwana continent; and that heralded the start of the main glaciation which lasted for a very long time, continuing on into the Early Permian. That was by far the most long-lived series of glacial events and development of their associated ice caps known in the whole Phanerozoic, and there are extensive Carboniferous glacial deposits in Brazil and Argentina (Fig. 9.2). The glacial rocks there range in age from the Serpukhovian to the Ghezelian (about 325–300 Ma), a span of more than 25 Myr, and that ice sheet may have extended over as far as North-East Africa in the Late Carboniferous (Fig. 16.5c). However, in the adjacent basins of South America, the glacial diamictites are earlier, ranging from the Tournaisian to the Late Visean (about 355–325 Ma). In Iran, the glaciogenic rocks are confined to two episodes in the Carboniferous, the first in the Bashkirian, and the second across the Carboniferous–Permian boundary from the Gzhelian to the Sakmarian (305–290 Ma). Although Fig. 9.7 shows much of the former Laurentian Craton as emergent land, sporadic melting of high-latitude ice caused much sea-level change, with the result that parts of that craton became sporadically flooded in a comparable way to the northern England sector of Laurussia.

In contrast to the end-Ordovician glaciation, an apparent long-term decline in pCO_2 (Figs. 16.2 and 16.7) at least led to favourable glacial conditions; but why did major glaciations start at around 330–320 Ma? Major orogenic events and uplift, and the opening and closure of seaways, have often been invoked as potential causes for major shifts in the Cenozoic global climate system (Fig. 16.3): one of the largest known seaways on Earth, the Rheic Ocean, separating Laurussia from Gondwana, closed dramatically during the Devonian (Fig. 8.1), and by the Mid–Late Carboniferous (Fig. 9.1) that former seaway was essentially closed and the Alleghanian–Variscan Orogeny enclosed a belt of about 7,500 km in length across the heart of Pangea.

Although there was extensive and prolonged glaciation in the higher latitudes of Gondwana, there seems little evidence that the equatorial regions were very much affected by those much colder climates at higher latitudes, indicating that the global temperature gradients must have been much more diverse (Fig. 16.5b) than in the preceding greenhouse periods, perhaps not dissimilar to the present day. In the Late Carboniferous, Central Pangea including North America and Europe was located near the Equator (Fig. 16.5b) and was at times covered by humid tropical rainforest, known as the Coal Forests. As the climate aridified, the rainforests collapsed and eventually were replaced by seasonally dry biomes (Sahney et al. 2010). In North America, an abrupt shift to more arid climates (Fig. 16.5b–d) is linked to the Late Carboniferous (Late Moscovian to Kasimovian) rainforest collapse, but the causal mechanism remains uncertain: one hypothesis is that aridification was triggered by a short-term but intense glacial phase which drove the global sea level to one of the lowest in Earth history (~310 Ma; Fig. 16.2), which is coincident with the most abrupt phase of vegetation change (Sahney et al. 2010).

Permian. There was substantial and significant global climate change at about 280 Ma (Sakmarian time) from icehouse (Fig. 16.5c) to greenhouse (Fig. 16.5d) conditions, which involved the final closure of the extraordinarily long-lived major Permian–Carboniferous glacial period. Subsequently the planet enjoyed or endured above-average temperatures until the end of the Permian, when there were the massive biological extinctions described in Chapter 11. Mass extinction is linked to the Siberian Traps which erupted at high latitudes between 58° N and 79° N.

The regions not within the glacial areas were nevertheless directly affected by the Permian–Carboniferous glacial episodes; for example, in North China the earlier Permian continental deposits are coal-bearing and include fluvial strata which are rich in plant fossils, in contrast to later fluvial red beds with many calcitic palaeosols. After the earliest Permian glacial interval was over in the Early Sakmarian, a substantial and steep climatic gradient was developed during the Asselian to Sakmarian along Gondwana's northern margin, seen in the contrasting brachiopod and fusuline faunas.

Triassic. The Triassic was a time of great continental emergence due to a combination of widespread orogenesis and low sea levels (Fig. 16.2g), particularly at the start of the period. The Triassic climates varied considerably, with consequent contractions and expansions of the various temperature zones; in particular, cooler conditions were reflected by the reduced ammonoid diversity in the higher latitudes in the northern hemisphere (the Boreal Realm) and in the southern hemisphere (the Himalayan Province) in the earliest Triassic at the start of the Induan. That was in contrast to the warmest temperatures and the largest numbers of ammonoids near the beginning of the Anisian only 4 Myr later, and there was sporadic fluctuation for the rest of the period (Zakharov et al., 2008). As can be deduced from various factors, including the differing leaf morphologies of the plants found at many horizons in Europe and elsewhere, the Triassic climate varied between arid and semi-arid (Fig. 16.5d).

In the exposed continental land areas, particularly in northern Pangea, there were widespread and substantial desert conditions, which resulted in wadi, dune, and lacustrine deposits known as the New Red Sandstone over much of Europe and North America. The falling sea levels and contemporary orogeny and uplift caused rejuvenation of the Palaeozoic massifs as land areas, and in the Early Triassic (Scythian and Anisian) there was widespread generation of fluvial conglomerates (the Bunter Pebble Beds in northern Europe).

In the later Triassic (the Norian), sea level rose and many parts of the cratons were flooded, particularly in northern Europe where the Zechstein Sea extended to Greenland and Svalbard at times. However, the transgressions and sea-level fluctuations were variable, and perhaps cyclical, and arid climates probably favoured the widespread dolomites and other evaporite rocks found over most of the Carnian and Norian carbonate platforms (Fig. 11.4).

The Triassic–Jurassic boundary saw the emplacement of the low-latitude CAMP LIP (Fig. 11.2) and consequent biotic extinctions, leading to acidification of the deep-ocean bottom water and a catastrophic release of greenhouse gases as well as poisonous sulphur compounds. It also led to an enormous amount of Aeolian dust being generated on the affected land areas, which was not only limited to Africa and North and South America but extended as far as Japan, and contributed greatly to the increase in cloud cover and the consequent effects on the climate (Ikeda et al., 2015). Carbon and oxygen isotope excursions from oysters from Britain in the same interval reflect relatively cool sea-surface temperatures at the start of the interval, but temperatures rose quickly by about 10 °C from 7–14 °C to 12–22 °C (Korte et al., 2009).

Jurassic. The beginning of the Jurassic saw major rises in sea level, with the result that, for example, nearly all of Laurasia, which had been a large area of land, became flooded in the latest Triassic (Rhaetian) at about 202 Ma. Within little more than a million years, during the Early Hettangian, parts of the region had become transformed into an archipelago of islands of various sizes within shelf seas, a situation which continued for most of the Jurassic.

Reflected in the sediments, the Jurassic saw substantial fluctuations in global climate. One of the most important was the Toarcian Oceanic Anoxic Event at about 183 Ma, when there was widespread bottom-water anoxia, global warming, a negative carbon isotope excursion of 5–7‰, and various biotic extinctions. Data from fossil wood indicate that unusually isotopically light carbon was present in the atmosphere during this event. Possible causes include outgassing of the Karroo–Ferrar LIP volcanism in Africa and East Antarctica, emplaced at intermediate southerly latitudes, and which might have disrupted ocean-water dynamics leading to the disassociation of methane hydrates, perhaps aided by astronomical forcing. However, for whatever reason, extensive organic shales with up to 18% organic carbon were deposited, not only in Europe, but also elsewhere, including South America. At higher latitudes there were fewer changes (Fig.12.4).

Cretaceous. Since it was such a long period, Cretaceous climates varied considerably. For most of the period it was much warmer than average, including some of the warmest climates in the whole Phanerozoic, with a temperature peak in the Turonian at about 90 Ma, and thus lush and varied vegetation flourished over all of the land areas at both poles. In contrast, although there is no evidence for Cretaceous glaciation, the Earth became significantly cooler near the end of the Campanian at about 70 Ma, to reach a Cretaceous minimum at the K-T boundary at the very end of the period at 65 Ma, although the temperatures rose again to become more equable during the subsequent Paleocene. In addition to the Deccan Traps (emplaced at latitudes of 20–28° S) in India and the effects of the Mexican meteorite impact, that cooling must also have played at least some part in the end-Cretaceous terminal events.

The Cretaceous saw the highest sea levels in the whole Phanerozoic (Fig. 16.2g), which peaked at about 95 Ma (Late Cenomanian). That was the result of a combination of very different factors, including the absence of polar ice caps, the high temperatures (which increased the volume of the sea water), and the larger-than-average sizes of the mid-ocean ridges. Thus, the total land area was much reduced, with transgressions over all the cratons, and only some 18% of the Earth's area was above sea level by comparison with 28% today (Figs. 13.8 and 13.9).

Paleogene and Neogene. The Late Paleocene–Eocene is characterised by many events of rapid change in atmospheric carbon dioxide and temperature. Global temperatures increased in the Late Paleocene and peaked over the Paleocene–Eocene boundary, an event known as the Paleocene–Eocene Thermal Maximum (PETM) or Eocene Thermal Maximum 1 (ETM1, Fig. 16.3) with global warmth and massive carbon addition. The PETM is characterised by a sharp negative excursion in the oxygen, carbon (Fig. 16.3), and nitrogen isotope ratios followed by a significant increase in phosphorus concentrations. Temperatures peaked to 30 °C in the Early Eocene (Ypresian: 55 Ma), as documented from molluscs in the Paris Basin, France. That temperature maximum led to increased humidity, reflected by increased kaolinite content in the sediments, and the associated anoxia was probably tied to a bloom of nitrogen-fixing cyanobacteria in the oceans.

The PETM is commonly associated with the North Atlantic Igneous Province (NAIP) which affected vast areas of Baffin Island, Greenland, the United Kingdom, Ireland, the Faroe Islands, and offshore regions. Volcanic activity started at around 62 Ma, linked to the impingement of the Iceland plume, and dated NAIP rocks peak between 56.2 and 53.5 Ma (Fig. 16.3), which is just prior to or during the initial opening of the north-east Atlantic. At about 55 Ma, the Iceland plume was located near the East Greenland margin but volcanism is recorded as far away as Disco in West Greenland (1,100 km), southwards to England (1,700 km) and in the more nearby offshore Norwegian margin. There, magmatic sills intruding organic-rich sediments have been dated to 55.6 ± 0.3 and 56.3 ± 0.4 Ma (U/Pb zircon), and within errors these ages overlap with the PETM. Svensen et al. (2004, 2010) maintain that PETM and global warming were triggered by the rapid release of greenhouse gases generated by heating of organic-rich sediments in the north-east Atlantic, rather than from dissociation of gas hydrates.

In the succeeding Lutetian, global average temperatures dropped back to about 20 °C, followed by transient warming in the Middle Eocene (Bartonian: 40 Ma) up to a Tertiary maximum of 32 °C (Huyghe et al., 2015). After that they declined steadily through the Eocene–Oligocene boundary at 34 Ma to reach a Paleogene low of about 12 °C, followed by a much smaller rise to 24 °C in the Early Oligocene at about 30 Ma.

There is no evidence for any ice caps at the poles during the earlier part of the Paleogene, including all of the Paleocene. However, during the steady cooling in the Eocene, there is the first evidence of glaciation, recorded in Antarctica in the Late Eocene at about 37 Ma. That Late Eocene cooling, with the coldest interval in the Early Oligocene at about 34 Ma, was echoed in the changes in vegetation and a mammalian faunal turnover. By the Miocene, there was a more marked temperature gradient from Equator to poles than previously.

During the Middle and Late Miocene, climatic differences between low and high latitudes increased significantly, reflecting the start of the polar cooling, and those changes obviously restricted the distributions of animals and plants with limited habitat tolerance. Between the polar and tropical regions, at middle to higher latitudes, there was an expansion of drier and more open environments with grasslands, although the forests became reduced in area.

Global sea level progressively fell during the Miocene (Fig. 16.2g), and the shallowness of the seas is demonstrated by the thick evaporites in many regions. However, further shallowing led to the Messinian Salinity Crisis, as the Tethys Ocean dwindled in the Mediterranean with thick deposits of halite and anhydrite, since the area was completely cut off from the open oceans.

The Plio-Pleistocene Glaciation. Although the Earth's temperature has gradually dwindled since the Early Paleocene, it was during the Pliocene, at about 2.7 Ma, that the first phase started of the remarkable Plio-Pleistocene glaciation (Fig.16.6c), which has probably not yet ended today. Although a hundred years ago it was concluded that there had been fewer than ten glacial episodes within it, through the analysis of complete cores both in deep-sea marine sediments and in the extensive ice caps at the higher latitudes it is now known that there have been at least 17 glacial periods alternating with a corresponding number of warmer periods termed 'interglacials', including the present. Perhaps it is worth mentioning here that the great variability observed in the current Cenozoic ice age is almost certainly enhanced by the better and more precise records of relatively recent events and changes when compared with other ice ages in the Precambrian, Ordovician, and Carboniferous–Permian.

The existence of ice sheets has a profound effect on both atmospheric and oceanic circulations, with climatic belts being pushed by up to 2,000 km towards the Equator, resulting in a marked narrowing of the Equatorial Zone during the glacial intervals. In addition, the desert areas become much more extensive during the glacial periods, even in areas some distance away from the ice caps, whilst in the interglacials the deserts are much reduced in size, enabling the enlargement of many areas fringing the erstwhile deserts which can support many more diverse faunal and floral communities and habitats.

Many hypotheses have been forwarded to explain the extensive Plio-Pleistocene glaciations: a decrease in atmospheric carbon dioxide may have played a role during the onset (as for the two Palaeozoic glaciations, Fig. 16.6a,b) but tectonic and/or oceanic events have also been invoked to explain the first ever known Phanerozoic Northern Glaciation. Greenland is a key player but needs to have sufficient latitude and altitude. There was probably no significant topography in East Greenland before 10 Ma, but Steinberger et al. (2014) proposed that the subsequent uplift was driven by the Iceland Plume, which accelerated at ~5 Ma, and which lifted the parts of the East Greenland margin closest to Iceland to elevations of more than 3 km above sea level. This uplift, combined with a northward drift of Greenland of ~6° over the past 60 Myr, true polar wander (~12°), and a decline in pCO_2, probably played a central role in preconditioning Greenland for widespread glaciation.

Endnote

The pages above will, we hope, have both informed and challenged the reader. We have tried to present the history of our planet, but this book is merely a snapshot of our present interpretations of the Earth's changes through time. The challenge for us all is to refine and illuminate further that very complex story, which reflects the progressive interactions between the effects of the eternal laws of physics, chemistry, and biology through very long periods of time. No doubt further insights will come not only through the acquisition of much fresh data, but also by inspired lateral thinking by many scientists who will re-interpret our collective view of the world and the processes through which it has developed. That new thinking should also generate new and improved models for the more accurate prediction of the evolution and margins of tectonic plates and also the variations and causes of global climates and the evolution of life, in the past, present, and future, as well as enabling the construction of even more realistic maps of the changing geographies during the Earth's long saga.

Appendix 1

Location of Phanerozoic Large Igneous Province (LIP) centres (today and at eruption time). **OP**, Oceanic Plateaus; CLIP, Continental LIP. LIPs are reconstructed according to a hybrid plate motion frame (moving hotspot frame back to 120 Ma and a true polar wander corrected palaeomagnetic frame before that time; Doubrovine et al. 2012; Torsvik et al. 2012).

Large Igneous Province		Age (Ma)	Type	Centre today		At eruption time	
				Lat.	Long.	Lat.	Long.
Columbia River Basalts	CRB	15	CLIP	46.0	241.0	47.7	−116.0
Ethiopia	ET	31	CLIP	10.0	39.5	5.5	36.1
North Atlantic Igneous Province	NAIP	62	CLIP	69.9	332.8	63.7	−14.3
Deccan Traps	D	65	CLIP	21.0	73.0	−14.8	53.6
Sierra Leone Rise	SL	73	**OP**	6.0	338.0	5.8	−32.4
Madagascar	M	87	CLIP/ **OP**	−26.0	46.0	−42.7	31.2
Broken Ridge	BR	95	**OP**	−30.0	96.0	−49.5	62.7
Hess Rise	HR	99	**OP**	34.0	177.0	5.4	−140.0
Central Kerguelen	CK	100	**OP**	−52.0	74.0	−48.9	59.2
Agulhas Plateau	AP	100	**OP**	−39.0	26.0	−53.2	−1.0
Nauru	N	111	**OP**	6.0	166.0	−23.3	−142.0
Southern Kerguelen	SK	114	**OP**	−59.0	79.0	−48.6	57.6
Rajhmahal Traps	R	118	CLIP	25.0	88.0	−37.9	60.7
Ontong Java/Manihiki combined	OJMP	123	**OP**	−6.8	167.7	−37.4	229.2
Wallaby Plateau	W	124	**OP**	−22.0	104.0	−37.5	75.5
Maud Rise	MR	125	**OP**	−65.0	3.0	−52.2	2.2
Bunbury Basalts	BB	132	CLIP	−34.0	115.0	−54.7	72.5
Paraná–Etendeka	PR	134	CLIP	−20.0	11.0	−31.8	−15.0
Gascoyne	G	136	**OP**	−23.0	114.0	−46.3	81.2
Magellan Rise	MR	145	**OP**	7.0	183.0	−1.1	−108.0
Shatsky Rise	SR	147	**OP**	34.0	160.0	7.3	−108.0
Argo Margin	AM	155	**OP**	−17.0	120.0	−43.4	82.9
Karroo	K	182	CLIP	−23.0	32.0	−37.0	−5.2
Central Atlantic Magmatic Province	CAMP	201	CLIP	27.0	279.0	16.8	−25.4
Siberian Traps	SBT	251	CLIP	65.0	97.0	62.5	44.2
Emeishan LIP	E	258	CLIP	26.6	104.0	−4.0	134.2
Panjal Traps	PT	285	CLIP	34.0	75.0	−42.9	59.5
Skagerrak Centred LIP	SCLIP	297	CLIP	57.5	9.0	9.8	−2.0
Yakutsk	Y	360	CLIP	63.0	130.0	48.2	348.4
Altai–Sayan	AS	400	CLIP	49.0	90.0	38.1	6.1
Kalkarindji	KA	510	CLIP	−16.3	133.3	21.5	0.4

Appendix 2

Some Mesozoic to Modern Panthalassic and Pacific Ocean plates.

Plate name	Plate ID	From	To	Entirely synthetic
Izanagi	926	Palaeozoic?	~55 Ma (subducted)	Yes
Phoenix	919	Palaeozoic?	~120 Ma (splits into 982, 983, 919, and 908)	Yes
▶ Manhiki	982	~120 Ma	~85 Ma (becomes part of 901)	Yes
▶ Hikurangi	983	~120 Ma	~85 Ma (becomes part of 901)	Yes
▶ Catequil	919	~120 Ma	~85 Ma (becomes part of 901/902)	Yes
▶ Chazca	908	~120 Ma	~85 Ma (becomes part of 902)	Yes
Farallon	902	Palaeozoic?	~23 Ma (splits into 911 and 924)	
▶ Nazca	911	~23 Ma	***Active***	
▶ Cocos	924	~23 Ma	***Active***	
Cache Creek	131	Palaeozoic	~140 Ma (subducted)	Yes
Pacific	901	~190 Ma	***Active***	
Kula	918	~83 Ma	~40 Ma (becomes part of 901)	
Vancouver	903	~52 Ma	~37 Ma (becomes 903)	
▶ Juan de Fuca	903	~37 Ma	***Active***	

Appendix 3 Orogenies

There are numerous names in the literature for both local and more widespread orogenies. We list here the ones cited in this book (with their chapter numbers), with brief descriptions.

Acadian. The Early Devonian orogeny affecting parts of the north-eastern USA, Canada, and Britain. 8.

Achalian. A Late Ordovician to Devonian event in South America, including the attachment of Cuyania to Gondwana. 8.

Alleghanian. A Late Palaeozoic multiple orogeny partly equivalent to the Variscan Orogeny and affecting parts of the USA and Canada as Laurussia accreted to Gondwana. 9.

Alpine. A major Late Cretaceous to modern event caused by the accretion of Africa to Europe. 13.

Andean. A major Eocene to modern event caused by the subduction of the eastern part of the Pacific Plate under western South America. 14.

Antler. A Late Devonian orogeny when the Roberts Mountains were thrust onto the western Laurussian Craton in the western USA. 8.

Benambran. Late Ordovician and Early Silurian multiple terrane accretion to Gondwana in eastern Australia. 7.

Bhimphelian. A Late Cambrian to Early Ordovician event of unknown cause in the Himalayan area. 6.

Brasiliano. A Late Precambrian and Early Cambrian event linking the cratons in South America, and part of the unification of Gondwana. 5.

Browns Fork. A local Permian orogeny in the Farewell Terrane of Alaska of unknown cause. 10.

Cadomian. A Late Precambrian and Early Cambrian event in southern Europe and north-western Africa, part of the unification of Gondwana. 5.

Caledonide or Caledonian. A major Middle Silurian to earliest Devonian event caused by the accretion of Laurentia to Avalonia–Baltica to form Laurussia, and affecting both margins of today's North Atlantic. 7.

Cimmerian. A Late Triassic and Jurassic event initially caused by the accretion of various blocks to each other in Central Asia, and which extended eastwards to South-East Asia. 11.

Cordilleran. A major Eocene to modern event caused by the subduction of and accretion of plates and terranes in the eastern Pacific under and beside western North America. 14.

Delamerian. Late Proterozoic to Middle Cambrian multiple terrane accretion to Gondwana in eastern Australia. 7.

East African. A Late Precambrian and Early Cambrian event between Africa, India, and Arabia, and part of the unification of Gondwana. 5.

Ellesmerian. Late Devonian to Early Carboniferous orogeny affecting parts of Arctic Canada, Greenland, and Svalbard. 8.

Eurekan. A Cretaceous to Paleogene event caused by the accretion of small terranes to the Arctic margin of North America, Greenland, and the Barents Sea and their subsequent adjustments. 14.

Famatinian. Ordovician multiple arc accretion to South America. 7.

Gondwanides. Nebulously defined Late Palaeozoic multiple orogenies around the southern continents caused by the accretion of various terranes to Gondwana. 10.

Hercynian. An alternative name for the Devonian and Carboniferous Variscan Orogeny of Europe. 8.

Himalayan. A major Eocene to modern event caused by the accretion of India to Laurasia. 14.

Hunter–Bowen. Permian orogeny in eastern Australia caused by the accretion of various terranes to Gondwana. 10.

Indosinian. Multiple Carboniferous to Triassic orogeny in South-East Asia caused by the accretion of various terranes to each other. 9.

Kanimblan. Middle Devonian to Carboniferous multiple terrane accretion to Gondwana in eastern Australia. 8.

Klakas. A local Late Silurian and Early Devonian orogeny affecting the Alexander Terrane in Alaska. 7.

Kuungan. A Late Precambrian and Early Cambrian event between India, eastern Antarctica, and western Australia, which formed part of the unification of Gondwana. 5.

Laramide. A Late Cretaceous event caused by the accretion of various terranes onto the western margin of North America. 13.

M'Clintock. Middle Ordovician deformation within the Pearya Terrane of Arctic Canada. 6.

Neoacadian. The Emsian to Famennian part of the Acadian Orogeny, affecting parts of the north-eastern USA and Canada. 7.

Ocloyic. Ordovician orogeny in the Eastern Cordillera and Puna areas of Argentina. 8.

Ouachita. Middle Carboniferous to Early Permian orogeny, in the southern USA, reflecting the union of Laurasia with Gondwana to form Pangea. 10.

Pampean. Late Precambrian and Early Cambrian events adjacent to the Famatinian and Pampean arc areas in south-western South America. 5.

Pan-African. A Late Precambrian and Early Cambrian event linking the cratons in Africa, and another part of the unification of Gondwana. 5.

Romanzov. A Late Silurian and Devonian orogeny affecting Alaska and north-western Canada. 7.

Ross. Multiple Palaeozoic orogeny when marginal terranes were accreted to Antarctica. 9.

Salinic (or Salinian). A local name for that part of the Caledonide Orogeny affecting the northern Appalachians and Newfoundland. 7.

Scandian. A local name for that part of the Caledonide Orogeny affecting Scandinavia. 7.

Shelvian. A minor latest Ordovician and earliest Silurian event in Britain, reflecting the Baltica–Avalonia oblique docking. 6.

Sonoma. A local Permian and Triassic orogeny in the Cordillera of western Laurussia caused by the accretion of various terranes onto the craton. 10.

Svalbardian. A minor Late Devonian orogeny uniting East and West Spitsbergen. 7.

Tabberabberan. Silurian and Devonian multiple terrane accretion to Gondwana in eastern Australia. 8.

Taconic. A latest Cambrian to Early Silurian series of events in the western Appalachians of the USA. 6.

Timanide. A Late Precambrian and Early Cambrian event uniting Baltica and the Timanian area now in northern Europe. 5.

Tyennan. Cambrian orogeny in Tasmania and south-eastern Australia. 5.

Uralian. The multiple orogeny along today's Ural Mountains, when the Kazakh terranes became progressively accreted to eastern Laurussia during the Late Palaeozoic. 9.

Variscan. Major Devonian and Carboniferous multiple orogeny in Europe caused by the accretion and amalgamation of numerous terranes as the Rheic Ocean progressively closed. 8.

Wales. A pre-Ordovician (probably Cambrian) orogeny affecting the Alexander Terrane of Alaska, USA. 7.

References

Abrajevitch, A., Van der Voo, R., Levashova, N.M. & Bazhenov, M.L. (2007). Paleomagnetic constraints on the paleogeography and oroclinal bending of the Devonian volcanic arc in Kazakhstan orocline. *Tectonophysics*, 441, 67–84.

Abrajevitch, A., Van der Voo, R., Bazhenov, M.L. et al. (2008). The role of the Kazakhstan orocline in the late Paleozoic amalgamation of Eurasia. *Tectonophysics*, 455, 61–76.

Allen, M.B., Alsop, G.I. & Zhemchuzhnikov, V.G. (2001). Dome and basin refolding and transpressive inversion along the Karatau Fault System, southern Kazakhstan. *Journal of the Geological Society, London*, 158, 83–95.

Almalki, K.A., Betts, P.G. & Ailleres, L. (2015). The Red Sea – 50 years of geological and geophysical research. *Earth-Science Reviews*, 147, 109–140.

Álvaro, J.J., Elicki, O., Rushton, A.W.A. & Shergold, J.H. (2003). Palaeogeographical controls on the Cambrian immigration and evolutionary patterns reported in the western Gondwana margin. *Palaeogeography, Palaeoclimatology, Palaeoecology*, 195, 5–35.

Alvey, A., Gaina, C., Kusznir, N.J. & Torsvik, T.H. (2008). Integrated crustal thickness mapping and plate reconstructions for the high Arctic. *Earth and Planetary Science Letters*, 274, 310–321.

Andersen, M.B., Elliott, T., Freymuth, H. et al. (2015). The terrestrial uranium isotope cycle. *Nature*, 517, 356–359.

Andersen, T.B., Jamtveit, B., Dewey, J.F. & Swensson, E. (1991). Subduction and eduction of continental crust: major mechanism during continent–continent collision and orogenic extensional collapse, a model based on the south Caledonides. *Terra Nova*, 3, 303–310.

Angiolini, L., Gaetani, M., Muttoni, G. et al. (2007). Tethyan oceanic currents and climate gradients 300 m.y. ago. *Geology*, 35, 1071–1074.

Arenas, R., Fernández, R.D., Martínez, S.S. et al. (2014). Two-stage collision: exploring the birth of Pangea in the Variscan terranes. *Gondwana Research*, 25, 756–763.

Armijo, R., Lacassin, R., Coudurier-Curveur, A. & Carrizo, D. (2015). Coupled tectonic evolution of Andean orogeny and global climate. *Earth-Science Reviews*, 143, 1–35.

Ashwal, L.D., Demaiffe, D. & Torsvik, T.H. (2002). Petrogenesis of Neoproterozoic granitoids and related rocks from the Seychelles: evidence for an Andean arc origin. *Journal of Petrology*, 43, 45–83.

Assumpção, M., Feng, M. Tassara, A. & Julia, J. (2013). Models of crustal thickness for South America from seismic refraction, receiver functions and surface wave tomography. *Tectonophysics*, 609, 82–96.

Astashkin, V.A., Pegel, T.V., Repina, L.N. et al. (1995). *The Cambrian System of the Foldbelts of Russia and Mongolia*. International Union of Geological Sciences Publications, 32.

Aubrey, M.P., Lucas, S.G. & Berggren, W.A. (eds.) (1998). *Late Paleocene–Early Eocene Climatic and Biotic Events in the Marine and Terrestrial Records*. New York: Columbia University Press.

Baarli, G.B., Johnson, M.E. & Antoshkina, A.L. (2003). Silurian stratigraphy and paleogeography of Baltica. *New York State Museum Bulletin*, 493, 3–34,

Badarch, G., Cunningham, W.D., & Windley, B.F. (2002). A new terrane subdivision for Mongolia: implications for the Phanerozoic crustal growth of central Asia. *Journal of Asian Earth Sciences*, 21, 87–110.

Bassett, M.G. & Cocks, L.R.M. (1974). A review of Silurian brachiopods from Gotland. *Fossils and Strata*, 3, 1–56.

Batkhishig, B, Noriyoshi, T. & Greg, B. (2010). Magmatism of the Shuteen Complex and Carboniferous subduction of the Gurvansaihan terrane, South Mongolia. *Journal of Asian Earth Sciences*, 37, 399–411.

Bazhenov, M.L., Collins, A.Q., Degtyarev, K.E. et al. (2003). Paleozoic northward drift of the North Tien Shan (Central Asia) as revealed by Ordovician and Carboniferous paleomagnetism. *Tectonophysics*, 366, 113–141.

Beck Jr., M.E. & Housen, B.A. (2003). Absolute velocity of North America during the Mesozoic from paleomagnetic data. *Tectonophysics*, 377, 33–54.

Becker, T.P., Thomas, W.A. & Gehrels, G.E. (2006). Linking Late Paleozoic sedimentary provenance in the Appalachian Basin to the history of the Alleghanian deformation. *American Journal of Science*, 306, 777–798.

Becker, T.W. & Boschi, I. (2002). A comparison of tomographic and geodynamic mantle models. *Geochemistry, Geophysics, Geosystems*, 3, doi:10.1029/2001GC000168.

Belasky, P., Stevens, C.H. & Hanger, R.A. (2002). Early Permian location of western North American terranes based on brachiopod, fusulinid and coral biogeography. *Palaeogeography, Palaeoclimatology, Palaeoecology*, 179, 245–266.

Belousov, V.I. (2007). The Upper Palaeozoic preflysch and overthrusting in the Türkstan–Alay ranges, southern Fergana. *Geotektonika*, 2007(5), 63–75 (in Russian).

Benedetto, J.L. (1998). Early Palaeozoic brachiopods and associated shelly faunas from western Gondwana: their bearing on the geodynamic history of the pre-Andean margin. In R.J. Pankhurst & C.W. Rapela (eds.), *The Proto-Andean Margin of Gondwana*. Geological Society, London, Special Publications, 142, pp. 57–83.

Benton, M.J. (1995). Diversification and extinction in the history of life. *Science*, 268, 52–58.

Benton, M.J. (2005). *Vertebrate Palaeontology*, 3rd edn. Oxford: Blackwell.

Benton, M.J. (2008). The end-Permian mass extinction events on land in Russia. *Proceedings of the Geologists' Association*, 119, 119–136.

Berner, R.A. (1997). The rise of plants and their effect on weathering and atmospheric CO_2. *Science*, 276, 544–546.

Berner, R.A., Lasaga, A.C. & Garrels, R.M. (1983). The carbonate–silicate geochemical cycle and its effect on atmospheric carbon dioxide over the past 100 million years. *American Journal of Science*, 283, 641–683.

Berra, F. & Angiolini, L. (2014). The evolution of the Tethys Region throughout the Phanerozoic: a brief tectonic reconstruction. In L. Marlow, C.C.G. Kendall & L.A. Yose (eds.), *Petroleum Systems of the Tethyan Region*. AAPG Memoir, 106, pp. 1–27.

Beuf, S., Bijou-Duval, V., De Charpal, O. et al. (1971). *Les Grés du Paléozoïque au Sahara*. Publications de l'institut français du pétrole, 18.

Biggin, A.J., Steinberger, B., Aubert, J. et al. (2012). Long term geomagnetic variations and whole-mantle convection processes. *Nature Geoscience*, 5, 526–533.

Bird, P. (2003). An updated digital model of plate boundaries. *Geochemistry, Geophysics, Geosystems*, 4, 1027, doi:10.1029/2001GC000252.

Biske, Y.S. & Seltmann, R. (2010). Paleozoic Tian-Shan as a transitional region between the Rheic and Urals–Turkestan oceans. *Gondwana Research*, 17, 602–613.

Blieck, A. & Cloutier, R. (2000). Biostratigraphical correlations of Early Devonian vertebrate assemblages of the Old Red Sandstone continent. *Courier Forschungsinstitut Senckenberg*, 223, 223–269.

Blodgett, R.B. & Stanley, G.D. (eds.) (2008). *The Terrane Puzzle: New Perspectives on Paleontology and Stratigraphy from the North American Cordillera*. Geological Society of America Special Paper, 442.

Bonev, N. (2006). Cenozoic tectonic evolution of the eastern Rhodope Massif (Bulgaria): basement structure and kinematics of syn- to postcollisional extensional deformation. In Y. Dilek & S. Pavlides (eds.), *Postcollisional Tectonics and Magmatism in the Mediterranean Region and Asia*. Geological Society of America Special Paper, 409, pp. 211–235.

Boschman, L.M., van Hinsbergen, D.J.J., Torsvik, T.H. et al. (2014). Kinematic reconstruction of the Caribbean region since the Early Jurassic. *Earth-Science Reviews*, 138, 102–136.

Boucot, A.J. (1975). *Evolution and Extinction Rate Controls*. Amsterdam: Elsevier.

Boucot, A.J. & Blodgett, R.B. (2001). Silurian–Devonian biogeography. In C.H.C. Brunton, L.R.M. Cocks & S.L. Long (eds.), *Brachiopods Past and Present*. London: Taylor and Francis, pp. 335–344.

Boucot, A.J., Johnson, J.G. & Talent, J.A. (1969). *Early Devonian Brachiopod Zoogeography*. Geological Society of America Special Paper, 119.

Boucot, A.J., Xu, C. & Scotese, C.R. (2013). *Phanerozoic Paleoclimate: An Atlas of Lithologic Indicators of Climate*. SEPM Concepts in Sedimentology and Paleontology, 11.

Bowring, S.A., Erwin, D.H., Jin, Y.G. et al. (1998). U/Pb zircon geochronology and tempo of the end-Permian mass extinction. *Science*, 280, 1039–1045.

Bowring, S.A. & Williams, I.S. (1999). Priscoan (4.00 ± 4.03 Ga) orthogneisses from northwestern Canada. *Contributions to Mineralogy and Petrology*, 134, 3–16.

Bradley, D.C. (2008). Passive margins through earth history. *Earth-Science Reviews*, 91, 1–26.

Braitenberg, C. (2015). Exploration of tectonic structures with GOCE in Africa and across-continents. *International Journal of Applied Earth Observation and Geoinformation*, 35, 88–95.

Brenchley, P.J. & Cocks, L.R.M. (1982). Ecological associations in a regressive sequence: the latest Ordovician of the Oslo–Asker district, Norway. *Palaeontology*, 25, 783–815.

Brenchley, P.J. & Rawson, P.F. (eds.) (2006). *The Geology of England and Wales*. The Geological Society, London.

Brew, G., Barazangi, M., Al-Maleh, A.K. & Sawaf, F. (2001). Tectonic and geologic evolution of Syria. *GeoArabia*, 6, 573–615.

Brown, D., Herrington, R. & Alvarez-Marron, J. (2011). Processes of arc–continent collision in the Uralides. In D. Brown & P.D. Ryan (eds.), *Arc–Continent Collision*. Berlin: Springer-Verlag, pp. 311–340.

Buiter, S.J.H. & Torsvik, T.H. (2007). Horizontal movements in the Eastern Barents Sea constrained by numerical models and plate reconstructions. *Geophysical Journal International*, 171, 1376–1389.

Buiter, S.J.H. & Torsvik, T.H. (2014). A review of Wilson Cycle plate margins: a role for mantle plumes in continental break-up along sutures? *Gondwana Research*, 26, 627–653, doi:10.1016/j.gr.2014.02.007.

Bullard, E.C., Everett, J.E. & Smith, A.G. (1965). The fit of the continents around the Atlantic. *Philosophical Transactions of the Royal Society*, A258, 41–51.

Burke, K. (2011). Plate tectonics, the Wilson Cycle, and mantle plumes: geodynamics from the top. *Annual Review of Earth and Planetary Sciences*, 39, 1–29.

Burke, K., Steinberger, B., Torsvik, T.H. & Smethurst, M.A. (2008). Plume Generation Zones at the margins of Large Low Shear Velocity Provinces on the core–mantle boundary. *Earth and Planetary Science Letters*, 265, 49–60.

Burke, K. & Torsvik, T.H. (2004). Derivation of Large Igneous Provinces of the past 200 million years from long-term heterogeneities in the deep mantle. *Earth and Planetary Science Letters*, 227, 531–538.

Burtman, V.S. (2008). Nappes of the southern Tien Shan. *Russian Journal of Earth Sciences*, 10(ES1006), 1–35.
Bussien, D., Gombojav, N., Winkler, W. et al. (2011). The Mongol–Okhotsk belt in Mongolia. *Tectonophysics*, 510, 132–150.
Cai, J.X. & Zhang, K.J. (2009). A new model for the Indochina and South China collision during the Late Permian to the Middle Triassic. *Tectonophysics*, 467, 35–43.
Calvès, G., Schwab, A.M., Huuse, M. et al. (2011). Seismic volcanostratigraphy of the western Indian rifted margin: the pre-Deccan igneous province. *Journal of Geophysical Research*, 116, B01101, doi:10.1029/2010JB000862.
Came, R.E., Eiler, J.M., Veizer, J. et al. (2007). Coupling of surface temperatures and atmospheric CO_2 concentrations during the Palaeozoic era. *Nature*, 449, 198–201.
Cavazza, W., Roure, F., Spakman, W., Stampfli, G.M. & Ziegler, P.A. (eds.) (2004). *The TRANSMED Atlas: The Mediterranean Region from Crust to Mantle*. Berlin: Springer.
Charvet, J., Shu, L. & Laurent-Charvet, S. (2007). Paleozoic structural and geodynamic evolution of eastern Tianshan (NW China): welding of the Tarim and Junggar plates. *Episodes*, 30, 162–186.
Chen, X., Zhou, Z. & Fan, J. (2010). Ordovician paleogeography and tectonics of the major paleoplates of China. In S.C. Finney and W.B.N. Berry (eds.), *The Ordovician Earth System*. Geological Society of America Special Paper, 466, pp. 85–104.
Chen, Z.Q., Shi, G.R. & Zhan L.P. (2003). Early Carboniferous athyridid brachiopods from the Qaidam Basin, northwest China. *Journal of Paleontology*, 77, 844–862.
Christiansen, J.L. & Stouge, S. (1999). Oceanic circulation as an element in palaeogeographical reconstructions: the Arenig (early Ordovician) as an example. *Terra Nova*, 11, 73–78.
Chulick, G.S., Detweiler, S. & Mooney, W.D. (2013). Seismic structure of the crust and uppermost mantle of South America and surrounding oceanic basins, *Journal of South American Earth Sciences*, 42, 260–276.
Clack, J.A. (2002). *Gaining Ground: The Origin and Evolution of Tetrapods*. Bloomington: Indiana University Press.
Cocks, L.R.M. (1972). The origin of the Silurian Clarkeia shelly fauna of South America, and its extension to West Africa. *Palaeontology*, 15, 623–630.
Cocks, L.R.M. (ed.) (1981). *The Evolving Earth*. Cambridge: Cambridge University Press.
Cocks, L.R.M. (2011). There's no place like home: Cambrian to Devonian brachiopods critically useful for analysing palaeogeography. *Memoir of the Association of Australasian Palaeontologists*, 41, 135–148.
Cocks, L.R.M. & Fortey, R.A. (1982). Faunal evidence for oceanic separations in the Palaeozoic of Britain. *Journal of the Geological Society, London*, 138, 465–478.
Cocks, L.R.M. & Fortey, R.A. (1988). Lower Palaeozoic facies and faunas round Gondwana. In M.G. Audley-Charles & A. Hallam (eds.), *Gondwana and Tethys*. Geological Society, London, Special Publications, 37, pp. 183–200.
Cocks, L.R.M. & Fortey, R.A. (2009). Avalonia: a long-lived terrane in the Lower Palaeozoic? In M.G. Bassett (ed.), *Early Palaeozoic Peri-Gondwana Terranes*. Geological Society, London, Special Publications, 325, pp. 141–155.
Cocks, L.R.M., Fortey, R.A. & Lee, C.P. (2005). A review of Lower and Middle Palaeozoic biostratigraphy in west peninsula Malaya and southern Thailand in its context within the Sibumasu Terrane. *Journal of Asian Earth Sciences*, 34, 703–717.
Cocks, L.R.M. & Rong, J. (2008). Earliest Silurian faunal survival and recovery after the end-Ordovician glaciation: evidence from the brachiopods. *Transactions of the Royal Society of Edinburgh Earth and Environmental Sciences*, 98, 291–301.
Cocks, L.R.M. & Torsvik, T.H. (2002). Earth geography from 500 to 400 million years ago: a faunal and palaeomagnetic review. *Journal of the Geological Society, London*, 159, 631–644.
Cocks, L.R.M. & Torsvik, T.H. (2005). Baltica from the Late Precambrian to mid-Palaeozoic times: the gain and loss of a terrane's identity. *Earth-Science Reviews*, 72, 39–66.
Cocks, L.R.M. & Torsvik, T.H. (2007). Siberia, the wandering northern terrane, and its changing geography through the Palaeozoic. *Earth-Science Reviews*, 82, 29–74.
Cocks, L.R.M. & Torsvik, T.H. (2011). The Palaeozoic geography of Laurentia and western Laurussia: a stable craton with mobile margins. *Earth-Science Reviews*, 106, 1–51.
Cocks, L.R.M. & Torsvik, T.H. (2013). The dynamic evolution of the Palaeozoic geography of eastern Asia. *Earth-Science Reviews*, 117, 40–79.
Cocks, L.R.M. & Verniers, J. (2000). Applicability of planktic and nektic fossils to palaeogeographic reconstructions. *Acta Universitatis Carolinae – Geologica*, 42, 399–400.
Cody, S., Richardson, J.E., Ruli, V. et al. (2010). The Great American Biotic Interchange revisited. *Ecography*, 33, 326–332.
Coffin, M.F. & Eldholm, O. (1994). Large igneous provinces: crustal structure, dimensions, and external consequences. *Reviews of Geophysics*, 32, 1–36.
Cohen, K.M., Finney, S.C., Gibbard, P.L. & Fan, J.-X. (2013; updated). The ICS International Chronostratigraphic Chart. *Episodes*, 36, 199–204.
Collier, J.S., Minshull, T.A., Haqmmond, J.O.S. et al. (2009). Factors influencing magmatism during continental breakup: new insights from a wide-angle seismic experiment across the conjugate Seychelles–Indian margins. *Journal of Geophysical Research*, 114, B03101.
Collinson, M.E. & Hooker, J.J. (2003). Paleogene vegetation of Eurasia: framework for mammalian faunas. *Deinsea*, 10, 41–84.
Colpron, M. & Nelson, J.L. (2006). *Palaeozoic Evolution and Metallogeny of Pericratonic Terranes at the Ancient Pacific*

Margin of North America. Geological Association of Canada Special Paper, 45.

Colpron, M. & Nelson, J.L. (2009). A Palaeozoic Northwest Passage: incursion of Caledonian, Baltican and Siberian terranes into eastern Panthalassa and the early evolution of the North American Cordillera. In P.A. Cawood & A. Kröner (eds.), *Earth Accretionary Systems in Space and Time*, Geological Society, London, Special Publications, 318, pp. 273–307.

Connelly, J.N., Bizzarro, M., Krot, A.N. et al. (2012). Absolute chronology and thermal processing of solids in the solar protoplanetary disk. *Science*, 338, 651–655.

Conrad, C.P., Steinberger, B. & Torsvik, T.H. (2014). Dynamic topography and sea level change inferred from dipole and quadrupole moments of plate tectonic reconstructions. American Geophysical Union Fall Meeting, San Francisco, Abstract.

Cook, H.E., Zhemchuzhnikov, V.G., Zempolich, W.G. et al. (2002). Devonian and Carboniferous platform facies in the Bolshoi Karatau, southern Kazakhstan: outcrop analogs for coeval carbonate oil and gas fields in the North Caspian Basin, western Kazakhstan. In W.G. Zempolich & H.E. Cook (eds.), *Paleozoic Carbonates of the Commonwealth of Independent States*. SEPM Special Publication, 74, pp. 81–122.

Cope, T., Ritts, B.D., Darby, B.J. et al. (2005). Late Paleozoic sedimentation on the northern margin of the North China Block: implications for regional tectonics and climate changes. *International Geology Review*, 47, 270–296.

Copper, P. (2002). Silurian and Devonian reefs. In W. Kiessling, E. Flügel & J. Golonka (eds.), *Phanerozoic Reef Patterns*. SEPM Special Publication, 72, pp. 181–238.

Copper, P. & Jin, J. (2015). Tracking the early Silurian post-extinction faunal recovery in the Jupiter Formation of Anticosti Island, eastern Canada: a stratigraphical revision. *Newsletters on Stratigraphy*, 48, 221–240.

Corfu, F., Polteau, S., Planke, S. et al. (2013). U–Pb geochronology of Cretaceous magmatism on Svalbard and Franz Josef Land, Barents Sea Large Igneous Province. *Geological Magazine*, 150, 1127–1135.

Courtillot, V., Davaille, A., Besse, J. & Stock, J. (2003). Three distinct types of hotspots in the Earth's mantle. *Earth and Planetary Science Letters*, 205, 295–308.

Courtillot, V., Gallet, Y., Le Mouël, J.-L., Fluteau, F. & Genevey, A. (2007). Are there connections between the Earth's magnetic field and climate? *Earth and Planetary Science Letters*, 253, 328–339.

Courtillot, V.E. & Renne, P.-R. (2003). On the ages of flood basalt events. *Comptes Rendus Geoscience*, 335, 113–140.

Dabard, M.P., Loi, A., Paris, F. et al. (2015). Sea-level curve for the Middle to early Late Ordovician in the Armorican Massif (western France): icehouse third-order glacio-eustatic cycles. *Palaeogeography, Palaeoclimatology, Palaeoecology*, 436, 96–111.

Dal Corso, J., Marzoli, A., Tateo, F. et al. (2014). The dawn of CAMP volcanism and its bearing on the end-Triassic carbon cycle disruption. *Journal of the Geological Society, London*, doi.org/10.1144/jgs2013-063.

Dalhquist, J.A., Pankhurst, R.J., Gaschnig, R.M. et al. (2013). Hf and Nd isotopes in Early Ordovician to Early Carboniferous granites in the Proto-Andean margin of Gondwana. *Gondwana Research*, 23, 1617–1630.

Dalziel, I.W.D. (1997). Neoproterozoic–Paleozoic geography and tectonics: review, hypothesis, environmental speculation. *Geological Society of America Bulletin*, 109, 16–42.

Darwin, C. (1859). *The Origin of Species*. London: John Murray.

Daukeev, S.Z., Uzhkenov, B.S., Miletenko, N.V. et al. (eds.) (2002). *Atlas of Lithology – Paleogeographical, Structural, Palinspastic and Geoenvironmental Maps of Central Eurasia*. Almaty: Scientific Research Institute of Natural Resources (in Russian).

Dawes, P.R. (2009). Precambrian–Palaeozoic geology of Smith Sound, Canada and Greenland: key constraint to palaeogeographical reconstructions of northern Laurentia and the North Atlantic region. *Terra Nova*, 21, 1–13.

Dean, W.T., Monod, O., Rickards, R.B. et al. (2000). Lower Palaeozoic stratigraphy and palaeontology, Karadire-Zirze area, Pontus Mountains, northern Turkey. *Geological Magazine*, 137, 555–582.

de Freitas, T.A. & Dixon, O.A. (1995). Silurian microbial buildups, Canadian Arctic. In C.L.V. Monty, D.W.J. Bosence, P.H. Bridges & B.R. Pratt (eds.), *Carbonate Mud-Mounds: Their Origin and Evolution*. International Association of Sedimentologists, Special Publication, 23, pp. 151–169.

Degtyarev, K.Y. & Ryazantsev, A.V. (2007). Cambrian arc–continent collision in the Palaeozoides of Kazakhstan. *Geotectonics*, 43, 63–86.

de Jong, K, Xiao, W., Windley, B.F. et al. (2006). Ordovician $^{40}Ar/^{39}Ar$ phengite ages from the blueschist-facies Ondor Sum subduction–accretion complex (Inner Mongolia) and implications for the Early Paleozoic history of continental blocks in China and adjacent areas. *American Journal of Science*, 306, 799–845.

Dewing, K., Harrison, J.C., Pratt, B.R. & Mayr, U. (2004). A probable late Neoproterozoic age for the Kennedy Channel and Ella Bay formations, northeastern Ellesmere Island and its implications for passive margin history in the Canadian Arctic. *Canadian Journal of Earth Sciences*, 41, 1013–1025.

Dhuime, B., Hawkesworth, C.J., Cawood, P.A. & Storey, C.D. (2012). A change in the geodynamics of continental growth 3 billion years ago. *Science*, 335,1334–1336.

Dhuime, B., Wuestefeld, A. & Hawkesworth, C.J. (2015). Emergence of modern continental crust about 3 billion years ago. *Nature Geoscience*, 8, doi:10.1038/NGEO2466.

Dickinson, W.R. (2000). Geodynamic interpretation of Paleozoic tectonic trends oriented oblique to the Mesozoic Klamath–Sierran continental margin in California. In M.J. Soreghan &

G.E. Gehrels (eds.), *Paleozoic and Triassic paleogeography and tectonics of western Nevada and Northern California*. Geological Society of America Special Paper, 347, pp. 200–245.

Dickinson, W.R. (2009). Anatomy and global context of the North American Cordillera. In S. Mahlburg Kay, V.A. Ramos & W.R. Dickinson (eds.), *Backbone of the Americas: Shallow Subduction, Plateau Uplift, and Ridge and Terrane Collision*. Geological Society of America Memoir, 204, pp. 1–29.

Dickinson, W.R. & Lawton, T.F. (2001). Carboniferous to Cretaceous assembly and fragmentation of Mexico. *Geological Society of America Bulletin*, 113, 1142–1160.

DiMichele, W.A., Montanez, I.P., Poulsen, C.J. & Tabor, N.J. (2009). Climate and vegetational regime shifts in the late Paleozoic ice age earth. *Geobiology*, 7, 200–226.

DiMichele, W.A., Gastaldo, R.A. & Pfefferkorn, H.W. (2005). Plant biodiversity partitioning in the Late Carboniferous and Early Permian and its implications for ecosystem assembly. *Proceedings of the California Academy of Sciences*, 56, Supplement 1(4), 32–49.

Dobretsov, N.L., Berzin, N.A., Buslov, M.M. (1995). Opening and tectonic evolution of the Paleo-Asian ocean. *International Geology Review*, 35, 335–360.

Dobretsov, N.L., Buslov, M.M. & Vernikovsky, V.A. (2003). Neoproterozoic to Early Ordovician evolution of the Paleo-Asian Ocean: implications to the break-up of Rodinia. *Gondwana Research*, 6, 143–159.

Dobretsov, N.L., Buslov, M.M., Zhimulev, F.I. et al. (2006). Vendian–early Ordovician geodynamic evolution and model for exhumation of ultrahigh- and high-pressure rocks from the Kokchetav subduction–collision zone. *Geologiya i Geofizika*, 47, 428–444 [in Russian].

Dodd, S.C., MacNiocaill, C. & Muxworthy, A.R. (2015). Long duration (>4 Ma) and steady-state volcanic activity in the early Cretaceous Paraná–Etendeka Large Igneous Province: new palaeomagnetic data from Namibia. *Earth and Planetary Science Letters*, 414, 16–29.

Domeier, M. (2015). A plate tectonic scenario in the Iapetus and Rheic oceans. *Gondwana Research*, doi:10.1016/j.gr.2015.08.003.

Domeier, M. & Torsvik, T.H. (2014). Plate tectonics in the late Paleozoic. *Geoscience Frontiers*, 5, 303–350.

Domeier, M., Van der Voo, R. & Torsvik, T.H. (2012). Paleomagnetism and Pangea: the Road to reconciliation. *Tectonophysics*, 514, 14–43.

Dornbos, S.Q. & Bottjer, D.J. (2000). Evolutionary paleoecology of the earliest echinoderms: helicoplacoids and the Cambrian substrate revolution. *Geology*, 28, 839–842.

Doubrovine, P.V., Steinberger, B. & Torsvik, T.H. (2012). Absolute plate motions in a reference frame defined by moving hotspots in the Pacific, Atlantic and Indian oceans. *Journal of Geophysical Research*, 117, B09101, doi:10.1029/2011JB009072.

Doubrovine, P.V. & Tarduno, J.A. (2004). Late Cretaceous paleolatitude of the Hawaiian Hot Spot: new paleomagnetic data from Detroit Seamount (ODP Site 883). *Geochemistry, Geophysics, Geosystems*, 5, Q11L04, doi:10.1029/2004GC000745.

Edwards, D., Cherns, L. & Raven, J.A. (2015). Could land-based early photosynthesizing ecosystems have bioengineered the planet in mid-Palaeozoic times? *Palaeontology*, 58, 803–837.

Egan, S.S., Mosar, J., Brunet, M.F. & Kangarli, T. (2009). Subsidence and uplift mechanisms within the South Caspian Basin: insights from the onshore and offshore Azerbaijan region. In M.-F. Brunet, M. Wilmsenj & W. Granath (eds.), *South Caspian to Central Iran Basins*. Geological Society, London, Special Publications, 312, pp. 219–240.

Eide, E.A. & Torsvik, T.H. (1996). Paleozoic supercontinent assembly, mantle flushing and genesis of the Kiaman Superchrons. *Earth and Planetary Science Letters*, 144, 389–402.

Eldholm, O. & Myhre, A.M. (1977). Hovgaard Fracture Zone. In *Årbok 1976* Oslo: Norsk Polarinstitutt, 195–208.

Elliott, D. (2013). The geological and tectonic evolution of the Transantarctic Mountains: a review. In M.J. Hambrey et al. (eds.), *Antarctic Palaeoenvironments and Earth-Surface Processes*. Geological Society, London, Special Publications, 381, pp. 7–35.

Embry, A.F. (1991). Middle–Upper Devonian clastic wedge of the Arctic islands. In H.P. Trettin (ed.) *Geology of the Innuitian Orogen and Arctic Platform of Canada and Greenland*. Ottawa: Geological Survey of Canada, Geology of Canada, 3, pp. 263–279.

Engen, Ø., Faleide, J.I. & Dyreng, T.K. (2008). Opening of the Fram Strait gateway: a review of plate tectonic constraints. *Tectonophysics*, 450, 51–69, doi:10.1016/j.tecto.2008.01.002.

Escayola, M.P., van Staal, C.R. & Davis, W.J. (2011). The age and tectonic setting of the Punoviscana Formation in northwestern Argentina: an accretionary complex related to early Cambrian closure of the Punoviscana Ocean and accretion of the Arequipa–Antofalla blocks. *Journal of South American Earth Sciences*, 32, 438–450.

Evans, D.A.D. (2013). Reconstructing pre-Pangean supercontinents. *Geological Society of America Bulletin*, 125, 1735–1751.

Evenchick, C.A., Davis, W.J., Bédard, J.H., Hayward, N. & Friedman, R.N. (2015). Evidence for protracted High Arctic Large Igneous Province magmatism in the central Sverdrup Basin from stratigraphy, geochronology, and paleodepths of saucer-shaped sills. *Geological Society of America Bulletin*, 127, 1366–1390, doi.org/10.1130/B31190.1

Faccenna, C., Becker, T.W., Auer, L. et al. (2014). Mantle dynamics in the Mediterranean, *Reviews of Geophysics*, 52, 283–332.

Faleide, J.I., Bjørlykke, K. & Gabrielsen, R.H. (2010). Geology of the Norwegian Continental Shelf. In K. Bjørlykke (ed.), *Petroleum Geoscience: From Sedimentary Environments to*

Rock Physics, Springer Science Business Media, pp. 467–499.

Fergusson, C.L. & Henderson, R.A. (2015). Early Palaeozoic continental growth in the Tasmanides of northeast Gondwana and its implications for Rodinia assembly and rifting. *Gondwana Research*, 28, 933–953.

Fielding, C.R., Frank, T.D. & Isbell, J.L. (eds.) (2008). *Resolving the Late Paleozoic Ice Age in Time and Space*. Geological Society of America Special Paper, 441, pp. 71–82.

Flecker, R., Krijgsman, W., Capella, W. et al. (2015). Evolution of the Late Miocene Mediterranean–Atlantic gateways and their impact on regional and global environmental change. *Earth-Science Reviews*, 150, 365–392.

Fortey, R.A. & Cocks, L.R.M. (2003). Palaeontological evidence bearing on global Ordovician–Silurian continental reconstructions. *Earth-Science Reviews*, 61, 245–307.

Fortey, R.A. & Cocks, L.R.M. (2005). Late Ordovician global warming: the Boda Event. *Geology*, 33, 405–408.

Francis, J.E., Ashworth, A., Cantrill, D.J. et al. (2008). 100 million years of Antarctic climate evolution: evidence from fossil plants. In A.K. Cooper, P.J. Barrett et al. (eds.), *Antarctica: A Keystone in a Changing World*. Washington, DC: National Academies Press, pp. 19–27,

Franke, W. (2006). The Variscan orogen in central Europe: construction and collapse. In D.G. Gee & R.A. Stephenson (eds.), *European Lithosphere Dynamics*. Geological Society, London, Memoir, 32, pp. 333–343.

Franke, W., Cocks, L.R.M. & Torsvik, T.H. (2016). Fresh insights into an old orogeny: the Variscan revisited. *Gondwana Research* (in press).

French, S.W. & Romanowicz, B. (2015). Broad plumes rooted at the base of the Earth's mantle beneath major hotspots. *Nature*, 525, 95–99.

Friend, P.F. & Williams, B.P.J. (eds.) (2000). *New Perspectives on the Old Red Sandstone*. Geological Society, London, Special Publications, 180.

Friis, E.M., Crane, P.R. & Pedersen, K.R. (2011). *Early Flowers and Angiosperm Evolution*. Cambridge: Cambridge University Press.

Froitzheim, N., Plašienka, D. & Schuster, R. (2008). Alpine tectonics of the Alps and western Carpathians. In T. McCann (ed.), *The Geology of Central Europe*. The Geological Society, London, pp. 1141–1232.

Gabrielse. H. & Yorath. C.J. (1992). *Geology of the Cordilleran Orogeny in Canada. The Geology of North America, Vol. G-2*. Ottawa: Geological Survey of Canada.

Gaetani, M. (1997). The Karakorum Block in Central Asia, from Ordovician to Cretaceous. *Sedimentary Geology*, 109, 339–359.

Gaetani, M., Angiolini, L, Ueno, K. et al. (2009). Pennsylvanian–Early Triassic stratigraphy in the Alborz Mountains (Iran). In M.-F. Brunet, M. Wilmsenj & W. Granath (eds.), *South Caspian to Central Iran Basins*. Geological Society, London, Special Publications, 312, pp. 79–128.

Gaina, C., Gernigon, L. & Ball, P. (2009). Palaeocene–Recent plate boundaries in the NE Atlantic and the formation of the Jan Mayen microcontinent. *Journal of the Geological Society, London*, 166, 601–616.

Gaina, C., Müller, D.R., Royer, J.-Y. et al. (1998). The tectonic history of the Tasman Sea: a puzzle with 13 pieces. *Journal of Geophysical Research*, 103, 12413–12433.

Gaina, C., Müller, D.R., Royer, J.-Y. & Symonds, P. (1999). Evolution of the Louisiade triple junction. *Journal of Geophysical Research*, 104, 12927–12939.

Gaina, C., Medvedev, S., Torsvik, T.H., Koulakov, I.Yu & Werner, S.C. (2013a). 4D Arctic: a glimpse into the structure and evolution of the Arctic in the light of new geophysical maps, plate tectonics and tomographic models. *Surveys in Geophysics*, 35, 1095–1122.

Gaina, C., Roest, W.R. & Muller, R.D. (2002). Late Cretaceous–Cenozoic deformation of northeast Asia. *Earth and Planetary Science Letters*, 197, 273–286.

Gaina, C., Torsvik, T.H., van Hinsbergen, D. et al. (2013b). The African Plate: a history of oceanic crust accretion and subduction since the Jurassic. *Tectonophysics*, 604, 4–25.

Gaina, C., van Hinsbergen, D. & Spakman, W. (2015). Tectonic interactions between India and Arabia since the Jurassic reconstructed from marine geophysics, ophiolite geology, and seismic tomography. *Tectonics*, 34(5), 875–906.

Garnero, E.J., Lay, T. & McNamara, A. (2007). Implications of lower-mantle structural heterogeneity for existence and nature of whole-mantle plumes. In G.R. Foulger & D.M. Jurdy (eds.), *Plates, Plumes and Planetary Processes*. Geological Society of America Special Paper, 430, pp. 79–101.

Gee, D.G. & Pease, V.I. (eds.) (2005). *The Neoproterozoic Timanide Orogen of Eastern Baltica*. Geological Society, London, Memoir, 30.

Gee. J.S. & Kent, D.V. (2007). Source of oceanic magnetic anomalies and the geomagnetic polarity time scale. *Treatise on Geophysics*, 5, 455–507.

Ghienne, J.F., Le Heron, D.P., Moreau, J. et al. (2007). The Late Ordovician sedimentary system of the North Gondwana platform. In M. J. Hambrey et al. (eds.), *Glacial Sedimentary Processes and Products*. International Association of Sedimentologists Special Publication, 39, pp. 297–319.

Ghienne, J.F., Monod, O., Kozlu, H. & Dean, W.T. (2010). Cambrian–Ordovician depositional sequences in the Middle East: a perspective from Turkey. *Earth-Science Reviews*, 101, 101–146.

Gibbons, W. & Moreno, T. (eds.) (2002). *The Geology of Spain*. The Geological Society, London.

Gibbons, A.D., Whittaker, J.M. & Müller, R.D. (2013). The breakup of East Gondwana: assimilating constraints from Cretaceous ocean basins around India into a best-fit tectonic model. *Journal of Geophysical Research*, 118, 1–15.

Gibling, M.R., Davies, N.S., Falcon-Lang, H.J. et al. (2014). Palaeozoic co-evolution of rivers and vegetation: a synthesis of current knowledge. *Proceedings of the Geologists' Association*, 125, 524–533.

Glass, L.M. & Phillips, D. (2006). The Kalkarindji continental flood basalt province: a new Cambrian large igneous province in Australia with possible links to faunal extinctions. *Geology*, 34, 461–464.

Glen, R.A. (2005). The Tasmanides of eastern Australia. In A.P.M. Vaughan, P.T. Leat & R.J. Pankhurst (eds.), *Terrane Processes at the Margins of Gondwana*. Geological Society, London, Special Publications, 246, pp. 23–96.

Glennie, K. (ed.) (2006). *Oman's Geological Heritage*, 2nd edn. Muscat: Petroleum Development Oman.

Goldreich, P. & Toomre, A. (1969). Some remarks on polar wandering. *Journal of Geophysical Research*, 74, 2555–2569.

Goodfellow, W.D., Cecile, M.P. & Leybourne, M.I. (1995). Geochemistry, petrogenesis and tectonic setting of Lower Paleozoic alkalic and potassic volcanic rocks, northern Canadian Cordilleran Miogeocline. *Canadian Journal of Earth Sciences*, 32, 1236–1254.

Greb, S.F., Pashin, J,C., Martino, R.L. & Eble, C.F. (2008). Appalachian sedimentary cycles during the Pennsylvanian: changing influences of sea level, climate, and tectonics. In C.R. Fielding, T.D. Frank & J.L. Isbell (eds.), *Resolving the Late Paleozoic Ice Age in Time and Space*. Geological Society of America Special Paper, 441, pp. 235–248.

Gulbranson, E.L., Ryberg, P.E., Decobeix, A.-L. et al. (2014). Leaf habit of Late Permian *Glossopteris* trees from high-palaeolatitude forests. *Journal of the Geological Society, London*, 171, 493–507.

Hall, R. (2012). Late Jurassic–Cenozoic reconstructions of the Indonesian margin and the Indian Ocean. *Tectonophysics*, 570–571, 1–41.

Hall, R. & Holloway, J.D. (eds.) (1998). *Biogeography and Geological Evolution of SE Asia*. Leiden: Backhuys.

Hallam, A. (1988). A reevaluation of Jurassic eustasy in the light of new data and the revised Exxon curve. In C.K. Wilgus et al. (eds.), *Sea-Level Changes: An Integrated Approach*. SPEM Special Publication, 42, pp. 261–273.

Haq, B.U. & Al-Qahtani, A.M. (2005). Phanerozoic cycles of sea-level change on the Arabian Platform. *GeoArabia*, 10, 127–160.

Haq, B.U. & Shutter, S.R. (2008). A chronology of Paleozoic sea-level changes. *Science*, 322, 64–68.

Harper, D.A.T., Mac Niocaill, C. & Williams, S.H. (1996). The palaeogeography of the early Ordovician Iapetus terranes: an integration of faunal and palaeomagnetic constraints. *Palaeogeography, Palaeoclimatology, Palaeoecology*, 121, 297–312.

Harper, D.A.T. & Servais, T. (eds.) (2013). *Early Palaeozoic Biogeography and Palaeogeography*. Geological Society, London, Memoirs, 38.

Hartz, E.H. & Torsvik, T.H. (2002). Baltica upside down: a new plate tectonic model for Rodinia and the Iapetus Ocean. *Geology*, 30, 255–258.

Hatcher, R.D., Thomas, W.A. & Viele, G.W. (eds.) (1989). *The Appalachian–Ouachita Orogeny in the United States: The Geology of North America, Vol. F-2*. Boulder: Geological Society of America.

Havlíček, V., Vaněk, J. & Fatka, O. (1994). Perunica microcontinent in the Ordovician (its position within the Mediterranean Province, series divisions, benthic and pelagic associations). *Sborník geologickych věd Geologie*, 46, 25–56.

Hawkesworth, C., Dhuime, B., Pietranik, A. et al. (2010). The generation and evolution of the continental crust. *Journal of the Geological Society, London*, 167, 229–248.

Hawkins, T., Smith, M.P., Herrington, R.J. et al. (2016). The geology and genesis of the iron skarns of the Turgai belt, northwestern Kazakhstan. *Ore Geology Reviews* (in press).

Heine, C., Zoethout, J. & Muller, R.D. (2013). Kinematics of the South Atlantic rift. *Solid Earth*, 4, 215–253.

Helbing, H. & Tiepolo, M. (2005). Age determination of Ordovician magmatism in NE Sardinia and its bearing on Variscan basement evolution. *Journal of the Geological Society, London*, 162, 689–700.

Hellinger, S.J. (1981). The uncertainties of finite rotations in plate tectonics. *Journal of Geophysical Research*, 86, 9312–9318.

Henriksen, N. (2008). *Geological History of Greenland: Four Billion Years of Earth Evolution*. Copenhagen: Geological Survey of Denmark.

Hervé, F., Calderón, M., Fanning, C.M. et al. (2013). Provenance variations in the Late Paleozoic accretionary complex of central Chile as indicated by detrital zircons. *Gondwana Research*, 23, 1122–1135.

Higgins, A.K., Gilotti, J.A. & Smith, M.P. (eds.) (2008). *The Greenland Caledonides: Evolution of the Northeast Margin of Laurentia*. Geological Society of America Memoir, 202.

Hildebrand, R.S. (2014). Geology, mantle tomography, and inclination corrected paleogeographic trajectories support westward subduction during Cretaceous orogenesis in the North American Cordillera. *Geoscience Canada*, 41, doi.org/10.12789/geocanj.2014.41.032.

Hoepffer, C., Soulaimani, A. & Piqué, A. (2005). The Moroccan Hercynides. *Journal of African Earth Sciences*, 43, 144–165.

Hoffman, P.F., Kaufman, A.J., Halverson, G.P. & Schrag, D.P. (1998). A Neoproterozoic Snowball Earth. *Science*, 281, 1342–1346.

Holmer, L.E., Popov, L.E., Koneva, S.P. & Bassett, M.G. (2001). *Cambrian–Early Ordovician Brachiopods from Malyi Karatau, the Western Balkash Region, and Tien Shan, Central Asia*. The Palaeontological Society, Special Papers in Palaeontology, 65.

Holz, M., França, A.B., Sousa, P.A. et al. (2010). A stratigraphic chart of the Late Carboniferous/Permian succession of the eastern border of the Paraná Basin, Brazil. *Journal of South American Earth Sciences*, 29, 381–389.

Husson, L., Conrad, C.P. & Faccenna, C. (2012). Plate motions, Andean orogeny, and volcanism above the South Atlantic convection cell. *Earth and Planetary Science Letters*, 317–318, 126–135.

Huyghe, D., Lartaud, F., Emmanuel, L., Merle, D. & Renard, M. (2015). Palaeogene climate evolution in the Paris Basin from oxygen stable isotope ($\delta^{18}O$) compositions of marine molluscs. *Journal of the Geological Society, London*, 172, 576–587.

Ikeda, M., Hori, R.S., Okada, Y. & Nakada, A. (2015). Volcanism and deep-ocean acidification across the end-Triassic extinction event. *Palaeogeography, Palaeoclimatology, Palaeoecology*, 440, 725–733.

IPCC (2013). Summary for Policymakers. In T.F. Stocker, D. Qin, G.-K. Plattner et al. (eds.), *Climate Change 2013: The Physical Science Basis. Contribution of Working Group I to the Fifth Assessment Report of the Intergovernmental Panel on Climate Change*. Cambridge: Cambridge University Press.

Isozaki, Y., Aoki, K., Nakama, T. & Yanai, S. (2010). New insight into a subduction-related orogeny: a reappraisal of the geotectonic framework and evolution of the Japanese islands. *Gondwana Research*, 18, 82–105.

Jaanusson, V. (1973). Aspects of carbonate sedimentation in the Ordovician of Baltoscandia. *Lethaia*, 6, 11–34.

James, K.H., Lorente, M.A. & Pindell, J.L. (eds.) (2009). *The Origin and Evolution of the Caribbean Plate*. Geological Society, London, Special Publications, 328.

Jenkyns, H.C. (2010). Geochemistry of oceanic anoxic events. *Geochemistry, Geophysics, Geosystems*, 11, Q03004, doi:10.1029/2009GC002788.

Jian, P., Liu, D., Kröner, A. et al. (2008). Time scale of an early to mid-Palaeozoic orogenic cycle of the long-lived Central Asian Orogenic Belt, Inner Mongolia of China: implications for continental growth. *Lithos*, 101, 233–259.

Jian, P., Liu, D., Kröner, A. et al. (2009a). Devonian to Permian plate tectonic cycle of the Paleo-Tethys Orogen in southwest China (I): geochemistry of ophiolites, arc/back-arc assemblages and within-plate igneous rocks. *Lithos*, 113, 748–766.

Jian, P., Liu, D., Kröner, A. et al. (2009b). Devonian to Permian plate tectonic cycle of the Paleo-Tethys Orogen in southwest China (II): insights from zircon ages of ophiolites, arc/back-arc assemblages and within plate igneous rocks and generation of the Emeishan CFB province. *Lithos*, 113, 767–784.

Joachimski, M.M., Breizig, S., Buggisch, W. et al. (2009). Devonian climate and reef evolution: insights from oxygen isotopes in apatite. *Earth and Planetary Science Letters*, 284, 599–609.

Johnson, D.P., Maillet, P.C. & Price, R. (1993). Regional setting of a complex backarc: New Hebrides Arc, northern Vanuatu–eastern Solomon Islands. *Geo-Marine Letters*, 13, 82–89.

Johnston, S.T. (2008). The Cordilleran Ribbon Continent of North America. *Annual Review of Earth and Planetary Sciences*, 36, 495–530.

Kay, S.M., Mpodozis, C. & Ramos, V.A. (2005). Andes. In R.C. Selley, L.R. M. Cocks & I.R. Plimer (eds.), *Encyclopedia of Geology, Volume 1*, Amsterdam: Elsevier, pp. 118–131.

Keller, B.M. & Predtechensky, N.N. (eds.) (1968). *Atlas of Lithology – Paleogeographical Maps of the U.S.S.R. Precambrian, Cambrian, Ordovician, Silurian, Volume 1*. Moscow: Ministry of Geology of the USSR (in Russian).

Kennan, L. & Pindell, J.L. (2009). Dextral shear, terrane accretion and basin formation in the Northern Andes: best explained by interaction with a Pacific-derived Caribbean Plate? In K.H. James, M.A. Lorente & J.L. Pindell (eds.), *The Origin and Evolution of the Caribbean Plate*. Geological Society, London, Special Publications, 328, pp. 487–531.

Kenrick, P. & Davis, P. (2004). *Fossil Plants*. London: The Natural History Museum.

Kenrick, P., Wellman, C.H., Schneider, H. & Edgecombe, G.D. (2012). A time-line for terrestrialization: consequences for the carbon cycle in the Palaeozoic. *Philosophical Transactions of the Royal Society*, B367, 519–536.

Kent, D.V. & Tauxe, L. (2005). Corrected Late Triassic latitudes for continents adjacent to the North Atlantic. *Science*, 307, 240–247.

Keppie, J.D. (2004). Terranes of Mexico revisited: a 1.3 billion year odyssey. *International Geology Review*, 46, 765–794.

Keppie, J.D., Dostal, J., Murphy, J.B. & Nance, R.D. (2008). Synthesis and tectonic interpretation of the westernmost Variscan orogeny in southern Mexico: from rifted Rheic margin to active Pacific margin. *Tectonophysics*, 461, 277–290.

Keppie, J.D., Nance, R.D., Dostal, J., Lee, J.K.W. & Ortega-Rivera, A. (2012). Constraints on the subduction erosion/extrusion cycle in the Paleozoic Acatlán Complex of southern Mexico: geochemistry and geochronology of the type Plaxtia Suite. *Gondwana Research*, 21, 1050–1065.

Kheraskova, T.N., Didenko, A.N., Bush, V.A. & Volozh, Y.A. (2003). The Vendian–Early Paleozoic history of the continental margin of eastern Paleogondwana, Paleoasian Ocean, and Central Asia Foldbelt. *Russian Journal of Earth Sciences*, 5, 165–184.

Khozyem, H., Adatte, T., Spangenberg, J.E. et al. (2013). Palaeoenvironmental and climatic changes during the Palaeocene–Eocene Thermal Maximum (PETM) at the Wadi Nukhul Section, Sinai, Egypt. *Journal of the Geological Society, London*, 170, 341–352.

Kirschvink, J.L. (1992). Late Proterozoic low-latitude global glaciation: the snowball Earth. In J.W. Schopf, C. Klein & D. Des Maris (eds.), *The Proterozoic Biosphere: A Multidisciplinary Study*. Cambridge: Cambridge University Press, pp. 51–52.

Knudsen, M.F. & Riisager, P. (2009). Is there a link between earth's magnetic field and low-latitude precipitation? *Geology*, 37, 71–74.

Kodama, K.P. (2009). Simplification of the anisotropy-based inclination correction technique for magnetite- and hematite-bearing rocks: a case study for the Carboniferous Glenshaw and Mauch Chunk formations, North America. *Geophysical Journal International*, 176, 467–477.

Kohn, M.J. (2014). Himalayan metamorphism and its tectonic implications. *Annual Review of Earth and Planetary Sciences*, 42, 381–419.

Koppers, A.A.P., Yamazaki, T., Geldmacher, J. et al. (2012). Limited latitudinal mantle plume motion for the Louisville hotspot. *Nature Geoscience*, 5, 911–917, doi:10.1038/NGEO1638.

Korte, C., Hesselbo, S.P., Jenkyns, H.C. et al. (2009). Palaeoenvironmental significance of carbon- and oxygen-isotope stratigraphy of marine Triassic–Jurassic boundary sections in SW Britain. *Journal of the Geological Society, London*, 166, 431–445.

Kravchinsky, V.A., Konstantinov, K.M. & Cogné, J.P. (2001). Palaeomagnetic study of Vendian and Early Cambrian rocks of South Siberia and Central Mongolia: was the Siberian platform assembled at the time? *Precambrian Research*, 110, 61–92.

Kröner, A. (ed.) (2015). *The Central Asian Orogenic Belt*. Stuttgart: Borntraeger.

Krstić, B., Maslarević, L., Ercegovać, M. & Dajić, S. (1999). Ordovician of the East-Serbian South Carpathians. *Acta Universitatis Carolinae – Geologica*, 43, 101–114.

Labails, C., Olivet, J.L., Aslanian, D. & Roest, W.R. (2010). An alternative early opening scenario for the Central Atlantic Ocean. *Earth and Planetary Science Letters*, 297, 355–368.

Landing, E., Rushton, A.A., Fortey, R.A. & Bowring, S.A. (2015). Improved geochronologic accuracy and precision for the ICS Chronostratigraphic Charts: examples from the late Cambrian–early Ordovician. *Episodes*, 38, 154–161.

Lapworth, C. (1879). On the tripartite classification of the Lower Palaeozoic rocks. *Geological Magazine*, 6, 1–15.

Lee, S., Choi, D.R. & Shi, G.R. (2010). Pennsylvanian brachiopods from the Geumcheon-Jangseong Formation, Pyeongan Supergroup, Taebaeksan Basin, Korea. *Journal of Paleontology*, 84, 417–443.

Lefebvre, B. & Fatka, O. (2003). Palaeogeographical and palaeoecological aspects of the Cambro-Ordovician radiation of echinoderms. *Palaeogeography, Palaeoclimatology, Palaeoecology*, 195, 73–97.

Lekic, V., Cottar, S., Dziewonski, A. & Romanowicz, B. (2012). Cluster analysis of global lower mantle tomography: a new class of structure and implications for chemical heterogeneity. *Earth and Planetary Science Letters*, 357, 68–77.

Leroy, S., Razin, P., Autin, J. et al. (2012). From rifting to oceanic spreading in the Gulf of Aden: a synthesis. *Arabian Journal of Geosciences*, 5, 859–901, doi:10.1007/s12517-011-0475-4.

Lethiers, F. & Crasquin-Soleau, S. (1995). Distributions des ostracodes et paléocurrantologie au Carbonifère terminal–Permien. *Geobios*, 18, 257–272.

Levashova, N.M., Van der Voo, R., Abrajevitch, A. & Bazhenov, M.L. (2009). Paleomagnetism of mid-Paleozoic subduction-related volcanics from the Chingiz Ridge in NE Kazakhstan: the evolving paleogeography of the MALGAMATING Eurasian composite continent. *Geological Society of America Bulletin*, 121, 555–573.

Leveridge, B.E & Shail, R.K. (2011). The marine Devonian stratigraphy of Great Britain. *Proceedings of the Geologists' Association*, 122, 540–567.

Li, J.Y. (2006). Permian geodynamic setting of Northeast China and adjacent regions: closure of the Paleo-Asian Ocean and subduction of the Paleo-Pacific Plate. *Journal of Asian Earth Sciences*, 26, 207–224.

Li, Z.X., Bogdanova, S.V., Collins, A.S. et al. (2008). Assembly, configuration, and break-up history of Rodinia: a synthesis. *Precambrian Research*, 160, 179–210.

Liu, L., Gurnis, M., Seton, M. et al. (2010). The role of oceanic plateau subduction in the Laramide orogeny. *Nature Geoscience*, 3, 353–357, doi:10.1038/NGEO829.

Liu, L. & Stegman, D.R. (2012). Origin of Columbia River flood basalt controlled by propagating rupture of the Farallon slab. *Nature*, 482, 386–390.

Lorenz, H., Männik, P., Gee, D. G. & Proskurnin, V. (2008). Geology of the Severnaya Zemlya Archipelago and new tectonic interpretation for the North Kara Terrane in the Russian high Arctic. *International Journal of Earth Sciences*, 97, 519–547.

Lowry, D.P, Poulsen, C.J., Horton, D.E. et al. (2014). Thresholds for Paleozoic ice sheet initiation. *Geology*, 42(7), 627–630.

Lyons, T.W., Reinhard, C.T. & Planavsky, N.J. (2014). The rise of oxygen in Earth's early ocean and atmosphere. *Nature*, 506, 307–315.

MacLeod, N., Rawson, P.E., Forey, P.L. et al. (1997). The Cretaceous–Tertiary biotic transition. *Journal of the Geological Society, London*, 154, 265–292.

Mac Niocaill, C., van de Pluijm, B.A. & Van der Voo, R. (1997). Ordovician paleogeography and evolution of the Iapetus Ocean. *Geology*, 25, 159–162.

Maloney, K.T., Clarke, G.L., Klepeis, K.A. & Quevedo, L. (2013). The Late Jurassic to present evolution of the Andean margin: drivers and the geological record. *Tectonics*, 32, 1049–1065.

Manankov, I.N., Shi, G.R. & Shen, S. (2006). An overview of Permian marine stratigraphy and biostratigraphy of Mongolia. *Journal of Asian Earth Sciences*, 26, 294–303.

Mander, L., Kürschner, W.M. & McElwain, J.C. (2013). Palynostratigraphy and vegetation history of the Triassic–Jurassic transition in East Greenland. *Journal of the Geological Society, London*, 170, 37–46.

Marcussen, C., Knudsen, C., Hopper, J.R. et al. (2015). Age and origin of the Lomonosov Ridge: a key continental fragment in Arctic Ocean reconstructions. *Geophysical Research Abstracts*, 17, EGU2015-10207-1.

Marshall, P.E., Widdowson, M. & Murphy, D.T. (2016). The Giant Lavas of Kalkarindji: rubbly pāhoehoe lava in an ancient

continental flood basalt province. *Palaeogeography, Palaeoclimatology, Palaeoecology*, 441, 22–37.
Martindale, R.C., Corsetti, F.A., James, N.P. & Bottjer, D.J. (2015). Paleogeographic trends in Late Triassic reef ecology from northeastern Panthalassa. *Earth-Science Reviews*, 142, 18–37.
McCall, G.J.H. (2006). The Vendian (Ediacaran) in the geological record: enigmas in geology's prelude to the Cambrian explosion. *Earth-Science Reviews*, 77, 1–229.
McCann, T. (ed.) (2008). *The Geology of Central Europe*. London: Geological Society.
McHone, J.G. (2002). Volatile emissions of Central Atlantic Magmatic Province basalts: mass assumptions and environmental consequences. In W.E. Hames, J.G. McHone, P.R. Renne & C. Ruppel (eds.), *The Central Atlantic Magmatic Province*. American Geophysical Union, Geophysical Monograph, 136, pp. 241–254.
McKenzie, P.M., Hughes, N.C., Myrow, P.M. et al. (2011). Trilobites and zircons link north China with the eastern Himalaya during the Cambrian. *Geology*, 39, 591–594.
McKerrow, W.S. & Cocks, L.R.M. (1976). Progressive faunal migration across the Iapetus Ocean. *Nature*, 263, 304–306.
McKerrow, W.S., Mac Niocaill, C., Ahlberg, P.E. et al. (2000). The Late Palaeozoic relations between Gondwana and Laurussia. In W. Franke, V. Haak, O. Oncken & D. Tanner (eds.), *Orogenic Processes: Quantification and Modelling in the Variscan Belt*. Geological Society, London, Special Publications, 179, 9–20.
McQuarrie, N. & van Hinsbergen, D.J.J. (2013). Retrodeforming the Arabia–Eurasia collision zone: age of collision versus magnitude of continental subduction. *Geology*, 41, 315–318.
Meert, J.G. (2003). A synopsis of events related to the assembly of eastern Gondwana, *Tectonophysics*, 362, 1–40.
Meert, J.G. (2012). What's in a name? The Columbia (Palaeopangea/Nuna) Supercontinent. *Gondwana Research*, 21, 987–993.
Meert, J.G. (2014). Strange attractors, spiritual interlopers and lonely wanderers: the search for pre-Pangæan supercontinents. *Geoscience Frontiers*, 5, 155–166.
Mei, S. & Henderson, C.M. (2001). Evolution of Permian conodont provincialism and its significance in global correlation and paleoclimate implication. *Palaeogeography, Palaeoclimatology, Palaeoecology*, 270, 217–260.
Meng, J. & McKenna, M.C. (1998). Faunal turnover of Palaeogene mammals from the Mongolian Plateau. *Nature*, 394, 364–367.
Metcalfe, I. (2006). Palaeozoic and Mesozoic tectonic evolution and palaeogeography of East Asian crustal fragments: the Korean Peninsula in context. *Gondwana Research*, 9, 24–46.
Metcalfe, I. (2011). Palaeozoic–Mesozoic history of SE Asia. In R. Hall, M.A. Cottamm & E.J. Wilson (eds.), *The SE Asian Gateway: History and Tectonics of the Australia–Asia Collision*. Geological Society, London, Special Publications, 355, pp. 7–35.
Michard, A., Saddiqi, O., Chalouan, A. & Lamotte, D. F. (eds.) (2008). *Continental Evolution: The Geology of Morocco*. Berlin: Springer-Verlag.
Mojzsis, S.J., Cates, N.L., Bleeker, W. et al. (2014). Component geochronology of the ca. 3960 Ma Acasta Gneiss. *Geochimica et Cosmochimica Acta*, 133, 68–96.
Montelli, R., Nolet, G., Dahlen, F. & Masters, G. (2006). A catalogue of deep mantle plumes: new results from finite-frequency tomography. *Geochemistry, Geophysics, Geosystems*, 7, Q11007, doi:10.1029/2006GC001248.
Moratti, G. & Chalouan, A. (eds.) (2006). *Tectonics of the Western Mediterranean and North Africa*. Geological Society, London, Special Publications, 262.
Moreno, T. & Gibbons, W. (eds.) (2007). *The Geology of Chile*. London: Geological Society.
Morgan, W.J. (1971). Convection plumes in the lower mantle. *Nature*, 230, 42–43.
Mortimer, N., Herzer, R.H., Gans, P.B., Parkinson, D.L. & Seward, D. (1998). Basement geology from Three Kings Ridge to West Norfolk Ridge, southwest Pacific Ocean: evidence from petrology, geochemistry and isotopic dating of dredge samples. *Marine Geology*, 148, 135–162.
Mortimore, R.N. (2011). A chalk revolution – what have we done to the chalk of England? *Proceedings of the Geologists' Association*, 122, 232–297.
Moulin, M., Aslanian, D. & Unternehr, P. (2010). A new starting point for the South and Equatorial Atlantic Ocean. *Earth-Science Reviews*, 98, 1–37.
Moullade, M. & Nairn, A.E.M. (eds.) (1983). *The Mesozoic, B. The Phanerozoic Geology of the World II*. Amsterdam: Elsevier.
Müller, R.D., Royer, J.-Y. & Lawver, L.A. (1993). Revised plate motions relative to the hotspots from combined Atlantic and Indian Ocean hotspot tracks. *Geology*, 21, 275–278.
Müller, R.D., Sdrolias, M., Gaina, C. & Roest, W.R. (2008). Age, spreading rates, and spreading asymmetry of the world's ocean crust. *Geochemistry, Geophysics, Geosystems*, 9, Q04006, doi:10.1029/2007GC001743.
Murphy, J.B, van Staal, C.R. & Keppie, J.D. (1999). Middle to Late Paleozoic Acadian orogeny in the northern Appalachians: a Laramide-style plume-related orogeny? *Geology*, 27, 653–656.
Musteikis, P. & Cocks, L.R.M. (2004). Strophomenide and orthotetide Silurian brachiopods from the Baltic region, with particular reference to Lithuanian boreholes. *Acta Palaeontologica Polonica*, 49. 455–482.
Myhre, A.M., Eldholm, O. & Sundvor, E. (1982). The margin between Senja and Spitsbergen fracture zones: implications from plate tectonics. *Tectonophysics*, 89(1–3), 33–50.
Nance, R.D., Keppie, J.D., Miller, B.V. et al. (2009). Palaeozoic palaeogeography of Mexico: constraints from detrital zircon age data. In J.B. Murphy, J.D. Keppie & A. Hynes (eds.), *Ancient Orogens and Modern Analogues*. Geological Society, London, Special Publications, 327, pp. 239–269.

Natal'in, B.A. & Şengör, A.M.C. (2005). Late Palaeozoic to Triassic evolution of the Turan and Scythian platforms: the pre-history of the Palaeo-Tethyan closure. *Tectonophysics*, 404, 175–202.

Nelson, J. & Colpron, M. (2007). Tectonics and metallogeny of the British Columbia, Yukon and Alaskan Cordillera: 1.8 Ga to the present. In W.D. Goodfellow (ed.), *Mineral Deposits of Canada*. Geological Association of Canada, Mineral Deposits Division, Special Publication, 5, pp. 755–791.

Nielsen, K.C. (2005). Ouachitas. In R.C. Selley, L.R.M. Cocks & I.R. Plimer (eds.), *Encyclopedia of Geology, Volume 4*, Amsterdam: Elsevier, pp. 61–71.

Nikishin, A.M., Ziegler, P.A., Stephenson, R.A. et al. (1996). Late Precambrian to Triassic history of the East European Craton: dynamics of sedimentary basin evolution. *Tectonophysics*, 268, 23–63.

Nokleberg, W.J., Parfenov, I.M., Monger, J.W.H. et al. (2000). *Phanerozoic Tectonic Evolution of the Circum-North Pacific*. US Geological Survey Professional Paper, 1626.

Oakey, G.N. & Damaske, D. (2006). Continuity of basement structures and dyke swarms in the Kane Basin region of central Nares Strait constrained by aeromagnetic data. *Polarforschung*, 74, 51–62.

Olierook, H.K.H., Merle, R.E., Jourdan, F. et al. (2015). Age and geochemistry of magmatism of the oceanic Wallaby Plateau and implications for the opening of the Indian Ocean. *Geology*, 43, 971–974.

O'Neill, C., Lenardic, A., Moresi, L. et al. (2007). Episodic Precambrian Subduction. *Earth and Planetary Science Letters*, 262, 552–562.

O'Neill, C., Müller, R.D. & Steinberger, B. (2005). On the uncertainties in hot spot reconstructions and the significance of moving hot spot reference frames. *Geochemistry, Geophysics, Geosystems*, 6, Q04003, doi:10.1929/2004GC000784.

Owen-Smith, T.M., Ashwal, L.D., Torsvik, T.H. et al. (2013). Seychelles alkaline suite records the culmination of Deccan Traps continental flood volcanism. *Lithos*, 182–183, 33–47.

Oxman, V.S. (2003). Tectonic evolution of the Mesozoic Verkhoyansk–Kolyma belt (NE Asia). *Tectonophysics*, 365, 45–76.

Parman, S.W. (2015). Time-lapse zirconography: imaging punctuated continental evolution. *Geochemical Perspective Letters*, 1, 43–52.

Parrish, J.T. (1982). Upwelling and petroleum source beds, with reference to the Paleozoic. *Bulletin – American Association of Petroleum Geologists*, 66, 750–774.

Pausata, F.S.R., Chafik, L., Caballero, R. & Battisti, D.S. (2015). Impacts of high-latitude volcanic eruptions on ENSO and AMOC. *Proceedings of the National Academy of Science*, 112, 13784–13788.

Pegel, T.V. (2000). Evolution of trilobite biofacies in Cambrian basins of the Siberian Platform. *Journal of Paleontology*, 74, 1000–1017.

Percival, I.G. (1991). Late Ordovician articulate brachiopods from central New South Wales. *Memoirs of the Society of Australasian Palaeontologists*, 12, 107–177.

Percival, I.G. & Glenn, R.A. (2007). Ordovician to earliest Silurian history of the Macquarie Arc, Lachlan Orogen, New South Wales. *Australian Journal of Earth Sciences*, 54, 143–165.

Peron-Pinvidic, G., Gernigon, L., Gaina, C. & Ball, P. (2012). Insights from the Jan Mayen system in the Norwegian-Greenland Sea: II. Architecture of a microcontinent. *Geophysics Journal International*, 191, 413–435.

Petterson, M.G., Babbs, T., Neal, C.R. et al. (1999). Geological–tectonic framework of Solomon Islands, SW Pacific: crustal accretion and growth within an intra-oceanic setting. *Tectonophysics*, 301, 35–60.

Pirajno, F., Mao, J., Zhang, Z. & Chai, F. (2008). The association of mafic–ultramafic intrusions and A-type magmatism in the Tian Shan and Altay orogens, NW China: implications for geodynamic evolution and potential for the discovery of new ore deposits. *Journal of Asian Earth Sciences*, 32, 165–183.

Plafker, G. & Berg, H.C. (eds.) (1994). *The Geology of Alaska. The Geology of North America, Vol. G-1*. Boulder: The Geological Society of America.

Popov, L.E., Bassett, M.G., Zhemchuzhnikov, V.G. et al. (2009). Gondwanan faunal signatures from early Palaeozoic terranes of Kazakhstan and central Asia. In M.G. Bassett (ed.), *Early Palaeozoic Peri-Gondwana Terranes*. Geological Society, London, Special Publications, 325, pp. 23–64.

Popov, L.E. & Cocks, L.R.M. (2017). Late Ordovician brachiopods from Kazakhstan, and their palaeogeography. *Acta Geologica Polonica* (in press).

Potter, A.W., Boucot, A.J., Bergström, S.M. et al. (1990). Early Paleozoic stratigraphic, paleogeographic, and biogeographic relations of the eastern Klamath belt, northern California. In D. S. Harwood & M.M. Miller (eds.), *Paleozoic and Early Mesozoic Paleogeographic Relations; Sierra Nevada, Klamath Mountains, and Related Terranes*. Geological Society of America Special Paper, 255, pp. 57–74.

Pownall, J.M., Hall, R. & Watkinson, I.M. (2013). Extreme extension across Seram and Ambon, eastern Indonesia: evidence for Banda slab rollback. *Solid Earth*, 4, 277–314.

Puchkov, V.N. (2009). The evolution of the Uralian orogen. In J.B. Murphy, J.D. Keppie & A. Hynes (eds.), *Ancient Orogens and Modern Analogues*. Geological Society, London, Special Publications, 327, pp. 161–195.

Qiao, L. & Shen, S. (2014). Global paleobiogeography of brachiopods during the Mississippian – response to the lobal tectonic configuration, ocean circulation, and climate changes. *Gondwana Research*, 26, 1173–1185.

Quintaville, M., Tongiorgi, M. & Gaetani, M. (2000). Lower to Middle Ordovician acritarchs and chitinozoans from northern Karakorum Mountains, Pakistan. *Rivista Italiana di Paleontologia e Stratigrafia*, 106, 3–18.

Radley, J.D. & Allen, P. (2012). The non-marine Lower Cretaceous Wealden strata of southern England. *Proceedings of the Geologists' Association*, 123, 235–385.

Ramos, V.A. & Folguera, A. (2009). Andean flat-slab subduction through time. In J.B. Murphy, J.D. Keppie & A. Hynes (eds.), *Ancient Orogens and Modern Analogues*, Geological Society, London, Special Publications, 327, pp. 31–54.

Rasmussen, C.M.Ø., Ullmann, C.V., Jakobsen, K.G. et al. (2016). Onset of main Phanerozoic marine radiation sparked by emerging Mid Ordovician icehouse. *Nature, Scientific Reports*, doi:10.1038/srep18884.

Raymo, M.E., Ruddiman, W.F. & Froelich, P.N. (1988). Influence of late Cenozoic mountain building on ocean geochemical cycles. *Geology*, 16, 649–653.

Rees, P.M. (2002). Land-plant diversity and the end-Permian mass extinction. *Geology*, 30, 827–830.

Rees, P.M., Noto, C.R., Parrish, J.M. & Parrish, J.T. (2004). Late Jurassic climates, vegetation and dinosaur distributions. *Journal of Geology*, 112, 643–653.

Reidel, S.P., Camp, V.E., Tolan, T.L. & Martin, B.S. (2013). The Columbia River flood basalt province: stratigraphy, areal extent, volume, and physical volcanology. In S.P. Reidel et al. (eds.), *The Columbia River Flood Basalt Province*, Geological Society of America Special Paper, 497, pp. 1–43.

Retallack, G.J. (2015). Silurian vegetation structure and density inferred from fossil soils and plants in Pennsylvania, USA. *Journal of the Geological Society, London*, 172, 693–709.

Ricci, J., Quidelleur, X., Pavlov, V. et al. (2013). New $^{40}Ar/^{39}Ar$ and K–Ar ages of the Viluy traps (Eastern Siberia): further evidence for a relationship with the Frasnian–Famennian mass extinction. *Palaeogeography, Palaeoclimatology, Palaeoecology*, 386, 531–540.

Ridd, M.F., Barber, A.J. & Crowe, M.J. (eds.) (2011). *The Geology of Thailand*. London: Geological Society.

Riefstahl, F., Estrada, S., Geissler, W.H. et al. (2013). Provenance and characteristics of rocks from the Yermak Plateau, Arctic Ocean: petrographic, geochemical and geochronological constraints. *Marine Geology*, 343, 125–145.

Ritsema, J. & Allen, R.M. (2003). The elusive mantle plume. *Earth and Planetary Science Letters*, 207, 1–12.

Roberts, N.M. & Spencer, C.J. (2014). The zircon archive of continent formation through time. In N.M.W. Roberts et al. (eds.), *Continent Formation through Time*. Geological Society, London, Special Publications, 389, pp. 197–225.

Rocha-Campos, A.C., Santos, P.R.D. & Canuto, J.R. (2008). Late Paleozoic glacial deposits of Brazil: Paraná Basin. In C.R. Fielding, T.D. Frank & J.L. Isbell (eds.), *Resolving the Late Paleozoic Ice Age in Time and Space*. Geological Society of America Special Paper, 441, pp. 97–114.

Rong, J., Boucot, A.J., Su, Y. & Strusz, D.L. (1995). Biogeographical analysis of Late Silurian brachiopod faunas, chiefly from Asia and Australia. *Lethaia*, 28, 39–60.

Rong, J., Chen, X., Su, Y. et al. (2003). Silurian paleogeography of China. *New York State Museum Bulletin*, 493, 243–298.

Rong, J. & Harper, D.A.T. (1988). A global synthesis of the latest Ordovician Hirnantian brachiopod fauna. *Transactions of the Royal Society of Edinburgh, Earth Sciences*, 79, 383–401.

Ross, C.A. & Ross, J.R.P. (1983). Late Paleozoic accreted terranes of western North America. In C.H. Stevens (ed.), *Pre-Jurassic Rocks in Western American Suspect Terranes*, Los Angeles: SEPM Pacific Section, pp. 7–22.

Royer, D.L. (2006). CO_2-forced climate thresholds during the Phanerozoic. *Geochimica et Cosmochimica Acta*, 70, 5665–5675.

Royer, D.L., Berner, R.A. & Park, J. (2007). Climate sensitivity constrained by CO_2 concentrations over the past 420 million years. *Nature*, 446, 530–532.

Rozman, K.S. (1978). Brachiopods of the Obikolon Beds. *USSR Academy of Sciences Siberian Branch Institute of Geology and Geophysics Transactions*, 397, 75–101 (in Russian).

Ruban, D.A., al-Husseini, M.L. & Iwasaki, Y. (2007). Review of Middle East plate tectonics. *GeoArabia*, 12, 35–56.

Ruddiman, W.F. (2014). *Earth's Climate Past and Future*, 3rd edn. New York: W.H. Freeman.

Rushton, A.W.A., Cocks, L.R.M. & Fortey, R.A. (2002). Upper Cambrian trilobites and brachiopods from Severnaya Zemlya, Arctic Russia, and their implications for correlation and biogeography. *Geological Magazine*, 139, 281–290.

Sahney, S., Benton, M.J. & Falcon-Lang, H.J. (2010). Rainforest collapse triggered Carboniferous tetrapod diversification in Euramerica. *Geology*, 38, 1079–1082.

Savage, R.J.G. & Long, M.R. (1986). *Mammal Evolution: An Illustrated Guide*. London: British Museum (Natural History).

Scarrow, J.H., Ayala, C. & Kimball, G.S. (2002). Insights into orogenesis: getting to the root of a continental-ocean–continent collision. *Journal of the Geological Society, London*, 159, 659–671.

Schallreuter, R. & Siveter, D.J. (1985). Ostracodes across the Iapetus Ocean. *Palaeontology*, 28, 577–598.

Schandelmeier, H. & Reynolds, P.O. (1997). *Palaeogeographic–Palaeotectonic Atlas of North-Eastern \Africa, Arabia, and Adjacent Areas*. Rotterdam: Balkema.

Schmid, S.M., Bernoulli, D., Fügenschuh, B. et al. (2008). The Alpine–Carpathian–Dinaridic orogenic system: correlation and evolution of tectonic units. *Swiss Journal of Geosciences*, 101, 139–183.

Scotese, C.R. & Barrett, S.F. (1990). Gondwana's movement over the South Pole during the Palaeozoic: evidence from lithological indicators of climate. In W.S. McKerrow & C.R. Scotese (eds.), *Palaeozoic Palaeogeography and Biogeography*. Geological Society, London, Memoir, 12, pp. 75–85.

Searle, M.P. (2013). *Colliding Continents: A Geological Exploration of the Himalaya, Karakoram, and Tibet*. Oxford: Oxford University Press.

Searle, M.P., Cherry, A.G., Ali, M.Y. & Cooper, D.J.W. (2014). Tectonics of the Musandam Peninsula and northern Oman

Mountains: from ophiolite obduction to continental collision. *GeoArabia*, 19, 135–174.

Sedgwick, A. & Murchison, R.I. (1835). On the Cambrian and Silurian systems, exhibiting the order in which the older sedimentary strata succeed each other in England and Wales. *The London and Edinburgh Philosophical Magazine and Journal of Science*, 7, 483–5.

Sedgwick, A. & Murchison, R.I. (1837). A classification of the old slate rocks of the north of Devonshire. *Report of the British Association for the Advancement of Science (for 1836)*, 95–96.

Sedlock, R.L. (2003). Geology and tectonics of the Baja California peninsula and adjacent areas. In S.E. Johnson, S.R. Paterson, J.M. Fletcher et al. (eds.), *Tectonic Evolution of Northwestern Mexico and the Southwestern USA*. Geological Society of America Special Paper, 374, pp. 1–42.

Selley, R.C., Cocks, L.R.M. & Plimer, I.R. (eds.) (2005). *Encyclopedia of Geology*. Amsterdam: Elsevier.

Şengör, A.M.C. & Atayman, S. (2009). *The Permian Extinction and the Tethys*. Geological Society of America Special Paper, 448.

Şengör, A.M.C. & Natal'in, B.A. (1996). Paleotectonics of Asia: fragments of a synthesis. In A. Yin & M. Harrison (eds.), *The Tectonic Evolution of Asia*. Cambridge: Cambridge University Press, pp. 486–646.

Sennikov, N.V. (2003). Ordovician events in Altai–Sayan–Kuznesty and Tuva basins and their influence on the sedimentary facies and marine biota (Siberia, Russia). *INSUGEO Serie Correlación Geológica*, 17, 461–465.

Seton, M., Gaina, C., Müller, R.D. & Heine, C. (2009). Mid-Cretaceous seafloor spreading pulse: fact or fiction? *Geology*, 37, 687–690.

Seton, M., Flament, N., Whittaker, J. et al. (2015). Ridge subduction sparked reorganization of the Pacific plate–mantle system 60–50 million years ago. *Geophysical Research Letters*, 42, 1732–1740.

Seton, M., Müller, R.D., Zahirovic, S. et al. (2012). Global continental and ocean basin reconstructions since 200 Ma. *Earth-Science Reviews*, 113, 212–270.

Shao, L., Zhang, P., Gayer, R.A. et al. (2003). Coal in a carbonate sequence stratigraphic framework: the Upper Permian Heshan Formation in central Guangxi, southern China. *Journal of the Geological Society, London*, 160, 285–298.

Shaw, J., Johnston, S.T., Gutiérrez-Alonso, G., et al. (2012). Oroclines of the Variscan orogeny of Iberia: paleocurrent analysis and paleogeographic implications. *Earth and Planetary Science Letters*, 329–330, 60–70.

Shelley, D. & Bossière, G. (2000). A new model for the Hercynian orogeny of Gondwanan France and Iberia. *Journal of Structural Geology*, 22, 757–776.

Shellnutt, J.G., Bhat, G.M., Brookfield, M.E., & Jahn, B.M. (2011). No link between the Panjal Traps (Kashmir) and the Late Permian mass extinctions. *Geophysical Research Letters*, 38, L19308.

Shen, S., Xie, J., Zhang, H. & Shi, G.R. (2009). Roadian–Wordian (Guadalupian, Middle Permian) global palaeobiogeography. *Global and Planetary Change*, 65, 166–181.

Shephard, G., Flament, N., Williams, S. et al. (2014). Circum-Arctic mantle structure and long-wavelength topography since the Jurassic. *Journal of Geophysical Research*, 119, 7889–7908, doi:10.1002/2014JB011078.

Shephard, G., Müller, R.D. & Seton, M. (2013). The tectonic evolution of the Arctic since Pangea breakup: integrating constraints from surface geology and geophysics with mantle structure. *Earth-Science Reviews*, 124, 148–183.

Shergold, J.H. (1988). Review of trilobite biofacies distributions at the Cambrian–Ordovician boundary. *Geological Magazine*, 125, 363–380.

Shergold, J.H. (1991). Late Cambrian (Payntonian) and Early Ordovician (Late Warendian) trilobite faunas of the Amadeus Basin Central Australia. *Bulletin of the Bureau of Mineral Resources, Geology and Geophysics*, 237, 15–75.

Shi, G.R. (2006). The marine Permian of east and northeast Asia: an overview of biostratigraphy, palaeobiogeography and palaeogeographical implications. *Journal of Asian Earth Sciences*, 26, 175–206.

Shirey, S.B. & Richardson, S.H. (2011). Start of the Wilson cycle at 3 Ga shown by diamonds from subcontinental mantle. *Science*, 333, 434–436.

Sigloch, K. & Mihalynuk, M. G. (2013). Intra-oceanic subduction shaped the assembly of Cordilleran North America. *Nature*, 496, 50–56, doi:10.1038/nature12019.

Silva, D.R.A., Mizusaki, A.M.P., Milani, E. & Pimentel, M. (2012). Determination of depositional age of Paleozoic and pre-rift supersequences of the Recôncavo Basin in northeastern Brazil by applying Rb–Sr radiometric dating technique to sedimentary rocks. *Journal of South American Earth Sciences*, 37, 13–24.

Sleep, N.H. (1997). Lateral flow and ponding of starting plume material. *Journal of Geophysical Research*, 102, 10001–10012.

Smelror, M., Petrov, O.V., Larssen, G.B. & Werner, S. (eds.) (2009). *Geological History of the Barents Sea*. Trondheim: Geological Survey of Norway.

Smith, M.P. & Rasmussen, J.A. (2008). Cambro-Silurian development of the Laurentian margin of the Iapetus Ocean in Greenland and related areas. In Higgins, A.K., Gilotti, J.A. & Smith, M.P. (eds.), *The Greenland Caledonides: Evolution of the Northeast Margin of Laurentia*. Geological Society of America Memoir, 202, pp. 137–167.

Sone, M. & Metcalfe, I. (2008). Parallel Tethyan sutures in mainland Southeast Asia: new insights for Palaeo-Tethys closure and implications for the Indosinian orogeny. *Comptes Rendus Geoscience*, 340, 166–179.

Song, S., Su, L., Niu, Y. et al. (2009b). Tectonic evolution of Early Paleozoic HP metamorphic rocks in the North Qilian Mountains, NW China: new perspectives. *Journal of Asian Earth Sciences*, 35, 334–353.

Spencer, A.M., Embry, A.F., Gautier, D.L. et al. (eds.) (2011). *Arctic Petroleum Geology*. Geological Society, London, Memoirs, 35.

Speranza, F., Minelli, L., Pignatelli, A. & Chiappini, M. (2012). The Ionian Sea: the oldest *in situ* ocean fragment of the world? *Journal of Geophysical Research*, 117, B12101, doi:10.1029–2012JB009475.

Stampfli, G.M. & Borel, G.D. (2004). The TRANSMED transects in space and time: constraints on the paleotectonic evolution of the Mediterranean domain. In W. Cavazza, F. Roure, W. Spakman, G.G. Stampfli & P.A. Ziegler (eds.), *The TRANSMED Atlas: The Mediterranean Region from Crust to Mantle*. Berlin: Springer, pp. 53–90.

Steinberger, B. (2000). Plumes in a convecting mantle: models and observations for individual hotspots. *Journal of Geophysical Research*, 105, 11,127–11,152.

Steinberger, B. & Gaina, C. (2007). Plate-tectonic reconstructions predict part of the Hawaiian hotspot track to be preserved in the Bering Sea. *Geology*, 35, 407–410.

Steinberger, B., Spakman, W., Japsen, P. & Torsvik, T.H. (2015). The key role of global solid-Earth processes in preconditioning Greenland's glaciation since the Pliocene. *Terra Nova*, 27, 1–8.

Steinberger, B., Sutherland, R. & O'Connell, R.J. (2004). Prediction of Emperor–Hawaii seamount locations from a revised model of global plate motion and mantle flow. *Nature*, 430, 167–173.

Steinberger, B. & Torsvik, T.H. (2008). Absolute plate motions and true polar wander. *Nature*, 452, 620–623.

Steinberger, B. & Torsvik, T.H. (2010). Toward an explanation for the present and past locations of the poles. *Geochemistry, Geophysics, Geosystems*, 11, Q06W06, doi:10.1929/2009GC002889.

Stemmerik, B. (2000). Late Palaeozoic evolution of the North Atlantic margin of Pangea. *Palaeogeography, Palaeoclimatology, Palaeoecology*, 161, 95–126.

Stern, R.J. (2008). Modern-style plate tectonics began in Neoproterozoic time: an alternative interpretation of Earth's tectonic history. In K. Condie & V. Pease (eds.), *When Did Plate Tectonics Begin?* Geological Society of America Special Paper, 440, pp. 265–280.

Stevens, C.H. & Stone, P. (2007). *The Pennsylvanian–Early Permian Bird Spring Carbonate Shelf, Southeastern California: Fusulinid Biostratigraphy, Paleogeographic Evolution, and Tectonic Implications*. Geological Society of America Special Paper, 429.

Stevens, L.G., Hilton, J., Bond, D.P.G. et al. (2011). Radiation and extinction patterns in Permian floras from North China as indicators for environmental and climate change. *Journal of the Geological Society, London*, 168, 607–619.

Stringer, C.B. (2002). Modern human origin: progress and prospects. *Philosophical Transactions of the Royal Society, London*, B357, 563–579.

Svensen, H., Planke, S. & Corfu, F. (2010). Zircon dating ties NE Atlantic sill emplacement to initial Eocene global warming. *Journal of the Geological Society, London*, 167, 433–436.

Svensen, H., Planke, S., Malthe-Sorenssen, A. et al. (2004). Release of methane from a volcanic basin as a mechanism for initial Eocene global warming. *Nature*, 429, 542–545.

Svensen, H., Hammer, Ø. & Corfu, F. (2015). Astronomically forced cyclicity in the Upper Ordovician and U–Pb ages of interlayered tephra, Oslo Region, Norway. *Palaeogeography, Palaeoclimatology, Palaeoecology*, 418, 150–159.

Szederkényi, T., Haas, N. & Hámor, G. (2012). Geology and history of evolution of the Tisza Unit. In J. Haas, (ed.), *Geology of Hungary*. Berlin: Springer-Verlag.

Tankard, A.J., Suárez-Soruco, R. & Welsink, H.J. (eds.) (1995). *Petroleum Basins of South America*. American Association of Petroleum Geologists Memoir, 6.

Tarduno, J., Bunge, H.-P., Sleep, N. & Hansen, U. (2009). The bent Hawaiian–Emperor hotspot track: inheriting the mantle wind. *Science*, 324, 50–53.

Tarduno, J.A., Duncan, R.A., Scholl, D.W. et al. (2003). The Emperor seamounts: southward motion of the Hawaiian hotspot plume in Earth's mantle. *Science*, 301, 1064–1069.

Tarduno, J.A., Brinkman, D.B., Renne, P.R. et al. (1998). Late Cretaceous Arctic volcanism: tectonic and climatic connections. In *American Geophysical Union Spring Meeting Abstracts*, Washington, DC: American Geophysical Union.

Tauxe, L. & Kent, D.V. (2004). A simplified statistical model for the geomagnetic field and the detection of shallow bias in paleomagnetic inclinations: was the ancient magnetic field dipolar? In J.E.T. Channell et al. (eds.), *Timescales of the Paleomagnetic Field, 145*, Washington, DC: American Geophysical Union, pp. 101–116.

Taylor, T.N., Taylor, E.L. & Krings, M. (2009). *Paleobotany: The Biology and Evolution of Fossil Plants*. Amsterdam: Academic Press.

Tejada-Lara, J.V., Salas-Gismondi, R., Pujos, F. et al. (2015). Life in Proto-Amazonia: Middle Miocene mammals from the Fitzcarrald Arch (Peruvian Amazonia). *Palaeontology*, 58, 341–378.

Tessensohn, F. & Henjes-Kunst, F. (2005). Northern Victoria Land terranes, Antarctica: far-travelled or local products? In A.P.M. Vaughan, P.T. Leat & R.J. Pankhurst (eds.), *Terrane Processes at the Margins of Gondwana*. Geological Society, London, Special Publications, 246, 275–291.

Torsvik, T.H. (2003). The Rodinia jigsaw puzzle. *Science*, 300, 1379–1381.

Torsvik, T.H., Amundsen, H., Hartz, E.H., et al. (2013). A Precambrian microcontinent in the Indian Ocean. *Nature Geoscience*, 6, 223–227.

Torsvik, T.H., Amundsen, H.E.F., Trønnes, R.G. et al. (2015). Continental crust beneath southeast Iceland. *Proceedings of the National Academy of Sciences*, 112, E1818–E1827, doi:10.1073/pnas.1423099112.

Torsvik, T.H. & Andersen, T.B. (2002). The Taimyr fold belt, Arctic Siberia: timing of pre-fold remagnetization and regional tectonics. *Tectonophysics*, 352, 335–348.

Torsvik, T.H., Burke, K., Steinberger, B. et al. (2010). Diamonds sourced by plumes from the core–mantle boundary. *Nature*, 466, 352–355.

Torsvik, T.H., Carter, L.M., Ashwal, L.D. et al. (2001). Rodinia refined or obscured: palaeomagnetism of the Malani Igneous Suite (NW India). *Precambrian Research*, 108, 319–333.

Torsvik, T.H. & Cocks, L.R.M. (2004). Earth geography from 400 to 250 Ma: a palaeomagnetic, faunal and facies review. *Journal of the Geological Society, London*, 161, 555–572.

Torsvik, T.H. & Cocks, L.R.M. (2005). Norway in space and time: a centennial cavalcade. *Norwegian Journal of Geology*, 85, 73–86.

Torsvik, T.H. & Cocks, L.R.M. (2009). The Lower Palaeozoic palaeogeographical evolution of the northeastern and eastern peri-Gondwanan margin from Turkey to New Zealand. In M.G. Bassett (ed.), *Early Palaeozoic Peri-Gondwana Terranes*. Geological Society, London, Special Publications, 325, pp. 3–21.

Torsvik, T.H. & Cocks, L.R.M. (2011). The Palaeozoic geography of central Gondwana. In D.J.J. van Hinsbergen, S.J.H. Buiter, T.H. Torsvik et al. (eds.), *The Formation and Evolution of Africa: A Synopsis of 3.8 Ga of Earth History*. Geological Society, London, Special Publications, 357, pp. 137–166.

Torsvik, T.H. & Cocks, L.R.M. (2012). From Wegener until now: the development of our understanding of Earth's Phanerozoic evolution. *Geologica Belgica*, 15, 181–192.

Torsvik, T.H. & Cocks, L.R.M. (2013). Gondwana from top to base in space and time. *Gondwana Research*, 24, 999–1030.

Torsvik, T.H., Gaina, C. & Redfield, T.F. (2008a). Antarctica and global paleogeography: from Rodinia, through Gondwanaland and Pangea, to the birth of the Southern Ocean and the opening of gateways. In A.K. Cooper, P. Barrett, H. Stagg et al. (eds.), *Antarctica, a Keystone in a Changing World*, Washington DC: National Academies Press, pp. 125–129.

Torsvik, T.H., Müller, R.D., Van der Voo, R. et al. (2008b). Global plate motion frames: towards a unified model. *Reviews of Geophysics*, 46, RG3004, doi:10.1029/2007RG000227.

Torsvik, T.H. & Rehnström, E.F. (2003). The Tornquist Sea and Baltica–Avalonia docking. *Tectonophysics*, 362, 67–82.

Torsvik, T.H., Rousse, S., Labails, C. & Smethurst, M.A. (2009). A new scheme for the opening of the South Atlantic Ocean and the dissection of an Aptian salt basin. *Geophysical Journal International*, 177, 1315–1333.

Torsvik, T.H., Smethurst, M.A., Burke, K. & Steinberger, B. (2006). Large Igneous Provinces generated from the margins of the Large Low Velocity Provinces in the deep mantle. *Geophysical Journal International*, 167, 1447–1460.

Torsvik, T.H., Smethurst, M.A., Meert, J.G. et al. (1996). Continental break-up and collisions in the Neoproterozoic and Palaeozoic: a tale of Baltica and Laurentia. *Earth-Science Reviews*, 40, 229–258.

Torsvik, T.H., Steinberger, B., Cocks, L.R.M. & Burke, K. (2008c). Longitude: linking Earth's ancient surface to its deep interior. *Earth and Planetary Science Letters*, 276, 273–282.

Torsvik, T.H., Van der Voo, R., Preeden, V. et al. (2012). Phanerozoic polar wander, palaeogeography, and dynamics. *Earth-Science Reviews*, 114, 325–368.

Torsvik, T.H., Van der Voo, R., Doubrovine, P.V. et al. (2014). Deep mantle structure as a reference frame for movements in and on the Earth. *Proceedings of the National Academy of Sciences*, 111, 24, 8735–8740.

Trettin, H.P. (1998). Pre-Carboniferous geology of the northern part of the Arctic Islands. *Geological Survey of Canada Bulletin*, 425, 1–401.

Tucker, R.D., Ashwal, L.D. & Torsvik, T.H. (2001). U–Pb geochronology of Seychelles granitoid: Neoproterozoic construction of a Rodinia continental fragment. *Earth and Planetary Science Letters*, 187, 27–38.

Tucker, R.D. & McKerrow, W.S. (1995). Early Palaeozoic chronology: a review in light of new U–Pb zircon ages from Newfoundland and Britain. *Canadian Journal of Earth Sciences*, 32, 368379.

Tull, J.F., Barineau, C.L., Mueller, P.A. & Wooden, J.L. (2007). Volcanic arc emplacement onto the southernmost Appalachian Laurentian shelf: characteristics and constraints. *Geological Society of America Bulletin*, 119, 261–274.

Tomurtogoo, O., Windley, B.F., Kröner, A., Badarch, G., Liu, D.Y. (2005). Zircon age and occurrence of the Adaatsag ophiolite and Muron shear zone, central Mongolia: constraints on the evolution of the Mongol–Okhotsk ocean, suture and orogen. *Journal of the Geological Society, London*, 162, 125–134.

Ustaszewski, K., Schmid, S.M., Fügenschuh, B. et al. (2008). A map-view restoration of the Alpine–Carpathian–Dinaridic system for the Early Miocene. *Swiss Journal of Geosciences*, 101, 273–294.

van der Meer, D.G., Spakman, W., van Hinsbergen, D.J.J. et al. (2010). Towards absolute plate motions constrained by lower mantle slab remnants. *Nature Geoscience*, 3, 36–40.

van der Meer, D.G., Torsvik, T.H., Spakman, W. et al. (2012). Intra-Panthalassa Ocean subduction zones revealed by fossil arcs and mantle structure. *Nature Geoscience*, 5, 215–219, doi:10.1038/NGEO1401.

van der Meer, D.G., Zeebe, R., van Hinsbergen, D.J.J. et al. (2014). Long-term trends in atmospheric CO_2 levels over the past 250 million years driven by plate tectonic volcanic degassing. *Proceedings of the National Academy of Sciences*, 111, 4380–4385, doi:10.1073/pnas.1315657111.

Van der Voo, R. (1993). *Paleomagmatism of the Atlantic, Tethys and Iapetus Oceans*. Cambridge: Cambridge University Press.

Van der Voo, R., van Hinsbergen, D.J.J., Domeier, M. et al. (2015). Latest Jurassic–earliest Cretaceous oroclinal closure of the

Mongol–Okhotsk Ocean and implications for Mesozoic Central Asian plate reconstructions. In T.H. Anderson, A.N. Didenko, C.L. Johnson, A.I. Khanchuk & J.H. MacDonald Jr. (eds.), *Late Jurassic Margin of Laurasia: A Record of Faulting Accommodating Plate Rotation*, Geological Society of America Special Paper, 513, doi:10.1130/2015.2513(19).

van Hinsbergen, D.J.J., Buiter, S.J.H., Torsvik, T.H., et al. (eds.) (2011). *The Formation and Evolution of Africa: A Synopsis of 3.8 Ga of Earth History*. Geological Society, London, Special Publications, 357.

van Hinsbergen, D.J.J., Lippert, P.C., Dupont-Nivet, G. et al. (2012). Greater India Basin hypothesis and a two-stage Cenozoic collision between India and Asia. *Proceedings of the National Academy of Sciences*, 109, 7659–7664, doi:10.1073/pnas.1117262109.

Van Roy, P., Daley, A.C., Briggs, D.E.G. (2015). Anomalocaridid trunk limb homology revealed by a giant filter-feeder with paired flaps. *Nature*, 522, 77–80.

van Staal, C.R., Whalen, J.B., McNicol, V.J. et al. (2007). The Notre Dame Arc and the Taconic Orogeny in Newfoundland. In R.D. Hatcher et al. (eds.), *4-D Framework of Continental Crust*. Geological Society of America Memoir, 200, pp. 511–552.

van Staal, C.R., Whalen, J.B., Vaquero, P.V. et al. (2009). Pre-Carboniferous episodic accretion-related orogenesis along the Laurentian margin of the northern Appalachians. In J.B. Murphy, J.D. Keppie & A. Hynes (eds.), *Ancient Orogens and Modern Analogues*. Geological Society, London, Special Publications, 327, pp. 271–316.

Van Wagoner, N.A., Leybourne, N.I., Dadd, K.A. et al. (2002). Late Silurian bimodal volcanism of southwestern New Brunswick, Canada: products of continental extension. *Geological Society of America Bulletin*, 114, 400–418.

Vaughan, A.P.M., Leat, P.J. & Pankhurst, R.J. (eds.) (2005). *Terrane Processes at the Margins of Gondwana*. Geological Society, London, Special Publications, 246.

Veevers, J.J. (2004). Gondwanaland from 650–500 Ma assembly through 320 Ma merger in Pangea to 185–100 Ma breakup: supercontinental tectonics via stratigraphy and radiometric dating. *Earth-Science Reviews*, 68, 1–132.

Veizer, J., Godderis, Y. & François, L.M. (2000). Evidence for decoupling of atmospheric CO_2 and global climate during the Phanerozoic eon. *Nature*, 408, 698–701.

Villas, E., Vizcaïno, D., Álvaro, J.J., Destombes, J. & Vennin, E. (2006). Biostratigraphic control of the latest-Ordovician glaciogenic unconformity in Alnif (Eastern Anti-Atlas, Morocco), based on brachiopods. *Geobios*, 39, 727–737.

Vine, F.J. & Matthews, D.H. (1963). Magnetic anomalies over oceanic ridges. *Nature*, 199, 947–949.

Vissers, R.L.M. &. Meijer, P.Th. (2012). Iberian Plate kinematics and Alpine collision in the Pyrenees. *Earth-Science Reviews*, 114, 61–83.

Walderhaug, H.J., Eide, E.A., Scott, R.A., Inger, S. & Golionko, E.G. (2005). Palaeomagnetism and $^{40}Ar/^{39}Ar$ geochronology from the South Taimyr igneous complex, Arctic Russia: a Middle–Late Triassic magmatic pulse after Siberian flood-basalt volcanism. *Geophysical Journal*, 163, 501–517.

Waldron, J.W.F., Schofield, D.I, White, C.E. & Barr, S.M. (2013). Cambrian successions of the Meguma terrane, Nova Scotia, and Harlech Dome, North Wales: dispersed fragments of a peri-Gondwanan basin? *Journal of the Geological Society, London*, 168, 83–98.

Waldron, J.W.F. & van Staal, C.R. (2001). Taconian orogeny and the accretion of the Dashwoods block: a peri-Laurentian microcontinent in the Iapetus Ocean. *Geology*, 29, 811–814.

Wang, B., Chen, Y., Zhan, S. et al. (2007). Primary Carboniferous and Permian paleomagnetic results from the Yili Block (NW China) and their implications on the geodynamic evolution of Chinese Tianshan Belt. *Earth and Planetary Science Letters*, 263, 288–308.

Watkins, R. (1994). Evolution of Silurian pentamerid communities in Wisconsin. *Palaios*, 9, 488–499.

Webby, B.D., Paris, F., Droser, M.L. & Percival, I.G. (eds.) (2004). *The Great Ordovician Biodiversification Event*. New York: Columbia University Press.

Wegener, A. (1912). Die Entstehung der Kontinente. *Dr. A. Petermanns Mitteilungen aus Justus Perthes geographischer Anstalt*, 58, 185–195, 253–256, 305–309.

Wegener, A. (1915). *Die Entstehung der Kontinente und Ozeane*. Brunswick: Vieweg.

Wellman, C.H. & Strother, P.K. (2015). The terrestrial biota prior to the origin of land plants (Embryophytes): a review of the evidence. *Palaeontology*, 58, 601–627.

Wignall, P.B. (2007). The end-Permian mass extinction: how bad did it get? *Geobiology*, 5, 303–309.

Wignall, P.B. & Bond, D.P.G. (2008). The end-Triassic and Early Jurassic mass extinction records of the British Isles. *Proceedings of the Geologists' Association*, 119, 73–84.

Wilde, S.A., Valley, J.W., Peck, W.H., Graham, C.M. (2001). Evidence from detrital zircons for the existence of continental crust and oceans on the Earth 4.4 Gyr ago. *Nature*, 409, 175–178.

Willem, C., Windley, B.F. & Stampfli, G.M. (2012). The Altaids of Central Asia: a preliminary innovative review. *Earth-Science Reviews*, 113, 303–341.

Wilson, J.T. (1963). A possible origin of the Hawaiian islands. *Canadian Journal of Physics*, 41, 863–870.

Windley, B.F., Alexeiev, D., Xiao, W. et al. (2007). Tectonic models for accretion of the Central Asian Orogenic Belt. *Journal of the Geological Society, London*, 164, 31–47.

Wright, A.J., Young, G.C., Talent, J.A. & Laurie, J.R., (2000). Paleobiogeography of Australasian faunas and floras. *Memoirs of the Association of Australasian Palaeontologists*, 23, 1–515.

Wright, J.E. & Wyld, S.J. (2006). Gondwanan, Iapetan, Cordilleran interactions: a geodynamic model for the Paleozoic tectonic evolution of the North American Cordillera. In J. Haggart,

R.J. Enkin & J.W.H. Monger (eds.), *Paleogeography of the North American Cordillera*. Geological Association of Canada Special Paper, 46, pp. 377–408.

Wu, F.Y., Sun, D.Y., Ge, W.C. et al. (2011). Geochronology of the Phanerozoic granitoids in northeastern China. *Journal of Asian Earth Sciences*, 41, 1–30.

Xiao, W., Han, C., Uan, C., et al. (2008). Middle Cambrian to Permian subduction-related accretionary orogenesis of Northern Xinjiang, NW China: implications for the tectonic evolution of central Asia. *Journal of Asian Earth Sciences*, 32, 102–117.

Xiao, W., Windley, B.F., Hao, J. & Li, J. (2002). Arc-ophiolite obduction in the Western Kunlun Range (China): implications for the Palaeozoic crustal evolution of central Asia. *Journal of the Geological Society, London*, 159, 517–528.

Xiao, W., Windley, B.F., Yong, Y. et al. (2009a). Early Palaeozoic to Devonian multiple-accretionary model for the Qilian-Shan, NW China. *Journal of Asian Earth Sciences*, 35, 323–333.

Xiao, W., Windley, B.F., Yuan, C., et al. (2009b). Paleozoic multiple subduction–accretion processes of the southern Altaids. *American Journal of Science*, 309, 221–270.

Yan, Z., Xiao, W., Windley, B.F., Wang, Z.Q. & Li, J.L. (2010). Silurian clastic sediments in the North Qilian Shan, NW China: chemical and isotopic constraints on their forearc provenance with implications for the Paleozoic evolution of the Tibetan Plateau. *Sedimentary Geology*, 231, 98–114.

Yanev, S., Göncüoğlu, M.C., Gedik, I. et al. (2006). Stratigraphy, correlations and palaeogeography of Palaeozoic terranes of Bulgaria and NW Turkey: a review of recent data. In A.H.F. Robertson & D. Mountrakis (eds.), *Tectonic Development of the Eastern Mediterranean Region*. Geological Society, London, Special Publications, 260, pp. 421–430.

Yang, J., Cawood, P.A., Du, Y. et al. (2014). A sedimentary archive of tectonic switching from Emeishan Plume to Indosinian orogenic sources in SW China. *Journal of the Geological Society, London*, 171, 269–280.

Yolkin, E.A., Sennikov, N.V., Bakharev, N.K. et al. (2003). Silurian paleogeography along the southwest margin of the Siberian continent: Altai–Sayan folded area. *New York State Museum Bulletin*, 493, 299–322.

Young, G.C. (1990). Devonian vertebrate distribution patterns and cladistics analysis of palaeogeographic hypotheses. In W.S. McKerrow & C.R. Scotese (eds.), *Palaeozoic Palaeogeography and Biogeography*. Geological Society, London, Memoir, 12, pp. 243–255.

Young, G.C. & Janvier, P. (1999). Early–middle Palaeozoic vertebrate faunas in relation to Gondwana dispersion and Asian accretion. In I. Metcalfe (ed.), *Gondwana Dispersion and Accretion*. Rotterdam: Balkema, pp. 115–140.

Yue, Y., Liao, J.G. & Graham, S.A. (2001). Tectonic correlation of Beishan and Inner Mongolia orogens and its implications for the palinspastic reconstruction of North China. In M.S. Hendrix & G.A. Davis (eds.), *Paleozoic and Mesozoic Tectonic Evolution of Central and Eastern Asia: From Continental Assembly to Intracontinental Deformation*. Geological Society of America Memoir, 194, pp. 101–116.

Zachos, J.C., Dickens, G.R. & Zeebe, R.E. (2008). An early Cenozoic perspective on greenhouse warming and carbon-cycle dynamics. *Nature*, 451, 279–283.

Zachos, J., Pagani, M., Sloan, L., Thomas, E. & Billups, K. (2001). Trends, rhythms, and aberrations in global climate 65 Ma to present. *Science*, 292, 688–693.

Zakharov, Y.D., Popov, A.M. & Blakov, A.S. (2008). Late Permian to Middle Triassic palaeogeographic differentiation of key ammonoid groups: evidence from the former USSR. *Polar Research*, 27, 441–468.

Zanchi, A., Poli, S., Fumagalli, P. & Gaetani, M. (2000). Mantle exhumation along the Trich Mir Fault Zone NW Pakistan: pre-mid-Cretaceous accretion of the Karakorum terrane to the Asian margin. In M.A. Khan et al. (eds.), *Tectonics of the Nanga Parbat Syntaxis and the Western Himalaya*. Geological Society, London, Special Publications, 170, pp. 237–252.

Zhan, R., Rong, Y., Percival, I.G. & Liang Y. (2011). Brachiopod biogeographic change during the Early to Middle Ordovician in South China. *Memoirs of the Society of Australasian Palaeontologists*, 41, 273–287.

Zhang, L., Ai, Y., Li, X. et al. (2007). Triassic collision of western Tianshan orogenic belt, China: evidence from SHRIMP U–Pb dating of zircon from HP/UHP eclogitic rocks. *Lithos*, 96, 266–280.

Zhang, L., Qin, K. & Xian, W. (2008). Multiple mineralisation events in the eastern Tienshan district, NW China: isotopic geochronology and geological significance. *Journal of Asian Earth Sciences*, 32, 236–246.

Zhang, W., Chen, P. & Palmer, A.R. (2003). *Biostratigraphy of China*. Beijing: Science Press.

Zhao, G.C., Sun, M., Wilde, S.A., Li, S.Z. (2004). A Paleo-Mesoproterozoic supercontinent: assembly, growth and breakup. *Earth-Science Reviews*, 67, 91–123.

Zhou, D., Graham, S.A., Chang, E.Z., Wang, B. & Hacker, B. (2001). Paleozoic tectonic amalgamation of the Chinese Tian Shan: evidence from a transect along the Dushanzi–Kuqa Highway. In M.S. Hendrix & G.A. Davis (eds.), *Paleozoic and Mesozoic Tectonic Evolution of Central and Eastern Asia: From Continental Assembly to Intracontinental Deformation*. Geological Society of America Memoir, 194, pp. 23–46.

Zhou, J., Wilde, S.A., Zhao, G.C. et al. (2010). Was the easternmost segment of the Central Asian Orogenic Belt derived from Gondwana or Siberia: an intriguing dilemma? *Journal of Geodynamics*, 50, 300–317.

Zhou, Z. & Dean, W.T. (eds.) (1996). *Phanerozoic Geology of Northwest China*. Beijing: Science Press.

Zhu, D., Zhao, Z., Niu, Y. et al. (2013). The origin and pre-Cenozoic evolution of the Tibetan Plateau. *Gondwana Research*, 23, 1429–1454.

Zhu, Y., Guo, X., Song, B., et al. (2009). Petrology, Sr–Nd–Hf isotopic geochemistry and zircon chronology of the Late Palaeozoic volcanic rocks in the southwestern Tianshan Mountains, Xinjiang, NW China. *Journal of the Geological Society, London*, 166, 1085–1099.

Ziegler, A.M., Cocks, L.R.M. & Bambach, R.K. (1968). The composition and structure of Lower Silurian marine communities. *Lethaia*, 1, 1–27.

Ziegler, A.M., Hulvey, M.I. & Rowley, D.B. (1997). Permian world topography and climate. In L.P. Martini (ed.), *Late Glacial and Post-Glacial Environmental Changes: Quaternary, Carboniferous, Proterozoic*. Oxford: Oxford University Press, pp. 111–146.

Ziegler, M.A. (2001). Late Permian to Holocene paleofacies evolution of the Arabian Plate and its hydrocarbon occurrences. *GeoArabia*, 6, 445–504.

Ziegler, P.A. (1989). *Evolution of Laurussia: A Study in Late Palaeozoic Plate Tectonics*. Dordrecht: Kluwer.

Ziegler, P.A. (1990). *Geological Atlas of Western and Central Europe*, 2nd edn. The Hague: Shell and London: Geological Society.

Žigaitė, Ž. & Blieck, A. (2006). Palaeobiogeographical significance of early Silurian thelodonts from central Asia and southern Siberia. *Geologiska Föreingens i Stockholm Förhandlingar*, 128, 203–206.

Zonenshain, L.P., Kuzmin, M.I. & Natapov, L.M. (1990). *Geology of the USSR: A Plate Tectonic Synthesis*. American Geophysical Union Geodynamics Series, 21.

Zurevinski, S.E., Heaman, L.M. & Creaser, R.A. (2011). The origin of Triassic/Jurassic kimberlite magmatism, Canada: two mantle sources revealed from the Sr–Nd isotopic composition of groundmass perovskite. *Geochemistry, Geophysics, Geosystems*, 12, Q09005, doi:10.1029/2011GC003659.

Index

ERA	PERIOD		EPOCH	STAGE	AGE (Ma)
					252
PALAEOZOIC	PERMIAN		Lopingian	CHANGHSINGIAN	254
				WUCHIAPINGIAN	260
			Guadalupian	CAPITANIAN	265
				WORDIAN	269
				ROADIAN	272
			Cisuralian	KUNGURIAN	279
				ARTINSKIAN	290
				SAKMARIAN	296
				ASSELIAN	299
	CARBONIFEROUS	PENNSYL-VANIAN	LATE	GZHELIAN	304
				KASIMOVIAN	307
			MIDDLE	MOSCOVIAN	315
			EARLY	BASHKIRIAN	323
		MISSIS-SIPPIAN	LATE	SERPUKHOVIAN	331
			MIDDLE	VISEAN	347
			EARLY	TOURNAISIAN	359
	DEVONIAN		LATE	FAMENNIAN	372
				FRASNIAN	383
			MIDDLE	GIVETIAN	388
				EIFELIAN	393
			EARLY	EMSIAN	408
				PRAGIAN	411
				LOCHKOVIAN	419
	SILURIAN		PRIDOLI		423
			LUDLOW	LUDFORDIAN	426
				GORSTIAN	427
			WENLOCK	HOMERIAN	430
				SHEINWOODIAN	433
			LLANDOVERY	TELYCHIAN	439
				AERONIAN	441
				RHUDDANIAN	444
	ORDOVICIAN		LATE	HIRNANTIAN	445
				KATIAN	453
				SANDBIAN	458
			MIDDLE	DARRIWILIAN	467
				DAPINGIAN	470
			EARLY	FLOIAN	~478
				TREMADOCIAN	487
	CAMBRIAN		FURONGIAN	STAGE 10	490
				JIANGSHANIAN	
				PAIBIAN	
			Epoch 3	GUZHANGIAN	494
				DRUMIAN	505
				STAGE 5	509
			Epoch 2	STAGE 4	514
				STAGE 3	521
			TERRENEUVIAN	STAGE 2	529
				FORTUNIAN	541

	EON	ERA	PERIOD	AGE (Ma)
				541
PRECAMBRIAN	PROTEROZOIC	NEOPRO-TEROZOIC	EDIACARAN	635
			CRYOGENIAN	850
			TONIAN	1000
		MESOPRO-TEROZOIC	STENIAN	1200
			ECTASIAN	1400
			CALYMMIAN	1600
		PALEOPRO-TEROZOIC	STATHERIAN	1800
			OROSIRIAN	2050
			RHYACIAN	2300
			SIDERIAN	2500
	ARCHEAN	NEOARCHEAN		2800
		MESO-ARCHEAN		3200
		PALEO-ARCHEAN		3600
		EOARCHEAN		4000
	HADEAN			

Precambrian and Palaeozoic time divisions, with dates in millions of years (Ma) at the base of each, with modifications (in red) near the Cambrian–Ordovician boundary.

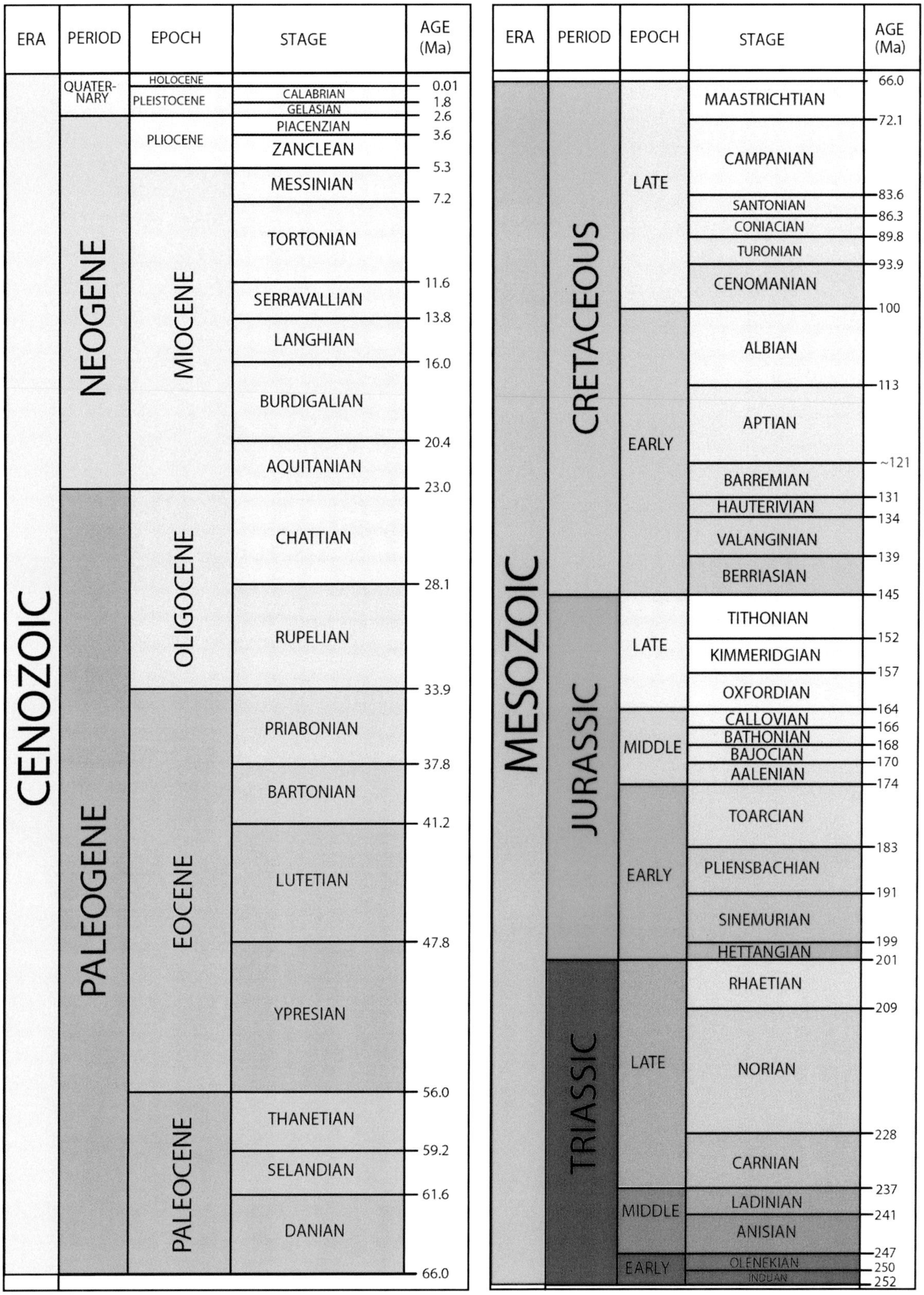

ERA	PERIOD	EPOCH	STAGE	AGE (Ma)
CENOZOIC	QUATERNARY	HOLOCENE		0.01
		PLEISTOCENE	CALABRIAN	1.8
			GELASIAN	2.6
	NEOGENE	PLIOCENE	PIACENZIAN	3.6
			ZANCLEAN	5.3
		MIOCENE	MESSINIAN	7.2
			TORTONIAN	11.6
			SERRAVALLIAN	13.8
			LANGHIAN	16.0
			BURDIGALIAN	20.4
			AQUITANIAN	23.0
	PALEOGENE	OLIGOCENE	CHATTIAN	28.1
			RUPELIAN	33.9
		EOCENE	PRIABONIAN	37.8
			BARTONIAN	41.2
			LUTETIAN	47.8
			YPRESIAN	56.0
		PALEOCENE	THANETIAN	59.2
			SELANDIAN	61.6
			DANIAN	66.0

ERA	PERIOD	EPOCH	STAGE	AGE (Ma)
				66.0
MESOZOIC	CRETACEOUS	LATE	MAASTRICHTIAN	72.1
			CAMPANIAN	83.6
			SANTONIAN	86.3
			CONIACIAN	89.8
			TURONIAN	93.9
			CENOMANIAN	100
		EARLY	ALBIAN	113
			APTIAN	~121
			BARREMIAN	131
			HAUTERIVIAN	134
			VALANGINIAN	139
			BERRIASIAN	145
	JURASSIC	LATE	TITHONIAN	152
			KIMMERIDGIAN	157
			OXFORDIAN	164
		MIDDLE	CALLOVIAN	166
			BATHONIAN	168
			BAJOCIAN	170
			AALENIAN	174
		EARLY	TOARCIAN	183
			PLIENSBACHIAN	191
			SINEMURIAN	199
			HETTANGIAN	201
	TRIASSIC	LATE	RHAETIAN	209
			NORIAN	228
			CARNIAN	237
		MIDDLE	LADINIAN	241
			ANISIAN	247
		EARLY	OLENEKIAN	250
			INDUAN	252

Mesozoic and Cenozoic time divisions, continuing upwards from the opposite page, with new modification (in red) at the Cretaceous (Barremian–Aptian) boundary. From Landing et al. (2015).

Printed in the United States
by Baker & Taylor Publisher Services